AF734876

Applications of Spinel Nano-Ferrites in Health, Environmental Sustainability, and Safety

This book presents the foundational concepts of spinel nano-ferrites, their synthesis techniques, and their cutting-edge applications in various fields. The book begins with an introduction to spinel nano-ferrites, discussing their unique properties and historical context. It then delves into various synthesis techniques, such as sol-gel methods, co-precipitation, and hydrothermal processes, highlighting the influence of these methods on the resultant structures and morphologies. The book also reviews the electrical and magnetic characteristics of spinel nano-ferrites, demonstrating their potential in electronic and magnetic devices. The role of spinel nano-ferrites in the health sector is explored in depth, detailing their use as antibacterial and antifungal agents and their innovative applications in magnetic resonance imaging (MRI), cancer detection, targeted drug delivery, and hyperthermia treatment. The chapters further examine the environmental applications of spinel nano-ferrites, including their effectiveness in pollution remediation, water purification, dye degradation, and the detection and removal of heavy metal ions and microplastics from wastewater. Toward the end, the book explores the development of gas sensors and humidity sensors using spinel nano-ferrites, as well as devices designed for human and environmental safety. This book is intended for researchers and students of nanotechnology, nanobiotechnology, healthcare professionals, and environmental scientists.

Key Features:

- Covers foundational concepts, unique properties, and cutting-edge applications of spinel nano-ferrites
- Explores various synthesis methods such as sol-gel, co-precipitation, and hydrothermal processes, and their impact on structures and morphologies of spinel nano-ferrites
- Discusses the electrical and magnetic characteristics of spinel nano-ferrites
- Examines the potential of spinel nano-ferrites as antibacterial and antifungal agents, and their roles in MRI, cancer detection, and drug delivery
- Highlights the effectiveness of spinel nano-ferrites in pollution remediation, water purification, and dye degradation

Nanotechnology Advances and Applications
About the Series

This book series explores the broader spectrum of applications of nanoparticles across biomedical, agricultural, and environmental sciences. Focusing on the biomedical applications of nanoparticles, the books cover their role in targeted drug delivery systems, advancements in nanoscale imaging for diagnostics, innovations in regenerative medicine and tissue engineering, and cutting-edge research in immunotherapy and vaccine development. In the realm of agricultural applications of nanoparticles, the series focuses on their use as nano-fertilizers, nano-pesticides, and in soil health monitoring techniques, sustainable farming practices, and the environmental benefits of reducing agricultural waste and chemical usage. The volumes cover the technologies behind antimicrobial packaging, freshness indicators, and the environmental impact of nano-enabled packaging solutions. The series also explores applications of nanoparticles in environmental sciences, including advanced filtration systems for air and water purification, efficient removal of heavy metals and pollutants through nano-adsorbents, and the development of sustainable energy solutions with biofuel cells. Each volume also addresses the ethical, safety, and regulatory aspects of nanotechnology, providing a balanced and comprehensive perspective.

Titles

1. Green Nanobiotechnology, 1st Edition (9781032820453)
Edited By Atul Thakur, Preeti Thakur, Deepa Suhag and S M Paul Khurana

2. Advancements in Nanobiology 1st Edition (9781032750491)
Edited by Atul Thakur, Preeti Thakur and Deepa Suhag

3. Applications of Spinel Nano-Ferrites in Health, Environmental Sustainability, and Safety (9781032904108)
Edited by Atul Thakur and Preeti Thakur

https://www.routledge.com/Nanotechnology-Advances-and-Applications/book-series/NAA

Applications of Spinel Nano-Ferrites in Health, Environmental Sustainability, and Safety

Edited by
Atul Thakur and Preeti Thakur

CRC Press
Taylor & Francis Group
Boca Raton London New York

CRC Press is an imprint of the
Taylor & Francis Group, an **informa** business

Designed cover image: Shutterstock

First edition published 2025
by CRC Press
2385 NW Executive Center Drive, Suite 320, Boca Raton FL 33431

and by CRC Press
4 Park Square, Milton Park, Abingdon, Oxon, OX14 4RN

CRC Press is an imprint of Taylor & Francis Group, LLC

© 2025 selection and editorial matter, Atul Thakur and Preeti Thakur; individual chapters, the contributors

Reasonable efforts have been made to publish reliable data and information, but the author and publisher cannot assume responsibility for the validity of all materials or the consequences of their use. The authors and publishers have attempted to trace the copyright holders of all material reproduced in this publication and apologize to copyright holders if permission to publish in this form has not been obtained. If any copyright material has not been acknowledged please write and let us know so we may rectify in any future reprint.

Except as permitted under U.S. Copyright Law, no part of this book may be reprinted, reproduced, transmitted, or utilized in any form by any electronic, mechanical, or other means, now known or hereafter invented, including photocopying, microfilming, and recording, or in any information storage or retrieval system, without written permission from the publishers.

For permission to photocopy or use material electronically from this work, access www.copyright.com or contact the Copyright Clearance Center, Inc. (CCC), 222 Rosewood Drive, Danvers, MA 01923, 978-750-8400. For works that are not available on CCC please contact mpkbookspermissions@tandf.co.uk

Trademark notice: Product or corporate names may be trademarks or registered trademarks and are used only for identification and explanation without intent to infringe.

ISBN: 9781032904108 (hbk)
ISBN: 9781032940984 (pbk)
ISBN: 9781003568940 (ebk)

DOI: 10.1201/9781003568940

Typeset in Times
by KnowledgeWorks Global Ltd.

Contents

Preface vii
About the Editors ix
Contributors xi

Chapter 1 Introduction to Spinel Nano-Ferrites 1
Preeti Thakur, Shivani Jangra, Fayu Wan, Blaise Ravelo, and Atul Thakur

Chapter 2 Synthesis Techniques for Spinel Nano-Ferrites 13
Atul Thakur, Preeti Thakur, Shivangi Sangwan, Dinesh Kumar, Ritesh Verma, and Jean-Luc Mattei

Chapter 3 Structures and Morphologies Attained by Spinel Nano-Ferrites 27
Preeti Thakur, Atul Thakur, Shivani Jangra, Irina Edelmana, and Sergey Ovchinnikov

Chapter 4 Electrical and Magnetic Characteristics of Spinel Nano-Ferrites 41
Preeti Thakur, Neetu Dhanda, An-Cheng Aidan Sun, and Atul Thakur

Chapter 5 Role of Spinel Nano-Ferrites in the Health Sector 48
Preeti Thakur, Shivani Jangra, and Atul Thakur

Chapter 6 Spinel Nano-Ferrites as Antibacterial Agents 61
Kavita Yadav, Atul Thakur, Preeti Thakur, Amitender Singh, and Ravi Kant Singh

Chapter 7 Antifungal Activity of Spinel Nano-Ferrites 78
Atul Thakur, Arti Rao, Muhammad Ehsan, Ritesh Verma, and Preeti Thakur

Chapter 8 Cancer Detection and Diagnosis with Spinel Nano-Ferrites 91
Atul Thakur, Preeti Thakur, Anand Salvi, Manish Shandilya, Renuka Kanojia, and Hui-Min David Wang

Chapter 9 Spinel Nano-Ferrites for Targeted Drug Delivery 105
Manish Shandilya, Renuka Kanojia, Anand Salvi, Larissa V. Panina, Preeti Thakur, and Atul Thakur

Chapter 10 Benefits of Spinel Nano-Ferrites in Catalysis 117
Pooja, Nisha Yadav, and Gyandshwar Kumar Rao

Chapter 11 Water Purification Using Spinel Nano-Ferrites 125
Preeti Thakur, Atul Thakur, Neetu Dhanda, and An-Cheng Aidan Sun

Chapter 12 Role of Spinel Nano-Ferrites for Dye Degradation 138
Vivek Srivastava

Chapter 13 Spinel Nano-Ferrites for Heavy Metal Ions Detection and Their Removal ... 149
Almas Ishaque, Sumbul Raza, and Shahab A. A. Nami

Chapter 14 Spinel Nano-Ferrites for Theranostic and Diagnostic Biomedical Applications ... 162
Akash A. Podutwar, Purva U. Chandorkar, Aniruddha R. Chabukswar, and Swati C. Jagdale

Chapter 15 A New Approach: Spinel Nano-Ferrites for Wastewater-Based Microplastics ... 171
Jyoti Bhattacharjee and Subhasis Roy

Chapter 16 Safety Measures Based on Spinel Nano-Ferrites ... 184
Vivek Srivastava

Chapter 17 Spinel Nano-Ferrites for High-Frequency Power Applications ... 194
Vivek Verma, Adarsh Singh, Manish Kumar Kansal, and Neelam Singh

Chapter 18 Devices for Human Safety Based on Spinel Nano-Ferrites ... 200
Preeti Thakur, Atul Thakur, Amitender Singh, and Kavita Yadav

Chapter 19 Future Perspectives of Spinel Nano-Ferrites ... 213
Arti Rao, Preeti Thakur, and Atul Thakur

Index ... 225

Preface

In recent years, the field of nanotechnology has witnessed remarkable advancements. The book *Applications of Spinel Nano-Ferrites in Health, Environmental Sustainability, and Safety* serves as a comprehensive guide, exploring the multifaceted applications of nanomaterials in improving human health, controlling environmental pollutants, and ensuring safety.

Spinel nano-ferrites, characterized by their unique properties at the nanoscale, offer tremendous potential for addressing some critical challenges. In the health sector, these materials have opened new possibilities for targeted drug delivery systems, cancer detection, and in the pharmaceutical field. Nano-ferrite materials such as nano-adsorbents, nano-photocatalysts, and nanosensors can be employed in pollution remediation offering innovative solutions for water purification, dye degradation, and air and pollutant detection. Spinel nano-ferrites have emerged as valuable tools in ensuring safety in today's world, sensors that detect and monitor hazardous substances. These nano-ferrite materials have proven their efficacy in safeguarding individuals and the environment.

This book aims to provide the reader with a comprehensive understanding of the diverse applications of spinel nano-ferrites in health, environment, and safety. Contributions from leading experts in the field have shared their insights, research findings, and practical applications and will encourage further exploration of the immense possibilities that these nanomaterials offer.

Happy reading!

The Editors

About the Editors

Prof. Atul Thakur is currently working as Director at Amity Institute of Nanotechnology, Amity University Haryana, Gurugram, India. Dr. Thakur is also a Professor at Nanjing University of Information Science & Technology and Jiangsu appointed Professor. He is also appointed as Adjunct Professor at Yuan Ze University. He is rated among the top 2% of researchers by Stanford University's ranking. During his postdoctorate in the Electronics Department of the University of Brest, France for two years, he worked on the size reduction of patch antennas. Thereafter, he worked as a postdoctorate researcher in the Physics Department of the world-renowned Taiwan National University, Taiwan, working on high-density memory storage systems and devices. The different areas of his research are ferrites for microwave applications, high-frequency applications, radar-absorbing materials, water purification, antenna miniaturization, high-density memory storage systems, sensor applications, metamaterials, and agricultural applications. Dr. Atul Thakur has published more than 275 research papers in international, peer-reviewed, SCI- and Scopus-indexed journals, and has filed more than 55 patents to date. He has edited books published by Springer, Elsevier, and CRC Press, and others. He has successfully completed several government-funded projects sponsored by ANR, DRDO, MNRE, DAE, and the Royal Academy of Engineering, UK.

Prof. Preeti Thakur is a Professor of Physics and currently working as Head of the Amity Institute of Nanotechnology, Amity University Haryana, Gurugram, India. She is figured among the top 2% of researchers by Stanford University's ranking. She is a postdoctorate from Taiwan National University, Taiwan and a gold medalist in her Masters; she has completed her Ph.D. in Physics from Himachal Pradesh University, Shimla. Her areas of interest are spinel ferrite synthesis and characterization for radar-absorbing materials, sensors, high-frequency materials, and application of magnetic nano-ferrite materials for wastewater treatment and agriculture purposes. She has published more than 260 articles in SCI/Scopus-indexed international journals, has filed 50+ patents, and is a reviewer of international indexed journals. She is an editorial board member of the *Journal of Magnetism and Magnetic Materials*. She has successfully completed research projects from government-funded agencies such as the Department of Science and Technology and the Royal Academy of Engineering. She has edited books published by Springer, Elsevier, and CRC Press, and others, and has published several book chapters.

Contributors

Jyoti Bhattacharjee
Department of Chemical Engineering
University of Calcutta
Kolkata, India

Aniruddha R. Chabukswar
Department of Pharmaceutical Sciences
School of Health Sciences and Technology
Dr Vishwanath Karad MIT World Peace University, MIT Campus
Pune, India

Purva U. Chandorkar
Department of Pharmaceutical Sciences
School of Health Sciences and Technology
Dr Vishwanath Karad MIT World Peace University, MIT Campus
Pune, India

Neetu Dhanda
Department of Physics
Amity University Haryana
Gurugram, India

Irina Edelmana
Kirensky Institute of Physics
Federal Research Center KSC, Siberian Branch
Russian Academy of Sciences
Krasnoyarsk, Russia

Muhammad Ehsan
Departement Informatique
CY Cergy-Paris Université, École CY-Tech
Pau, France

Almas Ishaque
Department of Industrial Chemistry, Faculty of Science
Aligarh Muslim University
Aligarh, India

Swati C. Jagdale
Department of Pharmaceutical Sciences
School of Health Sciences and Technology
Dr Vishwanath Karad MIT World Peace University, MIT Campus
Pune, India

Renuka Kanojia
Department of Chemistry, Biochemistry and Forensic Sciences
Amity School of Applied Sciences
Amity University Haryana
Gurugram, India

Manish Kumar Kansal
Department of Physics
Hindu College, University of Delhi
Delhi, India

Dinesh Kumar
Amity Institute of Nanotechnology
Amity University Haryana
Gurugram, India
and
DesignInnova Pvt Ltd.
New Delhi, India

Jean-Luc Mattei
Lab-STICC, CNRS
University of Brest
Brest, France

Shahab A. A. Nami
Department of Industrial Chemistry, Faculty of Science
Aligarh Muslim University
Aligarh, India

Sergey Ovchinnikov
Kirensky Institute of Physics
Federal Research Center KSC, Siberian Branch
Russian Academy of Sciences
Krasnoyarsk, Russia

Larissa V. Panina
National University of Science and Technology MISIS
Moscow, Russia

Akash A. Podutwar
Department of Pharmaceutical Sciences
School of Health Sciences and Technology
Dr Vishwanath Karad MIT World Peace University, MIT Campus
Pune, India

Pooja
Department of Chemistry, Biochemistry and Forensic Sciences
Amity University Haryana
Gurugram, India

Gyandshwar Kumar Rao
Department of Chemistry, Biochemistry and Forensic Sciences
Amity University Haryana
Gurugram, India

Blaise Ravelo
School of Electronics and Information Engineering
Nanjing University of Information Science & Technology
Nanjing, China

Sumbul Raza
Department of Industrial Chemistry, Faculty of Science
Aligarh Muslim University
Aligarh, India

Arti Rao
Amity Institute of Nanotechnology
Amity University Haryana
Gurugram, India

Subhasis Roy
Department of Chemical Engineering
University of Calcutta
Kolkata, India

Anand Salvi
Department of Chemistry, Biochemistry and Forensic Sciences
Amity School of Applied Sciences
Amity University Haryana
Gurugram, India

Shivangi Sangwan
Amity Institute of Nanotechnology
Amity University Haryana
Gurugram, India

Manish Shandilya
Department of Chemistry, Biochemistry and Forensic Sciences
Amity School of Applied Sciences
Amity University Haryana
Gurugram, India

Shivani Jangra
Amity Institute of Nanotechnology
Amity University Haryana
Gurugram, India

Adarsh Singh
Department of Physics
Hindu College, University of Delhi
Delhi, India

Amitender Singh
Amity Institute of Nanotechnology
Amity University Haryana
Gurugram, India

Neelam Singh
Department of Physics
Hansraj College, University of Delhi
Delhi, India

Ravi Kant Singh
Amity Institute of Biotechnology
Amity University
Noida, India

Vivek Srivastava
Mathematics and Basic Sciences–Chemistry
NIIT University
Neemrana, India

An-Cheng Aidan Sun
Department of Chemical Engineering and Materials Science
Yuan Ze University
Taoyuan, Taiwan

Atul Thakur
Nanjing University of Information Science & Technology, Nanjing, China
Nanotechnology Wing, Innovative Science Research Society, Shimla, India
nanoLatticeX LLP, Manesar, Gurugram, India
Amity Institute of Nanotechnology, Amity University Haryana, Gurugram, India

Preeti Thakur
Nanotechnology Wing, Innovative Science Research Society, Shimla, India
nanoLatticeX LLP, Manesar, Gurugram, India
Amity Institute of Nanotechnology, Amity University Haryana, Gurugram, India

Ritesh Verma
Department of Physics
Graphic Era (Deemed to be) University
Dehradun, Uttarakhand, India

Vivek Verma
Department of Physics
Hindu College, University of Delhi
Delhi, India

Fayu Wan
School of Electronics and Information Engineering
Nanjing University of Information Science & Technology
Nanjing, China

Hui-Min David Wang
Graduate Institute of Biomedical Engineering
National Chung Hsing University
Taichung, Taiwan

Kavita Yadav
Department of Physics
G.C.W. Gurawara
Rewari, Haryana, India

Nisha Yadav
Department of Chemistry, Biochemistry and Forensic Sciences
Amity University Haryana
Gurugram, India

1 Introduction to Spinel Nano-Ferrites

Preeti Thakur and Shivani Jangra*
Amity Institute of Nanotechnology, Amity University Haryana, Gurugram, India

Fayu Wan and Blaise Ravelo
School of Electronics and Information Engineering, Nanjing University of Information Science & Technology, Nanjing, China

Atul Thakur
Amity Institute of Nanotechnology, Amity University Haryana, Gurugram, India
Nanotechnology Wing, Innovative Science Research Society, Shimla, India
nanoLatticeX LLP, Manesar, Gurugram, India
Nanjing University of Information Science & Technology, Nanjing, China

1.1 INTRODUCTION

1.1.1 Historical Background

Ferrites have a fascinating history that spans millennia and incorporates contributions from several cultures. Lodestone, a naturally occurring magnetic mineral, was utilized for navigation in ancient China beginning in the 4th century BCE, and it is thought that this practice led to the usage of ferrite-like materials in compasses [1]. The foundation for understanding magnetism and the behaviour of magnetic materials was laid by these early compasses. The magnetic qualities of lodestone were also known to the Greeks and Romans [2]. The word "magnet" comes from the name of the lodestone-rich ancient Greek city of Magnesia. They saw that lodestone fragments had the ability to draw iron and organize themselves in a north-south orientation. The research of magnets and magnetic materials continued in medieval Europe. Researchers like William Gilbert and Peter Peregrinus made substantial advancements to our understanding of magnetism. Gilbert established the groundwork for comprehending magnetic domains and magnetic induction in his book "De Magnete," which was published in 1600 [3]. Gilbert characterized the Earth as a huge magnet. The terms "ferrite" and "ferromagnetism" were first used to refer to a group of ceramic materials having iron oxide as their main constituent around the beginning of the 20th century. These materials displayed distinctive ferrimagnetism and other magnetic characteristics. Magnetic moments in ferrimagnetic materials are oriented in opposite directions and provide a net magnetic moment. Due to its use in radar technology, ferrite materials attracted a lot of attention during World War II [4]. Radar systems and communication technologies advanced, thanks to the usage of ferrite cores in microwave devices and antennas. Following the war, ferrite materials were widely used in a variety of industries, such as telecommunication, magnetic recording, and television sets. They were extremely useful in the developing electronics industry because of their capacity to operate at high frequencies. Ferrite materials are still crucial parts of electronics and telecommunications in the present era [5]. They serve as the magnetic cores in electrical circuits as well as in transformers, inductors, and microwave devices. Furthermore, they are essential to data storage technology like hard drives. Specialized ferrite materials with tailored qualities for certain applications have been developed as a result of advancements in materials science. Researchers are still looking into new applications for ferrite materials in cutting-edge technology. The development and use of ferrite materials in numerous technical fields may be traced back to the progressive accumulation of information and scientific discoveries in the realm of magnetism. Ferrites are essential parts of many modern electronic and electromagnetic devices.

1.1.2 Early Applications

- Compasses are one of the ferrite-like materials and have the most important applications. Lodestone, a magnetic mineral that occurs naturally, was utilized as a compass needle by ancient Chinese sailors. Essentially, lodestone is magnetite, a type of ferrite, in its native state. These lodestone compasses were essential for marine exploration and navigation.
- The healing and protective powers of magnets were revered in a few prehistoric cultures. Natural ferrite materials like lodestone were used to make magnetic jewellery, including bracelets and amulets. Both medical and superstitious reasons were given for wearing these things [6].

* Corresponding author.

DOI: 10.1201/9781003568940-1

- People's fascination with magnetism led to the creation of early magnetic toys and novelties. For instance, magnetic "games" were well-liked among people in the 19th century. These often-involved small ferrite magnets that could attract and repel each other, leading to various entertaining effects.
- *Scientific Instruments:* As knowledge of magnetism and ferrite materials increased, these materials found use in the scientific apparatus. For instance, William Gilbert detailed how lodestone may be used in scientific experiments to examine magnetic phenomena in his 17th-century work "De Magnete."
- During the 19th century, ferrite materials were utilized in telegraphy and telegraphic apparatus. In order to produce dependable signalling devices, telegraph systems relied on the magnetic characteristics of substances like iron and ferrite.
- Early radio receivers relied heavily on ferrite-core antennas for their operation. To capture and magnify radio signals, the ferrite rod antenna, often referred to as a loop stick antenna, was a crucial part of crystal radio sets [7, 8].
- Radar technology used ferrite materials, which were developed during World War II. The development of more sophisticated radar systems for military applications was made possible using ferrite cores in microwave devices and waveguide components.

These early uses for ferrite materials as shown in Figure 1.1 demonstrate their significance in a variety of historical situations, from communication and entertainment to navigation and scientific discovery. Ferrite materials have become essential in contemporary electronics and telecommunications because of the advancement of technology and our growing understanding of ferrite materials over time [9].

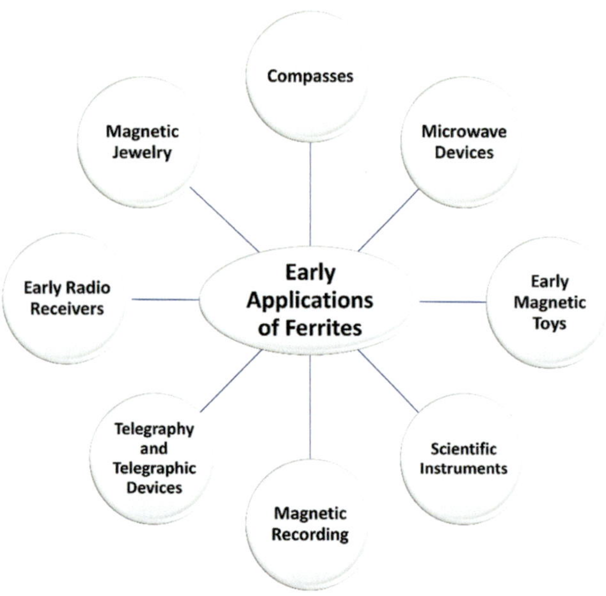

FIGURE 1.1 Some early applications of ferrites.

1.2 EVOLUTION OF FERRITE SCIENCE

Ferrite, derived from the Latin word "ferrum" (meaning iron), is a crystalline phase of iron, central to the development of materials science and metallurgy. Its evolution is a story of discovery, understanding, and innovation that has significantly shaped modern industries and technologies.

- Ferrite's story begins in the early 19th century when scientists started exploring the crystalline nature of materials. Pioneers like René-Just Haüy and William Lawrence Bragg laid the foundation for crystallography—the study of crystalline structures. Ferrite was first identified as a distinct crystalline structure within iron alloys [10].
- *The Role of Ferrite in Metallurgy* (*Late 19th Century*): The late 19th century marked significant advancements in metallurgy, coinciding with the Industrial Revolution. Engineers and metallurgists recognized that ferrite had a profound impact on the mechanical properties of iron and steel. Its presence made the material softer and more ductile, which had far-reaching implications for structural engineering, transportation, and construction [11].
- *Phase Diagrams and Transformations* (*Early 20th Century*): The early 20th century saw the development of phase diagrams, graphical representations of how materials transform into different phases at varying temperatures and compositions. The iron-carbon phase diagram was pivotal in understanding the behaviour of ferrites. It revealed that ferrite could exist in various forms, including alpha and delta ferrite, depending on temperature and carbon content [12].
- *Ferrite in Welding* (*Mid-20th Century*): The mid-20th century brought significant advancements in welding technology, and understanding the role of ferrite in welding became critical. The ferrite content in welds profoundly affected mechanical properties and corrosion resistance. Controlling ferrite content became essential for ensuring weld quality and durability in various applications [13].
- *Ferritic Stainless Steels* (*1950s–1960s*): During the 1950s and 1960s, researchers developed ferritic stainless steels, which contain a high proportion of ferrite. These alloys offered excellent corrosion resistance and found applications in kitchen appliances, automotive exhaust systems, and more. This marked a pivotal moment in harnessing ferrite's unique properties for specific needs.

- *Advancements in Materials Science (Late 20th Century)*: The late 20th century saw advancements in materials science that allowed for a deeper understanding of ferrite at the atomic level. Techniques like transmission electron microscopy (TEM) and electron diffraction were employed to study ferrite's microstructure and crystallography. This knowledge contributed to the design of advanced materials, including high-strength low-alloy (HSLA) steels and dual-phase steels, which combine ferrite with other phases for enhanced performance [14].
- *Modern Applications and Beyond (21st Century)*: In the 21st century, ferrite continues to play a vital role in various applications. Its unique combination of properties, including excellent magnetic properties, electrical conductivity, and corrosion resistance, makes it indispensable in fields like electronics, telecommunications, and renewable energy. Ferrite magnets, for example, are used in electric motors, transformers, and magnetic resonance imaging (MRI) machines.
- Moreover, researchers are exploring the potential of ferrite in emerging technologies such as spintronics, where the electron's spin is harnessed for information storage and processing. Ferrite nanoparticles are being investigated for their role in targeted drug delivery systems and as contrast agents in medical imaging.
- *Ferrite in Magnetic Materials and Electronics*: Ferrite's magnetic properties have made it a cornerstone of magnetic materials and electronics. In the mid-20th century, ferrite-based materials played a crucial role in the development of electronic components, such as transformers, inductors, and antennas. The unique magnetic properties of ferrite, including high magnetic permeability and low electrical conductivity, made it ideal for these applications.
- Ferrite cores are commonly used in transformers to efficiently transfer electrical energy from one coil to another while minimizing energy loss. Additionally, ferrite-based inductors are essential components in electronic circuits, serving to store and regulate electrical energy [15, 16].
- *Ferrite in Telecommunications*: Ferrite's magnetic properties have also found widespread use in the telecommunications industry. Ferrite materials are employed in microwave devices, such as circulators and isolators, to control the direction of electromagnetic waves. Circulators, for instance, allow signals to flow in a specific sequence and are critical components in radar systems and satellite communications.
- The unique property of ferrimagnetism—exhibited by some ferrite materials—enables the isolation of signals in one direction while allowing them to pass freely in another, making ferrite-based devices indispensable in the field of telecommunications.
- *Ferrite in Power Electronics*: Ferrite materials have become essential in power electronics, where they are used in a variety of applications, including power transformers and inductors. These applications require materials that can efficiently handle high-frequency signals and magnetic fields while minimizing energy loss. Ferrite cores in transformers and inductors provide the necessary magnetic characteristics to achieve these goals.
- Moreover, ferrite materials are employed in switch-mode power supplies (SMPS), which are widely used in electronic devices like computers and mobile phones. Ferrite-based components in SMPS help regulate voltage and current, ensuring efficient and stable power delivery to electronic devices.
- *Ferrite in the Automotive Industry*: The automotive industry relies on ferrite-based components for various applications. Ferrite magnets are used in electric power steering systems, helping to generate the magnetic fields required to assist in steering without the need for hydraulic fluids. These magnets are chosen for their durability and cost-effectiveness.
- Additionally, ferrite cores are used in ignition coils and spark plugs to manage electrical energy efficiently. Their ability to withstand high temperatures and electrical stress makes them ideal for these critical automotive components.
- *Ferrite in Renewable Energy*: As the world seeks cleaner energy sources, ferrite materials have found applications in renewable energy technologies. Ferrite magnets are used in the generators of wind turbines to convert mechanical energy from the rotating blades into electrical energy. These magnets are valued for their high magnetic flux density and stability in harsh environmental conditions.
- Ferrite-based inductors and transformers are also utilized in solar power inverters, which convert direct current (DC) generated by solar panels into alternating current (AC) for use in homes and the electrical grid. The efficiency and reliability of these components are crucial for the success of solar energy systems [17].
- *Cutting-Edge Research and Emerging Applications*: In recent years, research in ferrite materials has expanded into cutting-edge areas with the potential to revolutionize various industries.
  i. *Spintronics and Magnetic Data Storage*: Ferrite materials are explored for their potential in spintronics, a field that utilizes the electron's spin in addition to its charge for

information processing and storage. Ferrite nanoparticles and thin films have shown promise for spintronic applications, offering the possibility of higher-density magnetic data storage and faster data processing [18].

ii. *Biomedical Applications*: Ferrite nanoparticles are being investigated for their role in biomedical applications. They can be used as contrast agents in MRI to enhance imaging quality. Additionally, researchers are exploring ferrite nanoparticles for targeted drug delivery systems, where they can be guided to specific locations in the body for precise drug release.

iii. *Water Treatment and Environmental Remediation*: Ferrite-based materials are being studied for their potential in water treatment and environmental remediation. They can be used to remove contaminants, such as heavy metals, from water sources, contributing to clean and safe drinking water.

iv. *Ferrite in Quantum Technologies*: Ferrite materials are also entering the realm of quantum technologies. Researchers are exploring their use in quantum computing and quantum communication systems, leveraging their unique properties to process and transmit quantum information [19, 20].

1.3 MAGNETIC PROPERTIES OF FERRITES

1.3.1 Magnetic Domains

Magnetic domains in ferrite are regions within the material where the magnetic moments of atoms or ions align in a specific direction. These domains play a crucial role in understanding the magnetic behaviour of ferrites and have significant implications for their magnetic properties and applications.

Here are key aspects of magnetic domains in ferrite:

i. *Magnetic Moment Alignment*
   - Magnetic moments arise from the intrinsic magnetic properties of the atoms or ions within the ferrite crystal lattice. These moments can be thought of as tiny individual magnets, each with a north and south pole.
   - In an unmagnetized or demagnetized state, the magnetic moments in a ferrite material are randomly oriented. This means that the net magnetization of the material is close to zero because the magnetic moments cancel each other out [21].

ii. *Formation of Magnetic Domains*
   - As an external magnetic field is applied to the ferrite material, it starts to influence the orientation of the magnetic moments. Under the influence of the external field, some magnetic moments will tend to align with the field direction.
   - This alignment is not uniform across the entire material. Instead, it occurs in localized regions, creating small regions with coordinated magnetic moments. These regions are what we refer to as magnetic domains [22].

iii. *Role of Magnetic Domains*
   - Magnetic domains help explain how materials can exhibit magnetization despite having individual magnetic moments that cancel each other out in the absence of an external field.
   - Within each domain, the magnetic moments are aligned, resulting in a net magnetization within that domain. However, between domains, the magnetic moments may be oriented differently or even in opposite directions.

iv. *Magnetic Domain Wall*
   - The boundaries or interfaces between adjacent magnetic domains are called magnetic domain walls. These domain walls are regions where the magnetic moments gradually transition from the orientation in one domain to the orientation in the neighbouring domain.
   - Domain walls are dynamic and can move within the material under the influence of external magnetic fields. This movement of domain walls plays a critical role in the process of magnetization and demagnetization of ferrite materials [23].

v. *Hysteresis Loop*
   - The behaviour of magnetic domains and domain walls is reflected in the hysteresis loop of a ferrite material. A hysteresis loop shows how the magnetization of the material changes with the applied magnetic field.
   - As the external magnetic field is increased, the domain walls can move, and magnetic domains can align more with the field. This results in an increase in the material's overall magnetization.
   - The process of magnetization follows the ascending branch of the hysteresis loop. When the external field is reduced, the domain walls may move again, causing the material to retain some level of magnetization even after the field is removed. This is known as remanence [24].

vi. *Applications and Control of Magnetic Domains*
   - Understanding and controlling magnetic domains are essential in various applications. Engineers and scientists manipulate magnetic domains to optimize the magnetic properties of ferrite materials for specific uses.
   - In magnetic recording media (e.g., hard drives and magnetic tapes), for instance, the arrangement of magnetic domains is carefully

controlled to store and retrieve data effectively [25].

- In power electronics, such as transformers and inductors, the behaviour of magnetic domains influences the efficiency and performance of these components.

vii. *Factors Affecting Magnetic Domains*

- Several factors can influence the behaviour of magnetic domains in ferrite materials, including temperature, material composition, and the presence of external magnetic fields.
- At elevated temperatures, thermal energy can disrupt the alignment of magnetic moments, affecting the stability of magnetic domains.
- The composition of the ferrite, including the type and concentration of magnetic ions, can influence the size and stability of domains [26].
- External magnetic fields can reorient magnetic domains, either aligning them with the field (magnetization) or causing domain wall motion (demagnetization).

These domains are critical to understanding the magnetic behaviour of ferrite and play a central role in various applications, from magnetic storage to power electronics. Control and manipulation of magnetic domains are essential for optimizing the performance of ferrite materials in specific technological and industrial contexts.

1.3.2 Magnetic Susceptibility

Magnetic susceptibility is a measure of how much a material responds to an applied magnetic field by becoming magnetized. It quantifies the degree to which a material can be magnetized and is a key property in understanding the magnetic behaviour of materials, including ferrites. Magnetic susceptibility, denoted by the symbol χ (chi), is a dimensionless quantity that characterizes a material's response to an applied magnetic field. It quantifies how easily a material becomes magnetized when subjected to an external magnetic field [27]. Mathematically, magnetic susceptibility is defined as the ratio of the magnetization (M) of a material to the applied magnetic field (H), often expressed as: $\chi = \frac{M}{H}$

Magnetic susceptibility can be categorized into three main types based on the behaviour of materials in response to an applied magnetic field:

i. *Paramagnetic Susceptibility*: Paramagnetic materials have positive magnetic susceptibility. When subjected to an external magnetic field, the magnetic moments within paramagnetic materials align in the direction of the field, causing the material to become weakly magnetized.

ii. *Diamagnetic Susceptibility*: Diamagnetic materials have negative magnetic susceptibility. When exposed to an external magnetic field, diamagnetic materials develop small induced magnetic moments in the opposite direction of the applied field. This causes a weak repulsion from the field.

iii. *Ferromagnetic and Ferrimagnetic Susceptibility*: Ferromagnetic and ferrimagnetic materials exhibit a very high positive magnetic susceptibility. They can become strongly magnetized when exposed to a magnetic field and retain that magnetization even after the field is removed. Ferrites typically fall into this category [28].

The magnetic susceptibility of ferrites is influenced by several factors, including as follows:

- *Material Composition*: The type and concentration of magnetic ions (e.g., iron, nickel, manganese) in the ferrite composition significantly affect its susceptibility.
- *Crystal Structure*: The crystal structure of the ferrite, such as spinel or garnet structures, influences its magnetic properties, including susceptibility.
- *Temperature*: Magnetic susceptibility is temperature dependent. As temperature increases, thermal energy can disrupt the alignment of magnetic moments, affecting susceptibility [29].
- *External Magnetic Field*: The presence of an external magnetic field can temporarily alter the magnetic susceptibility of ferrites, especially during the magnetization process.

Magnetic susceptibility is a crucial property in various applications of ferrites:

- *Electronics and Telecommunications*: Ferrites are commonly used in electronic components, such as inductors, transformers, and magnetic cores, where their high positive susceptibility allows for efficient magnetic field control and energy transfer.
- *Magnetic Recording*: In data storage devices like hard disk drives (HDDs) and magnetic tapes, the magnetic susceptibility of the recording media determines the ability to write, store, and retrieve data [30].
- *Medical Imaging*: Ferrite-based contrast agents in MRI take advantage of their magnetic susceptibility to enhance image contrast and visualize specific tissues or structures within the body.

Magnetic susceptibility is typically measured using magnetic susceptibility meters or susceptometers. The units of magnetic susceptibility are dimensionless, as it is a ratio of magnetization to magnetic field strength. Magnetic susceptibility is a critical property that characterizes the response of ferrites to external magnetic fields. Ferrites, especially ferromagnetic and ferrimagnetic ones, exhibit high positive susceptibility, making them valuable in a wide range of applications where magnetic control, energy transfer, and data storage are essential. Understanding and controlling the magnetic susceptibility of ferrites are fundamental aspects of optimizing their performance in various technological and industrial contexts.

1.4 TYPES OF FERRITES

Ferrites are a class of ceramic compounds primarily composed of iron oxide (Fe_2O_3 or Fe_3O_4) along with other metal oxides. They exhibit a wide range of magnetic properties, making them suitable for various applications [31]. There are several types of ferrites, each with specific properties and applications. Some common types of ferrites are illustrated in Figure 1.2.

1.4.1 Soft Ferrites

Soft ferrites are a specific type of ferrite material known for their low coercivity and high magnetic permeability. These properties make them ideal for applications where magnetization needs to change rapidly with the applied magnetic field [32]. Soft ferrites are typically used in high-frequency applications and are characterized by their ability to efficiently transfer magnetic energy while minimizing energy loss. Here are some key characteristics and applications of soft ferrites:

- *Low Coercivity*: Soft ferrites have a low coercivity, which means they can be magnetized and demagnetized easily with relatively low applied magnetic fields. This property allows them to respond quickly to changes in the external magnetic field.
- *High Magnetic Permeability*: Soft ferrites exhibit high magnetic permeability, which enables them to effectively concentrate magnetic flux within their structure. This is crucial for applications where magnetic fields need to be confined or guided.

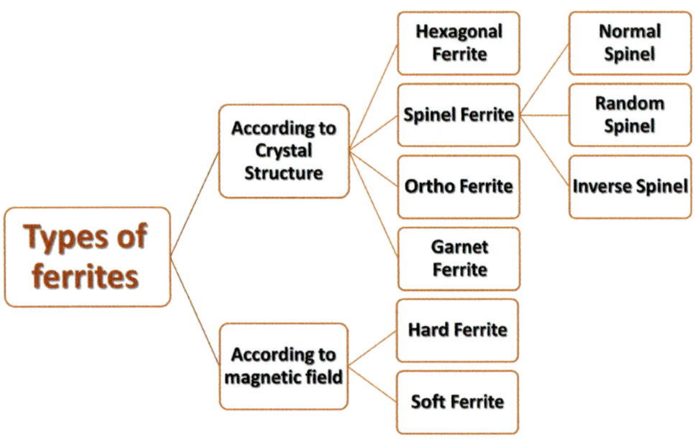

FIGURE 1.2 Types of ferrites.

- *Low Core Loss*: Soft ferrites have low core loss, which means they dissipate minimal energy in the form of heat when subjected to alternating magnetic fields. This makes them highly efficient in high-frequency applications.
- *Good Electrical Insulation*: Soft ferrites are electrically insulating materials, making them suitable for applications where electrical isolation is necessary.
- *Frequency Range*: Soft ferrites are typically used in applications ranging from a few hundred hertz (Hz) to several megahertz (MHz) and beyond. They are commonly found in the radiofrequency (RF) and microwave frequency bands [33, 34].

Common soft ferrite materials include manganese-zinc ferrites (MnZn) and nickel-zinc ferrites (NiZn). Here are some notable applications of soft ferrites:

i. *Transformers*: Soft ferrites are used as transformer cores in high-frequency power converters, such as those found in SMPS and voltage converters. They help efficiently transfer electrical energy between primary and secondary windings.
ii. *Inductors*: Soft ferrites serve as the core material in inductors used in various electronic circuits. They help store and control energy in applications like filtering, signal conditioning, and impedance matching.
iii. *Electromagnetic Interference (EMI) Suppression*: Soft ferrite beads and cores are placed on cables and wires to suppress high-frequency EMI. They absorb and dissipate EMI noise, preventing it from affecting sensitive electronic components.
iv. *High-Frequency Magnetic Components*: Soft ferrites are used in various high-frequency magnetic components, including chokes, baluns, and RF filters. They help shape and direct magnetic fields in RF and microwave circuits.
v. *Noise Suppression in Electronics*: Soft ferrite components are commonly used in electronic circuits to reduce EMI and noise, ensuring that electronic devices operate with minimal interference [35, 36].

Overall, soft ferrites are essential in high-frequency and high-efficiency applications where rapid magnetization changes and effective magnetic energy transfer are required, while simultaneously minimizing energy losses.

1.4.2 Hard Ferrites

Hard ferrites, also known as permanent ferrite magnets, are a specific type of ferrite material characterized by their high coercivity and ability to retain magnetization even after the applied magnetic field is removed [37]. These properties make them ideal for permanent magnet applications where a stable and strong magnetic field is required. Here are some key characteristics and applications of hard ferrites:

- *High Coercivity*: Hard ferrites have a high coercivity, which means they require a strong external magnetic field to demagnetize them. This property allows them to maintain their magnetization for extended periods.
- *Stability*: Hard ferrites are highly stable and do not easily lose their magnetic properties over time. This makes them suitable for long-term applications.
- *Permanent Magnetization*: They are used to create permanent magnets that can generate a constant magnetic field without the need for an external power source.
- *Resistant to Demagnetization*: Hard ferrites are resistant to demagnetization due to factors like temperature variations, mechanical shock, and external magnetic fields, making them reliable in a wide range of conditions.
- *Moderate Magnetic Strength*: While hard ferrites provide a strong and stable magnetic field, they are not as strong as some other types of permanent magnets like neodymium magnets. However, they are sufficient for many applications.
- *Low Cost*: Hard ferrites are relatively inexpensive to produce, making them cost-effective for various applications.

Common hard ferrite materials include strontium ferrite ($SrFe_{12}O_{19}$) and barium ferrite ($BaFe_{12}O_{19}$). Here are some notable applications of hard ferrites [38–40]:

i. *Electric Motors and Generators*: Hard ferrite permanent magnets are commonly used in electric motors and generators, providing the magnetic field required for rotational motion and power generation in appliances, automotive components, and industrial machinery.
ii. *Speakers and Audio Devices*: Hard ferrite magnets are used in the magnetic assemblies of speakers and headphones to convert electrical signals into sound waves.
iii. *Magnetic Separators*: In industries such as mining and recycling, hard ferrite magnets are used in magnetic separators to separate ferrous materials from non-ferrous materials.
iv. *Refrigerator Door Seals*: Hard ferrite magnets are used in the door seals of refrigerators to create an airtight seal and keep the door closed.
v. *Automotive Sensors*: They find applications in automotive sensors, such as wheel speed sensors and position sensors, where a stable magnetic field is needed for accurate measurements.
vi. *Magnetic Toys and Educational Materials*: Hard ferrite magnets are used in various educational toys and materials, including building blocks and magnetic puzzles.
vii. *Security Systems*: Hard ferrite magnets are used in security systems, such as magnetic door contacts, to detect the opening or closing of doors and windows.
viii. *Magnetic Latches and Closures*: They are used in magnetic latches and closures for cabinets, doors, and various household items.

Hard ferrites are valued for their reliability, stability, and cost-effectiveness in permanent magnet applications where a consistent and durable magnetic field is essential.

1.5 MAGNETIC HYSTERESIS

Magnetic hysteresis is a phenomenon that occurs when a material, such as ferrite, is magnetized and demagnetized. It describes the relationship between the magnetic field strength (H) applied to the material and its resulting magnetization (M). Ferrites, due to their unique magnetic properties, exhibit hysteresis behaviour [41]. This section deals with how ferrites are magnetized and demagnetized (Figures 1.3 and 1.4), and how magnetic hysteresis plays a role in this process.

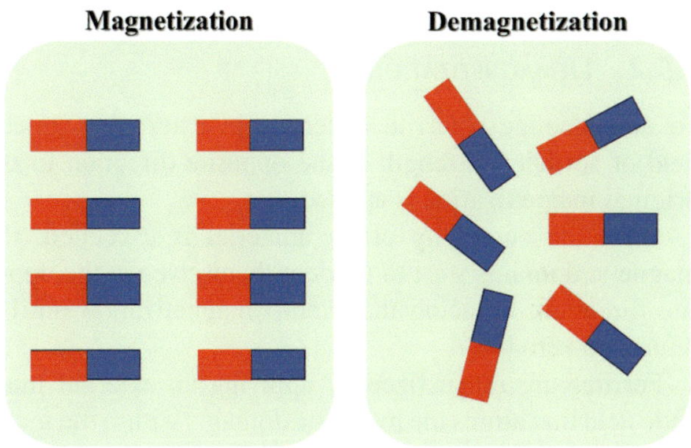

FIGURE 1.3 Magnetization and demagnetization of ferromagnetic and non-ferromagnetic materials.

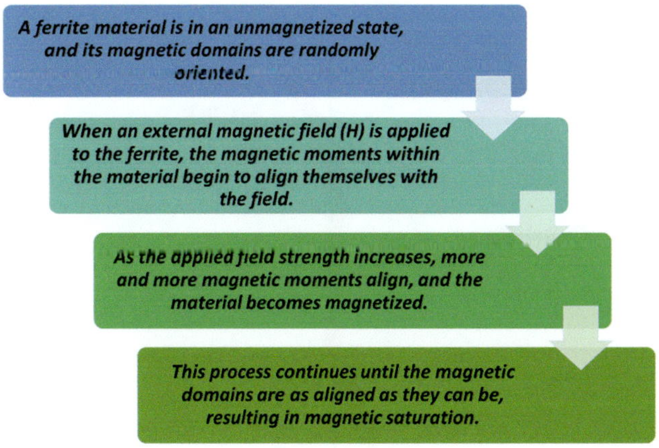

FIGURE 1.4 How magnetic saturation takes place in any material.

1.5.1 Magnetization (Magnetic Saturation)

1.5.1.1 Hysteresis Loop

As the external magnetic field is varied, the magnetization of the ferrite changes. This relationship is typically represented by a hysteresis loop on a magnetization curve as shown in Figure 1.5 [42]. When the external magnetic field is reduced from its maximum value (in the opposite direction), the magnetization of the ferrite does not immediately return to zero. Instead, it follows a path called the "demagnetization curve" within the hysteresis loop. This behaviour is due to the retention of some residual magnetization, which makes ferrites useful in applications where permanent magnets are needed [43].

1.5.1.2 Coercivity and Remanence

- *Within the hysteresis loop, two key points are defined*: coercivity (H_c) and remanence (B_r).
- Coercivity (H_c) is the strength of the reverse magnetic field required to completely demagnetize the material.
- Remanence (B_r) is the residual magnetization remaining in the material after the external field is removed.

1.5.2 Demagnetization

To demagnetize a ferrite material, an external magnetic field of sufficient strength in the opposite direction to the original magnetization is applied.

Once the coercivity of the material is exceeded, the magnetic domains start to reorient themselves in the opposite direction, reducing the overall magnetization until it returns to zero [44].

Ferrites are magnetized by applying an external magnetic field that aligns the magnetic domains within the material. Magnetic hysteresis describes the relationship between the applied field and the resulting magnetization, including how the material retains some magnetization after the field is removed. Demagnetization is achieved by applying an opposing magnetic field that exceeds the coercivity of the ferrite, causing the magnetic domains to realign and the material to return to its unmagnetized state.

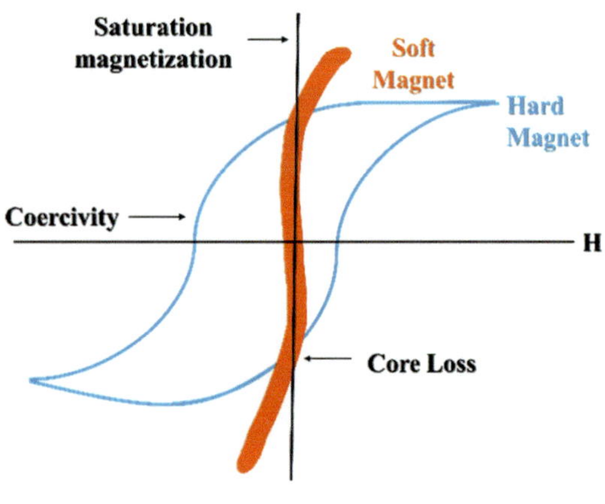

FIGURE 1.5 Magnetic hysteresis loop.

1.6 APPLICATIONS OF FERRITES THROUGH HISTORY

Ferrites have played a significant role in various applications throughout history, ranging from early telecommunications to emerging technologies like radar technology.

- *Early Telecommunications*
  - i. The early 20th-century ferrite cores were used in telecommunication devices such as radio antennas and transformers. They helped in tuning and improving the efficiency of radio transmission and reception.
  - ii. Ferrite rods and beads were utilized in radio antennas to enhance the selectivity and sensitivity of receivers, allowing for better signal reception and noise reduction.
  - iii. Ferrite transformers and inductors were used in telephone lines and early switching systems to manage the flow of electrical signals and improve signal quality [45–47].
- *Magnetic Recording*
  - i. Ferrite materials have been instrumental in the development of magnetic recording technology. Ferrite heads, used in early magnetic tape and disk drives, played a crucial role in reading and writing data.
  - ii. Ferrite-based magnetic storage media, such as ferrite-core memory, were used in early computers as a form of random-access memory (RAM) before the advent of semiconductor-based memory technologies [48].
- *Radar Technology*
  - i. During World War II, radar technology became pivotal for military operations. Ferrite materials were used in the construction of microwave circulators and isolators in radar systems (Figure 1.6).

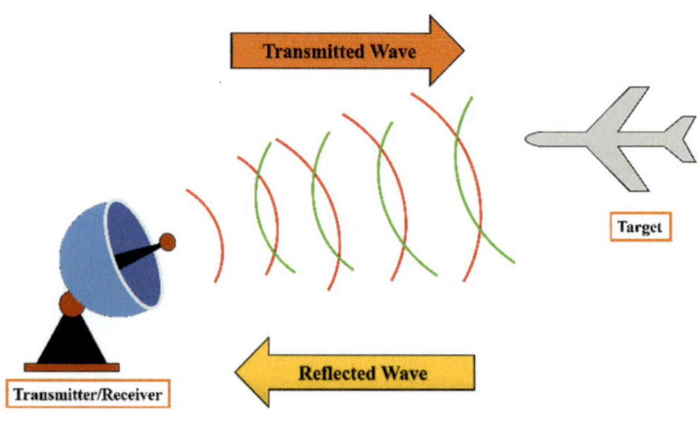

FIGURE 1.6 Working of radars.

   ii. Ferrite circulators and isolators allowed radar systems to transmit high-frequency microwave signals in one direction while isolating the receiver from the transmitter to prevent damage from reflected energy [45].

- *Microwave Devices*
   i. Ferrites continue to be essential in modern microwave devices. They are used in components like circulators, isolators, and phase shifters in radar systems, satellite communication, and microwave ovens [49].
   ii. Ferrite-based circulators are used to route microwave signals efficiently and protect sensitive microwave sources from reflected power [50].
- *Emerging Technologies*
- In contemporary times, ferrite materials are still in use in emerging technologies. They play a role in the development of high-frequency electronic devices, 5G technology, and wireless power transfer systems.
- Ferrites are utilized in miniaturized components for mobile devices, enabling efficient power management and signal processing.

Overall, ferrites have a rich history of contributing to the evolution of telecommunications, magnetic recording, radar technology, and other emerging technologies. Their unique magnetic properties and versatility continue to make them indispensable in a wide range of applications in the modern world.

1.7 MODERN APPLICATIONS OF FERRITES

Ferrites continue to have a wide range of modern applications due to their unique magnetic properties. Here are some of the key contemporary applications of ferrites:

i. *Electromagnetic Interference (EMI) Suppression*: Ferrite beads and cores are used to suppress EMI in electronic circuits. They are placed on cables and wires to absorb and dissipate high-frequency noise, ensuring that electronic devices operate without interference [51].

ii. *Transformers and Inductors*: Ferrite cores are commonly used in transformers and inductors for power electronics applications. They help to efficiently transfer and store electrical energy while minimizing energy loss and EMI [52].

iii. *Magnetic Sensors*: Ferrite-based magnetic sensors are used in various applications, such as automotive systems (e.g., wheel speed sensors and position sensors), industrial automation, and consumer electronics (e.g., compasses and proximity sensors).

iv. *Microwave Devices*: Ferrite materials are integral to the operation of microwave devices like circulators, isolators, and phase shifters. They help control the direction of microwave signals and protect sensitive components from reflected power [53].

v. *Data Storage*: While newer technologies like solid-state drives dominate the data storage market, some specialized applications, such as long-term archival storage, still use magnetic tape. Ferrite heads and magnetic media play a role in these systems.

vi. *Antennas*: Ferrite-loaded antennas are used in various wireless communication systems to improve antenna performance, reduce interference, and enhance signal reception.

vii. *RFID* (*Radio-Frequency Identification*): Ferrite materials are used in RFID tags and readers, allowing for wireless identification and tracking of objects in applications like inventory management, access control, and logistics [54–56].

viii. *Magnetic Resonance Imaging* (*MRI*): In the medical field, ferrite materials are used as part of the gradient coil systems in MRI machines, helping to create precise magnetic field gradients for imaging [57].

ix. *Magnetic Water Treatment*: Some applications employ ferrite magnets to treat water. These magnets are claimed to alter the properties of water and reduce scale buildup in pipes and appliances.

x. *Automotive Electronics*: Ferrites are used in various automotive electronic systems, including ignition systems, sensors, and power electronics, to improve performance and reduce EMI.

xi. *Wireless Power Transfer*: Ferrite materials are used in wireless charging systems, such as Qi wireless chargers for smartphones, to improve energy transfer efficiency and reduce EMI [58].

xii. *5G Technology*: In the development of 5G technology, ferrites are used in antennas and other components to enable high-frequency signal processing and transmission [59].

xiii. *Consumer Electronics*: Ferrites are incorporated into a range of consumer electronics, including speakers, headphones, and power adapters, to enhance performance and reduce interference [60].

Ferrites continue to find new applications and play a vital role in various industries due to their magnetic characteristics, ability to operate at high frequencies, and their capacity to manage EMI. As technology advances, the versatility and utility of ferrite materials are likely to expand even further.

1.8 NOTABLE DISCOVERIES AND INNOVATIONS OF FERRITES

Ferrites have a rich history of notable discoveries and innovations that have contributed significantly to various fields of science and technology. The following are some significant ferrites-related discoveries and innovations:

i. *Initial Discovery* (*1915*): The discovery of ferrites is often attributed to Kōichi Kajiwara, a Japanese scientist who first synthesized ferrite materials in 1915. His work laid the foundation for future research and applications [61].

ii. *Ferrimagnetism* (*1930s*): The understanding of ferrimagnetism, a property unique to ferrite materials, emerged in the 1930s. It was realized that ferrimagnetic materials have a spontaneous magnetization even in the absence of an external magnetic field, distinguishing them from ferromagnetic materials [62].

iii. *Magnetic Recording* (*1940s*): Ferrite heads and magnetic tapes were introduced in the 1940s, revolutionizing data storage and playback in devices such as tape recorders and computer storage systems. Ferrite materials enabled more reliable and efficient recording and playback of audio and data.

iv. *Radomes* (*1940s*): During World War II, the development of radar technology led to the use of ferrite materials in radomes. Radomes are protective coverings that allow radar signals to pass through while protecting the radar antenna from environmental factors. Ferrites played a crucial role in improving radar system performance [63].

v. *Microwave Devices* (*1950s*): The 1950s saw the rise of microwave technologies, and ferrite materials became essential components in microwave devices such as circulators, isolators, and phase shifters. These devices enabled the efficient control and manipulation of microwave signals.

vi. *Ferrite-Core Memory* (*1950s–1960s*): Ferrite-core memory was a key innovation in early computer technology. It served as a form of RAM and was widely used in computers during the 1950s and 1960s. Ferrite cores stored binary data by magnetizing or demagnetizing tiny ferrite rings [64].

vii. *Telecommunications* (*20th Century*): Ferrite materials have been pivotal in telecommunications, improving the performance of antennas, transformers, and inductors. They helped in signal transmission, reception, and noise reduction in various communication systems.

viii. *Emerging Technologies* (*Modern Era*): Ferrites continue to find applications in emerging technologies such as 5G wireless communication, wireless power transfer, and RFID systems. Their unique magnetic properties are harnessed to enhance performance and efficiency in these cutting-edge applications.

ix. *Medical Imaging* (*MRI*): Ferrite materials are used in MRI machines to create precise magnetic field gradients, allowing for detailed and non-invasive imaging of the human body. This innovation has revolutionized medical diagnosis and research.

x. *Energy Efficiency* (*Modern Era*): Ferrites are increasingly used in power electronics and renewable energy systems to improve energy efficiency and reduce EMI. They play a vital role in modern electric vehicles (EVs) and renewable energy generation systems [65, 66].

These discoveries and innovations highlight the versatility and importance of ferrite materials across a wide range of scientific and technological domains. They continue to be at the forefront of research and development, driving advancements in various industries.

1.9 CONCLUSION

In conclusion, the exploration of ferrite, its history, and fascinating magnetism has shed light on a material that has played a vital role in the development of technology and our understanding of magnetic phenomena. From its origins as a naturally occurring mineral to its engineered forms used in modern electronics, ferrite has continuously evolved, showcasing its remarkable properties in a variety of applications. The historical journey of ferrite has revealed the ingenuity of scientists and engineers who harnessed its unique magnetic characteristics for diverse purposes. Whether in the early experiments of Michael Faraday or the mass production of ferrite core during World War II, the story of ferrite mirrors the advancements in science and technology throughout history. Therefore, the intricate magnetism exhibited by ferrite materials has deepened our comprehension of the fundamental principles of magnetism. The discovery of ferrimagnetism and the role of magnetic domains within ferrite crystals have enriched our understanding of magnetic behaviour, enabling the development of more efficient and versatile magnetic materials. Ferrite continues to be a crucial component in a wide range of applications, from telecommunications to electric motors and transformers. Its low cost, high magnetic permeability, and resistance to electrical currents make it an indispensable material in the design and manufacturing of various electronic devices. Hence, the study of ferrite not only offers a historical perspective on the evolution of materials science but also underscores the enduring importance of this magnetic material in shaping our modern technological landscape. As we continue to push the boundaries of scientific knowledge and technological innovation, ferrite remains a valuable resource with the potential for even greater discoveries and applications in the future.

REFERENCES

1. Harris, V. G. (2022). Societal benefits of ferrites: Historical, scientific, and commercial breakthroughs. *Modern Ferrites: Basic Principles, Processing and Properties*, 1, 9–20.
2. Sugimoto, M. (1999). The past, present, and future of ferrites. *Journal of the American Ceramic Society*, 82(2), 269–280.
3. Herlihy, E. A. (1999). Studies of ferrite, borosilicate and intercalation materials (Doctoral dissertation, University College Cork).
4. Dormann, J. L., Fiorani, D., & Tronc, E. (1997). Magnetic relaxation in fine-particle systems. *Advances in Chemical Physics*, 98, 283–494.
5. Fuller, A. J. B. (1987). *Ferrites at Microwave Frequencies (No. 23)*. IEE Electromagnetic Waves Series 23, Peter Peregrinus Ltd. UK, Institution of Electrical Engineers.

6. Néel, L. (1948). Magnetic properties of ferrites: Ferrimagnetism and antiferromagnetism. *Ann. Phys. Paris* 3, 137–198.
7. Button, K. J. (1984). Microwave ferrite devices: The first ten years. *IEEE Transactions on Microwave Theory and Techniques*, 32(9), 1088–1096.
8. Albanese, G. (1977). Recent advances in hexagonal ferrites using nuclear spectroscopic methods. *Le Journal de Physique Colloques*, 38(C1), C1–85.
9. Reddy, C. G., Manorama, S. V., & Rao, V. J. (2000). Preparation and characterization of ferrites as gas sensor materials. *Journal of Materials Science Letters*, 19, 775–778.
10. Bengochea, R., López, B., & Gutiérrez, I. (1998). Effect of Retained Strain on the Microstructural Evolution during the Austenite to Ferrite Transformation. In *Materials Science Forum* (Vol. 284, pp. 201–208). Trans Tech Publications Ltd, Switzerland.
11. Thakur, A., Thakur, P., & Khurana, S. P. (2022). *Synthesis and Applications of Nanoparticles* (pp. 1–17). Singapore: Springer.
12. Beladi, H., Kelly, G. L., Shokouhi, A., & Hodgson, P. D. (2004). The evolution of ultrafine ferrite formation through dynamic strain-induced transformation. *Materials Science and Engineering: A*, 371(1-2), 343–352.
13. Liu, M., Song, C., & Cui, Z. (2021). Crystallographic texture evolution and martensite transformation in the strain hardening process of a ferrite-based low-density steel. *Journal of Materials Science & Technology*, 78, 247–259.
14. Pathak, T. K., Vasoya, N. H., Lakhani, V. K., & Modi, K. B. (2010). Structural and magnetic phase evolution study on needle-shaped nanoparticles of magnesium ferrite. *Ceramics International*, 36(1), 275–281.
15. Bellusci, M., Canepari, S., Ennas, G., Barbera, A. L., Padella, F., Santini, A., … & Varsano, F. (2007). Phase evolution in synthesis of manganese ferrite nanoparticles. *Journal of the American Ceramic Society*, 90(12), 3977–3983.
16. Thakur, P., Taneja, S., Sindhu, D., Lüders, U., Sharma, A., Ravelo, B., & Thakur, A. (2020). Manganese zinc ferrites: A short review on synthesis and characterization. *Journal of Superconductivity and Novel Magnetism*, 33, 1569–1584.
17. Eghbali, B., & Abdollah-Zadeh, A. (2006). Influence of deformation temperature on the ferrite grain refinement in a low carbon Nb–Ti microalloyed steel. *Journal of Materials Processing Technology*, 180(1–3), 44–48.
18. Azizi, A., & Sadrnezhaad, S. K. (2010). Effects of annealing on phase evolution, microstructure and magnetic properties of mechanically synthesized nickel-ferrite. *Ceramics International*, 36(7), 2241–2245.
19. Swaminathan, R., Woods, J., Calvin, S., Huth, J., & McHenry, M. E. (2006). Microstructural evolution model of the sintering behaviour and magnetic properties of NiZn ferrite nanoparticles. *Advances in Science and Technology*, 45, 2337–2344.
20. Abe, F., & Nakazawa, S. (1992). Microstructural evolution and creep behaviour of bainitic, martensitic, and martens ite–ferrite dual phase Cr–2W steels. *Materials Science and Technology*, 8(12), 1063–1069.
21. Rikukawa, H. (1982). Relationship between microstructures and magnetic properties of ferrites containing closed pores. *IEEE Transactions on Magnetics*, 18(6), 1535–1537.
22. Shokrollahi, H., & Janghorban, K. (2007). Influence of additives on the magnetic properties, microstructure and densification of Mn–Zn soft ferrites. *Materials Science and Engineering: B*, 141(3), 91–107.
23. Bhalla, N., Taneja, S., Thakur, P., Sharma, P. K., Mariotti, D., Maddi, C., … & Thakur, A. (2021). Doping independent work function and stable band gap of spinel ferrites with tunable plasmonic and magnetic properties. *Nano Letters*, 21(22), 9780–9788.
24. Sato, T., Iijima, T., Seki, M., & Inagaki, N. (1987). Magnetic properties of ultrafine ferrite particles. *Journal of Magnetism and Magnetic Materials*, 65(2–3), 252–256.
25. Thakur, P., & Thakur, A. (2022). Nanomaterials, Their Types and Properties. In *Synthesis and Applications of Nanoparticles* (pp. 19–44). Singapore: Springer Nature Singapore.
26. Franco, A. (2010). High temperature magnetic properties of cobalt ferrite nanoparticles. *Applied Physics Letters*, 96(17).
27. Punia, P., Bharti, M. K., Dhar, R., Thakur, P., & Thakur, A. (2022). Recent advances in detection and removal of heavy metals from contaminated water. *ChemBioEng Reviews*, 9(4), 351–369.
28. Bharti, M. K., Chalia, S., Thakur, P., & Thakur, A. (2021). Effect of lanthanum doping on microstructural, dielectric and magnetic properties of $Mn_{0.4}Zn_{0.6}Cd_{0.2}La_xFe1._{8-x}O_4$ ($0.0 \leq x \leq 0.4$). *Journal of Superconductivity and Novel Magnetism*, 34(10), 2591–2600.
29. Ichiyanagi, Y., Kubota, M., Moritake, S., Kanazawa, Y., Yamada, T., & Uehashi, T. (2007). Magnetic properties of Mg-ferrite nanoparticles. *Journal of Magnetism and Magnetic Materials*, 310(2), 2378–2380.
30. Liu, X., Wang, J., Gan, L. M., & Ng, S. C. (1999). Improving the magnetic properties of hydrothermally synthesized barium ferrite. *Journal of Magnetism and Magnetic Materials*, 195(2), 452–459.
31. Chahar, D., Thakur, P., Sun, A. C. A., & Thakur, A. (2023). Investigation of structural, electrical and magnetic properties of nickel substituted Co–Zn nanoferrites. *Journal of Materials Science: Materials in Electronics*, 34(10), 901.
32. Punia, P., Dhar, R., Ravelo, B., Trukhanov, A. V., Panina, L. V., Thakur, P., & Thakur, A. (2021). Microstructural, optical and magnetic study of Ni–Zn nanoferrites. *Journal of Superconductivity and Novel Magnetism*, 34(8), 2131–2140.
33. Kwon, H. J., Shin, J. Y., & Oh, J. H. (1994). The microwave absorbing and resonance phenomena of Y-type hexagonal ferrite microwave absorbers. *Journal of Applied Physics*, 75(10), 6109–6111.
34. Iida, K., Minachi, Y., Masuzawa, K., Kawakami, M., Nishio, H., & Taguchi, H. (1999). High-performance ferrite magnets: M-type Sr-ferrite containing lanthanum and cobalt. *Journal of the Magnetics Society of Japan*, 23(4_2), 1093–1096.
35. Thakur, A., Thakur, P., Kumar, D., & Sharma, P. B. (2022). Electromagnetic Characterizations of Nanomaterials. In *Synthesis and Applications of Nanoparticles* (pp. 255–265). Singapore: Springer Nature Singapore.
36. Jing, W., Hong, Z., Shuxin, B., Ke, C., & Changrui, Z. (2007). Microwave absorbing properties of rare-earth elements substituted W-type barium ferrite. *Journal of Magnetism and Magnetic Materials*, 312(2), 310–313.
37. Rana, K., Thakur, P., Tomar, M., Gupta, V., & Thakur, A. (2018). Investigation of cobalt substituted M-type barium ferrite synthesized via Co-precipitation Method for Radar Absorbing Material in Ku-band (12–18 GHz). *Ceramics International*, 44, 6370–6375.
38. Shaula, A. L., Pivak, Y. V., Waerenborgh, J. C., Gaczyñski, P., Yaremchenko, A. A., & Kharton, V. V. (2006). Ionic conductivity of brownmillerite-type calcium ferrite under oxidizing conditions. *Solid State Ionics*, 177(33–34), 2923–2930.

39. Nakamura, T., & Hankui, E. (2003). Control of high-frequency permeability in polycrystalline (Ba, Co)-Z-type hexagonal ferrite. *Journal of Magnetism and Magnetic Materials*, 257(2–3), 158–164.
40. Li, B. W., Shen, Y., Yue, Z. X., & Nan, C. W. (2007). Influence of particle size on electromagnetic behavior and microwave absorption properties of Z-type Ba-ferrite/polymer composites. *Journal of Magnetism and Magnetic Materials*, 313(2), 322–328..
41. Lu, H. Y., Zhu, J. G., & Hui, S. R. (2007). Measurement and modelling of thermal effects on magnetic hysteresis of soft ferrites. *IEEE Transactions on Magnetics*, 43(11), 3952–3960.
42. Suhag, D., Thakur, A., & Thakur, P. (eds) (2023). *Integrated Nanomaterials and Their Applications* (pp. 1–499). Singapore: Springer Nature Singapore. https://doi.org/10.1007/978-981-99-6105-4
43. Narang, S. B., & Pubby, K. (2021). Nickel spinel ferrites: A review. *Journal of Magnetism and Magnetic Materials*, 519, 167163.
44. Taneja, S., Thakur, P., Kumar, D., Slimani, Y., Thakur, A. (2024). Comparison of RE doping (Nd and Sm) on structural and magnetic properties of nickel-zinc-bismuth nano ferrites. *Journal of Rare Earths*, 42, 534–542.
45. Setiawan, A., Maulana, Y. Y., Sulaeman, Y., Praludi, T., & Taryana, Y. (2017, October). Design of 3 GHz stripline ferrite circulator for radar applications. In *2017 International Conference on Radar, Antenna, Microwave, Electronics, and Telecommunications (ICRAMET)* (pp. 154–157). IEEE.
46. Matters-Kammerer, M., Mackens, U., Reimann, K., Pietig, R., Hennings, D., Schreinemacher, B., … & Martiny, C. (2006). Material properties and RF applications of high k and ferrite LTCC ceramics. *Microelectronics Reliability*, 46(1), 134–143.
47. Roess, E. (1982). Soft magnetic ferrites and applications in telecommunication and power converters. *IEEE Transactions on Magnetics*, 18(6), 1529–1534.
48. Morisako, A., Naka, T., Ito, K., Takizawa, A., Matsumoto, M., & Hong, Y. K. (2002). Properties of Ba-ferrite/AlN double layered films for perpendicular magnetic recording media. *Journal of Magnetism and Magnetic Materials*, 242, 304–310.
49. Harris, V. G., Geiler, A., Chen, Y., Yoon, S. D., Wu, M., Yang, A., … & Vittoria, C. (2009). Recent advances in processing and applications of microwave ferrites. *Journal of Magnetism and Magnetic Materials*, 321(14), 2035–2047.
50. Baba, P. D., Argentina, G., Courtney, W., Dionne, G., & Temme, D. (1972). Fabrication and properties of microwave lithium ferrites. *IEEE Transactions on Magnetics*, 8(1), 83–94.
51. Urabe, J., Fujii, K., Dowaki, Y., Jito, Y., Matsumoto, Y., & Sugiura, A. (2006). A method for measuring the characteristics of an EMI suppression ferrite core. *IEEE Transactions on Electromagnetic Compatibility*, 48(4), 774–780.
52. Tang, S. C., Hui, S. Y., & Chung, H. H. (2002). Evaluation of the shielding effects on printed-circuit-board transformers using ferrite plates and copper sheets. *IEEE Transactions on Power Electronics*, 17(6), 1080–1088.
53. Amiri, S., & Shokrollahi, H. (2013). The role of cobalt ferrite magnetic nanoparticles in medical science. *Materials Science and Engineering: C*, 33(1), 1–8.
54. Tatarchuk, T., Bououdina, M., Judith Vijaya, J., & John Kennedy, L. (2017). Spinel ferrite nanoparticles: synthesis, crystal structure, properties, and perspective applications. In *Nanophysics, Nanomaterials, Interface Studies, and Applications. NANO2016,* Springer Proceedings in Physics, vol 195. (pp. 305–325) Springer, Cham.
55. Nishioka, Y., Maeshima, O., Uno, T., & Adachi, S. (1999). FDTD analysis of resistor-loaded bow-tie antennas covered with ferrite-coated conducting cavity for subsurface radar. *IEEE Transactions on Antennas and Propagation*, 47(6), 970–977.
56. Zhu, H., Lai, S., & Dai, H. (2007, September). Solutions of metal surface effect for hf rfid systems. In *2007 International Conference on Wireless Communications, Networking and Mobile Computing* (pp. 2089–2092). IEEE.
57. Umut, E., Coşkun, M., Pineider, F., Berti, D., & Güngüneş, H. (2019). Nickel ferrite nanoparticles for simultaneous use in magnetic resonance imaging and magnetic fluid hyperthermia. *Journal of Colloid and Interface Science*, 550, 199–209.
58. Sharma, P., Sharma, A., Sharma, M., Bhalla, N., Estrela, P., Jain, A., … & Thakur, A. (2017). Nanomaterial fungicides: In vitro and in vivo antimycotic activity of cobalt and nickel nanoferrites on phytopathogenic fungi. *Global Challenges*, 1(9), 1700041.
59. Li, J., Tsai, M., Chiu, R., He, E., Hsieh, A., Tsai, M. F., & Wang, Y. P. (2020, June). EMI shielding technology in 5G RF system in package module. In *2020 IEEE 70th Electronic Components and Technology Conference (ECTC)* (pp. 931–937). IEEE.
60. Tejeda, A., Carretero, C., Boys, J. T., & Covic, G. A. (2016). Ferrite-less circular pad with controlled flux cancelation for EV wireless charging. *IEEE Transactions on Power Electronics*, 32(11), 8349–8359.
61. Uchida, K., Nonaka, T., Kikkawa, T., Kajiwara, Y., & Saitoh, E. (2013). Longitudinal spin Seebeck effect in various garnet ferrites. *Physical Review B*, 87(10), 104412.
62. Rado, G. T., Wright, R. W., & Emerson, W. H. (1950). Ferromagnetism at very high frequencies. III. Two mechanisms of dispersion in a ferrite. *Physical Review*, 80(2), 273.
63. Parsa, A., Kodera, T., & Caloz, C. (2010). Ferrite based non-reciprocal radome, generalized scattering matrix analysis and experimental demonstration. *IEEE Transactions on Antennas and Propagation*, 59(3), 810–817.
64. Punia, P., Thakur, P., Kumar, R., Syal, R., Dhar, R., & Thakur, A. (2022). Synthesis and characterization of Ca substituted Ni-Zn nanoferrites-microstructural, magnetic and dielectric analysis. *Journal of Alloys and Compounds*, 928, 167248.
65. Thakur, P., Sharma, P., Mattei, J. L., Queffelec, P., Trukhanov, A. V., Trukhanov, S. V., & Thakur, A. (2018). Influence of cobalt substitution on structural, optical, electrical and magnetic properties of nanosized lithium ferrite. *Journal of Materials Science: Materials in Electronics*, 29, 16507–16515.
66. Verma, R., Thakur, P., Chauhan, A., Jasrotia, R., & Thakur, A. (2023). A review on MXene and its' composites for electromagnetic interference (EMI) shielding applications. *Carbon*, 208, 170–190.

2 Synthesis Techniques for Spinel Nano-Ferrites

Atul Thakur and Preeti Thakur*
Amity Institute of Nanotechnology, Amity University Haryana, Gurugram, India
Nanotechnology Wing, Innovative Science Research Society, Shimla, India
Nano LatticeX, Amity University Haryana, Gurugram, India

Shivangi Sangwan
Amity Institute of Nanotechnology, Amity University Haryana, Gurugram, India

Dinesh Kumar
Amity Institute of Nanotechnology, Amity University Haryana, Gurugram, India
and
DesignInnova Pvt Ltd., New Delhi, India

Ritesh Verma
Graphic Era (Deemed to be) University Dehradun, Uttarakhand, India

Jean-Luc Mattei
Lab-STICC, CNRS, University of Brest, Brest, France

2.1 INTRODUCTION

A significant category of materials known as ferrite has the potential to be used in integrated circuitry, transformer cores, and magnetic recording. Because ferrites possess novel features and can be used in a variety of technological applications, the study of ferrites has garnered a significant amount of interest from the scientific community [1–3]. This is especially true when the particle size reaches the nanoscale scale. One of the areas that is seeing the most activity is the study of nanocrystalline materials. In today's world, research in the field of material science is focused on developing new materials that have improved qualities as well as unique methods of synthesis in order to keep up with the ever-increasing demand from the technological industry [4–6]. When compared with their bulk counterparts, ferrite nanoparticles (NPs) have fundamentally different magnetic and electrical properties, which makes the study of these particles extremely interesting. It has been discovered that nano-oxides, due to the electrical qualities they possess, make for excellent catalysts, in comparison to the conventional oxides that correspond to them. Nanomaterials are currently the focus of much study because of the vast number of uses that can be found for them and the intriguing features that they possess [7–9]. Ferrites are a class of ferromagnetic compounds that often make use of magnetic NPs. Their chemical formula is MFe_2O_4, where M can represent any one of numerous divalent metallic elements such as Mg, Fe, Co, Ni, or Cu [10–12]. Magnetic NPs of spinel ferrites are of great interest in fundamental science, especially for addressing the fundamental relationships between magnetic properties and their crystal chemistry and structure [13]. Crystal chemistry is the study of how the internal structure, chemical make-up (chemical formula), and physical characteristics of minerals are interconnected with one another. In recent years, research has been conducted on spinel ferrites for the purpose of determining their relevant electrical and magnetic properties as well as potential applications in the fields of information storage systems, magnetic bulk cores, magnetic fluids, microwave absorbers, and medical diagnostics [14–16]. The synthesis and magnetic structure characterization of spinel metastable nanoferrites have been investigated with a great deal of interest and a great deal of attention has been focused on the preparation and characterization of superparamagnetic metal oxide NPs of spinel ferrites, $MeFe_2O_4$ (metal Me = Co, Mg, Mn, Zn, etc.) [17–19]. Spinel ferrite systems have a complex crystal chemistry, which provides significant potential for understanding and fine-tuning the superparamagnetic characteristics of NPs through the use of chemical manipulation. For the most part, researchers made use of a number of different synthesis procedures in order to create metal ferrite. These processes include combustion, hydrothermal, co-precipitation, and citrate precursor and other techniques like sol-gel [20].

* Corresponding author.

DOI: 10.1201/9781003568940-2

2.2 SYNTHESIS TECHNIQUES OF SPINEL FERRITES

2.2.1 Sol-Gel Auto-Combustion Process for Spinel Ferrite Synthesis

This technique (Figure 2.1) utilizes an exothermic and self-sustaining thermally induced anionic redox reaction of xerogel. The xerogel is generated from an aqueous solution that contains the required metal salts (oxidizer) and organic complexant (reductant). Burning of reactant mixes in a combustion reaction. In most cases, the ratio of complexant to salts is computed according to the valences of the reactive elements in order to achieve a ratio of oxidizer to reductant that is equal to 1. Because they provide a source of water-soluble NO^{-}_{3} oxidant that may be used at low temperatures, nitrate salts are favored for usage as precursors in chemical synthesis. In certain processes, the metal nitrates and the complexant are mixed together immediately without the addition of water. This process involves vigorous swirling and heating. Since metal nitrates have the property of hygroscopicity, they readily take on the state of slurry when exposed to moisture. The flash-combustion method refers to this particular variation of the sol-gel auto-combustion process [21]. In order to prepare the precursor solution, the metal salts, such as nitrates, chlorides, or acetates, are first dissolved in a suitable solvent, such as water, alcohol, or a mixture of the two. Next, the solution is heated to a temperature that is high enough to completely evaporate the solvent. In order to create a sol-gel precursor, a chelating agent such as citric acid or ethylene glycol is added to the mixture. In order to facilitate gelation, the sol-gel precursor goes through the heating process. This can be accomplished by heating the food on a hot plate, in an oven, or by subjecting it to irradiation from a microwave. The gel is first dried, and then it is calcined at a high temperature in order to eliminate the chelating agent as well as any other impurities and to generate the spinel ferrite NPs. Kumar et al. [22] studied that the $Ni_{1-x}Cu_xFe_2O_4$ ferrites (x = 0.0, 0.1, 0.2, 0.3, 0.4, 0.5, 0.6, 0.7, 0.8, 0.9, and 1.0) have been synthesized using the sol-gel process in order to obtain a homogeneous crystal structure, and they have then been sintered at a high temperature. Chen et al. [23] used sol-gel method, which was utilized in the preparation of $NiFe_2O_4$ NPs, and polyacrylic acid (PAA) was used as the chelating agent.

2.2.2 Co-precipitation Process for Spinel Ferrite Synthesis

The co-precipitation method is a straightforward process that can be utilized in order to fabricate ferrite NPs at lower temperatures. In addition, the co-precipitated NPs are suitable for magnetic characterization since they have a high purity and do not agglomerate. On the other hand, because the size of the crystallites may be controlled, the co-precipitation approach can be utilized for the purpose of determining how the magnetic characteristics vary with crystallite size. The production of ferrite NPs can be accomplished by the co-precipitation approach (Figure 2.2), which is characterized by its ease of use and adaptability. It requires the use of a base in order to accomplish the simultaneous precipitation of metal ions from an aqueous solution. The metal salts should be dissolved in an appropriate solvent. In most cases, distilled water is employed as the solvent; however, ethanol and methanol are two other examples of solvents that can be utilized. In most cases,

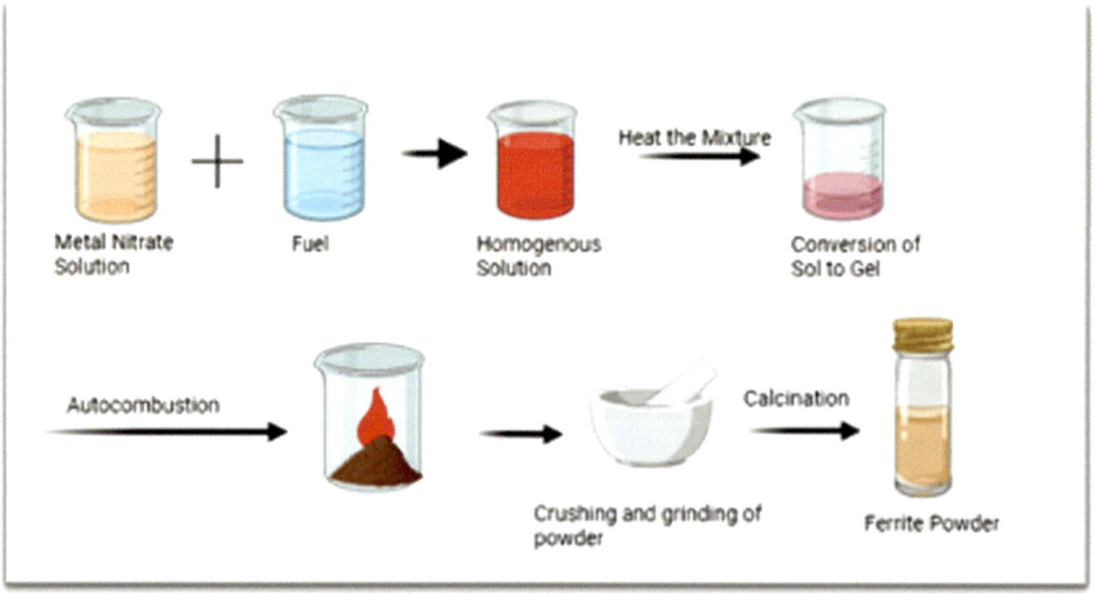

FIGURE 2.1 Diagrammatic representation of synthesis of spinel ferrites via Sol-Gel auto-combustion method.

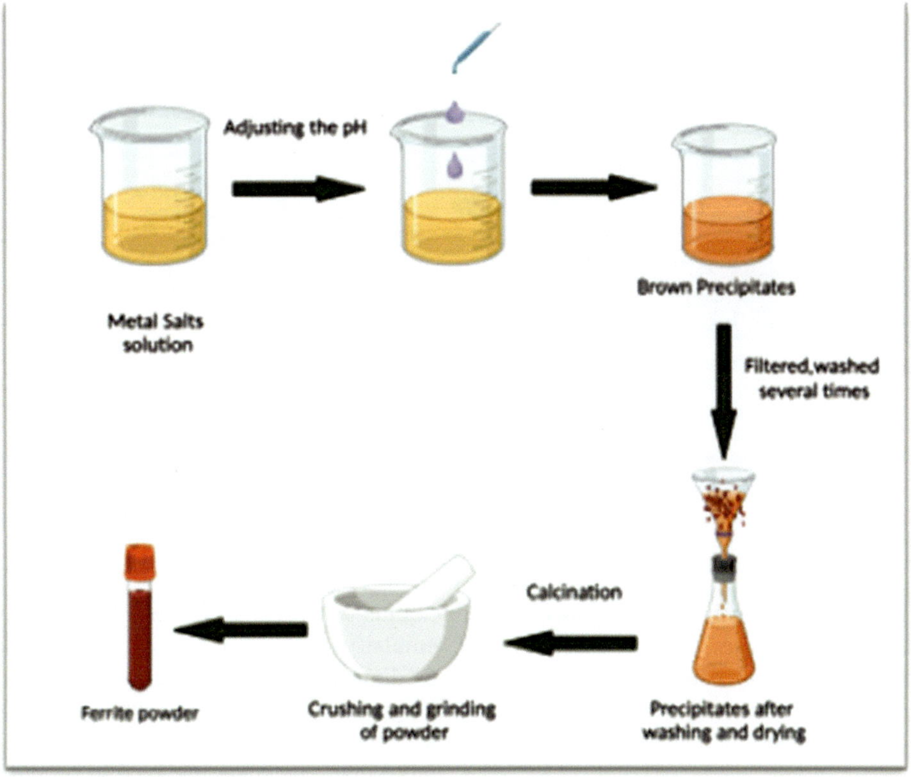

FIGURE 2.2 Diagrammatic representation of synthesis of spinel ferrites via co-precipitation method.

nitrates or chlorides are present in the metal salts. Make the necessary adjustments to bring the pH of the solution up to the required level. The particle size of the ferrites as well as their shape will be influenced by the pH of the solution. For the co-precipitation of ferrites, a pH in the range of 10–12 is commonly utilized. If you want the metal ions to precipitate out of the solution, add a base. In most cases, either sodium hydroxide (NaOH) or potassium hydroxide (KOH) is used as the base. Drop by drop, the base is added while continuously forcefully swirling the solution. Maintain constant agitation of the solution for the allotted amount of time. The ferrite particles are able to expand and aggregate as a result of this. Distilled water and ethanol should be used to wash the precipitate. The ferrites are cleansed of any contaminants as a result of this process. At a higher temperature, the precipitation should be dried. This eliminates all trace of the solvent and results in the precipitate being converted into a powder. Shaikh et al. [24] used the co-precipitation method to create fine particles of the $Co_{1-x}Ni_xFe_{1.9}Mn_{0.1}O_4$ system, with x being equal to 0.2, 0.4, 0.6, and 0.8, respectively, whereas by using the co-precipitation approach, Shahraki et al. [25] successfully produced zinc ferrite ($ZnFe_2O_4$) NPs without undergoing any subsequent calcination; finally Rahman et al. [26] used co-precipitation approach to create NPs of soft magnetic spinel ferrite ZnMg-ferrite ($Zn_{1-x}Mg_xFe_2O_4$, where x ranges from 0.0 to 40.0, 0.1 to 0.2, 0.3 to 0.4, and 0.5).

2.2.3 Solvothermal Process for Spinel Ferrite Synthesis

The ease with which solvothermal synthesis can be performed, in addition to the fact that it enables the production of highly crystalline products at temperatures that is significantly lower than those required by other procedures, which are just two of the many reasons why it is superior to other methods. There are many more reasons why it is superior to other methods, but these are just two of the most important ones. In addition, it is possible to precisely regulate the chemical composition, and uniform nanopowders can be produced using this method. As a result of this, the particle form and size can be controlled, for which the approach is able to control the growth of crystals, and it is adequate enough to manufacture samples on a large scale. In order to start the solvothermal synthesis of ferrites (Figure 2.3), you would typically start by dissolving metal salts or precursors in a solvent such as ethanol. This would be the first step in the process. After an extra half an hour of stirring, the mixture is ready to use. Then the mixture is transferred into an autoclave constructed of stainless steel, where it will be heated for a period of time in a vessel that is completely sealed while being subjected to high temperatures and pressures. After that, the lid of the autoclave is removed. Upon being washed with ethanol on a number of times, the material that was obtained is dried in an oven at a temperature of 80°C.

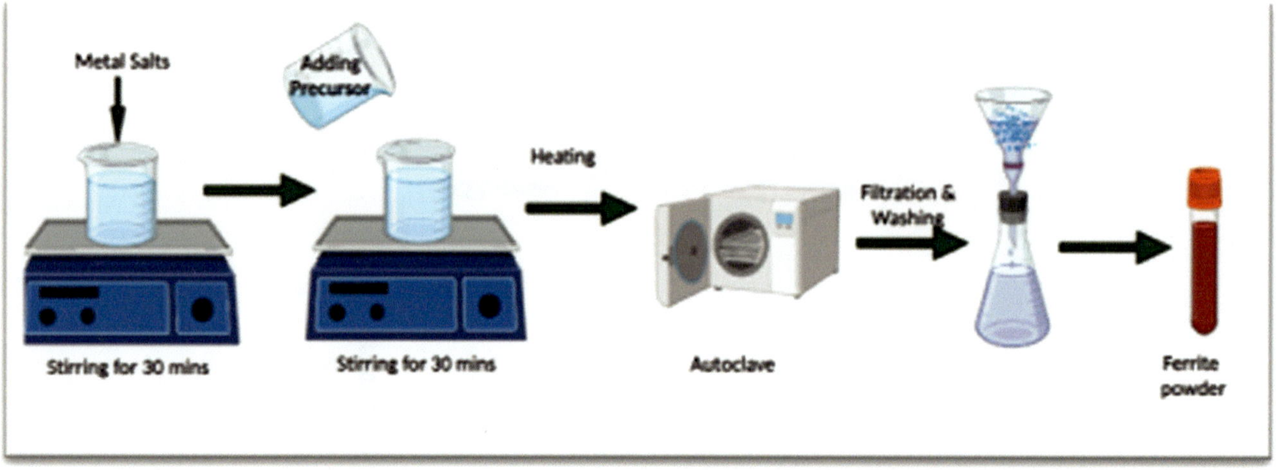

FIGURE 2.3 Diagrammatic representation of synthesis of spinel ferrites via Solvothermal method.

As Deng Ni et al. [27] shown, nanocomposites of nickel-zinc ferrites, also known as Ni-Zn ferrites, were produced using a solvothermal technique, whereas Chaudhuri et al. [28] used solvothermal approach to produce well-crystallized bismuth ferrite NPs without the assistance of any mineralizer.

2.2.4 Microemulsion Process for Spinel Ferrite Synthesis

An example of a type of liquid dispersion system known as a microemulsion is one that is composed of two immiscible liquids (often oil and water) that are held together by an emulsifying agent. It is possible to employ it as a reaction medium in order to create ferrites (Figure 2.4), which are magnetic materials that have a variety of applications in the fields of electronics, medicine, and catalysis. There is more than one stage involved in producing ferrites by the use of microemulsions. First, the microemulsion is made by vigorously swirling together the oil phase, which is often a hydrocarbon, and the water phase, which is typically an aqueous solution containing a surfactant. This is done while the oil and water phases are being mixed

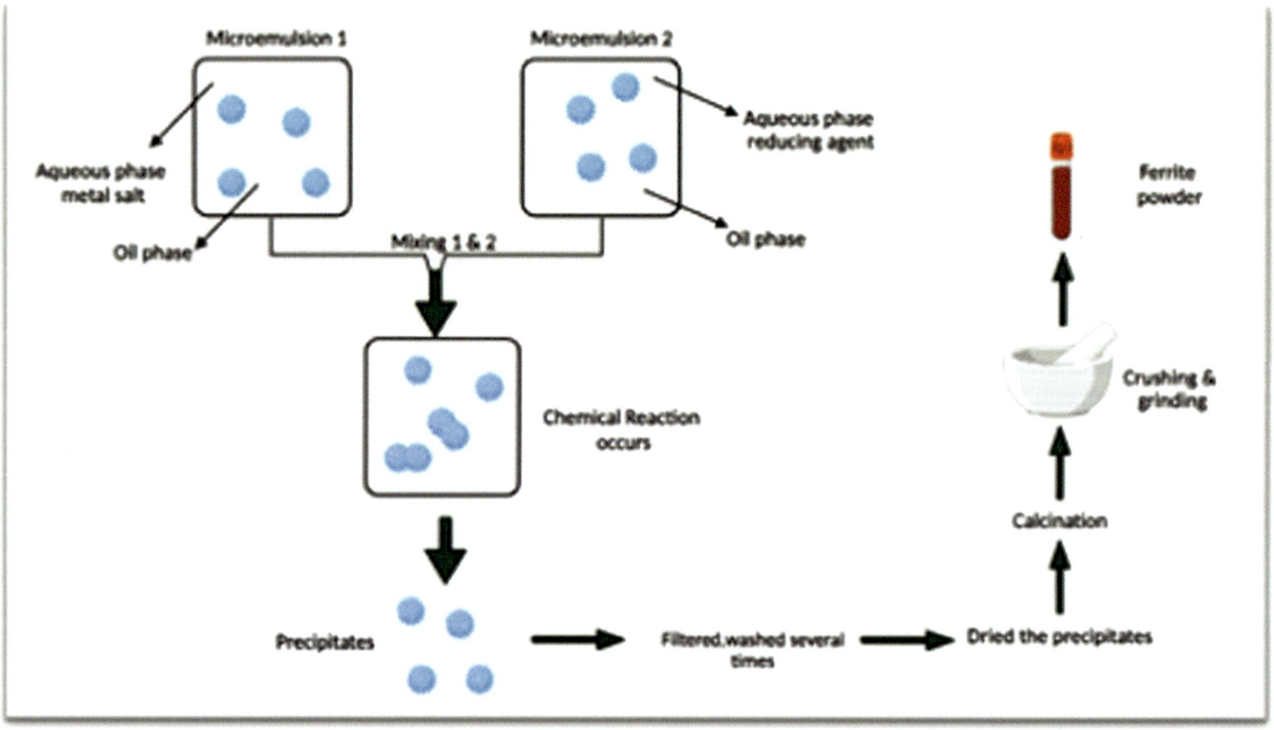

FIGURE 2.4 Diagrammatic representation of synthesis of spinel ferrites via Microemulsion method.

together. This leads to the production of a liquid system that is transparent and unshakeable. After that, the microemulsion will have the predecessors of the ferrite material, which are typically metal salts, dissolved in it. The required chemical make-up and physical characteristics of the ferrite should guide the selection of precursors. For instance, zinc ferrite ($ZnFe_2O_4$) can be manufactured by combining zinc nitrate and iron (III) chloride in the appropriate proportions. Next, the microemulsion that contains the precursors goes through a thermal treatment, which is normally carried out at a temperature ranging from 100 to 200°C, in order to cause the creation of ferrite NPs. Modifying the reaction conditions—for example, the temperature, the amount of time, and the concentration of the precursors—allows one to exert influence over the NPs' morphology and size. In the final step of the process, the ferrite NPs are extracted from the microemulsion by employing procedures such as centrifugation or filtration. As Rajjab Ali et al. [29] shown, Ca–Ni co-substituted samples of nanocrystalline spinel ferrites with the chemical formula $Mg_{1-x}Ca_xNi_yFe_{2-y}O_4$ (x=0.0–0.6, y=0.0–1.2) were produced using the microemulsion process, and then those were annealed at 700°C for 7 hours. Azhar Khan et al. [30] used microemulsion process to produce Tb^{3}-doped nanocrystalline zinc ferrites, which had a nominal composition of $Zn_{1-x}T_{b-x}Fe_2O_4$ (x=0.40, 0.03, 0.06, 0.09, 0.12, and 0.15). The nanocrystalline zinc ferrites were then annealed at 600°C for 8.5 hours.

2.2.5 Sonochemical Process for Spinel Ferrite Synthesis

The process of sonochemical synthesis of ferrites (Figure 2.5) uses ultrasound sounds to help ferrite NPs form in a solution. To make ferrite NPs using sonochemistry, ultrasound waves are applied to a fluid that has metal salts and a reducing agent in it. Ultrasound waves make cavitation bubbles in the solution, which help metal ions break down into NPs. Pick the right metal salts to use as building blocks for the ferrite you want to make. Which precursors you use will depend on the type of ferrite mixture you want to make. Choose a solvent that can successfully break down the precursors. Water, ethanol, and other organic liquids are often used as solvents in sonochemical synthesis. To make a uniform solution, dissolve the metal salts (precursors) in the liquid of your choice. It is possible to do this by stirring the mixture at the right temperature until all the precursors are gone. The solution needs a surfactant added to it. During the sonochemical process, surfactants help keep the particles stable and stop them from sticking together. It is important to pick a surfactant that works well with the liquid and doesn't get in the way of making the ferrite you want. Move the solution to a sonication generator or an ultrasonic bath. The fluid is then put in contact with very strong ultrasound waves. The ultrasonic sounds make cavitation bubbles in the solution, which causes high temperatures and pressures to form in certain areas. These very harsh conditions help ferrite NPs form and grow

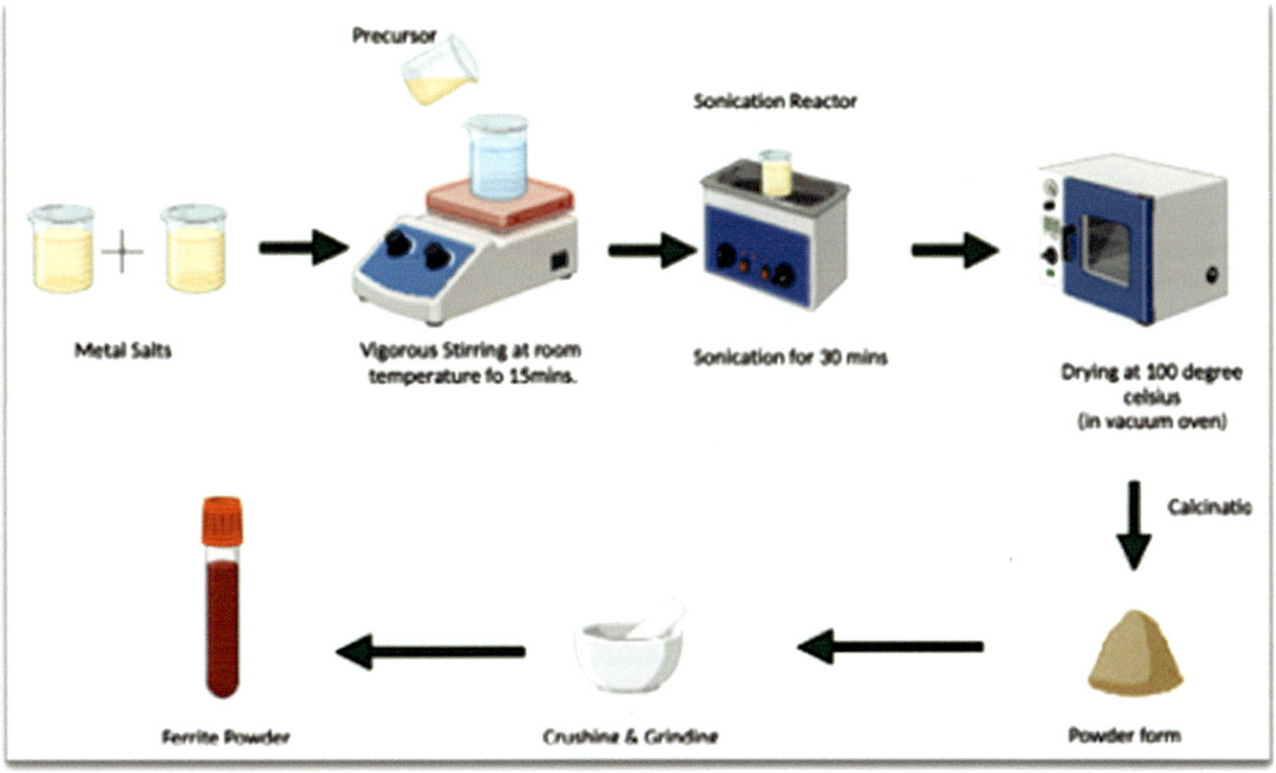

FIGURE 2.5 Diagrammatic representation of synthesis of spinel ferrites via Sonochemical method.

very quickly. How long and how strong the sonication lasts depend on things like the concentration of the precursor, the particle size that is wanted, and the type of ferrite that is being made. Sonochemical treatment can last anywhere from a few minutes to a few hours. Heat the solution to get rid of the solvent after the sonochemical process. Putting the solution on a hot plate or in an oven set to a certain temperature will do this. During the drying process, a solid powder with ferrite NPs can be made. The powder needs to be calcined after the solvent has been fully evaporated. For a certain amount of time, the powder is heated in a kiln at high temperatures (usually above 500°C) during calcination. This step helps to make the ferrite NPs more stable and crystallized. Before going for characterization of the powder, let it cool to room temperature. Almessiere et al. [31] take NPs of the formula $Co_{0.3}Ni_{0.5}Mn_{0.2}Eu_xFe_{2-x}O_4$, where $0.00 \leq x \leq 0.10$ (hereafter called CNMEuF) were made using a sonochemical method and an ultrasonic homogenizer with a frequency of 20 kHz and a power of 70 W.

2.2.6 Solid-State Process for Spinel Ferrite Synthesis

A common way to make ferrite materials with clear magnetic qualities is through solid-state synthesis of ferrites (Figure 2.6). Mixing metal oxides or carbonates in certain stoichiometric amounts, grinding them into a fine powder, and burning them up in a furnace are the steps in the process. During the heating process, the metal oxides or carbonates mix with each other, making a single-phase ferrite material. Which metal oxides or carbonates are used in the synthesis process depends on the qualities of the ferrite material that is being made. Iron oxide (Fe_2O_3) is often used as the iron source. Other metal oxides, like zinc oxide (ZnO), manganese oxide (MnO), and nickel oxide (NiO), can also be used to make different kinds of ferrites. Making sure that the metal oxides or carbonates are evenly spread out in the powder mixture is very important during the mixing process. Usually, this is done by grinding the combination with a ball mill or some other machine. The powder that you get should have particles that are all the same size and shape. Usually, the heating process takes place in a boiler that is hotter than 1000°C. The type of ferrite material being made determines the exact temperature and length of time needed to heat it. If you heat the metal oxides or carbonates, they combine to make a single-phase ferrite material. After that, the material is slowly cooled to room temperature so that it doesn't crack or have any other flaws. It is common to use solid-state synthesis of ferrites to make ferrite magnets, microwaves, and other electrical parts because it can make high-quality, single-phase materials with clear magnetic properties. Mishra et al. [32] used the standard solid-state reaction approach to produce nanostructured manganese ferrites ($MnFe_2O_4$) with diameters ranging from 45 to 30 nm. This was accomplished by doping the material with Ti^{4+} ions.

2.2.7 Microbial-Assisted Process for Spinel Ferrite Synthesis

The production of NPs can be helped by microbes through a technique known as microbial-assisted synthesis of NPs (Figure 2.7). The fact that this approach is both economical and kind to the environment has contributed to its meteoric rise in popularity over the past few years. In order to convert metal ions into their corresponding NPs, the process calls for the utilization of a wide variety of microorganisms such as bacteria, fungus, and algae. These organisms perform the function of reducing agents. Microorganisms are responsible for the production of certain enzymes, which have the ability to convert metal ions into NPs by contributing energy and lowering power requirements. NP synthesis can be accomplished through either extracellular or intracellular means. Both approaches are viable. In the extracellular approach, the microorganisms secrete enzymes that decrease metal ions outside of the cell, but in the intracellular way, the reduction process takes place inside the cell. The extracellular method is the more common of the two. Find a microbe that can create NPs and

FIGURE 2.6 Diagrammatic representation of synthesis of spinel ferrites via solid-state method.

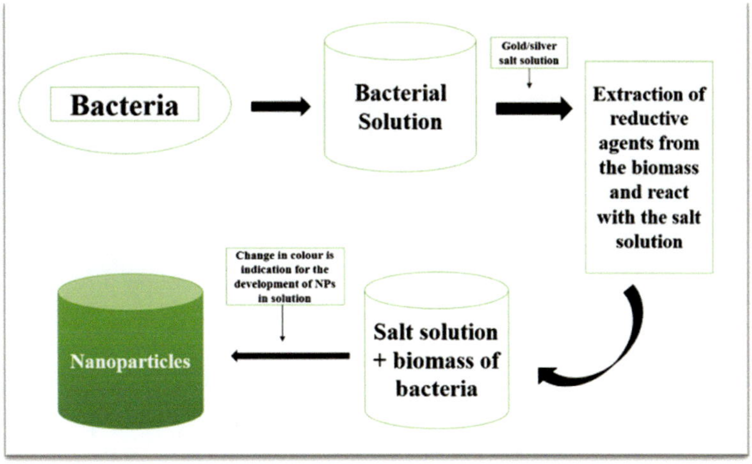

FIGURE 2.7 Diagrammatic representation of synthesis of spinel ferrites via Microbial-assisted method.

has the ability to decrease metal ions, such as bacteria, fungi, or algae, and use it. Because the ability of various microbes to produce NPs differ, it is essential to select the appropriate species. Create a culture medium that can support the growth of the microorganism of your choice while also offering the appropriate nutrients and environmental factors. In order to achieve this goal, it may be necessary to make adjustments to the temperature, pH, and other environmental conditions. Place the microbe of choice into the culture medium, and then let it to develop while maintaining strict control over its environment. This can be accomplished by including a negligible quantity of the microorganism in the medium and then incubating the mixture under the suitable conditions. The culture medium should be supplemented with the desired metal ions, such as silver, gold, or iron. These metal ions will be taken up by the microorganisms, which will then kick off the reduction process. The microorganisms produce enzymes that function as reducing agents, thereby transforming the metal ions into the appropriate NPs of each element. It is possible for this reduction process to take place either extracellularly (outside the cell) or intracellularly (inside the cell), depending on the microbe and method that are utilized. When the synthesis is finished, the NPs have to be extracted from the growth medium and then purified before they can be used. The processes of centrifugation, filtration, and ultracentrifugation are all methods that can be used to accomplish this goal. Mubraiz et al. [33] synthesized cobalt oxide (Co_3O_4) NPs by utilizing plant extracts from the *Aerva javanica* plant, bacterial isolates from the rhizospheric soil of *Potentilla atrosanguinea*, *Swertia petiolata*, and *Senecio chrysanthemoides*, as well as fungus *Fusarium oxysporum*. Zhang et al. [34] gave an overview of the current state of knowledge about the synthesis of NPs by microorganisms such as bacteria, fungus, actinomycetes, and yeast, as well as their effect on the processes that occur within microbiology.

2.2.8 Microwave-Assisted Process for Spinel Ferrite Synthesis

Spinel ferrites are a category of magnetic materials. Microwave-assisted synthesis is a process that speeds up the synthesis of spinel ferrites (Figure 2.8). Spinel ferrite synthesis can benefit from the use of microwave-assisted synthesis in a number of ways, including a reduction in the amount of time needed for reactions, an increase in crystallinity and purity, and more control over particle size and morphology. However, it is essential to carefully adjust the reaction parameters in order to guarantee the quality of the product that is wanted and to prevent any potential problems, such as overheating or the degradation of the sample. The synthesis of spinel ferrite begins with the selection of acceptable metal precursors, hence the first stage in this process is the selection of precursors. In most cases, these precursors consist of metal salts or metal-organic compounds that contain the needed metal ions (such as Fe, Co, Ni, etc.) in addition to the other essential elements. The selected precursors are combined in an appropriate solvent or medium to produce a homogeneous solution or suspension, which is the next step in the preparation of the reaction mixture. In order to exert control over the particle size, morphology, or any number of other qualities, this mixture might also contain additives or surfactants. After this step, the reaction mixture is subjected to irradiation with microwaves. Microwaves are a type of electromagnetic wave that can produce heat by inducing rapid rotation in polar molecules present in a reaction mixture. This results in a localized heating effect. The irradiation from the microwaves causes the reaction mixture to heat up very quickly, which encourages the nucleation and development of spinel ferrite particles. In comparison to more traditional techniques of heating, the targeted heating and rapid energy transfer both contribute to the faster reaction kinetics. During the

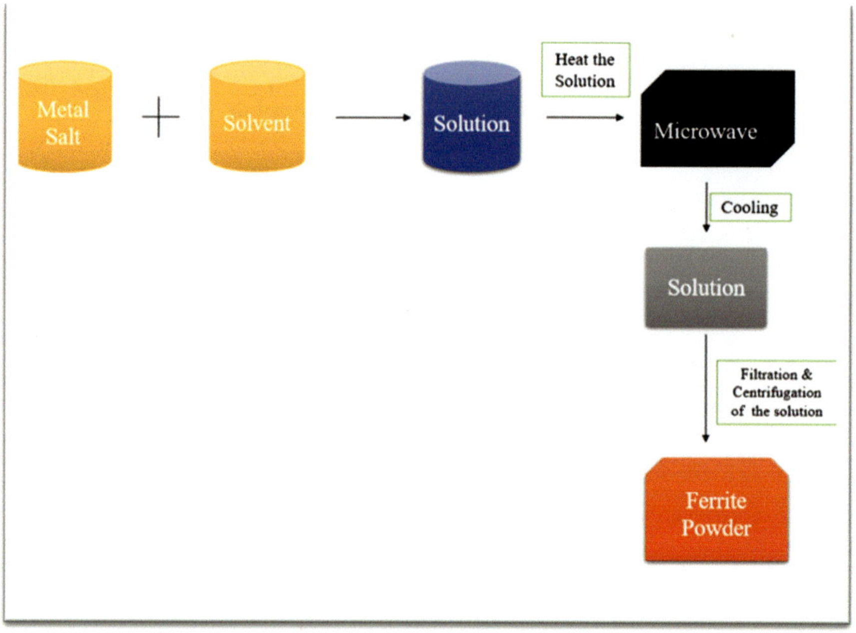

FIGURE 2.8 Diagrammatic representation of synthesis of spinel ferrites via Microwave-Assisted method.

microwave-assisted synthesis, a number of different reaction parameters can be adjusted in order to acquire the properties that are required. In this context, "parameters" refers to the power and duration of the microwave irradiation, the concentration of the precursors, the composition of the solvent, and the presence of any additives. After the desired amount of time has passed for the reaction, the microwave irradiation is turned off, and the mixture that was being used in the reaction is allowed to cool down before the product is recovered. After that, methods like filtration and centrifugation are utilized in order to separate the product that was produced from the medium or solvent that was used. In order to obtain the finished spinel ferrite product, it is subsequently washed and dried. The nanostructured M(II)-substituted $M_xFe_{3-x}O_4$ magnetic ferrites (M = Fe, Mn, or Co; x = 0–1) spherolites have been effectively synthesized by a microwave-assisted reflux process in EG solvent without the inclusion of any surfactants. This method is convenient, robust, and saves time when compared to other methods. The birth and growth of ferrite nanostructures is incredibly fast, and it happens within 30 minutes after heating ferrite with a microwave, As Solano et al. [35] shown, the well-defined synthesis of high-quality spinel ferrite magnetic NPs made of MFe_2O_4 (M = Mn, Fe, Co, Ni, Zn, and Cu), with diameters ranging from 0.1 to 1 nm, below 10 nm, based on facile and experimental evidence, has been described as a solvothermal or microwave-assisted one-pot process that is efficient in heating processes and procedures. The co-precipitation process [36] was used to produce $MnZnFe_2O_4$ ferrite NPs, and the heating device used was a microwave oven set to a temperature of 100°C.

2.2.9 Green Synthesis Process for Spinel Ferrite Synthesis

In the context of material synthesis, the term "green synthesis" refers to an ecologically friendly strategy that minimizes the use of potentially harmful chemicals while emphasizing the use of sustainable technologies [37–39]. Ferrites are a type of magnetic material that can be manufactured through the use of environmentally friendly synthesis methods [40–42]. The utilization of sustainable and non-hazardous precursors is an essential part of green synthesis. When it comes to ferrites, frequent green precursors include metal salts or organic compounds having the required metal ions. These can be used in the manufacturing process (Figure 2.9). In order to make a precursor solution, the desired green precursors are first dissolved in an appropriate solvent. The solvent that is utilized should not be harmful to humans or animals and should not be poisonous. The precursor solution is then infused with a stabilizing agent, which allows for more control over the size, shape, and stability of the synthesized ferrites. Stabilizing agents can be organic molecules like surfactants or polymers, and they help govern the nucleation and development of ferrite NPs by inhibiting or promoting these processes, respectively. In order to facilitate the creation of ferrite NPs, the reaction parameters, which include temperature, pressure, and reaction time, are tuned to their optimum levels. After the reaction is finished, the produced ferrite NPs are extracted from the reaction mixture using methods such as centrifugation or filtration. This step occurs after the reaction has been completed. After the NPs have been obtained, they are cleaned with the proper solvent to get rid of any contaminants or precursors that have

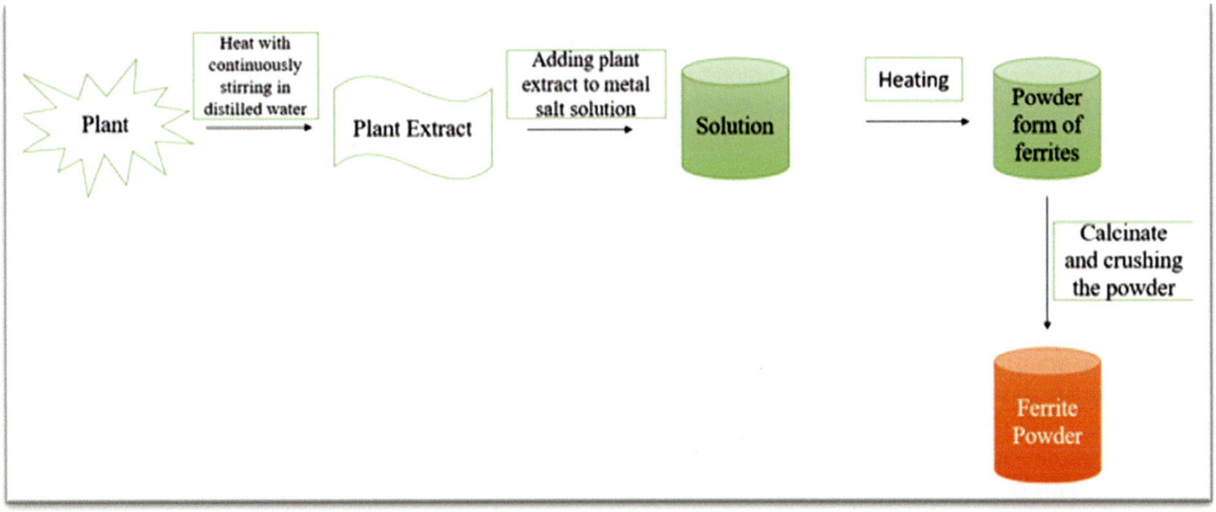

FIGURE 2.9 Diagrammatic representation of synthesis of spinel ferrites via Green Synthesis method.

not yet reacted. In order to achieve a powder that is solid, the washed NPs are finally dried under carefully monitored circumstances. Dhanda et al. [43] used the auto-combustion approach to prepare nickel-doped cobalt nano-ferrites ($Ni_xCo_{1-x}Fe_2O_4$, where x = 0.0 to 1.0 with step size 0.2), and the fuel for this method was aloe vera extract.

2.2.10 Ceramic Processing for Spinel Ferrite Synthesis

Ceramic processing of synthesis of ferrites is the process of applying techniques that are employed in ceramic processing to synthesize ferrite powders, which can subsequently be used to create ferrite ceramics. This process is also known as ceramic processing of synthesis of ferrites as shown in Figure 2.10. The preparation of the raw materials is the very first step in the manufacturing of ceramics. When it comes to ferrites, the primary components are metal oxides like iron oxide (Fe_2O_3) as well as metal carbonates or nitrates. These primary components are chosen with great care and then weighed out to get the correct stoichiometric ratio. After the quantities of the raw components have been determined, they are combined in order to create a consistent batch. Depending on the particular demands

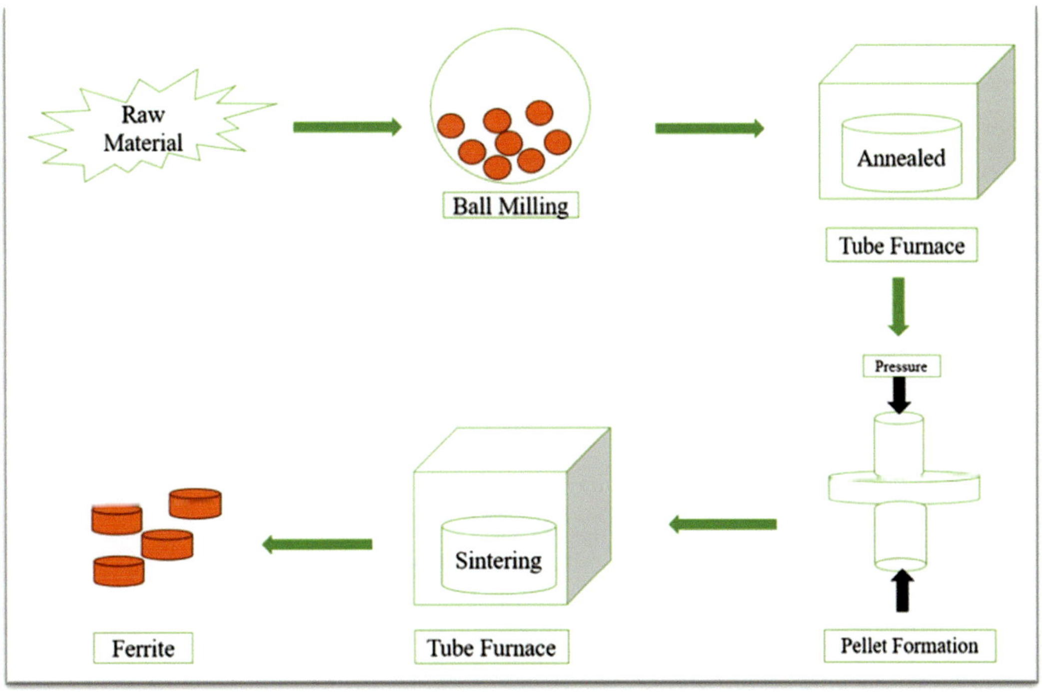

FIGURE 2.10 Diagrammatic representation of synthesis of spinel ferrites via Ceramic processing method.

of the synthesis process, this can be accomplished through the use of a variety of techniques, such as dry mixing (ball milling) or wet mixing (attrition milling), among others. Calcination is the process that is used on the powder after it has been combined and heated to a very high temperature. This stage involves heating the powder combination in a controlled atmosphere in order to promote chemical reactions and transform the raw ingredients into the required ferrite phase. The heat is applied to the mixture while it is in powder form. Temperature and time of calcination are fine-tuned to get optimum results in terms of achieving a full chemical transformation and the production of the crystal structure that is required. Following calcination, the calcined powder will normally be milled in order to minimize the size of the particles and increase the homogeneity of the product. This stage contributes to the achievement of a fine and uniform particle size distribution, both of which are essential for getting the desired qualities in the finished product. After the milling process, the powder is molded into the required shape. Pressing, extrusion, and slip-casting are three common methods that are utilized when ferrites are shaped. Extrusion requires driving the powder combination through a die to obtain a continuous shape, whereas pressing includes compacting the powder under high pressure to form green bodies. The milled powder and the liquid binder are combined into a slurry in the slip-casting process, and then the resulting mixture is poured into a mold to make the desired shape. After being molded, the green bodies are dried in order to remove any moisture and binders that may still be present. In order to prevent the green bodies from breaking or otherwise becoming deformed, this is normally done slowly and under carefully regulated conditions. Sintering is a process that involves drying the green bodies and then heating them to a very high temperature. The green bodies are heated at a temperature that is lower than their melting point during the sintering process in order to induce densification and grain growth. In order to manufacture a ferrite product with the correct density and mechanical qualities, the sintering process must be carried out for the appropriate amount of time and subjected to the appropriate temperature. After the sintering procedure, the ferrite components may undergo further processing steps such as grinding, polishing, or coating in order to achieve the surface finish and dimensional precision that are needed. Chen et al. [44] use ball-milling-assisted ceramic technique to produce all of the M-type hexagonal ferrites $Sr_{1-x}Co_xNd_xFe_{12-x}O_{19}$ (x = 0, 0.08, 0.16, and 0.24) samples.

2.2.11 Citrate Precursor for Spinel Ferrite Synthesis

Synthesizing spinel ferrites using the citrate precursor method (Figure 2.11) is a common practice because it is efficient and inexpensive. The weighing and mixing of the metal salts in the required stoichiometric ratio is the first step in the citrate precursor method. This step is performed in a separate container. Precursors are often metal salts, and common examples include iron nitrate, zinc nitrate, and cobalt nitrate. After the metal salts have been weighed, they are dissolved in distilled water to create a solution that is consistent throughout. In order to act as a chelating agent and form complexes with the metal ions in the metal salt

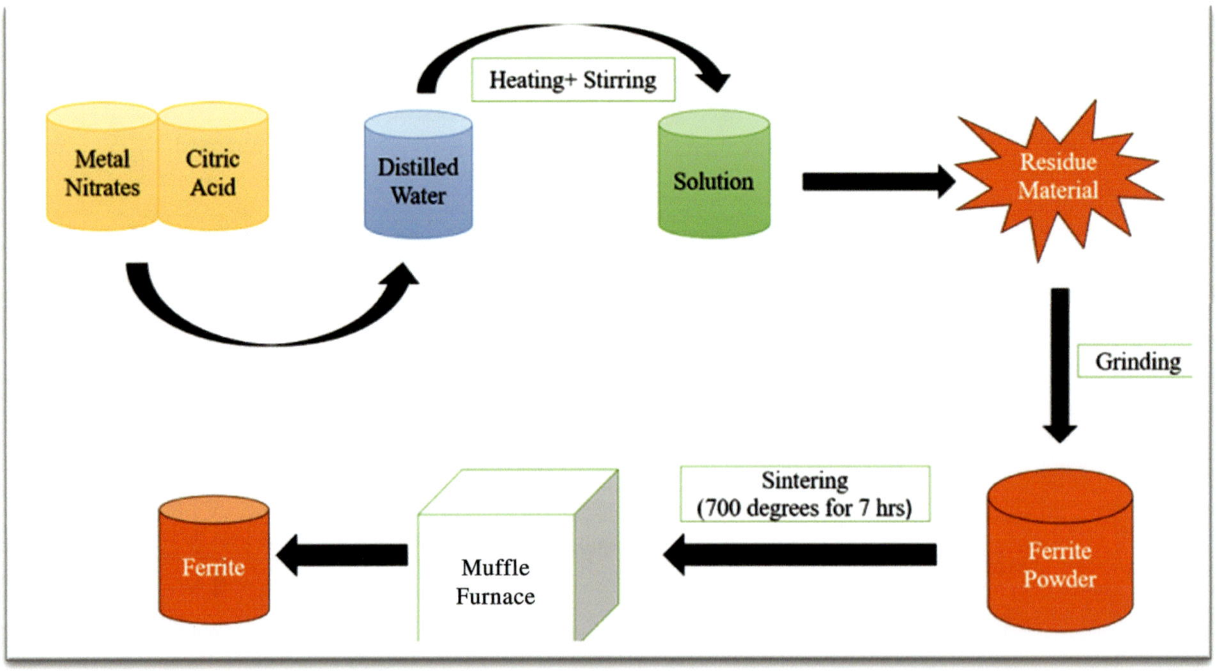

FIGURE 2.11 Diagrammatic representation of synthesis of spinel ferrites via Citrate precursor method.

solution, citric acid is added to the metal salt solution. By adding citric acid, one is able to better control the nucleation and growth of ferrite NPs, as well as prevent the creation of contaminants or other phases. After that, the solution of citric acid and metal salt is heated and agitated for a number of hours in order to produce a precursor that resembles a gel. The citric acid-metal salt solution goes through a number of different chemical events when it is heated. These reactions include dehydration, decarboxylation, and oxidation-reduction reactions, and the end result is the production of a stable gel. The precursor that has the consistency of a gel is then dried at a low temperature in order to remove any trace of water content and obtain a powder that is solid. After the powder has been dried, it is calcined at a high temperature (usually above 500°F) to stimulate the creation of spinel ferrite NPs. The temperature of the calcination and the amount of time it takes are both regulated so that the precursor is completely transformed into the ferrite phase that is needed. After being calcined, the powder is milled in order to achieve a fine particle size distribution and increase the homogeneity of the product. After being milled into a powder, the material can be formed into the appropriate form by employing a variety of processes, such as pressing, extrusion, or slip-casting. After the green bodies have been shaped, they go through a process called sintering, which involves high temperatures. The green bodies are heated at a temperature that is lower than their melting point during the sintering process in order to induce densification and grain growth. To achieve the necessary density and mechanical qualities in the finished ferrite product, the sintering temperature and time are carefully controlled. Ahmed et al. [45] used the citrate precursor approach in the production of nanometric nickel copper ferrites with the formula $Ni_{1-x}Cu_xFe_2O_4$, whereas Verma et al. [46] used the citrate precursor method, a mixture of manganese-zinc and nickel-zinc ferrites with a composition of $Mn_{0.2}Ni_{0.8-x}Zn_xFe_2O_4$ that has successfully been synthesized. The values for x range from 0.4, 0.5, and 0.6.

2.2.12 Polyol Process for Spinel Ferrite Synthesis

The polyol procedure is one of the most common ways that spinel ferrites are produced in synthetic form (Figure 2.12). The process of making polyols begins with the selection of metal precursors that are appropriate for the procedure. For the most part, metal salts including iron chloride, cobalt chloride, and zinc chloride are utilized as precursors in the manufacture of ferrite. As the medium for the reaction, we will be using a polyol solvent such as glycerol or ethylene glycol. During the synthesis process, the polyol solvent that is used should function as a reducing agent in addition to having a high level of solubility for the metal precursors. In order to ensure that the polyol is completely dissolved, the solvent is heated to a predetermined temperature. After heating the polyol solvent, the metal precursors, often in the form of metal chlorides or nitrates, are added to the mixture. The metal precursors are able to dissolve in the polyol solvent, which results in the formation of a solution that is homogeneous. Stabilizing agent or surfactant is added to the metal precursor solution in order to control the particle size and prevent agglomeration of the produced ferrite NPs. This is done in order to prevent agglomeration. Oleic acid and

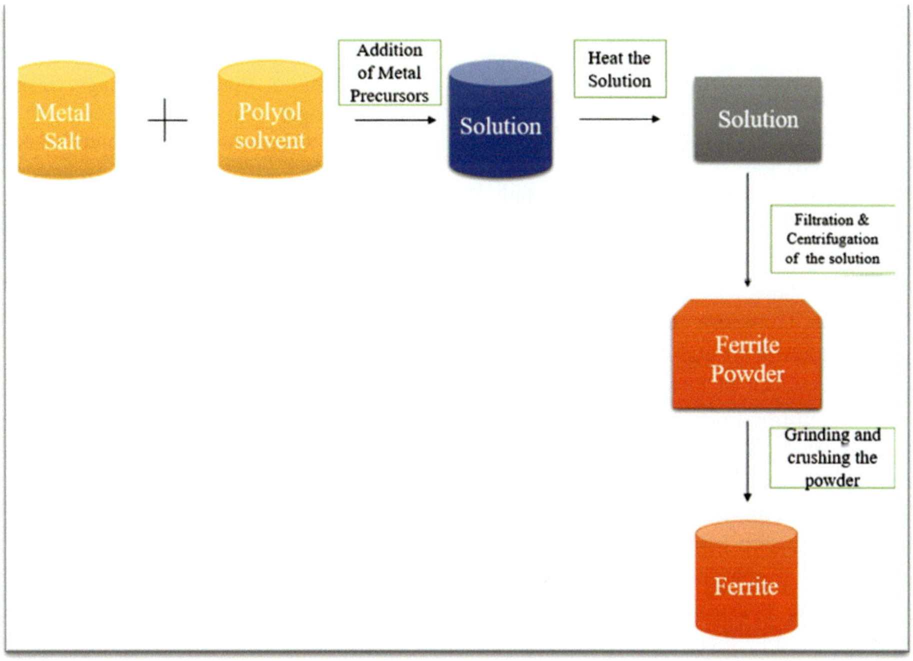

FIGURE 2.12 Diagrammatic representation of synthesis of spinel ferrites via Polyol process method.

polyvinylpyrrolidone (also known as PVP) are two examples of commonly employed stabilizing agents. Controlling the nucleation and development of the NPs is made easier by the presence of the stabilizing agent. In order to create a reducing atmosphere, the metal precursor solution is subjected to further heating until it reaches a higher temperature. In the presence of the stabilizing agent, the reduction of metal ions takes place, which ultimately results in the production of metal nuclei. The metal nuclei that were produced in the step before this one go through additional phases of growth and annealing. The temperature and the length of time that this process lasts are both meticulously managed in order to facilitate the formation of spinel ferrite NPs while simultaneously preventing an excessive amount of particle growth or aggregation. After the synthesis reaction has been completed, the resultant ferrite NPs are extracted from the reaction mixture in order to preserve their purity. In most cases, centrifugation or filtration is used to accomplish this goal. After the NPs have been produced, they are rinsed in the appropriate solvent, such as ethanol or water, in order to get rid of any lingering contaminants or unreacted precursors. In order to achieve a powder that is solid, the washed NPs are finally dried under carefully monitored circumstances. As J. Töpfer et al. [47] shown, in a medium consisting of triethylene glycol, nanocrystalline Fe_3O_4 and Mn-Zn ferrite powders with crystallite diameters of 10 and 7 nm, respectively, and limited size distribution were produced, whereas Ibrahim et al. [48] used the polyol approach to prepare nano-sized samples of single-phase cobalt ferrite. Ethylene glycol was used both as a high boiling point solvent and as a reducing agent during the preparation process.

2.3 CONCLUSION

Spinel ferrites can be produced by a number of different synthesis techniques, including the solid-state reaction, the sol-gel approach, the co-precipitation method, the hydrothermal method, and the microwave-assisted synthesis technique. The choice of method is determined by a number of parameters, including the features of the ferrite that are sought, the amount of money that is spent, and how easily it can be synthesized. Each approach has both advantages and disadvantages. Synthesizing spinel ferrites often involves a solid-state reaction, which calls for high temperatures and extended reaction periods. However, this process is quite popular. The sol-gel method is a procedure that is not overly complicated, can be carried out at low temperatures, and enables one to exert fine-grained control over the ferrite's chemical make-up. The production of highly homogeneous particles with a limited size distribution can also be accomplished by the widely used method of co-precipitation. The production of spinel ferrites with high levels of crystallinity and purity requires the employment of high pressure and temperature during the hydrothermal synthesis process. The microwave-assisted synthesis is a relatively new method that enables the speedy and efficient synthesis of spinel ferrites. This method was developed in the 1990s.

REFERENCES

1. P. Thakur et al., "Mg-Zn ferrite nanoparticles: A brief review on synthesis, characterizations and applications," *ChemBioEng Rev*, vol. 10, pp. 711–736, 2023, doi: 10.1002/cben.202200063
2. P. Punia, M. K. Bharti, R. Dhar, P. Thakur, A. Thakur, "Recent advances in detection and removal of heavy metals from contaminated water," *ChemBioEng Rev*, vol. 9, pp. 351–369, 2022, doi: 10.1002/cben.202100053
3. P. Thakur, D. Chahar, S. Taneja, N. Bhalla, A. Thakur, "A review on MnZn ferrites: Synthesis, characterization and applications," *Ceram Int*, vol. 46, pp. 15740–15763, 2020, doi: 10.1016/j.ceramint.2020.03.287
4. A. A. H. El-Bassuony, "A comparative study of physical properties of Er and Yb nanophase ferrite for industrial application," *J Supercond Nov Magn*, vol. 31, no. 9, pp. 2829–2840, Sep. 2018, doi: 10.1007/s10948-017-4543-1
5. M. Kumar, H. Singh Dosanjh, Sonia, J. Singh, K. Monir, and H. Singh, "Review on magnetic nanoferrites and their composites as alternatives in waste water treatment: Synthesis, modifications and applications," *Environ Sci: Water Res Technol*, vol. 6, no. 3. Royal Society of Chemistry, pp. 491–514, Mar. 01, 2020, doi: 10.1039/c9ew00858f
6. D. Suhag, A. Thakur, P. Thakur (eds) (2023) "Integrated Nanomaterials and their Applications". *Springer Nature Singapore.* pp 1–499. https://doi.org/10.1007/978-981-99-6105-4
7. Tovstolytkin, A. I., Kulyk, M. M., Kalita, V. M., Ryabchenko, S. M., Zamorskyi, V. O., Fedorchuk, O. P., ... & Belous, A. G, "Nickel-zinc spinel nanoferrites: Magnetic characterization and prospects of the use in self-controlled magnetic hyperthermia," *J Magnet Mater*, vol. 473, pp. 422–427, 2019.
8. K. Sharma, N. Aggarwal, N. Kumar, and A. Sharma. "A review paper: Synthesis techniques and advance application of Mn-Zn nano-ferrites." *Mater Today: Proc*, vol. 72, 2023. https://doi.org/10.1016/j.matpr.2022.12.088
9. E. Ahilandeswari, R. R. Kanna, and K. Sakthipandi, "Synthesis of neodymium-doped barium nanoferrite: Analysis of structural, optical, morphological, and magnetic properties," *Phys B: Condensed Matter*, vol. 599, p. 412425, 2020.
10. N. Dhanda, P. Thakur, A.-C. Aidan Sun, and A. Thakur, "Structural, optical and magnetic properties along with antifungal activity of Ag-doped Ni-Co nanoferrites synthesized by eco-friendly route," *J Magn Mater*, vol. 572, p. 170598, 2023.
11. S. Kumari, N. Dhanda, A. Thakur, V. Gupta, S. Singh, R. Kumar, S. Hameed, and P. Thakur, "Nano Ca–Mg–Zn ferrites as tuneable photocatalyst for UV light-induced degradation of rhodamine B dye and antimicrobial behavior for water purification," *Ceram Int*, vol. 49, no. 8, pp. 12469–12480, 2023.
12. N. Dhanda, S. Kumari, R. Kumar, D. Kumar, A.-C. A. Sun, P. Thakur, and A. Thakur, "Influence of Ni over magnetically benign Co ferrite system and study of its structural, optical, and magnetic behaviour," *Inorg Chem Commun*, vol. 151, p. 110569, 2023.
13. D. S. Mathew and R. S. Juang, "An overview of the structure and magnetism of spinel ferrite nanoparticles and their synthesis in microemulsions," *Chem Eng J*, vol. 129, no. 1–3, pp. 51–65, May 01, 2007, doi: 10.1016/j.cej.2006.11.001.
14. A. Thakur, A. Pathania, A. Salvi, S. Kharbanda, N. Dhanda, M. Shandaliya, F. Wan, and P. Thakur. "Biomedical Applications of 2D MXene Nanocomposites: A Review." *ChemBioEng Reviews*.

15. M. Amiri, M. Salavati-Niasari, and A. Akbari, "Magnetic nanocarriers: Evolution of spinel ferrites for medical applications," *Adv Colloid Interface Sci*, vol. 265, pp. 29–44, 2019.
16. K. K. Kefeni, T. A. M. Msagati, T. T. I. Nkambule, and B. B. Mamba, "Spinel ferrite nanoparticles and nanocomposites for biomedical applications and their toxicity," *Mater Sci Eng: C*, vol. 107, p. 110314, 2020.
17. P. Thakur, D. Chahar, S. Taneja, N. Bhalla, and A. Thakur, "A review on MnZn ferrites: Synthesis, characterization and applications," *Ceram Int*, vol. 46, no. 10, pp. 15740–15763, 2020.
18. Ch. S.L.N. Sridhar, Ch S. Lakshmi, G. Govindraj, S. Bangarraju, L. Satyanarayana, and D. M. Potukuchi, "Structural, morphological, magnetic and dielectric characterization of nano-phased antimony doped manganese zinc ferrites," *J Phys Chem Solids*, vol. 92, pp. 70–84, 2016.
19. Thakur, A., Thakur, P., Khurana, S.M.P. (eds) (2022) *Synthesis and Applications of Nanoparticles*. Springer Nature Singapore. pp 1–544. Singapore. https://doi.org/10.1007/978-981-16-6819-7
20. N. Dhanda *et al.*, "Influence of Ni over magnetically benign Co ferrite system and study of its structural, optical, and magnetic behaviour," *Inorg Chem Commun*, vol. 151, May 2023, doi: 10.1016/j.inoche.2023.110569.
21. A. Sutka and G. Mezinskis, "Sol-gel auto-combustion synthesis of spinel-type ferrite nanomaterials," *Front Mater Sci*, vol. 6, no. 2, pp. 128–141, Jun. 2012, doi: 10.1007/s11706-012-0167-3
22. G. R. Kumar, K. V. Kumar, and Y. C. Venudhar, "Synthesis, structural and magnetic properties of copper substituted nickel ferrites by sol-gel method," *Mater Sci Appl*, vol. 03, no. 02, pp. 87–91, 2012, doi: 10.4236/msa.2012.32013
23. D.-H. Chen and X.-R. He, "Synthesis of nickel ferrite nanoparticles by sol-gel method," 2001.
24. P. A. Shaikh, R. C. Kambale, A. V. Rao, and Y. D. Kolekar, "Structural, magnetic and electrical properties of Co-Ni-Mn ferrites synthesized by Co-precipitation method," *J Alloys Compd*, vol. 492, no. 1–2, pp. 590–596, Mar. 2010, doi: 10.1016/j.jallcom.2009.11.189.
25. R. Raeisi Shahraki, M. Ebrahimi, S. A. Seyyed Ebrahimi, and S. M. Masoudpanah, "Structural characterization and magnetic properties of superparamagnetic zinc ferrite nanoparticles synthesized by the coprecipitation method," *J Magn Mater*, vol. 324, no. 22, pp. 3762–3765, Nov. 2012, doi: 10.1016/j.jmmm.2012.06.020.
26. S. Rahman, K. Nadeem, M. Anis-Ur-Rehman, M. Mumtaz, S. Naeem, and I. Letofsky-Papst, "Structural and magnetic properties of ZnMg-ferrite nanoparticles prepared using the co-precipitation method," *Ceram Int*, vol. 39, no. 5, pp. 5235–5239, Jul. 2013, doi: 10.1016/j.ceramint.2012.12.023.
27. N. Deng, L. Zhou, X. Peng, X. Wang, and H. Ge, "Preparation and characterization of nickel-zinc ferrites by a solvothermal method," *Xiyou Jinshu Cailiao Yu Gongcheng/Rare Metal Mater Eng*, vol. 44, no. 9, pp. 2126–2131, Sep. 2015, doi: 10.1016/s1875-5372(16)30010-8
28. A. Chaudhuri, S. Mitra, M. Mandal, and K. Mandal, "Nanostructured bismuth ferrites synthesized by solvothermal process," *J Alloys Compd*, vol. 491, no. 1–2, pp. 703–706, 2010, doi: 10.1016/j.jallcom.2009.11.049
29. R. Ali *et al.*, "Structural, magnetic and dielectric behavior of Mg1-xCa xNiyFe2-yO4 nano-ferrites synthesized by the micro-emulsion method," *Ceram Int*, vol. 40, no. 3, pp. 3841–3846, Apr. 2014, doi: 10.1016/j.ceramint.2013.08.024.
30. M. Azhar Khan *et al.*, "High frequency dielectric response and magnetic studies of $Zn_{1-x}Tb_xFe_2O_4$ nanocrystalline ferrites synthesized via micro-emulsion technique," *J Magn Mater*, vol. 360, pp. 188–192, Jun. 2014, doi: 10.1016/j.jmmm.2014.02.059.
31. M. A. Almessiere *et al.*, "Sonochemical synthesis and physical properties of $Co_{0.3}Ni_{0.5}Mn_{0.2}EuxFe_{2-x}O_4$ nano-spinel ferrites," *Ultrason Sonochem*, vol. 58, Nov. 2019, doi: 10.1016/j.ultsonch.2019.104654.
32. S. Mishra, T. K. Kundu, K. C. Barick, D. Bahadur, and D. Chakravorty, "Preparation of nanocrystalline MnFe2O4 by doping with Ti^{4+} ions using solid-state reaction route," *J Magn Mater*, vol. 307, no. 2, pp. 222–226, Dec. 2006, doi: 10.1016/j.jmmm.2006.04.005.
33. N. Mubraiz, A. Bano, T. Mahmood, and N. Khan, "Microbial and plant assisted synthesis of cobalt oxide nanoparticles and their antimicrobial activities," *Agronomy*, vol. 11, no. 8, Aug. 2021, doi: 10.3390/agronomy11081607.
34. X. Zhang, S. Yan, R. D. Tyagi, and R. Y. Surampalli, "Synthesis of nanoparticles by microorganisms and their application in enhancing microbiological reaction rates," *Chemosphere*, vol. 82, no. 4. Elsevier Ltd, pp. 489–494, 2011, doi: 10.1016/j.chemosphere.2010.10.023.
35. E. Solano *et al.*, "Facile and efficient one-pot solvothermal and microwave-assisted synthesis of stable colloidal solutions of MFe2O4 spinel magnetic nanoparticles," *J Nanopart Res*, vol. 14, no. 8, Aug. 2012, doi: 10.1007/s11051-012-1034-y.
36. L. Zhenyu, X. Guangliang, and Z. Yalin, "Microwave assisted low temperature synthesis of MnZn ferrite nanoparticles," *Nanoscale Res Lett*, vol. 2, no. 1, pp. 40–43, Jan. 2007, doi: 10.1007/s11671-006-9027-3.
37. Q. Y. Tamboli, S. M. Patange, Y. K. Mohanta, R. Sharma, and K. R. Zakde, "Green synthesis of cobalt ferrite nanoparticles: An emerging material for environmental and biomedical applications," *J Nanomater*, vol. 2023, 2023.
38. M. M. Naik, H. S. Bhojya Naik, G. Nagaraju, M. Vinuth, K. Vinu, and R. Viswanath, "Green synthesis of zinc doped cobalt ferrite nanoparticles: Structural, optical, photocatalytic and antibacterial studies," *Nano-Struct Nano-Objects*, vol. 19, p. 100322, 2019.
39. T. Tatarchuk, A. Shyichuk, Z. Sojka, J. Gryboś, M. Naushad, V. Kotsyubynsky, M. Kowalska, S. Kwiatkowska-Marks, and N. Danyliuk, "Green synthesis, structure, cations distribution and bonding characteristics of superparamagnetic cobalt-zinc ferrites nanoparticles for Pb (II) adsorption and magnetic hyperthermia applications," *J Mol Liq*, vol. 328, p. 115375, 2021.
40. A. Bilal, J. K. Kasi, A. K. Kasi, M. Bokhari, S. Ahmed, and S. W. Ali, "Environment friendly synthesis of nickel ferrite nanoparticles using *Brassica oleracea* var. capitate (green cabbage) as a fuel and their structural and magnetic characterizations," *Mater Chem Phys*, vol. 290, p. 126483, 2022.
41. N. M. Deraz, "Facile and eco-friendly route for green synthesis of magnesium ferrite nano particles," *Sci Sinter*, vol. 52, no. 1, pp. 53–65, 2020.
42. A. R. Chavan, P. P. Khirade, S. B. Somvanshi, S. V. Mukhamale, and K. M. Jadhav. "Eco-friendly green synthesis and characterizations of $CoFe_{2-x}Al_xO_4$ nanocrystals: Analysis of structural, magnetic, electrical, and dielectric properties." *J Nanostruct Chem*, vol. 11, pp. 1–13, 2021).
43. N. Dhanda *et al.*, "Green-synthesis of Ni-Co nanoferrites using aloe vera extract: Structural, optical, magnetic, and antimicrobial studies," *Appl Organomet Chem*, vol. 37, no. 7, Jul. 2023, doi: 10.1002/aoc.7110

44. W. Chen, W. Wu, C. Zhou, S. Zhou, M. Li, and Y. Ning, "Structural and magnetic properties evolution of Co-Nd substituted M-type hexagonal strontium ferrites synthesized by ball-milling-assisted ceramic process," *J Electron Mater*, vol. 47, no. 3, pp. 2110–2119, Mar. 2018, doi: 10.1007/s11664-017-6021-8.
45. M. A. Ahmed, S. F. Mansour, and M. Afifi, "Structural and electrical properties of nanometric NiCu ferrites synthesized by citrate precursor method," *J Magn Mater*, vol. 324, no. 1, pp. 4–10, Jan. 2012, doi: 10.1016/j.jmmm.2011.07.010.
46. A. Verma and R. Chatterjee, "Effect of zinc concentration on the structural, electrical and magnetic properties of mixed Mn-Zn and Ni-Zn ferrites synthesized by the citrate precursor technique," *J Magn Mater*, vol. 306, no. 2, pp. 313–320, Nov. 2006, doi: 10.1016/j.jmmm.2006.03.033.
47. J. Töpfer and A. Angermann, "Nanocrystalline magnetite and Mn-Zn ferrite particles via the polyol process: Synthesis and magnetic properties," *Mater Chem Phys*, vol. 129, no. 1–2, pp. 337–342, Sep. 2011, doi: 10.1016/j.matchemphys.2011.04.025.
48. A. M. Ibrahim, M. M. A. El-Latif, and M. M. Mahmoud, "Synthesis and characterization of nano-sized cobalt ferrite prepared via polyol method using conventional and microwave heating techniques," *J Alloys Compd*, vol. 506, no. 1, pp. 201–204, Sep. 2010, doi: 10.1016/j.jallcom.2010.06.177.

3 Structures and Morphologies Attained by Spinel Nano-Ferrites

Preeti Thakur, Atul Thakur and Shivani Jangra*
Amity Institute of Nanotechnology, Amity University Haryana, Gurugram, India

Irina Edelmana and Sergey Ovchinnikov
Kirensky Institute of Physics, Federal Research Center KSC, Siberian Branch, Russian Academy of Sciences, Krasnoyarsk, Russia

3.1 INTRODUCTION

A type of magnetic material known for its unique combination of structural, electrical, and magnetic characteristics is ferrites. These ceramics, which consist of iron oxide and other metal oxides, are unique in the field of materials research because of their crystalline structure. The peculiar crystal structure known as spinel, which underpins ferrites' special qualities and uses, is responsible for their magnetic properties.

Spinel structure is essentially a face-centered cubic (FCC) lattice of oxygen ions with metal cations dispersed among octahedral and tetrahedral locations. This configuration, which is essential to the nature of ferrites, determines their applicability for a wide range of technical applications and produces their magnetic behavior. There are three main kinds of ferrites that are worth mentioning: Hematite (α-Fe_2O_3), maghemite (γ-Fe_2O_3), and magnetite (Fe_3O_4). With its various iron oxidation states, magnetite is a strong and naturally magnetic mineral [1, 2]. As a byproduct of magnetite alteration, maghemite has distinct magnetic characteristics. Conversely, hematite is highly valued for its red hue and is frequently used in a variety of industrial settings.

This chapter begins with a thorough understanding of ferrites, covering their morphological characteristics and intricate structure. Our goal is to discover the principles that control the behavior and usefulness of these materials by decoding their architecture at the atomic and nanoscale levels. Our voyage covers the various aspects that lead to the identification of ferrites, ranging from flaws and surface features to crystallography. Ferrites are much more important than just being magnetic materials. Magnetic cores in transformers, microwaves, magnetic recording media, biological imaging agents, and catalysts are just a few examples of the wide range of applications for magnetic materials [3]. We have paved the way for an investigation that will not only deepen our understanding of ferrites but also create new opportunities for creativity and progress in the field of materials science and technology.

* Corresponding author.

3.1.1 Significance of Studying Structural and Morphological Properties

Understanding the morphological and structural properties of ferrites is essential since it provides a wealth of knowledge that is required for both fundamental research and practical applications. Some significant ramifications of researching these qualities are described in the following sections.

3.1.1.1 Applications and Magnetism's Behavior

3.1.1.1.1 Adapting Magnetic Properties to Needs

The crystal structure of ferrite has a significant impact on its magnetic behavior. Magnetic properties can be tailored for use in magnetic recording media, transformers, and microwave devices, among other applications, by looking at the structural components [4].

3.1.1.2 Improvement of Electrical Properties

3.1.1.2.1 Electrical Conductivity

The electronic structure and crystal lattice imperfections of ferrite affect its electrical conductivity. Gaining the most out of these qualities is crucial when using them in applications like sensors and electronic devices.

3.1.1.2.2 Characteristics at the Nanoscale

Nanotechnology and Nanoscale Control: At the nanoscale, ferrites often exhibit distinguishing properties. The study of morphology is crucial to controlling particle size, shape, and surface qualities, which is necessary for the development of nanotechnological applications, including nanomedicine and nanoelectronics.

3.1.1.3 Chemical Reactivity and Catalysis

3.1.1.3.1 Surface characteristics

Surface Characteristics: Defects and surface structure have a big impact on catalytic activity [5]. Analyzing these anatomical features helps develop effective catalysts for chemical reactions, water filtration, and environmental restoration.

DOI: 10.1201/9781003568940-3

3.1.1.4 Biomedical Applications

3.1.1.4.1 Biocompatibility

To ensure biocompatibility in biomedical applications, morphological knowledge is essential. Ferrites' size and surface properties along with the magnetic traits [6] influence how well they interact with magnetic resonance imaging (MRI) contrast agents and tailored drug delivery systems.

3.1.1.5 Durability and Strength of the Material

3.1.1.5.1 Mechanical Characteristics

The crystal structure of ferrites affects their mechanical strength and durability. For applications where structural integrity is critical, such as transformer magnetic cores, an understanding of these properties is crucial [7].

3.1.1.6 Optimization of Fabrication and Synthesis

3.1.1.6.1 Methods of Synthesis

Morphological properties are directly impacted by the synthesis methods employed. Analyzing these traits enables manufacturing procedures to be improved to produce desired outcomes, like precisely sized or tailored nanoparticles.

3.1.1.7 Sustainability and the Environment's Consequences

3.1.1.7.1 Toxicity and Environmental Considerations

It is critical to understand the morphological and surface characteristics of ferrites when assessing their environmental impact, especially when used in large quantities [8]. This knowledge facilitates the development of sustainable and environmentally favorable.

3.1.1.7.2 Technological Advancement and Innovation

Material Engineering: It is easier to invent and develop new technologies when one is aware of the structural and morphological characteristics. This could lead to the creation of more efficient machinery, improved medical diagnostics, and innovations in a wide range of industrial fields [9].

3.2 CRYSTAL STRUCTURE

3.2.1 Introduction to Crystallography of Ferrites

The intricate crystallographic structure of ferrites, a kind of magnetic ceramic, is responsible for their extraordinary characteristics. The study of atomic and molecular crystal structures, or crystallography, is a crucial lens through which we may examine the electrical properties, magnetic behavior, and general usefulness of these materials [10]. We take a voyage into the crystallography of ferrites, revealing the fundamental ideas that underpin their distinct configurations and aid in their varied uses.

3.2.1.1 Crystal Structure: Dissecting the Spinel Arrangement

A defining feature of the crystallography of ferrites is the spinel structure. This unusual configuration consists of a FCC lattice of oxygen ions, in which metal cations are positioned at strategic tetrahedral and octahedral positions [11]. The characteristics of various ferrite phases, including magnetite, maghemite, and hematite, are determined by the interaction of iron oxide (Fe_3O_4) and other metal oxides inside this framework.

Comprehending the spinel structure is essential in understanding electrical and magnetic properties of ferrites. The distinct magnetic behavior displayed by these materials is influenced by the interatomic distances, the distribution of cations, and the occupancy of crystallographic sites [12].

3.2.1.2 Crystallographic Impurities and Point Defects

Furthermore, the function of point defects in the ferrite lattice is shown via crystallography. The ideal crystal structure is broken by vacancies, interstitials, and substitutional dopants, which affect the mechanical, electrical, and magnetic characteristics. The investigation of crystallographic imperfections sheds light on how these flaws affect ferrites' overall performance in a range of settings.

3.2.1.3 Crystallography and Synthesis Techniques

Synthesis techniques are essential in figuring out crystallographic properties of ferrites. Whether utilizing hydrothermal synthesis, co-precipitation, or sol-gel methods, each technique gives the crystal structure a unique identity. Through an awareness of these synthesis-structure correlations, scientists can modify ferrite crystallography to satisfy needs, optimizing characteristics.

3.2.1.4 Crystallography-Guided Applications: Electronics to Medicine

The uses of ferrites are impacted by crystallography in a significant way. Customized materials with specific electrical and magnetic characteristics can be created in the field of electronics through the modification of crystal structures [13]. Furthermore, in medicinal applications, such as tailored drug delivery systems and MRI contrast agents, control over crystallographic characteristics is crucial to compatibility and efficacy.

The magnetic appeal and functional diversity of ferrites are based on a rich tapestry of atomic configurations that we learn about when we study their crystallography [14]. This investigation lays the groundwork for a more thorough comprehension of the structural subtleties that underpin the technological advancements and applications made possible by these magnetic ceramics.

3.2.2 Crystal Structure of Different Ferrite Phases (Spinel Structure, Hexagonal Structure)

Ferrites can have a variety of crystal forms, the most prevalent of which is the spinel structure. However, differences in crystal structures may arise based on the ferrite. Here, we investigate the crystal structures of various ferrite phases,

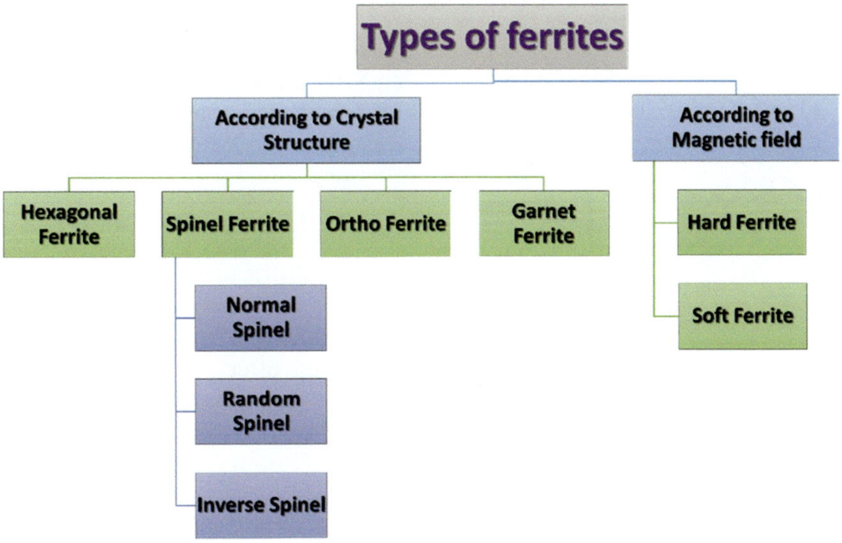

FIGURE 3.1 Types of ferrites.

with particular attention to the hexagonal and spinel structures (Figure 3.1):

- *Spinel Structure*: The most common crystal structure seen in ferrites is the spinel structure. It is made up of oxygen ions arranged in a FCC configuration. Within the FCC lattice, tetrahedral and octahedral sites are home to metal cations, among them iron [15]. Ferrites with the spinel structure have the general chemical formula AB_2O_4, where A and B stand for metal cations.

  Example: Iron ions occupy both tetrahedral and octahedral positions in magnetite (Fe_3O_4), which contributes to its special magnetic characteristics. Maghemite (γ-Fe_3O_4): Depending on the circumstances of synthesis, maghemite can show structural variants in addition to its typical spinel structure.
- *Hexagonal Structure*: Certain ferrites, such as hexagonal ferrites (Figure 3.2), adopt a hexagonal crystal lattice instead of following the spinel structure. Metal cations occupy locations within the hexagonal layers that oxygen ions form in the structure [16]. Hexagonal ferrites have a variety of chemical formulas but most frequently have the form MFe12O1–, where M stands for a divalent metal cation.

  Example: Barium Hexaferrite (BaFe12O19): This hexagonal ferrite is well-known for its application in microwave and magnetic recording medium. One of the reasons for its anisotropic magnetic characteristics is the hexagonal shape.

These disparate crystal formations are essential to customizing characteristics of ferrites for uses. The selection

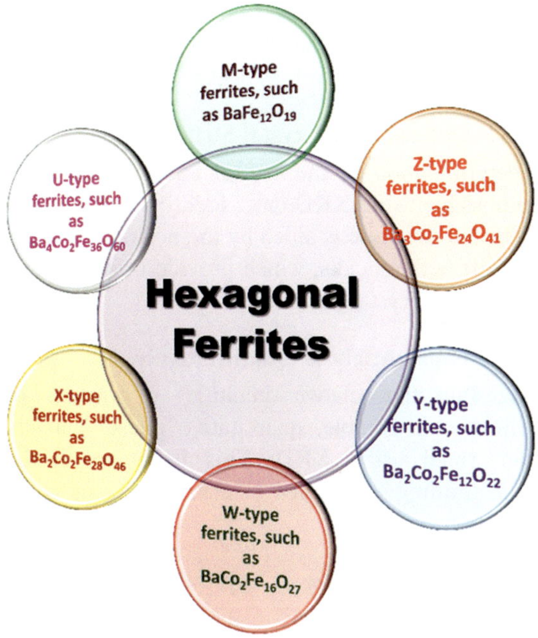

FIGURE 3.2 Classification of hexagonal ferrites.

of crystal structure is crucial in the design and engineering of ferrite materials since it affects electrical conductivity, magnetic behavior, and other characteristics.

3.3 METHODS OF STRUCTURAL ANALYSIS

3.3.1 X-Ray Diffraction

One effective method for determining the crystallographic structure of materials, including ferrites, is X-ray diffraction (XRD) examination. The arrangement of atoms in a crystal lattice, the size of the crystal, and the strain inside

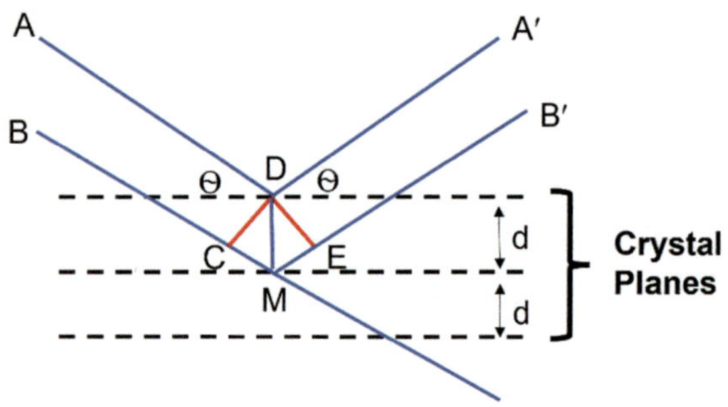

FIGURE 3.3 Schematic illustration of Bragg's Law of diffraction.

the material are all revealed by XRD. The interaction of X-rays with a crystal lattice of the material is necessary for XRD to occur. Diffraction patterns are produced when X-rays collide with crystals due to constructive and destructive interference (Figure 3.3) [17]. It is possible to measure the angles and intensities of diffracted X-rays, which yields details on the crystal structure. XRD is used to determine which crystal phases are in a sample of ferrite. The distinct diffraction peaks exhibited by each crystal phase facilitate the identification of various ferrite phases, such as magnetite and maghemite, as well as possible contaminants.

3.3.1.1 Determining Crystal Structure

The determination of the crystal structure of ferrites is made possible using XRD data. Identification of the crystallographic structure is aided by the positions and intensities of diffraction peaks, which provide information about the spacing of crystal planes [18].

3.3.1.2 Phase Analysis Quantitatively

To ascertain the relative quantities of various phases contained in a sample, quantitative phase analysis can be performed using XRD. This is especially helpful when examining the degree of crystallinity or in ferritin combinations.

3.3.1.3 Strain and Size of Crystallites

The size of the crystallites in the material can be determined by analyzing XRD data. Diffraction peak broadening is correlated with crystallite size, and peak form indicates the existence of defects or strain.

3.3.1.4 Analysis of Texture

Texture analysis is made possible by XRD, which provides details about the preferred orientation of crystallites inside a sample. When thinking about anisotropic qualities of materials, this is significant.

3.3.1.5 Analysis of Residual Stress

A residual stress of material can be evaluated using XRD. Stress can be determined by measuring the size of the shift in diffraction peak positions, which is an indicator of stress [19].

3.3.1.6 Instrumentation

A sample holder, a detector, and an X-ray source make up an XRD instrument in most cases. For improved resolution, modern devices may make use of synchrotron radiation or rotating anode X-ray generators.

3.3.2 Neutron Diffraction

Neutron diffraction analysis is another powerful technique used for investigating the structural properties of materials, including ferrites. Neutron diffraction complements XRD and offers unique advantages, particularly in the study of magnetic materials like ferrites. Neutron diffraction is based on the interaction of neutrons with the atomic nucleus. Neutrons, being uncharged, interact primarily with the atomic nuclei, providing complementary information to XRD [20]. The scattering intensity and pattern reveal details about the atomic arrangement within the crystal lattice.

3.3.2.1 Sensitivity to Light Elements

Neutrons are particularly sensitive to light elements such as hydrogen, which makes them valuable for studying materials with hydrogen-containing components, common in ferrites.

3.3.2.2 Magnetic Structure Determination

Neutron diffraction is especially powerful for determining magnetic structures in materials, including ferrites [21]. It can provide information about the magnetic moments and their orientations within the crystal lattice.

3.3.2.3 Complementary to X-ray Diffraction

Neutron diffraction is often used in conjunction with XRD to obtain a more comprehensive understanding of the crystal structure of materials. While X-rays are sensitive to electron density, neutrons are sensitive to nuclear and magnetic structures.

3.3.2.4 Investigation of Hydrogen Positions

Neutron diffraction is useful for determining the positions of hydrogen atoms within the crystal lattice. This is valuable in ferrites, where hydrogen can play a role in the crystal structure [22].

3.3.2.5 Isotopic Sensitivity

Neutron diffraction is sensitive to isotopic substitution, providing a means to selectively study specific elements within a material.

3.3.2.6 Temperature and Pressure Studies

Neutron diffraction is often employed in studies involving temperature and pressure variations. It allows researchers to investigate structural changes under different conditions, providing insights into the behavior of ferrites.

3.3.2.7 Instrumentation

Neutron diffraction experiments are typically conducted at specialized facilities known as neutron sources, which can be research reactors or spallation sources [23]. Neutron

TABLE 3.1
Strengths and Limitations of X-Ray Diffraction Technique

Strength	Limitations
Effective and quick (less than 20 minutes) method for identifying unknown minerals	The ideal material for identifying an unknown is homogeneous and single-phase
In most situations, it delivers a clear mineral identification	Access to a standard reference file of inorganic chemical is required
Small sample preparation is needed	2% of the sample is the detection limit for the mixed materials
XRD units are generally accessible	Indexing of the patterns for non-isometric crystal systems is difficult for unit cell measurements
Data interpretation is not too difficult	Peak overlay is possible for high-angle "reflections" and gets worse

detectors capture the scattered neutrons, and the data are analyzed to determine the crystal structure.

In summary, neutron diffraction analysis is a valuable tool for investigating the structural and magnetic properties of ferrites, providing detailed information about the atomic arrangements, magnetic moments, and other characteristics. It is particularly advantageous in situations where other techniques, such as XRD, may face limitations (Table 3.1).

3.3.3 Fourier Transform Infrared (FTIR) Spectroscopy

Fourier transform infrared (FTIR) spectroscopy is a potent analytical method for determining the molecular structure and chemical makeup of various materials, including ferrites. FTIR calculates how much infrared light a sample absorbs. The basic idea is to expose the sample to infrared light; the material's molecular vibrations are revealed by the absorbance spectrum that results (Figure 3.4).

3.3.3.1 Identification of Chemicals

Chemical composition of ferrites can be determined using FTIR, which looks for vibrational modes linked to the molecular groups of materials. Oxygen-containing functional groups, such as metal-oxygen bonds in ferrites, show distinctive absorption bands [24].

3.3.3.2 Description of Oxygen-Metal Bonds

Certain peaks in the FTIR spectrum are attributed to metal-oxygen bonding seen in ferrites. For example, the infrared spectrum can be used to examine the stretching vibrations of metal-oxygen bonds, which can reveal information about the composition of these bonds in ferrites.

3.3.3.3 Impurity Detection

In ferrite samples, FTIR can detect the presence of contaminants or secondary phases [25]. Variations in chemical composition or the presence of contaminants can be detected via absorption peaks that correspond to functional groups.

3.3.3.4 Analysis of Oxidation States

FTIR can be used to assess the oxidation state of iron in ferrites. The valence state of iron can be determined by looking for discrete infrared absorption bands, which are present in different oxidation states.

3.3.3.5 Examining the Surface

Information regarding surface characteristics of ferrites can be obtained via FTIR. It is possible to identify impurities and surface functional groups, which adds to a thorough understanding of the material.

3.3.3.6 Analytical Quantitative Methods

Researchers can use FTIR for quantitative analysis to find out how much of a particular functional group or component there is in a ferrite sample.

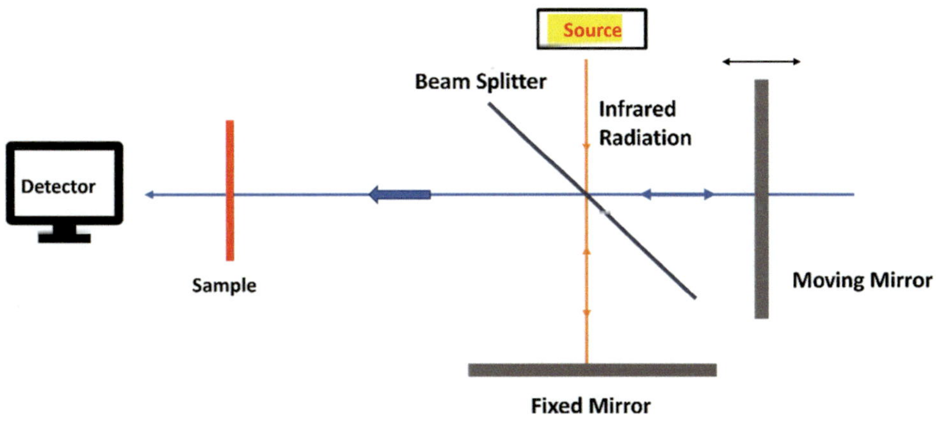

FIGURE 3.4 Working phenomena of Fourier transform infrared (FTIR) spectroscopy.

3.3.3.7 Research on Temperature Dependence

The thermal behavior of ferrites can be studied via FTIR analysis at various temperatures [26]. Phase transitions and structural alterations can be seen by observing variations in the FTIR spectrum with temperature.

3.3.3.8 Instrumentation

FTIR equipment using a Fourier transform mechanism provides enhanced spectrum resolution and sensitivity. Accessories for attenuated total reflectance (ATR) make it possible to analyze solid materials without requiring complicated sample preparation.

In summary, ferrites can be studied using FTIR spectroscopy, which is a flexible analytical method that can provide important details about the molecular structure, surface characteristics, and chemical makeup of ferrites. It is very helpful for characterizing materials, ensuring quality, and learning more about the functional groups found in ferrite samples.

3.4 MORPHOLOGICAL PROPERTIES

The structural attributes and traits pertaining to the size, shape, and surface structure of ferrite particles are referred to as the morphological properties of ferrites. Understanding these morphological properties is crucial for tailoring the performance of ferrites in various applications.

3.4.1 Size Distribution and Particle Size

i. *Nanostructure*: Ferrites frequently have dimensions on the nanoscale, and the size of the particles greatly affects their characteristics [27].
ii. *Size distribution*: The distribution of particle sizes in a sample influences the behavior and applications of the material.
   - *Morphological anisotropy*: Modifications in the shape of ferrite particles can impact their magnetic and physicochemical properties. Morphological anisotropy can play a major role in applications that need specific orientations of the magnetic moment.
   - *Surface Area*: Ferrite particles need to have a surface area in order to interact with their environment [28]. A substance's reactive and catalytic properties may be enhanced via increased surface areas.
   - The concept of "porosity" refers to the vacuum gaps that are present within ferrite particles. This property could have an impact on the substance's adsorption capacity, density, and catalytic activity.
   - *Material dispersion and Uniformity*: Ferrite particles tend to agglomerate into particle clusters or aggregate into clusters of agglomerates. Identifying these tendencies is necessary to control material dispersion and provide uniform properties.

TABLE 3.2
Size and Shape Distribution of Ferrite Nanoparticles

Size Distributions	Shape Distribution
Nanoscale dimensions	Spherical particles
Uniformity and monodispersed	Anisotropic shapes
Methods of size control	Agglomeration and aggregation

- *Crystal Structure*: Although the internal arrangement of atoms is more closely tied to crystal structure, exterior morphological qualities of ferrite are influenced by it [29]. Particle sizes and shapes might vary depending on the crystal structure.
- The external morphological features of ferrite are impacted by crystal structure, even though the internal arrangement of atoms is more intimately linked to it [29]. Depending on the crystal structure, different particles may have different sizes and shapes (Table 3.2).
- The technique of functionalizing or coating ferrite particles to alter their surface structure is known as surface modification. This could enhance the material's stability, biocompatibility, and capacity for chemical interactions.
- Ferrites are not limited to the usual particle shapes; they can also exist as nanowires or nanorods. These arrangements might exhibit unique magnetic and electrical properties.
- *Defects and Dislocations*: Surface flaws and dislocations can impact the mechanical, electrical, and magnetic characteristics of ferrite particles.
- *Surface Functionalization*: To improve the capabilities of ferrites for certain uses, like biomedicine or catalysis, their surfaces can be functionalized with groups or ligands [30].
- *Surface charge*: The stability of the particle of ferrite in solution and interactions with other charged species are influenced by their surface charge. Applications such as environmental cleanup and colloidal systems require this.
- *Morphological Variations in Various Settings*: Under a variety of circumstances, including variations in temperature, pressure, or exposure to distinct atmospheres, ferrite particles may experience morphological alterations [31]. Predicting the behavior of materials in a variety of contexts requires an understanding of these changes.

In conclusion, morphological characteristics are crucial in dictating how they behave in a variety of settings, including electronics, biomedical devices, and catalysis. Researchers frequently modify these characteristics to maximize the material's performance for purposes and improve its overall performance.

3.4.2 Agglomeration and Aggregation

Agglomeration can occur due to various factors, including van der Waals forces, magnetic attraction, and electrostatic forces. It is often influenced by the surface properties of the ferrite particles, including their charge and chemistry.

3.4.2.1 Impact on Properties

Agglomeration can alter the magnetic, electrical, and mechanical properties of ferrites [32]. The collective behaviour of agglomerated particles differs from that of individual particles.

3.4.2.2 Processing Challenges

Agglomerates can pose challenges during processing and manufacturing, affecting the homogeneity of the material.

3.4.2.3 Prevention and Control

By adding repulsive forces between the particles, surface modification of ferrite particles might lessen the likelihood of agglomeration [12]. Agglomeration can be avoided by stabilizing the dispersion of ferrite particles in a medium using surfactants or dispersants.

The process of forming larger structures or clusters of agglomerates, which produce even larger assemblies of ferrite particles, is referred to as aggregation. Agglomeration is driven by the same factors that drive agglomeration, but it takes place when particles that have already grouped together engage and bind to form larger structures.

3.4.2.4 Macroscopic Properties

Compared to single particles or tiny agglomerates, aggregated ferrite structures can have substantially different properties.

3.4.2.5 Controlling Process Parameters

The degree of aggregation can be influenced by carefully regulating processing parameters such as temperature, pressure, and solvent selection [33], applying stabilizing material to ferrite particles.

3.5 INFLUENCE OF SYNTHESIS TECHNIQUE ON MORPHOLOGY

The morphological aspects of ferrites, including their size, shape, surface characteristics, and crystallinity, are largely determined by the synthesis procedure used to create them [34]. Different synthesis techniques can be applied, each of which contributes to unique morphological characteristics.

- *Co-Precipitation Method*: This process creates a precursor by co-precipitating metal cations from a solution. The precursor is subsequently annealed to produce the final ferrite.

  Impact on the Morphology

  i. Frequently produces homogeneous, fine particles.
  ii. Precursor concentration and reaction conditions can be changed to alter particle size.
  iii. Because nanoscale particle formation is frequent, it is appropriate for applications that call for small particle sizes.
- *Sol-Gel Method*: This method forms the final ferrite by turning a sol, or colloidal dispersion of particles, into a gel and then heating it.

  Impact on Morphology

  i. Facilitates the production of very pure nanoscale particles.
  ii. Particle shape and size can be controlled during the gel-forming step.
  iii. Fit for generating thin films of ferrite with certain morphologies.
- *Hydrothermal Synthesis*: This process entails the reaction of precursors in an aqueous medium at high temperatures and pressures.

  Impact on Morphology

  i. Promotes the development of well-defined-shaped nanoscale particles.
  ii. Particle morphology and size are influenced by pressure and temperature conditions.
  iii. Fit for achieving morphologies and crystal phases.
- *Solid-State Reaction*: Solid-state reaction involves the direct heating and reaction of metal oxides or other precursor compounds.

  Influence on Morphology

  i. Typically results in larger particle sizes compared to wet chemical methods.
  ii. Particle size is influenced by factors such as temperature and duration of the reaction.
  iii. Shape is often less controlled compared to wet chemical methods.
- *Microemulsion Method*: Microemulsion involves the use of surfactant-stabilized microdroplets to control the synthesis of nanomaterials.

  Influence on Morphology

  i. Enables the production of nanoscale particles with controlled size and shape.
  ii. Surfactants contribute to the prevention of agglomeration and control of morphology.
  iii. Suitable for obtaining monodisperse nanoparticles.
- *Combustion Synthesis*: Combustion synthesis involves the combustion of a mixture of precursors, often aided by fuel-rich conditions.

  Influence on Morphology

  i. Rapid reaction kinetics can lead to the formation of fine particles.
  ii. Particle morphology is influenced by combustion conditions.
  iii. Suitable for producing nanoscale ferrite powders with high surface areas.

It is essential to comprehend how synthesis methods affect ferrite morphology to modify its properties for certain uses [35–38]. Depending on the required shape and the planned application of the ferrite material, researchers can choose the best synthesis technique.

3.6 CHARACTERIZATION TECHNIQUES FOR MORPHOLOGY

3.6.1 Scanning Electron Microscopy

One of the most effective methods for describing the morphology of ferrite materials is scanning electron microscopy, or SEM (Figure 3.5). High-resolution imaging is made possible by SEM, which offers comprehensive data on surface characteristics, particle size, shape, and distribution [39, 40].

3.6.1.1 Preparing the Sample

Ferrite samples must be positioned correctly on SEM stubs or conductive mounts. The sample surface is frequently covered with a small layer of conductive material, such as carbon or gold, to improve conductivity and lessen charging effects [41]. This coating is essential for insulating samples.

3.6.1.2 Instrument Adjustment

- *Voltage and Current*: Depending on the sample's characteristics, the SEM instrument is calibrated for the proper voltage and current settings.
- *Working Distance*: To maximize depth of field and focus, the working distance is adjusted.

3.6.1.3 Image Acquisition

- *Overview Images*: Low-magnification images provide an overview of the entire sample, helping to identify areas of interest [42].
- *High-Magnification Images*: High-magnification images are captured to study individual particles, surface features, and structural details.

3.6.1.4 Particle Size Analysis

- *Measurement*: Particle sizes are frequently determined directly from SEM images using software tools [43].
- *Distribution Analysis*: To determine the range and variability, the distribution of particle sizes within the sample is examined.

3.6.1.5 EDS/EDX Elemental Analysis

- *Integration with SEM*: To offer elemental analysis of the ferrite particles, energy-dispersive X-ray spectroscopy (EDS/EDX) can be merged with SEM.
- *Mapping*: To display the spatial distribution of the various elements in the sample, elemental maps can be created.

3.6.1.6 Topographical Analysis

- *Surface Roughness*: SEM images can be used to analyze the surface roughness of ferrite particles.
- *Feature Identification*: Topographical features, such as pores or cracks, can be identified and studied.

3.6.1.7 3D Imaging

- *Tilt Series*: Tilt series acquisition or other advanced imaging techniques can be employed for three-dimensional reconstructions of the ferrite surface.
- *Software Reconstruction*: Dedicated software is used to create 3D models from the acquired SEM images.

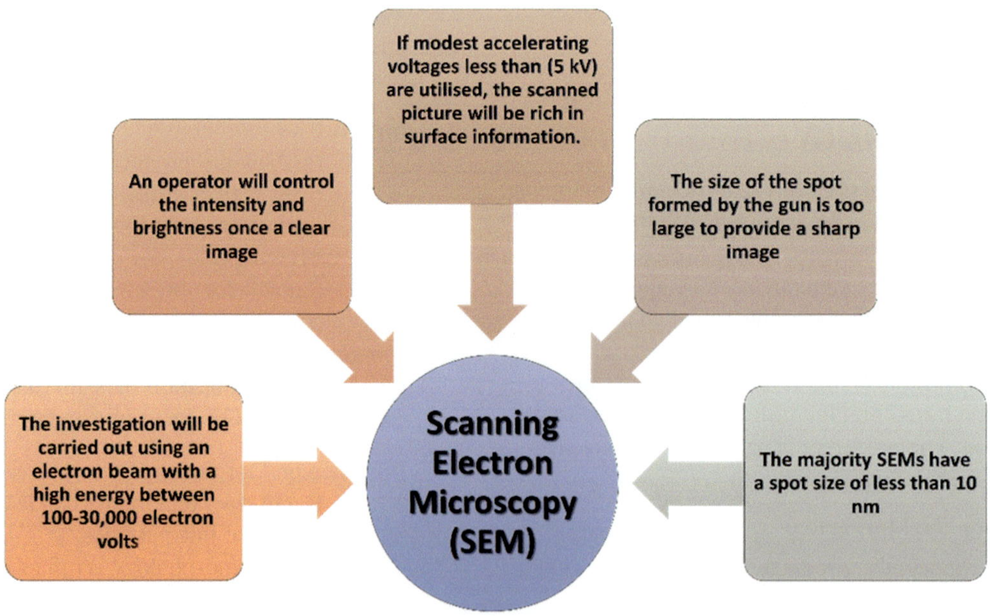

FIGURE 3.5 Scanning electron microscopy working procedure.

3.6.1.8 Cross-Sectional Imaging

- *Preparation*: Cross-sectional imaging involves preparing samples with polished cross-sections for SEM analysis [44].
- *Internal Structure*: Cross-sectional SEM images reveal details about the internal structure and layering of the ferrite sample.

3.6.1.9 Image Analysis and Reporting

- *Software Tools*: Various image analysis software tools are used for quantifying parameters, measuring dimensions, and extracting meaningful data.
- *Reporting*: Results and findings from the SEM analysis are compiled into a report, often including images, measurements, and any relevant observations.

SEM analysis is a versatile tool for studying the morphology of ferrites, providing essential insights for material characterization, quality control, and optimization of synthesis processes [45].

3.6.2 Transmission Electron Microscopy

Transmission electron microscopy (TEM) is a powerful technique for studying the morphology of materials at the nanoscale. TEM provides detailed information about the internal and surface structure of ferrites, offering high-resolution imaging and elemental analysis [46].

Ferrite samples are prepared as ultrathin sections using techniques like ultramicrotomy or focused ion beam (FIB) milling. The ultrathin sections are mounted on TEM grids for analysis. The TEM instrument is calibrated for the appropriate voltage and current settings [47]. Fine-tuning of focus and astigmatism is performed to optimize image quality.

3.6.2.1 Electron Beam Interaction

- *Transmission of Electrons*: The electron beam passes through the thin specimen, and the transmitted electrons form the TEM image.
- *Interaction with Specimen*: Electrons interact with the specimen, leading to scattering, diffraction, and absorption, providing contrast in the image.

Bright-field imaging provides information about the density variations in the specimen. Differences in electron density within the ferrite sample contribute to image contrast [48]. Dark-field imaging enhances contrast by selectively imaging scattered electrons. Features with different crystal orientations or defects can be highlighted. High-resolution TEM allows for the visualization of individual lattice planes and atomic arrangements. The arrangement of atoms in the crystal lattice of ferrites can be observed at a high level of detail. Energy-dispersive X-ray spectroscopy (EDS/EDX) is often integrated with TEM for elemental analysis [49]. A series of images are acquired at different tilt angles, and software reconstructs a 3D model. Various software tools are used for image analysis, including measurements of particle sizes, lattice spacings, and other morphological features. Results and findings from TEM analysis are compiled into a report, often including images, diffraction patterns, and any relevant observations.

TEM analysis provides detailed insights into the nanoscale morphology of ferrites, aiding in the understanding of crystal structure, defects, and elemental composition [50]. It is a crucial tool for materials research, contributing to the development and optimization of ferrite-based materials for diverse applications.

3.6.3 Atomic Force Microscopy

Atomic force microscopy (AFM) is a powerful technique for studying the morphology of materials at the nanoscale. Unlike conventional optical microscopy, AFM operates by scanning a sharp tip over the surface of a sample, measuring the interaction forces between the tip and the sample.

Ferrite samples are typically deposited onto a suitable substrate, such as a glass slide or silicon wafer [51]. The substrate should have a flat surface to facilitate accurate AFM imaging. AFM tips are typically made of materials like silicon or silicon nitride and have sharp tips that scan the sample surface. The choice of cantilever depends on factors such as the stiffness and resonance frequency, which are matched to the properties of the ferrite sample. The laser beam used to detect the cantilever's deflection is aligned properly. The spring constant of the cantilever is calibrated to ensure accurate force measurements. In contact mode, the tip maintains constant contact with the sample surface while scanning. The deflection of the cantilever is monitored as the tip scans the sample, producing a topographical image of the surface. In non-contact mode, the tip oscillates close to, but not in direct contact with, the sample surface [52]. Changes in oscillation amplitude are measured, providing information about the sample's topography and surface properties.

Measurement of force as the tip interacts with the sample is known as force spectroscopy. Analyzing force-indentation curves can be used to investigate material qualities, such as elasticity and hardness.

3.6.3.1 Excellent-Resolution Imaging

Individual particles and surface characteristics can be imaged in detail because of excellent lateral and vertical resolution of AFM [53]. Defects and crystal formations are among the small details that high-resolution imaging can make visible.

3.6.3.2 Data analysis

To enhance features, AFM images are treated with software techniques like filtering and flattening. Parameters including particle height, roughness, and lateral dimensions can all be quantitatively analyzed. The results of an AFM study are compiled into a report

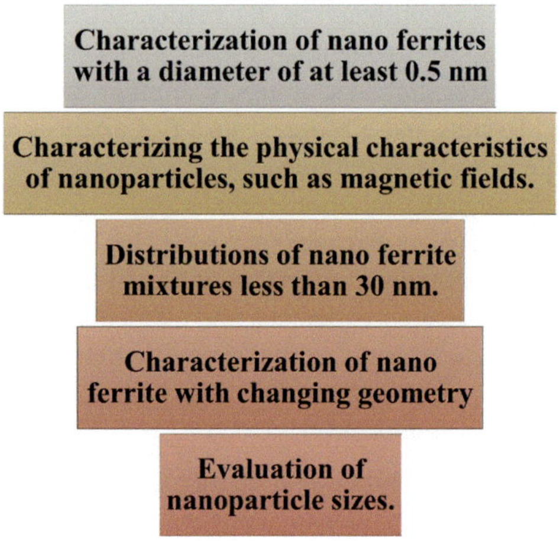

FIGURE 3.6 Advantages of AFM characterization for nanoferrites.

that often contains images, measurements, and any relevant observations.

Morphology can be studied with the help of AFM analysis, which can reveal information about the material's characteristics, particle sizes, and surface characteristics [54]. It is an effective method for characterizing materials and is essential to comprehending and improving ferrite-based materials for a range of uses. Figure 3.6 presents some of the advantages of AFM nanoferrite characterization.

3.6.4 Dynamic Light Scattering

Dynamic light scattering (DLS), often referred to as photon correlation spectroscopy (PCS), is a method for determining the size distribution of particles in a suspension. It's crucial to remember that DLS mostly offers data on hydrodynamic diameter, which is determined by the size, shape, and surrounding medium of the particles [55]. Even though electron microscopy techniques are more frequently utilized for morphological investigations, DLS can nevertheless provide insightful information about the size distribution of ferrite particles.

To create a stable suspension, ferrite samples must be well disseminated in a suitable solvent. Particle aggregation should be prevented, if possible, by utilizing surfactants or dispersants. DLS equipment is calibrated to operate at that wavelength. A laser beam is directed at the suspended ferrite particles, and the scattered light is detected at a fixed angle. The Brownian motion of particles in the suspension causes fluctuations in the scattered light intensity. The correlation function is calculated based on the temporal fluctuations in scattered light intensity. The decay of the correlation function provides information about the dynamics of particles in the suspension. DLS primarily reports the hydrodynamic diameter, which is influenced by both the size and shape of the particles. The PDI indicates the breadth of the size distribution. A lower PDI suggests a more monodisperse sample. The intensity-weighted size distribution plot shows the relative contribution of particles of different sizes to the overall scattering intensity. DLS can also measure the zeta potential of particles in suspension [56]. Zeta potential provides insights into the electrostatic stability of the suspension. DLS analysis can be performed at different temperatures to study the temperature-dependent behavior of the ferrite suspension. Changes in size distribution with temperature can indicate the stability of the particles in suspension. DLS can be used for dynamic studies, providing real-time information about changes in particle size or stability under specific conditions.

3.6.4.1 Data Analysis

DLS data are typically analyzed using dedicated software. Results, including size distribution plots, PDI values, and zeta potential measurements, are compiled into a report.

DLS is particularly useful for studying the size distribution and stability of particles in solution, making it applicable for assessing the dispersion and hydrodynamic behaviors of ferrites in liquid media. While it may not provide detailed morphological information, it offers valuable insights into the dynamic behaviors of ferrite particles in a suspension.

3.7 APPLICATIONS OF STRUCTURAL AND MORPHOLOGICAL INSIGHTS

Advances in materials science and engineering depend heavily on structural and morphological insights because they provide a deep understanding of the structure, composition, and properties of materials at the atomic, molecular, and nanoscale levels. These discoveries, which are frequently made possible by methods like electron microscopy, spectroscopy, and XRD, enable scientists and engineers to customize materials for uses, sparking advancements across a wide range of industries. We investigate the uses of structural and morphological knowledge in a wide range of fields in this thorough investigation.

- *Magnetic Substances and Equipment*: To create magnetic materials with certain properties, a thorough understanding of the crystal structure is essential. Understanding structural relationships aids in the development of magnetic materials for hard drives and other data storage devices as well as electromagnetic parts like transformers.
- *Telecommunications and electronics*: Creating electronic components with the best possible electrical and magnetic properties is made easier by an understanding of morphology [57]. In the field of telecommunications, the creation of ferrite components for use in circulators, isolators, microwave devices, and other electronic applications is guided by structural and morphological insights.

- *Catalysis*: Developing efficient catalysts with increased catalytic activity requires an understanding of crystal structures. Catalysis for industrial processes and environmental remediation depends heavily on structural understanding, which affects the effectiveness of chemical reactions and pollutant removal.
- *Applications in Biomedicine*: MRI, to customize ferrite nanoparticles for biomedical imaging applications, structural and morphological understanding is essential. These discoveries guide the investigation of ferrite nanoparticles as MRI contrast agents, perhaps leading to better medical imaging diagnoses.
- *Magnetic Hyperthermia*: The design of ferrite nanoparticles used in magnetic hyperthermia for cancer treatment depends heavily on the intricacies of crystallography and magnetic characteristics. Ferrite nanoparticles that preferentially kill cancer cells when subjected to an alternating magnetic field are developed using morphological insights [57].
- *Environmental Remediation:* Effective adsorbents for eliminating impurities from water are designed with structural considerations in mind. Ferrites are used in wastewater treatment to help remove heavy metals by using their structural features as a guide.
- *Energy Conversion and Storage*: The comprehension of crystallography and morphology is crucial to maximize the efficiency of ferrite-based materials in energy storage applications. Ferrites are used in supercapacitors and lithium-ion batteries, where their specially designed structures improve electrochemical performance.
- *Gas Sensing*: To maximize the surface characteristics of ferrites used in gas sensors, morphological information is crucial. Certain morphologies of ferrites aid in the creation of extremely selective and sensitive gas sensors for environmental monitoring.
- *Sensing Technologies*: Accurate and dependable magnetic sensor designs are aided by structural and morphological understanding [58]. Ferrite-based sensors are used in many different technologies, such as navigation systems and location sensing.
- *Electronic Devices*: The development of ferrite-based semiconductor materials requires a fundamental understanding of crystal structures. Because of their specific crystallographic characteristics, ferrites are used in integrated circuits and electrical devices.
- *Coatings and Thin Films*: Corrosion Protection: The design of ferrite coatings for corrosion protection is guided by morphological insights. In a variety of industries, ferrite-based coatings improve corrosion resistance due to the influence of their surface shape.
- *Ceramic Materials*: For ferrite ceramics to be tailored for usage in microwave dielectrics, structural features are essential. Based on crystallography, ferrite ceramics are used in electronic devices where dielectric characteristics are essential.
- *Inks and Pigments*: Particle Morphology, the formulation of inks and pigments is aided by morphological insights into ferrite particles. Printing processes use ferrite-based inks and pigments, which are impacted by their morphology.

To sum up, the foundation of materials research and development comprises structural and morphological insights that impact advancements in a wide range of applications. With the help of these discoveries, scientists and engineers will be able to maximize the ferrite material's qualities, which will progress the fields of electronics, medicine, environmental preservation, and other fields. The future of materials and their uses will be greatly influenced by structural and morphological discoveries as long as technology keeps developing.

3.8 FUTURE TRENDS AND EMERGING TECHNIQUES

The future trends of ferrites are shaped by ongoing research and technological advancements that aim to enhance their properties, expand their applications, and address emerging challenges. Future research will focus on designing ferrites with tailored magnetic properties, such as high saturation magnetization, low coercivity, and tuneable magnetic anisotropy, for applications in data storage, sensors, and magnetic devices. Continued exploration of nanostructured ferrites includes nanoparticles and their nanocomposites, harnessing their size-dependent effects will further enhance performance in magnetic hyperthermia, and enable novel applications in biomedicine and catalysis. The trend toward multifunctionality involves combining magnetic, electrical, and optical properties in a single ferrite material. This opens avenues for applications in sensors, energy conversion, and advanced electronic devices. The future will likely see increased emphasis on green synthesis methods, such as sustainable and environmentally friendly routes, to produce ferrite materials, aligning with the growing demand for sustainable and eco-conscious materials. The integration of ferrite materials into flexible and wearable electronics enables the development of bendable electronic components, sensors, and devices for applications in healthcare, communications, and consumer electronics. Research into using ferrites for energy harvesting from ambient sources, such as vibrations and waste heat, to generate electrical power for low-power electronic devices. Continued efforts to improve the electrochemical performance of ferrites for supercapacitor applications, contribute to energy storage solutions. Integration of artificial intelligence (AI) and machine learning in materials research accelerates the discovery and optimization of ferrite materials with desired properties [59].

These future trends reflect the dynamic nature of materials science and engineering, where ongoing research and technological innovations continually expand the possibilities and applications of ferrite materials. The convergence of these trends is expected to drive advancements in electronics, healthcare, energy, and various other fields, unlocking new opportunities for the use of ferrites in diverse technologies and applications.

3.9 CONCLUSION

In conclusion, the exploration of the structural and morphological properties of ferrites unveils a rich tapestry of characteristics that underpin their diverse applications across numerous technological domains. Morphologically, ferrites manifest in various forms, ranging from bulk materials to nanoscale particles, each offering unique advantages and challenges. The size and shape of ferrite particles profoundly influence their magnetic behaviors, introducing phenomena such as superparamagnetism in nanoscale dimensions. This morphological diversity opens avenues for tailored applications, from magnetic data storage to biomedical devices, where the manipulation of size imparts specific functionalities. Understanding the structure and morphology of ferrites is paramount for optimizing their performance in specific applications. Tailoring these properties through advanced synthesis methods, such as sol-gel processes or co-precipitation techniques, empowers researchers and engineers to fine-tune ferrite characteristics for enhanced efficiency in diverse technologies. Advancements in characterization techniques, such as XRD, electron microscopy, and spectroscopy, have illuminated the microscopic world of ferrites, enabling precise analysis of their structural features. As we delve deeper into the structural and morphological landscape of ferrites, the synergistic interplay between theoretical insights and experimental advancements continues to propel the field forward. Ongoing research endeavors strive to unravel the nuances of ferrite behaviors at the atomic and nanoscale levels, unlocking new possibilities for innovation and extending the boundaries of their applications. In essence, the structural and morphological properties in ferrites are not merely a scientific pursuit but a gateway to harnessing the full potential of these magnetic materials. The journey into their microscopic realms not only enriches our understanding of fundamental principles but also fuels the quest for engineering solutions that amplify the impact of ferrites in emerging technologies.

REFERENCES

1. Srivastava, R., & Yadav, B. C. (2012). Ferrite materials: Introduction, synthesis techniques, and applications as sensors. *International Journal of Green Nanotechnology*, 4(2), 141–154.
2. Adam, J. D., Davis, L. E., Dionne, G. F., Schloemann, E. F., & Stitzer, S. N. (2002). Ferrite devices and materials. *IEEE Transactions on Microwave Theory and Techniques*, 50(3), 721–737.
3. Suhag, D., Thakur, A., & Thakur, P. (eds) (2023). *Integrated Nanomaterials and Their Applications*. Springer Nature Singapore. pp 1–499. https://doi.org/10.1007/978-981-99-6105-4
4. Toledo, J. A., Valenzuela, M. A., Bosch, P., Armendariz, H., Montoya, A., Nava, N., & Vazquez, A. (2000). Effect of Al^{3+} introduction into hydrothermally prepared $ZnFe_2O4$. *Applied Catalysis A: General*, 198(1–2), 235–245.
5. Dhiman, M., Chudasama, B., Kumar, V., Tikoo, K. B., & Singhal, S. (2019). Augmenting the photocatalytic performance of cobalt ferrite via change in structural and optical properties with the introduction of different rare earth metal ions. *Ceramics International*, 45(3), 3698–3709.
6. Peng, Z., Fu, X., Ge, H., Fu, Z., Wang, C., Qi, L., & Miao, H. (2011). Effect of Pr^{3+} doping on magnetic and dielectric properties of Ni–Zn ferrites by "one-step synthesis". *Journal of Magnetism and Magnetic Materials*, 323(20), 2513–2518.
7. Thakur, A., Thakur, P., & Khurana, S. P. (2022). *Synthesis and Applications of Nanoparticles* (pp. 1–17). Singapore: Springer.
8. Thakur, P., Taneja, S., Sindhu, D., Lüders, U., Sharma, A., Ravelo, B., & Thakur, A. (2020). Manganese zinc ferrites: A short review on synthesis and characterization. *Journal of Superconductivity and Novel Magnetism*, 33, 1569–1584.
9. Button, K. J. (1984). Microwave ferrite devices: The first ten years. *IEEE Transactions on Microwave Theory and Techniques*, 32(9), 1088–1096.
10. Albanese, G. (1977). Recent advances in hexagonal ferrites using nuclear spectroscopic methods. *Le Journal de Physique Colloques*, 38(C1), C1–85.
11. Berggren, J. (1971). Refinement of the crystal structure of dicalcium ferrite, Ca2Fe2O5. *Acta Chemica Scandinavica*, 25, 3616–3624.
12. Chahar, D., Thakur, P., Sun, A. C. A., & Thakur, A. (2023). Investigation of structural, electrical and magnetic properties of nickel substituted Co–Zn nanoferrites. *Journal of Materials Science: Materials in Electronics*, 34(10), 901.
13. Hill, P. M., Peiser, H. S., & Rait, J. R. (1956). The crystal structure of calcium ferrite and β calcium chromite. *Acta Crystallographica*, 9(12), 981–986.
14. Li, J., Duan, Y., He, H., & Song, D. (2001). Crystal structure, electronic structure, and magnetic properties of bismuth-strontium ferrites. *Journal of Alloys and Compounds*, 315(1–2), 259–264.
15. Bid, S., & Pradhan, S. K. (2004). Characterization of crystalline structure of ball-milled nano-Ni–Zn-ferrite by Rietveld method. *Materials Chemistry and Physics*, 84(2–3), 291–301.
16. Bhalla, N., Taneja, S., Thakur, P., Sharma, P. K., Mariotti, D., Maddi, C., … & Thakur, A. (2021). Doping independent work function and stable band gap of spinel ferrites with tunable plasmonic and magnetic properties. *Nano Letters*, 21(22), 9780–9788.
17. Geller, S. (1956). Crystal structure of gadolinium orthoferrite, GdFeO3. *The Journal of Chemical Physics*, 24(6), 1236–1239.
18. Punia, P., Bharti, M. K., Dhar, R., Thakur, P., & Thakur, A. (2022). Recent advances in detection and removal of heavy metals from contaminated water. *ChemBioEng Reviews*, 9(4), 351–369.
19. Thakur, P., & Thakur, A. (2022). Nanomaterials, their types and properties. In *Synthesis and Applications of Nanoparticles* (pp. 19–44). Singapore: Springer Nature Singapore.

20. Kumar, L., Kumar, P., Narayan, A., & Kar, M. (2013). Rietveld analysis of XRD patterns of different sizes of nanocrystalline cobalt ferrite. *International Nano Letters*, 3, 1–12.
21. Modak, S., Ammar, M., Mazaleyrat, F., Das, S., & Chakrabarti, P. K. (2009). XRD, HRTEM and magnetic properties of mixed spinel nanocrystalline Ni–Zn–Cu-ferrite. *Journal of Alloys and Compounds*, 473(1–2), 15–19.
22. Mittal, V. K., Bera, S., Nithya, R., Srinivasan, M. P., Velmurugan, S., & Narasimhan, S. V. (2004). Solid state synthesis of Mg–Ni ferrite and characterization by XRD and XPS. *Journal of Nuclear Materials*, 335(3), 302–310.
23. Raghavender, A. T., Biliškov, N., & Skoko, Ž (2011). XRD and IR analysis of nanocrystalline Ni–Zn ferrite synthesized by the sol–gel method. *Materials Letters*, 65(4), 677–680.
24. Deepty, M., Srinivas, C., Kumar, E. R., Mohan, N. K., Prajapat, C. L., Rao, T. C., ... & Sastry, D. L. (2019). XRD, EDX, FTIR and ESR spectroscopic studies of co-precipitated Mn–substituted Zn–ferrite nanoparticles. *Ceramics International*, 45(6), 8037–8044.
25. Pradeep, A., & Chandrasekaran, G. (2006). FTIR study of Ni, Cu and Zn substituted nano-particles of MgFe2O4. *Materials Letters*, 60(3), 371–374.
26. Rathod, V., Anupama, A. V., Kumar, R. V., Jali, V. M., & Sahoo, B. (2017). Correlated vibrations of the tetrahedral and octahedral complexes and splitting of the absorption bands in FTIR spectra of Li-Zn ferrites. *Vibrational Spectroscopy*, 92, 267–272.
27. Taneja, S., Thakur, P., & Thakur, A. (2022). Effect of bismuth doped Ni-Zn nanoferrite on structural and magnetic properties. *Materials Today: Proceedings*, 67, 195–200.
28. Thakur, P., Kumari, S., Singh, S., Ravelo, B., Wan, F., & Thakur, A. (2023). Mg-Zn ferrite nanoparticles: A brief review on synthesis, characterizations, and applications. *ChemBioEng Reviews*, 10(5), 711–736.
29. Deraz, N. M., & Alarifi, A. (2012). Structural, morphological and magnetic properties of nano-crystalline zinc substituted cobalt ferrite system. *Journal of Analytical and Applied Pyrolysis*, 94, 41–47.
30. Qin, M., Shuai, Q., Wu, G., Zheng, B., Wang, Z., & Wu, H. (2017). Zinc ferrite composite material with controllable morphology and its applications. *Materials Science and Engineering: B*, 224, 125–138.
31. Alarifi, A., Deraz, N. M., & Shaban, S. (2009). Structural, morphological and magnetic properties of $NiFe_2O_4$ nano-particles. *Journal of Alloys and Compounds*, 486(1–2), 501–506.
32. Kadam, A. A., Shinde, S. S., Yadav, S. P., Patil, P. S., & Rajpure, K. Y. (2013). Structural, morphological, electrical and magnetic properties of Dy doped Ni–Co substitutional spinel ferrite. *Journal of Magnetism and Magnetic Materials*, 329, 59–64.
33. Bharati, V. A., Somvanshi, S. B., Humbe, A. V., Murumkar, V. D., Sondur, V. V., & Jadhav, K. M. (2020). Influence of trivalent Al–Cr co-substitution on the structural, morphological and Mössbauer properties of nickel ferrite nanoparticles. *Journal of Alloys and Compounds*, 821, 153501.
34. Atiq, S., Majeed, M., Ahmad, A., Abbas, S. K., Saleem, M., Riaz, S., & Naseem, S. (2017). Synthesis and investigation of structural, morphological, magnetic, dielectric and impedance spectroscopic characteristics of Ni-Zn ferrite nanoparticles. *Ceramics International*, 43(2), 2486–2494.
35. Akhtar, M. N., Khan, M. A., Ahmad, M., Nazir, M. S., Imran, M., Ali, A., ... & Murtaza, G. (2017). Evaluation of structural, morphological and magnetic properties of CuZnNi (CuxZn0.5–xNi0. 5Fe2O4) nanocrystalline ferrites for core, switching and MLCI's applications. *Journal of Magnetism and Magnetic Materials*, 421, 260–268.
36. Tatarchuk, T., Myslin, M., Mironyuk, I., Bououdina, M., Pędziwiatr, A. T., Gargula, R., ... & Kurzydło, P. (2020). Synthesis, morphology, crystallite size and adsorption properties of nanostructured Mg–Zn ferrites with enhanced porous structure. *Journal of Alloys and Compounds*, 819, 152945.
37. Sarwar, M., & Priestner, R. (1996). Influence of ferrite-martensite microstructural morphology on tensile properties of dual-phase steel. *Journal of Materials Science*, 31, 2091–2095.
38. Sonia, M. M. L., Anand, S., Vinosel, V. M., Janifer, M. A., Pauline, S., & Manikandan, A. (2018). Effect of lattice strain on structure, morphology and magneto-dielectric properties of spinel NiGdxFe_2– xO_4 ferrite nano-crystallites synthesized by sol-gel route. *Journal of Magnetism and Magnetic Materials*, 466, 238–251.
39. Garcia-Mateo, C., Jimenez, J. A., Lopez-Ezquerra, B., Rementeria, R., Morales-Rivas, L., Kuntz, M., & Caballero, F. G. (2016). Analyzing the scale of the bainitic ferrite plates by XRD, SEM and TEM. *Materials Characterization*, 122, 83–89.
40. Goto, K., Ito, M., & Sakurai, T. (1980). Studies on magnetic domains of small particles of barium ferrite by colloid-SEM method. *Japanese Journal of Applied Physics*, 19(7), 1339.
41. Allaedini, G., Tasirin, S. M., & Aminayi, P. (2015). Magnetic properties of cobalt ferrite synthesized by hydrothermal method. *International Nano Letters*, 5, 183–186.
42. Punia, P., Dhar, R., Ravelo, B., Trukhanov, A. V., Panina, L. V., Thakur, P., & Thakur, A. (2021). Microstructural, optical and magnetic study of Ni–Zn nanoferrites. *Journal of Superconductivity and Novel Magnetism*, 34(8), 2131–2140.
43. Thakur, A., Thakur, P., Kumar, D., & Sharma, P. B. (2022). Electromagnetic Characterizations of Nanomaterials. In *Synthesis and Applications of Nanoparticles* (pp. 255–265). Singapore: Springer Nature Singapore.
44. Naagar, M., Chalia, S., Wan, F., Panina, L.V., Thakur, P., Sharma, P.B., Thakur, A. (2024) Investigations on temperature-dependent magnetic properties and magnetic thermal stability of magnesium ferrite ($MgFe_2O_4$) nanoparticles. *Journal of Magnetism and Magnetic Materials*, 592, 171798.
45. David, S. A. (1981). *Ferrite Morphology and Variations in Ferrite Content in Austenitic Stainless-Steel Welds*. TN: Oak Ridge National Lab.
46. Son, S., Taheri, M., Carpenter, E., Harris, V. G., & McHenry, M. E. (2002). Synthesis of ferrite and nickel ferrite nanoparticles using radio-frequency thermal plasma torch. *Journal of Applied Physics*, 91(10), 7589–7591.
47. Sivakumar, P., Ramesh, R., Ramanand, A., Ponnusamy, S., & Muthamizhchelvan, C. (2011). Synthesis and characterization of nickel ferrite magnetic nanoparticles. *Materials Research Bulletin*, 46(12), 2208–2211.
48. Huo, J., & Wei, M. (2009). Characterization and magnetic properties of nanocrystalline nickel ferrite synthesized by hydrothermal method. *Materials Letters*, 63(13–14), 1183–1184.
49. Zhong, Y., Xiao, F., Zhang, J., Shan, Y., Wang, W., & Yang, K. (2006). In situ TEM study of the effect of M/A films at grain boundaries on crack propagation in an ultra-fine acicular ferrite pipeline steel. *Acta Materialia*, 54(2), 435–443.

50. Franco, A. (2010). High temperature magnetic properties of cobalt ferrite nanoparticles. *Applied Physics Letters*, 96(17).
51. Kester, E., Rabe, U., Presmanes, L., Tailhades, P., & Arnold, W. (1999). Measurement of mechanical properties of nanoscaled ferrites using atomic force microscopy at ultrasonic frequencies. *Nanostructured Materials*, 12(5–8), 779–782.
52. Cheng, F., Peng, Z., Xu, Z., Liao, C., & Yan, C. (1999). The sol-gel preparation and AFM study of spinel $CoFe_2O_4$ thin film. *Thin Solid Films*, 339(1–2), 109–113.
53. Punia, P., Thakur, P., Kumar, R., Syal, R., Dhar, R., & Thakur, A. (2022). Synthesis and characterization of Ca substituted Ni-Zn nanoferrites-microstructural, magnetic and dielectric analysis. *Journal of Alloys and Compounds*, 928, 167248.
54. Thakur, P., Sharma, P., Mattei, J. L., Queffelec, P., Trukhanov, A. V., Trukhanov, S. V., … & Thakur, A. (2018). Influence of cobalt substitution on structural, optical, electrical and magnetic properties of nanosized lithium ferrite. *Journal of Materials Science: Materials in Electronics*, 29, 16507–16515.
55. Gandhi, S., & Roy, I. (2019). Synthesis and characterization of manganese ferrite nanoparticles, and its interaction with bovine serum albumin: A spectroscopic and molecular docking approach. *Journal of Molecular Liquids*, 296, 111871.
56. Baldi, G., Bonacchi, D., Franchini, M. C., Gentili, D., Lorenzi, G., Ricci, A., & Ravagli, C. (2007). Synthesis and coating of cobalt ferrite nanoparticles: A first step toward the obtainment of new magnetic nanocarriers. *Langmuir*, 23(7), 4026–4028.
57. Darwish, M. S., Kim, H., Lee, H., Ryu, C., Lee, J. Y., & Yoon, J. (2019). Synthesis of magnetic ferrite nanoparticles with high hyperthermia performance via a controlled co-precipitation method. *Nanomaterials*, 9(8), 1176.
58. Gadkari, A. B., Shinde, T. J., & Vasambekar, P. N. (2010). Ferrite gas sensors. *IEEE Sensors Journal*, 11(4), 849–861.
59. Naito, Y., & Suetake, K. (1971). Application of ferrite to electromagnetic wave absorber and its characteristics. *IEEE Transactions on Microwave Theory and Techniques*, 19(1), 65–72.

4 Electrical and Magnetic Characteristics of Spinel Nano-Ferrites

*Preeti Thakur**
Nanotechnology Wing, Innovative Science Research Society, Shimla, India
nanoLatticeX LLP, Manesar, Gurugram, India
Amity Institute of Nanotechnology, Amity University Haryana, Gurugram, India

Neetu Dhanda
Amity University Haryana, Gurugram, India

An-Cheng Aidan Sun
Yuan Ze University, Taoyuan, Taiwan

Atul Thakur
Nanjing University of Information Science & Technology, Nanjing, China
Nanotechnology Wing, Innovative Science Research Society, Shimla, India
nanoLatticeX LLP, Manesar, Gurugram, India
Amity Institute of Nanotechnology, Amity University Haryana, Gurugram, India

4.1 INTRODUCTION

In the area of magnetic and electrical applications, spinel ferrite nanoparticles are candidates that could be potentially considered. They have a wide range of applications in the field of medical diagnostics as well as medication delivery systems [1]. Spinel ferrites are helpful in microwave devices, magnetic recording, and computer memories because of their low dielectric losses and strong electrical resistivities. The family of spinel ferrites includes magnesium ferrite, which is an n-type semiconducting soft material [2]. Magnesium ferrite has a number of uses in sensor technology, including thermal coagulation therapy, in which tumours are locally heated by the application of alternating magnetic fields [3]. In general, the structural formula for $MgFe_2O_4$ is written as $(Mg_{1-\lambda}Fe_{\lambda})[Mg_{\lambda}Fe_{2-\lambda}]O_4$, where square brackets and parentheses indicate cation sites of octahedral [B] and tetrahedral (A) coordination, respectively, and λ denotes the degree of inversion, which is defined as (A) sites fraction full by Fe^{3+} ions. Recently, milled nano-sized spinel ferrites have been shown to exhibit attractive electrical and magnetic properties. These features are distinct from those exhibited by their bulk counterparts. In comparison to those of bulk materials, nanosized ferrite particles have been found to exhibit a fluctuation in the Néel temperature that is proportional to high coercivity [4, 5], grain size [6], and improved magnetic moments [7, 8]. When the grain size of $Ni_{0.5}Zn_{0.5}Fe_2O_4$ spinel ferrite was decreased to 14 nm by ball milling (high-energy) [9], the Néel temperature was observed to increase from 538K to 611K in the bulk state. This change occurred from a starting point of 538K. Impedance spectroscopy has been used for over 50 years to study the bound dynamics or free ions in the interfacial or bulk areas of any liquid or solid medium, including semiconducting, ionic, dielectrics, and mixed electronic–ionic. This research has been carried out in order to better understand how these ions behave. A wide range of disordered solids, such as ionic, amorphous semiconductors, conducting (electronic) polymers, ion-conducting glasses, doped semiconductors or extremely defective crystals, and single crystals, have been observed to exhibit dispersion in conductivity [10–14]. In the case of $NiFe_2O_4$, Ponpandian et al. [15] have found that the dielectric loss (tanδ) and dielectric constant (ε') for milled samples are approximately two orders of magnitude less than the bulk. There are a few different findings on the magnetic characteristics of nanoscale $MgFe_2O_4$ that may be found in the published research [16–22]. There have only been a handful of researchers who have looked into the electrical characteristics of $MgFe_2O_4$ in its bulk form [23–25]. So in this chapter, the magnetic and electrical properties of spinel ferrites, for example, magnesium ferrite, are discussed.

4.2 ELECTRICAL STUDIES

4.2.1 Grain Size and DC Conductivity

The electrical conductivity of prepared samples, for example, magnesium ferrite is displayed in Figure 4.1 (Arrhenius plot) for the temperature range of 370K–500K. The data on the impedance were analysed with the non-linear least-squares (NLLS) fitting procedure, which allowed for the

* Corresponding author.

DOI: 10.1201/9781003568940-4

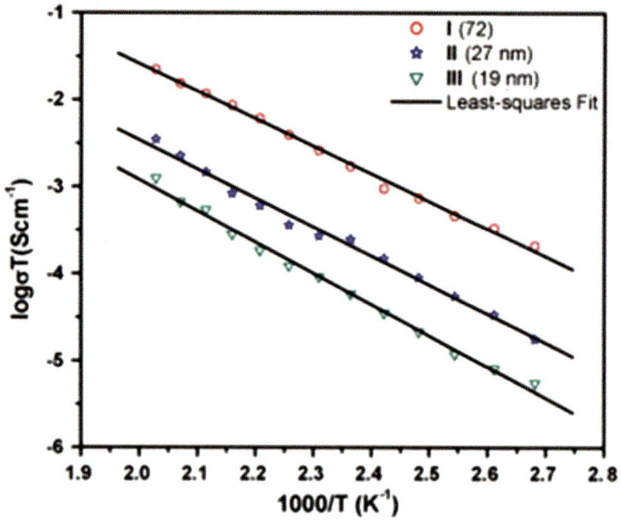

FIGURE 4.1 The Arrhenius plots for nano-crystalline $MgFe_2O_4$ spinel samples electrical conductivity. Equation (4.1) least-squares fit are shown by the solid lines. (Reproduced by permission from the Ref. [19], License number 5650280447042, copyright 2010, Elsevier.)

calculation of the resistance. In every instance, the relationship between the inverse temperature and the change in conductivity is nearly linear. Conductivity was found to increase with temperature for the sample, which is consistent with what was predicted based on the spinel ferrite's semiconducting behaviour, as seen earlier [9, 15].

It has been hypothesized that the electrons hopping from Fe^{2+} to Fe^{3+} ions at the octahedral [B] sites causes $MgFe_2O_4$ to have the properties of an n-type semiconductor [2]. The activation energy for electrical conductivity can be found by using the Arrhenius relation, which is written as

$$\sigma_T = \sigma_0 \exp\left[-\frac{E_a}{K_B T}\right] \tag{4.1}$$

In this equation, σ_0 represents the pre-exponential component with (Ω/cm) K dimensions, E_a represents the DC conductivity activation energy, and K_B represents the Boltzmann constant. The values of the activation energies were found to be 0.62 eV for $MgFe_2O_4$ sample, and the grain size was 72 nm [26]. It was hypothesized that the grain–grain connection's blocking nature was caused by the segmentation of the lattice mismatching, host lattice cations, dislocations, defects, and other such phenomena. In a normal situation, conductivity is determined by the combined influence of a number of parameters, including the Fe^{2+} ions concentration, the grain size, the perfection of the crystal structure, and the micro-structural uniformity. Usually, the ferrite's conductivity is mostly caused by the electron's transfer among the Fe^{2+} and Fe^{3+} ions that are available on equivalent lattice sites (octahedral site) [27, 28]. Additionally, it is known that the degree of covalency for the ions in the A-sites is higher than that of the cations in the B-sites [15], and the distance among the A-sites cations is greater than the distance among the B-sites cations. It is reasonable to anticipate that the holes and electrons mobility at A-sites will be lower than that at B-sites because of the difference in structure between the two types of sites and the higher degree of covalency found in A-sites. As a result of the lower conductivity of samples reducing the eddy current losses at high frequencies, the crushed ferrite samples investigated in this study are appropriate for use in high-frequency applications.

4.2.2 Dielectric Properties

Figure 4.2a and b illustrate, respectively, the frequency dependence of the real component of the dielectric constant (ε'), which is measured at 300K, and the dielectric loss factor (tanδ), which is measured at the same temperature. It can be seen from Figure 4.2a that the first goes up, and then it starts going down gradually with regard to the frequency of spinel ferrite. In a general sense, the atomic and electronic polarizations that are present in ferrite grains are the primary contributors to the dielectric constant [29, 30]. As the electrons hopping among the Fe^{3+} and Fe^{2+} ions are unable

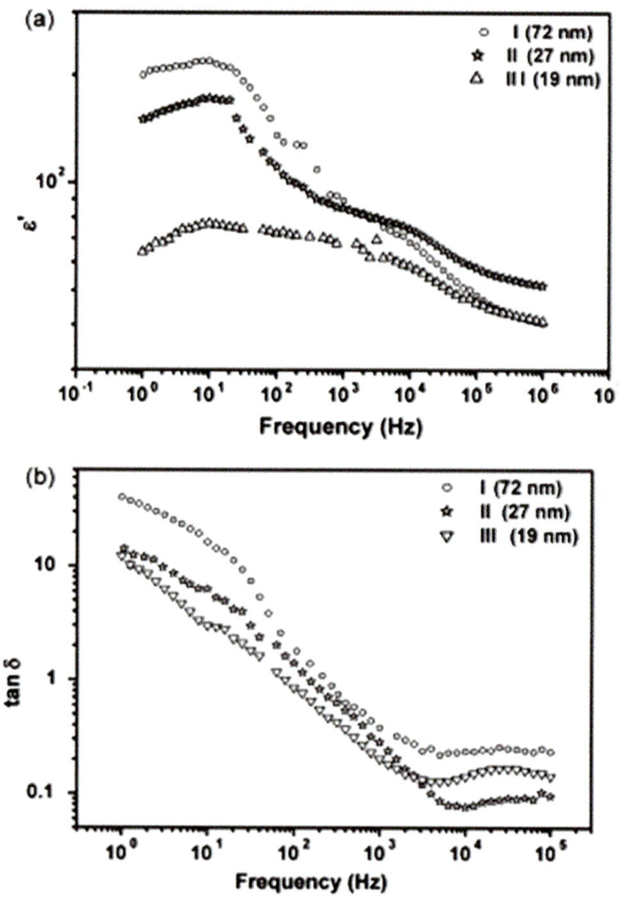

FIGURE 4.2 (a) The real component of the dielectric constant (ε) as a function of frequency for various $MgFe_2O_4$ grain sizes at 300K and (b) the dielectric loss factor (tanδ) for different $MgFe_2O_4$ grain sizes as a function of frequency at 300K. (Reproduced by permission from the Ref. [19], License number 5650280447042, copyright 2010, Elsevier.)

to follow the applied field beyond a certain frequency, the value falls as the frequency increases in normal behaviour.

In the instance of $MgFe_2O_4$, the presence of Mg^{2+} ions results in the generation of p-type and n-type carriers that are generated as a result of the exchange of electrons between Fe^{2+} and Fe^{3+} ions [31]. It is possible to interpret the abnormal dielectric behaviour as being owing to a combined contribution to the polarization made by the n-type carriers and the p-type carriers. In addition to the contributions made by the n-type charge carriers, the net polarization is also affected by the local displacements of p-type carriers in the same direction as the external electric field [32–34]. Because the mobility of p-type carriers is lower than that of n-type carriers, the contribution to the polarization made by the former will diminish more quickly and even at lower frequencies. This is because the p-type carriers are less mobile than the n-type carriers. As a result, there will be an increase in the net contribution at first, followed by a drop as the frequency increases. The location of the highest value will be determined by the ratio of p-type carriers to other types of carriers present in the material. The peak frequency increases in direct proportion to the number of p-type carriers present. In the instance of Cu–Ni ferrites, [32] has observed behaviour that is analogous to that described above. The shift in the Fe^{2+} ion concentration is the primary contributor to the change that occurs in ferrite. If there is a higher Fe^{2+} ions concentration, then the value of the dielectric constant will also be higher [35–37].

4.2.3 The Investigations of AC Conductivity and DC Resistivity

The alternating current (AC) conductivity and permittivity are experimentally and, respectively, represented as $\sigma'(\omega)\ltimes\omega^n$ and $\epsilon'(\omega)\ltimes\omega^{n-1}$, with $0<n\leq1$. This occurs above the dielectric loss peak frequency, which is denoted by the notation $\omega>\omega_p$. The conductivity of AC has been measured throughout a broad frequency spectrum and at a variety of temperatures [38–40]. Figure 4.3 illustrates the frequency dependence of the conductivity at a number of different temperatures. According to West and Almond [41], the AC conductivity of some conducting materials may be described as,

$$\sigma_{ac} = \sigma_{dc} + A\omega^n = \sigma_{dc}\left[1+\left(\frac{\omega}{\omega_p}\right)^n\right] \tag{4.2}$$

which is referred to as the universal power law. This law describes how the AC conductivity of some conducting materials behaves. NLLS fitting is used to fit the conductivity spectra at various temperatures to Equation (4.2) for all of the samples, and then the parameters p, n, and DC are derived from the study. In the typical power law, the frequency exponent, n, can take on values that are close to or slightly below unity and depend on the frequency. The equation that was just presented is a general one that may be used to describe

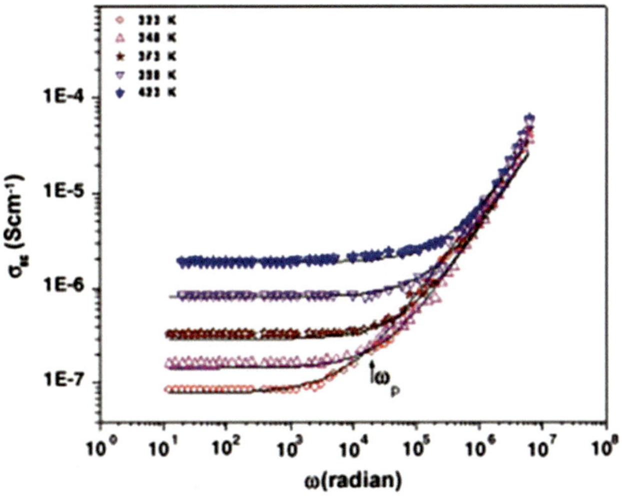

FIGURE 4.3 The $MgFe_2O_4$ sample (72 nm) frequency dependency of the AC conductivity. (Reproduced by permission from the Ref. [19], License number 5650280447042, copyright 2010, Elsevier.)

glasses, amorphous semiconductors, and electronic conductors alike. In an effort to determine where the power law dispersion comes from, numerous theoretical models have been proposed. Other non-Debye dielectric response functions, such as Havirliak–Negami, Cole–Davidson, and Cole–Cole [42–44], exhibit the power law characteristics as well. According to Almond and West's [41] research, the previously mentioned equation can also be written as

$$\sigma_{ac} = \sigma_{dc}\left(1+(\omega\tau)^n\right) \tag{4.3}$$

where τ denotes the relaxation period, the frequency-dependent conductivity data for the sample (72 nm) at a range of temperatures were modelled using the aforementioned equation and the results are depicted in Figure 4.3.

The material's DC conductivity is recognized using the plateaus that are frequency independent in the $\log_{10}\sigma$ vs. $\log_{10}\omega$ graphs. The high-frequency dispersion is responsible for the rise in bulk conductivity that occurs in the high-frequency band. From Figure 4.3, it can be observed that a frequency-independent plateau that occurs repeatedly throughout multiple decades of frequency for all ranges of temperature examined. However, because of an extra action that occurs within the grain, the increase in conductivity may only be observed at higher frequency ranges [45]. This manifests itself as an increase in conductivity that follows a power law relationship with increasing frequency. With a temperature range of 323K–873K and increments of 50K, the Keithley Electrometer Model 2401 used to perform the two probe I–V measuring procedure. DC electrical resistivity was calculated using Equations (4.4) and (4.5) and expressed as

$$R = \rho L/A \tag{4.4}$$

$$\rho = RA/L \tag{4.5}$$

where R stands for resistance, L for thickness, and A is the area of the pallet. The relationship between resistivity and temperature is given by Equation (4.6):

$$\rho_{DC} = \rho_O \exp(\nabla E / K_B T) \quad (4.6)$$

where ∇E is the activation energy, K_B is the Boltzmann constant, and ρ_{DC} is the electrical resistivity at infinity of temperature. The conductivity is said to be caused by the electrons' mobility. Fe^{2+} ions within the crystal lattice cause the mobility of charge carriers. A study has shown the Cr–Mg ($MgFe_{2-x}Cr_xO_4$) ferrite's temperature dependency on DC electrical resistivity [45]. Above 290°C, it has been found that ferrite samples have a semiconducting character, and that DC electrical resistivity decreases as temperature rises as shown in Figure 4.4. The metallic phase changes into the semiconducting phase at this temperature. All of the compounds have a DC resistivity that ranges between 106 and 109 cm. The compounds' resistivity values show that the elements with a single stable oxidation state are present at the B-site. Electrical resistivity is low in transition metal oxides when the compounds comprise cations of the same element positioned at the same location but with a valency difference of one [46]. The hopping mechanism makes it possible to see the conduction. Identifying Drift Mobility (μ_d) with the aid of the connection provided in Equation (4.7), the drift mobility d has been established [47].

$$\mu_d = 1/\eta e \rho_{DC} \quad (4.7)$$

where e stands for the electron's charge, while ρ_{DC} is the electrical resistivity, and η stands for the charge carrier concentration. Expression was used to compute the charge carrier concentration (4.8).

$$\eta = N_A \rho_b P_{Fe} / M \quad (4.8)$$

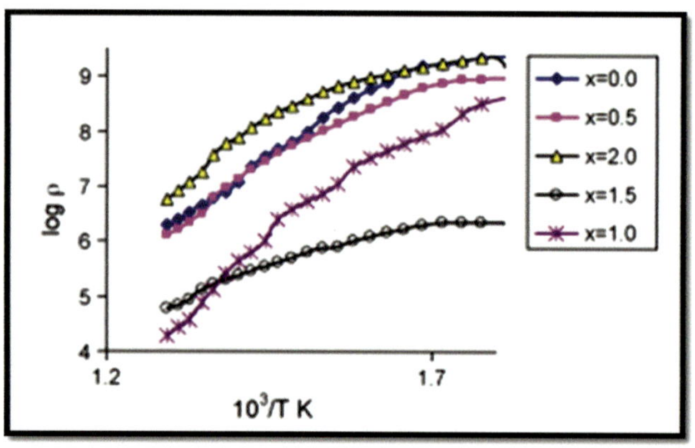

FIGURE 4.4 Variation of $MgFe_{2-x}Cr_xO_4$'s DC electrical resistivity (ρ) with temperature (K) for various compositions (x). (Reproduced by permission from the Ref. [16], License number 5530810514312, copyright 2009, Elsevier.)

where N_A is Avogadro's number (6.022×10^{23} mol^{-1}), ρ_b is bulk density, M is the sample's molecular weight, and P_{Fe} represents a number of the iron atoms in the composition. The quantity of iron atoms in the composition formula is given by xy^2. The heat activation process is what causes the charge carriers' upward trend in mobility [48].

The relationship between electrical resistivity and drift mobility is inversely correlated. The hopping of charge carriers is enhanced by a rise in temperature, which causes the drift mobility to increase with temperature [49]. By hopping Fe ions at the octahedral site, the conduction can be shown. Resistivity in this system increases for x = 2.0 but reduces up to x = 1.5 (Figure 4.4). According to reports, Fe ions only partially fill the A- and B-sites, while Cr ions prefer to occupy the octahedral (B) site. Most of the Mg^{2+} ions in $MgFe_2O_4$ reside on the B-site, and only tiny percentages migrate to the A-site. Fe ion concentration (at B-site) falls when Cr content rises. As a result, some Mg^{2+} ions move towards the A-site and some Fe^{3+} ions go towards the B-site [50]. The electrical conductivity consequently rises. It was discovered that as the Cr ion level rose, so did the activation energy of electric conduction. Due to the steady oxidation state of chromium in $MgCr_2O_4$, the resistivity rises once more. Its ionic structure is represented by the formula $Mg^{2+}[Cr^{3+}Cr^{3+}]O_4$. Therefore, the difficulty in transferring electrons between the ions at the B-site is evident, as is the high resistivity value. Additionally, it suggests that octahedral locations are devoid of ions with diverse oxidation states. This is in line with the finding that higher conductivity and lower electrical resistivity are related to lower activation energy.

4.2.3.1 Magnetic Studies

The magnetic nature of spinel ferrites is their most noteworthy characteristic. By exchange interactions mediated by oxygen anions, atoms at tetrahedral sites can orient their spins either anti-parallel or parallel towards the octahedral ions. Since the magnetic forces of the cations do not cancel out, sub-lattices alignment is often aligned in an anti-parallel fashion and leads to a ferrimagnetic alignment [51]. Getting the examined material's hysteresis curves is one technique to examine its magnetic characteristics. Figure 4.5 depicts an illustration of typical hysteresis behaviour.

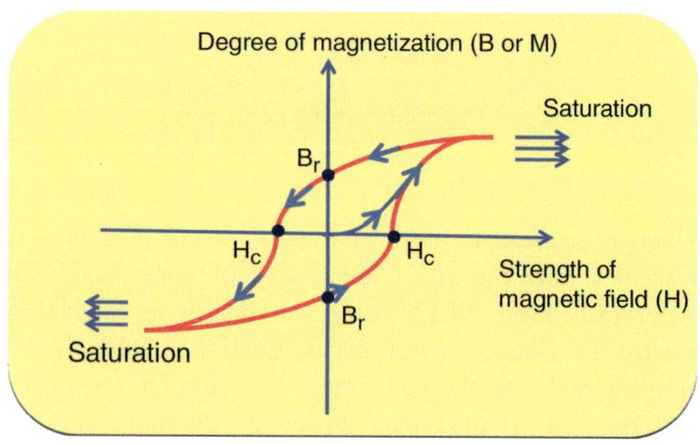

FIGURE 4.5 Typical magnetic hysteresis curve illustration.

The sample's magnetization when applied field is not present in the absence is referred to as remanence or remanent magnetization (Mr) [52]. At the moment when the hysteresis curve crosses the y axis, this value is acquired. This parameter indicates how much spontaneous magnetism is available in a material at the experiment's temperature. The highest level of magnetization a material may acquire under the influence of powerful magnetic field is known as saturation magnetization (Ms) [53].

In the crystalline framework, this characteristic is typically a useful indicator for the internal configuration of cations since it denotes the maximum magnetism of the material. When magnetization does not increase further even after increasing the applied field, then saturation is considered to have occurred. In the case of incomplete saturation, the word maximal magnetization (M*) might be used [54]. The field required to invert the magnetization direction is indicated by a sample's coercivity. Coercivity is frequently correlated with how challenging it is to move the magnetic domain barriers within a crystal. Hence, the coercivity of the material will be directly impacted by any procedure or characteristic that might prevent or promote the movement of these domain barriers (Figure 4.6). This metric may be impacted by a wide range of variables [5].

It is widely believed that structural parameters, such as crystallite size and the existence of defects, can strongly affect the amount of energy required for the domain walls motion. The absolute intensity of the coercive field is also determined by magnetic anisotropies, including form, tension, and magneto-crystalline. Each ferrite is classified as either a soft ferrite or a hard ferrite based on the coercivity value [33]. Low coercivity is present in soft ferrites, and the magnetization direction can be easily changed. This category includes the majority of spinel ferrites. Vibrating sample magnetometer (VSM) analysis was used to assess the magnetic behaviour of ferrites at ambient temperature with a maximum applied field of 1.5 (T). The following relationships have been used to derive the magnetic properties of prepared samples, including magneto-crystalline anisotropy constant (K), Bohr magnetic moments (μ_B), Y–K (Yafet and Kittel) angles, and μ_i is initial permeability.

$$\mu_B = M \times \frac{M_S}{5585 \times \rho_{X-Rav}} \tag{4.9}$$

$$K = H_C \times \frac{M_S}{0.96} \tag{4.10}$$

$$\mu_i = M_S^2 \times \frac{D}{K} \tag{4.11}$$

$$n_B = (6+x)Cos\alpha_{Y-K} - 5(1-x) \tag{4.12}$$

where D, M, n_B, and x stand for the synthetic samples' crystalline size, molecular weights, experimental magnetic moment, and sample's concentration values. Moreover, the ferrimagnetic order change from collinear to non-collinear order may be responsible for the decrease in these magnetic properties. The magnetic characteristics of the spinel ferrites were significantly impacted by the dopant element's ionic radii and other cations present. Consequently, the size of the bond lengths and ionic radii both affect the magnetic behaviour of nanoparticles. The electric or magnetic losses have a significant impact on the devices' performance. For example, VSMs were used by Easwari et al. to study the magnetic hysteresis loop and magnetic properties of the synthesized $MgFe_2O_4$ samples at room temperature. At room temperature, the sample shows ferromagnetic behaviour. It is well recognized that the sample shape, crystallinity, magnetization direction, and other factors affect a material's magnetic properties [55]. At 250°C, the Ms, Mr, and coercivity (Hc) of magnesium ferrite nanoparticles are 25.4 emu/g, 1.48 emu/g, and 27.7 G, respectively. Due to the spin canting of ferrite nanoparticles at their surface [56] and the replacement of a portion of Mg^{2+} in the octahedral sites by Fe^{2+} with a higher magnetic moment, the Ms of the sample is close to the reported bulk value. From the reported results, it is clear that the synthesis process also affects magnetic properties. It is being investigated how this method of preparation affects magnetic properties.

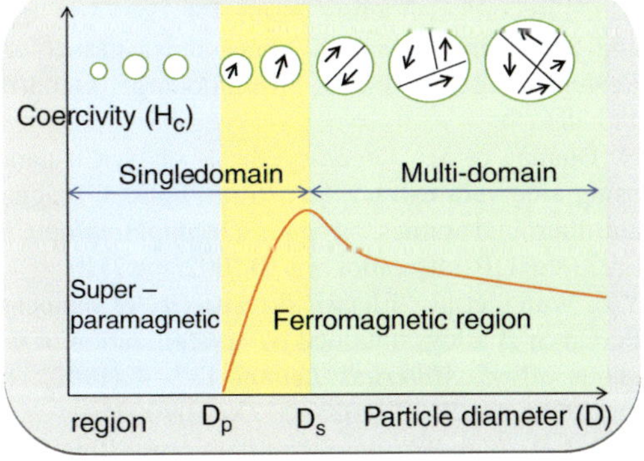

FIGURE 4.6 Coercivity is dependent on the size of the magnetic particles.

4.3 CONCLUSION

The study of magnetic and electrical properties of spinel ferrites are discussed in this chapter. It is observed that spinel ferrites with good electrical properties show high conductivity, and these electrical properties change with heating or annealing the sample. From the magnetic characteristics, it is observed that the spinel ferrites are significantly impacted by the dopant element's ionic radii and other cations present. Consequently, the size of the bond lengths and ionic radii both affect the magnetic behaviour of nanoparticles. The electric or magnetic losses have a significant impact on the devices' performance.

REFERENCES

1. P. Thakur et al. "Cobalt nanoferrites: A review on synthesis, characterization, and applications." *Journal of Superconductivity and Novel Magnetism* 35 (2022): 2639–2669, https://doi.org/10.1007/s10948-022-06334-1
2. D.K. Mahato, and S. Banerjee. "Dielectric characteristics of $MgFe_2O_4$ ferrite prepared by sol-gel auto-combustion method." *Materials Today: Proceedings* 4.4 (2017): 5525–5531, https://doi.org/10.1016/j.matpr.2017.06.008
3. K. Konishi et al. "Heating ferrite powder with AC magnetic field for thermal coagulation therapy." *Journal of Magnetism and Magnetic Materials*, 272–276 (2004): 2428–2429, https://doi.org/10.1016/j.jmmm.2003.12.591
4. N. Chinnasamy et al. "Grain size effect on the Néel temperature and magnetic properties of nanocrystalline NiFe2O4 spinel." *Journal of Magnetism and Magnetic Materials*, 238.2–3 (2002): 281–287, https://doi.org/10.1016/S0304-8853(01)00928-3
5. N. Dhanda, P. Thakur, and A. Thakur. "Green synthesis of cobalt ferrite: A study of structural and optical properties." *Materials Today: Proceedings* 73.Part 2 (2023): 237–240, ISSN 2214-7853, https://doi.org/10.1016/j.matpr.2022.07.202
6. Y. Shi, and J. Ding. "Strong unidirectional anisotropy in mechanically alloyed spinel ferrites." *Journal of Applied Physics* 90.8 (2001): 4078, https://doi.org/10.1063/1.1405834
7. S.A. Oliver et al. "Large zinc cation occupancy of octahedral sites in mechanically activated zinc ferrite powders." *Applied Physics Letters* 76.19 (2000): 2761–2763, https://doi.org/10.1063/1.126467
8. J.Z. Jiang, G.F. Goya, and H.R. Rechenberg. "Magnetic properties of nanostructured $CuFe_2O_4$." *Journal of Physics: Condensed Matter* 11.20 (1999): 4063, https://doi.org/10.1088/0953-8984/11/20/313
9. N. Sivakumar, et al. "Effect of mechanical milling on the electrical and magnetic properties of nanostructured $Ni_{0.5}Zn_{0.5}Fe_2O_4$." *Journal of Physics D: Applied Physics* 39.21 (2006): 4688, https://doi.org/10.1088/0022-3727/39/21/028
10. J.C. Dyre, and T.B. Schroder. "Universality of ac conduction in disordered solids." *Reviews of Modern Physics* 72.3 (2000): 873, https://doi.org/10.1103/RevModPhys.72.873
11. B. Roling, C. Martiny, and S. Murugavel. "Ionic conduction in glass: New information on the interrelation between the "jonscher behaviour" and the "nearly constant-loss behaviour" from broadband conductivity spectra." *Physical Review Letters* 87.8 (2001): 085901.
12. J.R. Macdonald, "Possible universalities in the ac frequency response of dispersed, disordered materials." *Journal of Non-Crystalline Solids* 210.1 (1997): 70–86. https://doi.org/10.1016/S0022-3093(96)00564-9
13. S. Kharbanda et al. "Multiferroic perovskite bismuth ferrite nanostructures: A review on synthesis and applications." *Journal of Magnetism and Magnetic Materials* 170569 (2023).
14. W.K. Lee, J.F. Liu, and A.S. Nowick. "Limiting behavior of ac conductivity in ionically conducting crystals and glasses: A new universality." *Physical Review Letters* 67.12 (1991): 1559, https://doi.org/10.1103/PhysRevLett.67.1559
15. N. Ponpandian, P. Balaya, and A. Narayanasamy. "Electrical conductivity and dielectric behaviour of nanocrystalline NiFe2O4 spinel." *Journal of Physics: Condensed Matter* 14.12 (2002): 3221, https://doi.org/10.1088/0953-8984/14/12/311
16. P. Hankare et al. "Synthesis, characterization and studies on magnetic and electrical properties of Mg ferrite with Cr substitution." *Materials Chemistry and Physics* 113 (2009) 233–238, https://doi.org/10.1016/j.matchemphys.2008.07.066
17. S. Kumari et al. (2022) "Structural, Morphological, Spectroscopic, and Magnetic Properties of Mg-Zn Nanoferrite for High-frequency Applications." In *International Conference on Nanotechnology: Opportunities and Challenges* (pp. 333–338). Singapore: Springer Nature Singapore.
18. Q. Chen, and Z.J. Zhang. "Size-dependent superparamagnetic properties of $MgFe_2O_4$ spinel ferrite nanocrystallites." *Applied Physics Letters* 73.21 (1998): 3156–3158, https://doi.org/10.1063/1.122704
19. N. Sivakumar et al. "Electrical and magnetic behaviour of nanostructured $MgFe_2O_4$ spinel ferrite." *Journal of Alloys and Compounds* 504.2 (2010): 395–402, https://doi.org/10.1016/j.jallcom.2010.05.125
20. V. Šepelák, et al. "Enhanced magnetisation in nanocrystalline high-energy milled $MgFe_2O_4$." *Scripta Materialia* 48.7 (2003): 961–966, https://doi.org/10.1016/S1359-6462(02)00600-0
21. M. Gateshki et al. "Structure of nanocrystalline $MgFe_2O_4$ from X-ray diffraction, Rietveld and atomic pair distribution function analysis." *Journal of Applied Crystallography* 38.5 (2005): 772–779, https://doi.org/10.1107/S0021889805024477
22. I. Bergmann, V. Šepelák, and K.D. Becker. "Preparation of nanoscale $MgFe_2O_4$ via non-conventional mechanochemical route." *Solid State Ionics* 177.19–25 (2006): 1865–1868, https://doi.org/10.1016/j.ssi.2006.04.002
23. D. Ravinder, and K. Latha. "Dielectric behaviour of mixed Mg–Zn ferrites at low frequencies." *Materials Letters* 41.5 (1999): 247–253, https://doi.org/10.1016/S0167-577X(99)00138-X
24. M.A. El Hiti. "Dielectric behaviour in Mg–Zn ferrites." *Journal of Magnetism and Magnetic Materials* 192.2 (1999): 305–313, https://doi.org/10.1016/S0304-8853(98)00356-4
25. N. Dhanda et al. "Influence of Ni over magnetically benign Co ferrite system and study of its structural, optical, and magnetic behaviour." *Inorganic Chemistry Communications* 151 (2023): 110569, https://doi.org/10.1016/j.inoche.2023.110569
26. N. Dhanda, P. Thakur, A.-C. Sun, and A. Thakur. "Structural, optical and magnetic properties along with antifungal activity of Ag-doped Ni-Co nanoferrites synthesized by eco-friendly route." *Journal of Magnetism and Magnetic Materials* 572 (2023): 170598, https://doi.org/10.1016/j.jmmm.2023.170598
27. G.A. Sawatzky, F. Van Der Woude, and A.H. Morrish. "Recoilless-fraction ratios for Fe 57 in octahedral and tetrahedral sites of a spinel and a garnet." *Physical Review* 183.2 (1969): 383, https://doi.org/10.1103/PhysRev.183.383
28. N. Dhanda et al. "Green-synthesis of Ni-Co nanoferrites using aloe vera extract: Structural, optical, magnetic, and antimicrobial studies." *Applied Organometallic Chemistry* (2023): e7110, https://doi.org/10.1002/aoc.7110
29. Y.Z. Wang, et al. "Electrical resistivity of nanocrystalline Fe-Cu-Si-B alloys obtained by crystallization of the amorphous alloy." *Materials Letters* 17.3–4 (1993): 152–154, https://doi.org/10.1016/0167-577X(93)90075-9
30. X.D. Liu, et al. "Properties of nanocrystalline Fe-Cu-Si-B alloys generated by crystallization of the amorphous alloy." *Physica B: Condensed Matter* 192.4 (1993): 345–350, https://doi.org/10.1016/0921-4526(93)90009-U

31. S. Kumari, N. Dhanda, A. Thakur, S. Singh, and P. Thakur "Investigation of calcium substitution on magnetic and dielectric properties of Mg–Zn nano ferrites." *Materials Chemistry and Physics* 297 (2023): 127394, https://doi.org/10.1016/j.matchemphys.2023.127394
32. A. Pathania, S. Bhardwaj, S.S. Thakur, J.L. Mattei, P. Quefelec, L.V. Panina, P. Thakur, and A. Thakur. "Investigation of structural, optical, magnetic and electrical properties of tungsten doped Ni–Zn nano-ferrites." *Physica B: Condensed Matter* 531 (2018): 45–50, https://doi.org/10.1016/j.physb.2017.12.008
33. D. Suhag, A. Thakur, and P. Thakur (eds) (2023) *Integrated Nanomaterials and their Applications* (pp. 1–499). Springer Nature Singapore, https://doi.org/10.1007/978-981-99-6105-4
34. C.G. Koops. "On the dispersion of resistivity and dielectric constant of some semiconductors at audiofrequencies." *Physical Review* 83.1 (1951): 121, https://doi.org/10.1103/PhysRev.83.121
35. D.C. Dube. "Study of Landau-Lifshitz-Looyenga's formula for dielectric correlation between powder and bulk." *Journal of Physics D: Applied Physics* 3.11 (1970): 1648, https://doi.org/10.1088/0022-3727/3/11/313
36. N. Ponpandian, and A. Narayanasamy. "Influence of grain size and structural changes on the electrical properties of nanocrystalline zinc ferrite." *Journal of Applied Physics* 92.5 (2002): 2770–2778, https://doi.org/10.1063/1.1498883
37. U. Van, and G. Legrand. "Dielectric properties of and conductivity in ferrites." *Proceedings of the IRE* 44.10 (1956): 1294–1303, https://doi.org/10.1109/JRPROC.1956.274952
38. S. Kumari et al. "Nano Ca–Mg–Zn ferrites as tuneable photocatalyst for UV light-induced degradation of rhodamine B dye and antimicrobial behavior for water purification." *Ceramics International* 49.8 (2023): 12469–12480, https://doi.org/10.1016/j.ceramint.2022.12.107
39. K. Prasad, K. Kumari, and K.L. Yadav. "Hopping type of conduction in $(Na_{0.5}Bi_{0.5})$ ZrO_3 ceramic." *Journal of Physics and Chemistry of Solids* 68.8 (2007): 1508–1514, https://doi.org/10.1016/j.jpcs.2007.03.023
40. M.A.L. Nobre, and S. Lanfredi. "Dielectric spectroscopy on $Bi_3Zn_2Sb_3O_{14}$ ceramic: An approach based on the complex impedance." *Journal of Physics and Chemistry of Solids* 64.12 (2003): 2457–2464, https://doi.org/10.1016/j.jpcs.2003.08.007
41. D.P. Almond, and A.R. West. "Mobile ion concentrations in solid electrolytes from an analysis of ac conductivity." *Solid State Ionics* 9 (1983): 277–282, https://doi.org/10.1016/0167-2738(83)90247-3
42. D.W. Davidson, and C.H. Robert. "Dielectric relaxation in glycerol, propylene glycol, and n-propanol." *The Journal of Chemical Physics* 19.12 (1951): 1484–1490, https://doi.org/10.1063/1.1748105
43. S. Havriliak, and S. Negami. "A complex plane representation of dielectric and mechanical relaxation processes in some polymers." *Polymer* 8 (1967): 161–210, https://doi.org/10.1016/0032-3861(67)90021-3
44. K.S. Cole, and R.H. Cole. "Dispersion and absorption in dielectrics I. Alternating current characteristics." *The Journal of Chemical Physics* 9.4 (1941): 341–351, https://doi.org/10.1063/1.1750906
45. D.P. Almond, G.K. Duncan, and A.R. West. "The determination of hopping rates and carrier concentrations in ionic conductors by a new analysis of ac conductivity." *Solid State Ionics* 8.2 (1983): 159–164, https://doi.org/10.1016/0167-2738(83)90079-6
46. M.K. Shobana, K. Kim, and J.H. Kim. "Impact of magnesium substitution in nickel ferrite: Optical and electrochemical studies." *Physica E: Low-Dimensional Systems and Nanostructures* (2018), https://doi.org/10.1016/j.physe.2018.12.013
47. V. Banihashemi et al. "Dual Ca–Zn substituted strontium hexaferrite; investigation of structural, magnetic and optical properties." Physica B 605 (2021): 412670, https://doi.org/10.1016/j.physb.2020.412670
48. H.J. Kardile et al. "Effect of Cd^{2+} doping on structural, morphological, optical, magnetic and wettability properties of nickel ferrite thin films. *Optik (Stuttg)* 207: 164462 (2020), https://doi.org/10.1016/j.ijleo.2020.164462
49. S. Kavitha, and M. Kurian. "Effect of zirconium doping in the microstructure, magnetic and dielectric properties of cobalt ferrite nanoparticles." *Journal of Alloys and Compounds* 799 (2019): 147–159, https://doi.org/10.1016/j.jallcom.2019.05.183
50. M.M. Naik et al. "Effect of aluminum doping on structural, optical, photocatalytic and antibacterial activity on nickel ferrite nanoparticles by sol–gel auto-combustion method." *Journal of Materials Science: Materials in Electronics* 29 (2018): 20395–20414, https://doi.org/10.1007/s10854-018-0174-y
51. A. Rafferty, T. Prescott, and D. Brabazon. "Sintering behaviour of cobalt ferrite ceramic." *Ceramics International* 34.1 (2008): 15–21, https://doi.org/10.1016/j.ceramint.2006.07.012
52. S.U.N. Ke et al. "Effect of sintering process on microstructure and magnetic properties of high frequency power ferrite." *Rare Metals* 25.6 (2006): 509–514.
53. W. Schiessl et al. "Magnetic properties of the $ZnFe_2O_4$ spinel." *Physical Review B* 53 (1996), 9143–9152, https://doi.org/10.1103/PhysRevB.53.9143
54. J. Philip et al. "Effect of thermal annealing under vacuum on the crystal structure, size, and magnetic properties of $ZnFe_2O_4$ nanoparticles." *Journal of Applied Physics* 102 (2007): 54305, https://doi.org/10.1063/1.2777168
55. S. Ayyappan et al. "Room temperature ferromagnetism in vacuum annealed $ZnFe_2O_4$ nanoparticles." *Applied Physics Letters* 96 (2010):143106, https://doi.org/10.1063/1.3374332
56. C.E.T. Rodríguez et al. "Oxygen-vacancyinduced local ferromagnetism as a driving mechanism in enhancing the magnetic response of ferrites." *Physical Review B* 89 (2014): 104411, https://doi.org/10.1103/PhysRevB.89.104411

5 Role of Spinel Nano-Ferrites in Health Sector

Preeti Thakur, Shivani Jangra, and Atul Thakur*
Amity Institute of Nanotechnology, Amity University Haryana, Gurugram, India

5.1 INTRODUCTION TO SPINEL FERRITES

Spinel ferrites represent a fascinating class of materials with versatile applications, driven by their distinctive crystal structure and magnetic properties. These ferrites, which are based on the structure of the mineral spinel, have attracted a lot of interest from various scientific and technical fields. Figure 5.1 depicts the different types of ferrites. An outline of the essential properties that make spinel ferrites attractive for a variety of applications is given in this introduction.

5.1.1 Spinel Structure

The distinct crystal structure of spinel ferrites, or the spinel structure, is the fundamental component of these materials. In this configuration, divalent and trivalent metal cations are arranged in a face-centred cubic (FCC) lattice of oxygen ions at tetrahedral and octahedral sites, respectively [1]. The unique arrangement adds to its interesting characteristics and stability.

5.1.2 Material

Spinel ferrites have the typical chemical formula AFe_2O_4, where "A" stands for a divalent metal cation (like cobalt, manganese, zinc, or nickel) [2]. The arrangement of these metal cations inside the spinel structure gives the ferrite particular properties that affect how it behaves in magnetic fields.

5.1.3 Magnetic Characteristics

- Ferromagnetism is one of the main magnetic characteristics of spinel ferrites. Adjacent magnetic moments in ferromagnetic materials naturally align parallel to one another. Because of this alignment, spinel ferrites have a net magnetic moment inside the material, which makes them perfect for magnetic applications.
- Spinel ferrites can achieve magnetic saturation, which means that when they encounter an external magnetic field, they can become completely

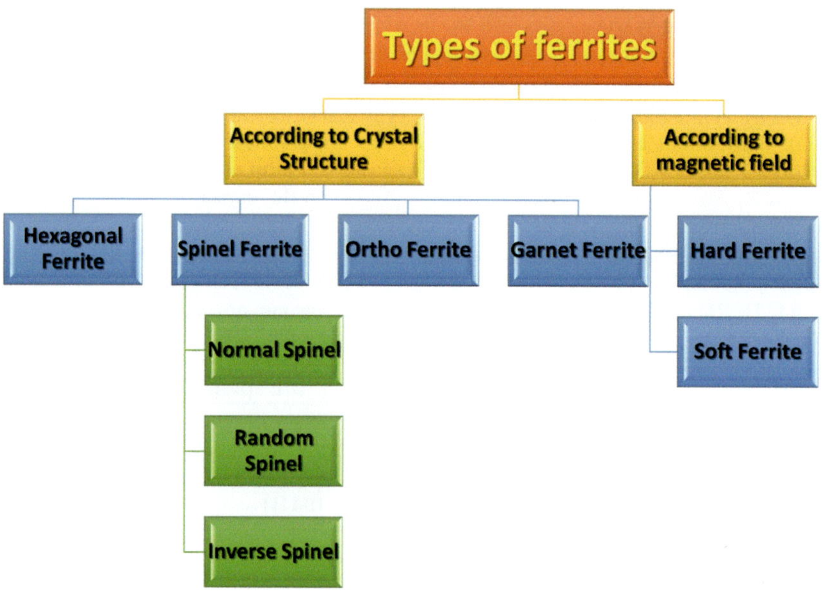

FIGURE 5.1 Types of ferrites.

* Corresponding author.

DOI: 10.1201/9781003568940-5

magnetized [3]. To ensure peak performance in devices like magnetic recording medium, this feature is essential.

- Spinel ferrites exhibit temperature-dependent magnetic properties. The critical value at which a material changes from a ferromagnetic to a paramagnetic state is represented by the Curie temperature. The material loses its spontaneous magnetism beyond the Curie temperature.
- When a magnetic field is applied, a material can experience magnetostriction, a phenomenon where the material changes shape [4]. This can happen to spinel ferrites. Actuators and sensors both use this characteristic.

5.1.4 Importance of Spinel Ferrites in Various Applications

These types of magnetic materials known as spinel ferrites, due to their distinctive crystal structure, have proven to be invaluable in a wide range of scientific and technological applications. They are useful and adaptable for a variety of applications, from electronics and telecommunications to biomedical devices, due to their remarkable magnetic characteristics, electrical conductivity, and biocompatibility. In this comprehensive exploration, we delve into the multifaceted importance of spinel ferrites, highlighting their role in shaping modern technologies and advancing scientific research.

5.1.4.1 Electronics and Telecommunications

These two industries are among the main ones in which spinel ferrites excel. Their strong electrical resistance and magnetic qualities make them perfect for parts like inductors and transformers. Because of their stability and effective energy transfer properties, spinel ferrites are used to create electronic circuits, power supplies, and telecommunications equipment [5]. They are employed in microwave devices because of their high magnetic permeability and low eddy current losses, which allow for component downsizing and improved performance.

5.1.4.2 Magnetic Recording Media

The development of magnetic recording media, including magnetic tapes and hard drives, has been greatly aided by spinel ferrites. For data storage applications, their magnetic stability and capacity to achieve magnetic saturation are essential. Because spinel ferrites have a high coercivity, recorded data can be guaranteed to be preserved and reliably recovered. Spinel ferrites aid in the advancement of more capable and effective magnetic recording technologies to be the need for data storage grows.

5.1.4.3 Biomedical Applications

Recently, spinel ferrites have emerged as a potentially useful material for biomedical applications, paving the way for novel approaches to treatment and diagnosis. For these applications, spinel ferrite nanoparticles are especially appealing due to their special blend of magnetic and biocompatible qualities.

5.1.4.4 Drug Delivery System

Nanoparticles of spinel ferrite are carriers in medication delivery systems. Because of their magnetic responsiveness, medications may be delivered precisely to the right places in the body, increasing therapeutic efficacy and reducing adverse effects [6]. Medications can be added to the surface of these nanoparticles, and an external magnetic field can regulate when the medications are released.

5.1.4.5 Magnetic Resonance Imaging (MRI)

Spinel ferrite nanoparticles serve as contrast agents in magnetic resonance imaging (MRI), a branch of medical imaging. Their magnetic characteristics affect the surrounding water molecules' relaxation durations, which enhances the sharpness and clarity of MRI scans. This application helps to improve disease monitoring and diagnosis accuracy.

5.1.4.6 Treatment for Hyperthermia

Magnetic hyperthermia features are observed in spinel ferrite nanoparticles. They produce heat when placed in an alternating magnetic field, and this heat can be used to cure cancer through a process called hyperthermia therapy. Treatments like chemotherapy and radiation therapy are more successful when they target and heat specific areas of malignant tissue.

5.1.4.7 Actuators and Sensors

Because of their magnetostrictive qualities, spinel ferrites are essential to the creation of actuators and sensors. The process known as magnetostriction occurs when a material undergoes a shape-changing reaction to an applied magnetic field. This characteristic is utilized in the design of actuators that provide accurate control in mechanical systems as well as sensors that identify magnetic fields [7]. Applications for the capacity of spinel ferrite to transform magnetic energy into mechanical motion range from vibration control systems to automobile sensors.

The significance of spinel ferrites in a variety of applications is evidence of their special blend of electrical, magnetic, and biocompatible qualities. Spinel ferrites continue to influence technological landscapes and make contributions to improvements that benefit society, from their fundamental role in electronics to their cutting-edge uses in healthcare sectors.

5.2 BIOCOMPATIBILITY AND BIOSAFETY

5.2.1 Biocompatibility of Spinel Ferrites for Medical Applications

5.2.1.1 In Vitro Cytotoxicity Studies

In vitro experiments are the first step towards comprehending the biocompatibility of spinel ferrites, as scientists show effects on different cell lines of these materials.

To find out how much spinel ferrites may impact cell viability, proliferation, and metabolic activity, cytotoxicity experiments are essential. Early in the new millennium, several investigations revealed that spinel ferrites—including magnetite—have minimal cytotoxicity. These substances typically cause cells to remain viable within reasonable bounds, which paves the way for their possible employment in biological applications [8]. In vitro experiments usually entail incubating cells with different concentrations of spinel ferrite nanoparticles and measuring things like oxidative stress indicators, membrane integrity, and cell survival.

5.2.1.2 In Vivo Biocompatibility Assessments

In vivo experiments are essential for comprehending the biocompatibility of spinel ferrites in intricate physiological settings as they go from cell cultures to living creatures. Studies on animals where spinel ferrite nanoparticles are injected or implanted shed light on the possible impacts on various tissues, organs, and systemic reactions. Early in the new millennium, research on spinel ferrites in vivo biocompatibility revealed few unfavourable impacts. To assess the dispersion of spinel ferrite nanoparticles, their accumulation in organs, and any indications of toxicity, animals—often rodents—were employed as models. These investigations yielded important insights into the safety profile of spinel ferrites in living things, hence bolstering the possibility of using them in therapeutic therapies.

5.2.1.3 Enhanced Biocompatibility by Surface Functionalization

Understanding how surface properties affect biocompatibility, scientists have investigated surface functionalization methods to customize spinel ferrites for certain biomedical uses. Surface modification is covering the nanoparticles with biocompatible substances, like surfactants or polymers, to increase their interactions with biological systems, decrease their toxicity, and improve their durability [9]. Spinel ferrites are surface functionalized for two reasons. Firstly, by forming a biocompatible surface that prevents the core material from coming into direct contact with living things, it lessens the possibility of negative consequences. Secondly, it creates opportunities for additional customization by enabling the targeting of ligands or the attachment of biomolecules. This adaptability is especially helpful in applications where controlled release and precise targeting are critical, such as medicine delivery.

5.2.1.4 Extended Stability in Biological Settings

One important component of the biocompatibility of spinel ferrites is their long-term stability in biological settings [10]. The purpose of stability studies is to ascertain whether these materials can be used for lengthy periods without degrading or producing hazardous byproducts while maintaining their structural, magnetic, and functional qualities. The stability of spinel ferrites across a range of biological fluids and conditions was evaluated in early research conducted in the 2000s. According to the results, these materials show strong stability, which makes them suitable for long-term use. The lack of discernible deterioration or changes in magnetic properties with time adds to spinel ferrites' attractiveness as long-term medical therapies.

5.2.1.5 Biological Compatibility and Immunological Response

Predicting the acceptance of the body of spinel ferrites requires an understanding of the reaction of immune system to them. Studies on immunology evaluate the interactions between these substances and the immune system, including any possible inflammatory reactions and immune cell activation. Early 2000s research revealed that spinel ferrites typically cause a little immune reaction. This observation is particularly important for situations where an overabundance of immune response may limit therapeutic outcomes, such as medication administration. Overall, the biocompatibility profile of spinel ferrites is influenced by their immunocompatibility, which facilitates their incorporation into medical technology without inducing unfavourable immune responses.

5.2.1.6 Applications in Biomedicine

The incorporation of spinel ferrites into a range of biomedical applications has been made possible by their biocompatibility. One of the most prominent uses is in diagnostic imaging, where magnetite nanoparticles and other special magnetic characteristics of spinel ferrites have shown to be advantageous as MRI contrast agents. These nanoparticles improve tissue and organ visibility, resulting in sharper, more detailed images that help in diagnosis. Spinel ferrites, particularly magnetite nanoparticles, have shown promise in the treatment of hyperthermia in the field of cancer therapy [11]. These nanoparticles can selectively target and destroy cancer cells because they produce localized heat when encountering an alternating magnetic field. In this regard, the biocompatibility of spinel ferrites is essential since it guarantees the security of these therapeutic procedures.

Targeted medication distribution is made possible by these magnetic qualities and biocompatibility [12]. Therapeutic medicines can be delivered to certain cells or tissues via functionalized spinel ferrite nanoparticles, reducing side effects and increasing treatment efficacy overall. Research on the biocompatibility of spinel ferrites will probably continue as technology develops, maybe revealing new possibilities and resolving any unmet difficulties. Continued research on spinel ferrites in medicine could lead to novel treatments and better patient outcomes a major advancement at the nexus of materials science and medicine.

5.2.2 Exploration of Their Safety Profiles in Biological Systems

Understanding the safety profiles of spinel ferrites in biological systems is crucial for assessing their viability in

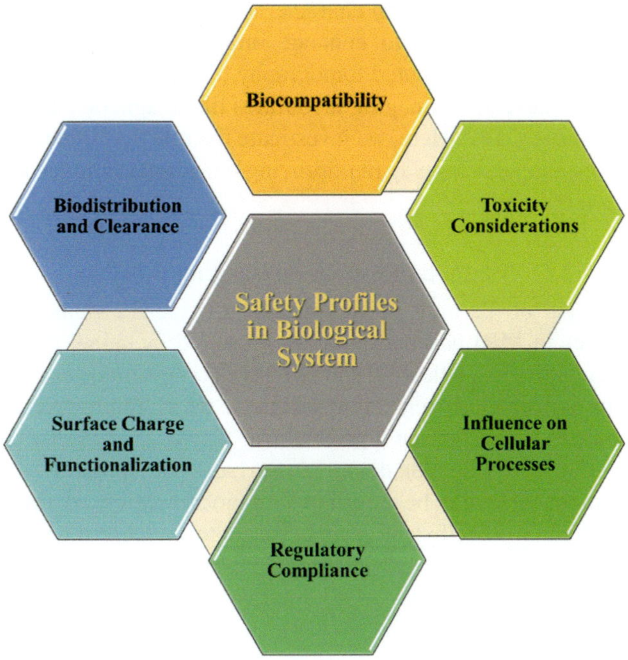

FIGURE 5.2 Safety profiles in biological systems in spinel ferrite nanoparticles.

medical applications (Figure 5.2). A comprehensive exploration involves examining various aspects, including cytotoxicity, genotoxicity, immunotoxicity, long-term effects, and the influence of surface characteristics [13]. Here, we delve into the safety profiles of spinel ferrites within biological systems.

5.2.2.1 Cytotoxicity

Cytotoxicity studies are fundamental in evaluating the impact of spinel ferrites on cell viability, proliferation, and metabolic activity. In vitro studies on various cell lines have consistently shown that spinel ferrites, including magnetite (Fe_3O_4), exhibit low cytotoxicity. Cells exposed to these materials generally maintain viability within acceptable limits, indicating a favourable biocompatibility profile.

5.2.2.2 Genotoxicity

Evaluations of genotoxicity seek to ascertain whether spinel ferrites cause harm to a genetic material of cell. According to preliminary studies, these substances typically show minimal genotoxicity [11]. The results of assays used to evaluate possible DNA damage, such as the comet assay and chromosomal aberration analysis, have shown little effect on the integrity of cellular DNA.

5.2.2.3 Long-Term Effects and Stability

The study on these two topics is crucial to determining whether spinel ferrites are safe to use for an extended amount of time. Studies have demonstrated that these materials remain stable in diverse biological settings, suggesting that they could have long-term medical uses. Their safety profile is improved by the lack of notable structural and magnetic property changes or degradation.

5.2.2.4 Clearance and Biological Fate

Monitoring the distribution, accumulation, and clearance of spinel ferrites within the body is necessary to comprehend their biological fate. According to studies, spinel ferrites appear to have adequate clearance, allaying worries about long-term buildup and possible negative consequences [14]. Their biological destiny is determined by surface functionalization, surface charge, and particle size.

5.2.2.5 Impact of Surface Functionalization

Surface properties are essential for how nanoparticles and biological systems interact. To improve spinel ferrites' safety and biocompatibility, surface functionalization with biocompatible substances like polymers or surfactants has been investigated. In addition to increasing stability, this method reduces direct contact between biological entities and the core material by creating a biocompatible interface.

5.2.2.6 Biocompatibility in Particular Uses

The application of spinel ferrites greatly influences their safety appearance [15]. Spinel ferrites are used as contrast agents for MRI in diagnostic imaging, and their biocompatibility is essential to avoid damaging the tissues being imaged. Surface functionalization is essential for maintaining safety and guaranteeing targeted delivery in drug delivery systems.

5.2.2.7 Mechanisms of Cellular Uptake

Understanding the cellular uptake and intracellular behaviour of spinel ferrites is crucial for elucidating their interactions with biological systems, particularly in the context of medical applications such as drug delivery and imaging. The cellular uptake mechanisms and subsequent intracellular fate of spinel ferrites are influenced by various factors, including particle size, surface charge, and surface functionalization. Determining the behaviour of spinel ferrites in living cells requires an understanding of the various mechanisms involved in their cellular uptake [16]. One important factor affecting the uptake of nanoparticles is their size. Smaller nanoparticles tend to be more cellularly absorbed than larger ones. Surface charge of spinel ferrites also has an effect; positively charged nanoparticles frequently exhibit enhanced cellular internalization because of interactions with negatively charged cell membranes.

Ensuring the safety of spinel ferrites in biological systems involves a multidimensional approach, considering biocompatibility, toxicity, immunogenicity, biodistribution, and regulatory compliance [17]. Ongoing research efforts, adherence to standardized testing protocols, and advancements in imaging technologies will collectively contribute to unlocking the full potential of spinel ferrites in biomedical applications while prioritizing safety and minimizing potential risks to living organisms.

5.3 DIAGNOSTIC IMAGING

5.3.1 Utilization of Spinel Ferrites in Magnetic Resonance Imaging (MRI) for Diagnostic Purposes

Spinel ferrites are great contrast agents for improving the visibility of tissues and organs in MRI scans because of their special magnetic properties. The use of spinel ferrites in MRI for diagnostic purposes involves the following important factors.

5.3.1.1 Contrast Enhancement

In MRI, spinel ferrites—particularly magnetite nanoparticles—act as contrast agents to increase tissue visibility [18]. The superparamagnetic behaviour of these nanoparticles increases the rates at which neighbouring water protons relax. Consequently, the MRI images become more contrasted, making it easier to distinguish between various tissues and pathological conditions.

5.3.1.2 T1 and T2 Relaxation Effects

Both T1 and T2 relaxation effects are influenced by the relaxation characteristics of water protons when spinel ferrites are present. The recovery of longitudinal magnetization, or T1 relaxation, affects the tissue brightness in MRI. Conversely, T2 relaxation modifies the contrast and sharpness of the images by causing transverse magnetization to decay. Depending on the imaging requirements, the magnetic properties of spinel ferrites can be optimized to maximize both T1 and T2 relaxation effects.

5.3.1.3 Biocompatibility and Safety

One of the most important aspects of using spinel ferrites for medical imaging is their biocompatibility. To make sure that these nanoparticles are safe to use in vivo, a great deal of research has been done. To improve biocompatibility and lower the possibility of adverse reactions, surface modifications and coatings are frequently used. This ensures the safety of patients receiving spinel ferrite contrast agents during MRI scans.

Molecular imaging and targeted imaging are made possible by the ability of spinel ferrites to bind targeting ligands, which permits targeted imaging in particular cells or tissues [19–21]. This is especially useful for molecular imaging since spinel ferrites can bind to specific disease-related biomarkers only. Through targeted enhancement of the specificity of MRI, pathological conditions can be precisely localized and detected early thanks to this approach.

5.3.2 Enhancement of Imaging Contrast and Resolution through Spinel Ferrite Nanoparticles

Optimizing contrast and resolution in different imaging modalities is a promising application of spinel ferrite nanoparticles in biomedical imaging. Due to their distinct magnetic and structural characteristics, these nanoparticles can be engineered to enhance imaging performance in methods like computed tomography (CT), MRI, as well as newly developed imaging modalities like magnetic particle imaging (MPI). Here, we investigate the role that spinel ferrite nanoparticles play in improving the contrast and resolution of imaging.

5.3.2.1 Magnetic Resonance Imaging (MRI)

Effects of T1 and T2 Relaxation: T1 and T2 relaxation times in MRI can be markedly impacted by spinel ferrite nanoparticles. Selecting spinel ferrite compositions enables the adjustment of relaxation durations, which improves contrast in various tissue types.

Surface Functionalization: Surfaces of spinel ferrite nanoparticles can be modified with biocompatible coatings, including polyethylene glycol (PEG), to increase their stability in physiological settings, lengthen their circulation durations, and improve their contrast in MRI.

5.3.2.1.1 Enhancement of Resolution

5.3.2.1.1.1 High Magnetization An enhanced signal-to-noise ratio (SNR) in MRI is a result of high magnetization of spinel ferrites [22]. Superior image clarity and spatial detail are made possible by this high SNR.

5.3.2.1.1.2 Targeted Imaging By adding targeting ligands to spinel ferrite nanoparticles, it is possible to specifically bind to biological markers or tissues. This enhances resolution by concentrating on areas of interest.

5.3.2.2 Contrast Enhancement in Computed Tomography (CT)

The X-ray attenuation attributes of spinel ferrite nanoparticles are improved by adding heavy metals such as bismuth [23]. CT imaging contrast is enhanced because of this increased attenuation.

5.3.2.2.1 Dual-Mode Imaging

Spinel ferrite nanoparticles can be used as contrast materials in both CT and MRI scans, offering supplementary data that enhances the imaging studies' ability to diagnose patients.

5.3.2.2.2 Enhancement of Resolution

5.3.2.2.2.1 Nanostructure Design Optimizing the dimensions and form of spinel ferrite nanoparticles enhances CT imaging resolution. More precise morphology and sizing of nanoparticles provide improved tissue penetration and clarity of imaging.

5.3.2.2.2.2 Biocompatible Coatings By covering spinel ferrite nanoparticles with biocompatible materials, one can improve imaging resolution by stabilizing the particles and ensuring a homogeneous distribution within biological tissues.

5.3.2.3 Enhancement in Magnetic Particle Imaging (MPI)

5.3.2.3.1 Properties of Superparamagnetic

Since spinel ferrite nanoparticles frequently display superparamagnetic activity, MPI is a perfect fit for them [24]. Superparamagnetic nanoparticles have quick magnetization and demagnetization processes which improve image contrast of MPI. By improving spatial resolution in MPI, nanoscale dimensions of a spinel ferrite nanoparticle enable comprehensive molecular imaging of biological components.

5.3.2.3.2 Enhancement of Resolution

Real-time monitoring of dynamic processes is made possible by capacity to function as tracers in MPI [25]. By altering the surface to make spinel ferrite nanoparticles more biocompatible, long-term in vivo use is made possible, allowing for high-resolution longitudinal research in MPI.

A flexible platform for improving imaging contrast and resolution across a range of biological imaging modalities is provided by spinel ferrite nanoparticles. Because of their magnetic and structural tuning capabilities as well as surface modifications for biocompatibility, spinel ferrite nanoparticles hold great promise as contrast agents for both established and developing imaging modalities. These nanoparticles will continue to be refined through ongoing research and advances in nanotechnology, bringing us one step closer to a new age of high-precision, non-invasive imaging in medical research and diagnostics.

5.4 DRUG DELIVERY SYSTEMS

5.4.1 Role of Spinel Ferrites in Drug Delivery Platforms

Targeted drug delivery, controlled release, and the possibility of external control over drug release kinetics are some of the benefits of using spinel ferrite nanoparticles into drug delivery systems (Figure 5.3) [26]. In this chapter, we discuss the several functions of spinel ferrites in medication delivery systems.

- *Magnetic Targeting*: Because spinel ferrite nanoparticles are inherently ferromagnetic or superparamagnetic, they have magnetic responsiveness. It makes it possible to concentrate and direct drug-loaded ferrite nanoparticles to target areas within the body by applying an external magnetic field [27, 28]. Drug-loaded spinel ferrite nanoparticles can be delivered to particular disease sites by using magnetic targeting, which enhances therapeutic efficacy and reduces systemic side effects.
- Spinel ferrite nanoparticles regulate drug release in response to external stimuli, such as temperature changes, pH shifts, or magnetic fields [29]. This feature is known as external stimulus-responsive release. The conditions of the target site can be catered for in this controlled release. Nanocarriers that react to external stimuli can be designed thanks to the magnetic properties of spinel ferrites. This versatility allows for the optimization of treatment regimens and on-demand medication release.
- *Magnetic Hyperthermia*: When subjected to an alternating magnetic field, spinel ferrite nanoparticles can produce heat. This phenomenon is known as magnetic hyperthermia. This characteristic is used in hyperthermia therapy, where the raised body temperature produces localized therapeutic effects that improve the effectiveness of drug delivery systems used to treat cancer [30]. When medication administration and hyperthermia therapy are combined, a synergistic strategy can be

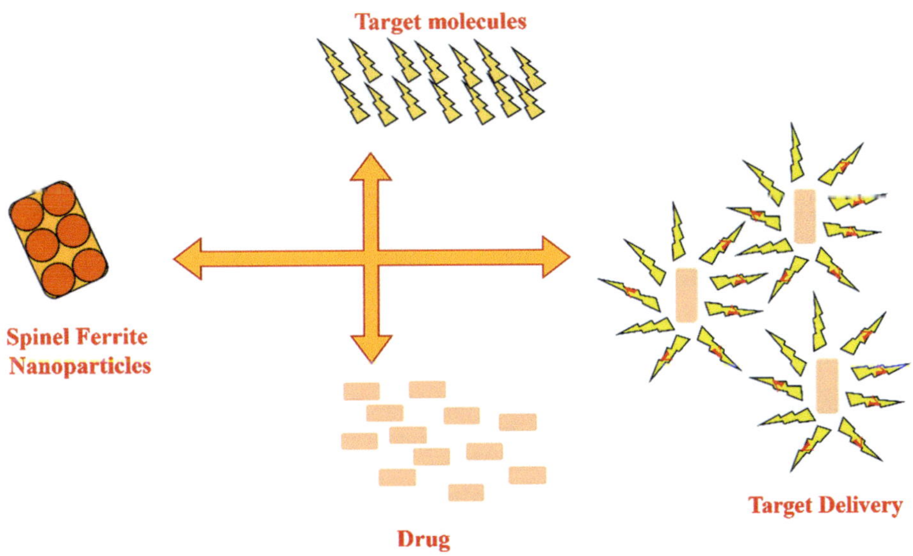

FIGURE 5.3 Drug delivery in spinel ferrite nanoparticles.

used. In this case, the heat produced by spinel ferrite nanoparticles improves drug penetration and cellular uptake, improving the therapeutic result.

Because they provide a flexible and responsive platform for targeted and regulated drug release, spinel ferrites are essential to advancing drug delivery systems. Because spinel ferrites combine surface changes, magnetic properties, and imaging capabilities, they are a promising prospect for the creation of novel and customized therapeutic techniques [31]. The potential of spinel ferrites to transform medication administration in a variety of medical applications is still worth exploring as research advances.

5.4.2 Nanoparticle-based Drug Carriers and Their Advantages in Targeted Therapy

These carriers, which are usually nanoscale in size, provide several benefits when precisely delivering medicinal substances to cells or tissues (Figure 5.4). The benefits of drug carriers based on nanoparticles for targeted therapy are examined here:

- *Targeted Drug Delivery*: It is possible to design nanoparticle carriers to only accumulate in cells or tissues [32]. Functionalization that minimizes off-target effects and increases therapeutic efficacy is made possible by ligands that recognize receptors or biomarkers. Because of the leaky vasculature, the enhanced permeability and retention (EPR) effect enables nanoparticles to passively collect in tumour tissues. Because nanoparticles selectively aggregate in tumour areas during cancer therapy, this passive targeting method is beneficial.
- *Enhancement of Drug Solubility*: Many medicinal substances are poorly soluble in water, particularly those used to treat cancer. Hydrophobic medications can be efficiently encapsulated, and their solubility is enhanced for efficient delivery by the hydrophobic environment that nanoparticle carriers create within their cores.
- *Gradual and Managed Release*: Therapeutic substances can be released gradually and under control thanks to nanoparticles [33, 34]. Release kinetics can be adjusted to produce sustained therapeutic benefits by varying the content and structure of the nanoparticle, which can eliminate the need for repeated administrations. Systems of stimuli-responsive nanoparticles react to environmental cues, such as temperature, enzyme activity, or pH. Therapeutic precision is improved by this responsiveness, which enables triggered medication release at the target spot.
- *Drug Molecule Protection*: Drug molecules are shielded by nanoparticle carriers from enzyme breakdown and early bloodstream clearance [35]. By prolonging the therapeutic circulation time and increasing its bioavailability, this protection improves its stability. By limiting systemic exposure to therapeutic substances, targeted drug delivery lowers toxicity and lessens off-target effects. This is especially helpful for treating chronic illnesses, as patient compliance depends on limiting side effects.
- *The Effect of Enhanced Permeability and Retention, or EPR*: Carriers based on nanoparticles can take use of the EPR phenomenon to concentrate in tumour tissues more preferentially [36]. This phenomenon is particularly useful for cancer treatment because it allows therapeutic medicines to be delivered to cancerous cells precisely while protecting healthy tissues.

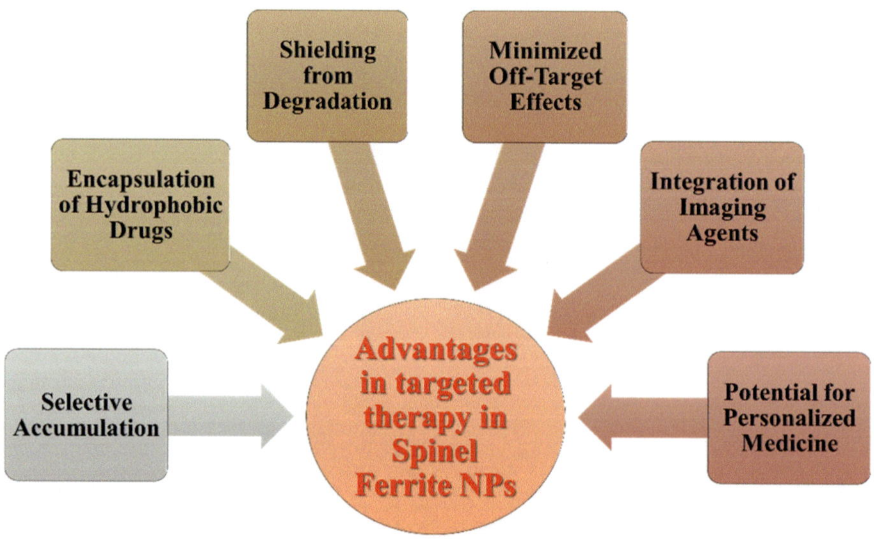

FIGURE 5.4 Advantages in targeted therapy in spinel ferrite nanoparticles.

- *Platforms with Multiple Functions*: Nanoparticles can be engineered to serve as multipurpose platforms that combine therapeutic and imaging agents for diagnostic applications. Real-time tracking of medication administration and treatment response is made possible by this theragnostic technique [37, 38]. Combination therapy is made possible by the co-delivery of several therapeutic substances, which is made possible by nanoparticles. This strategy can overcome drug resistance, increase synergistic effects, and boost overall treatment outcomes.
- *Customization for Disease*: Drug carriers based on nanoparticles can be tailored to the features of certain illnesses. This entails modifying ligand targeting, surface charge, and size to correspond with the specific characteristics of the target tissue or cells [39].
- *Surface Functionalization Ease*: A variety of targeting ligands and biomolecules can be functionalized onto the surfaces of nanoparticles. This adaptability makes it possible to design customized drug carriers that can identify tissue or cellular markers.
- *Patient-Specific Techniques*: Personalized medicine techniques are supported by the adaptability of medication carriers based on nanoparticles [28]. By considering differences in disease features and patient reactions, carriers that are customized to unique profile of each patient might maximize therapy success.

5.5 APPLICATION OF SPINEL FERRITES IN THE DEVELOPMENT OF MAGNETIC BIOSENSORS

Spinel ferrites possess superparamagnetism, a magnetic feature that renders them extremely responsive to alterations in their immediate surroundings. Magnetic biosensors use this sensitivity to identify changes brought about by the presence of target biomolecules [40]. The magnetic signal is altered when target biomolecules connect to spinel ferrite-based nanoparticles, which makes it possible to detect analytes with high sensitivity and selectivity.

Immunoassays are a common method for biomolecule detection. Spinal ferrite-based magnetic biosensors are used in these experiments. On the outermost layer of spinel ferrite nanoparticles, the antibodies particular to target biomolecules, such as proteins or antigens, are immobilized. The ability to detect biomolecules is made possible by the specific binding of these antibodies to their targets, which results in modifications to the magnetic characteristics. ELISA stands for enzyme-linked immunosorbent assay [41, 42]. By adding magnetic biosensors to ELISA platforms, one can increase responsiveness and identify enzymatic activities magnetically, indicating signs of the existence of biomolecules.

5.5.1 Nucleic Acid Detection

Spinel ferrites are essential for the detection of pathogen-derived nucleic acids, such as DNA or RNA. Magnetic pathogen identification can be made possible by the specific attraction of functionalized spinel ferrite nanoparticles to nucleic acid sequences.

5.5.2 PCR-Based Methodologies

The integration of magnetic biosensors into polymerase chain reaction (PCR) procedures allows increased nucleic acids to be magnetically separated for further detection and analysis.

5.5.3 Applications for Diagnosis

Spinel ferrite-based magnetic biosensors are appropriate for point-of-care testing (POCT), providing quick and on-site diagnoses [43]. This sensitivity of biosensors makes it possible to identify pathogens or biomarkers at low quantities, facilitating prompt and effective diagnosis. One important use is the identification of disease biomarkers, such as antigens unique to cancer or biomolecules suggestive of infectious illnesses. Magnetic biosensors aid in the early detection and tracking of illnesses.

In the creation of magnetic biosensors for the sensitive and focused identification of biomolecules and pathogens in diagnostic applications, spinel ferrites are essential. These biosensors have the potential to improve customized treatment and illness management by providing quick, point-of-care diagnostics [44, 45]. The incorporation of spinel ferrites into cutting-edge biosensing platforms shows promise for tackling challenging issues in healthcare and enhancing diagnostic precision as research progresses.

5.6 REGENERATIVE MEDICINE

5.6.1 Importance in Regenerative Medicine

5.6.1.1 Magnetic Hyperthermia

When exposure to an alternating magnetic field, spinel ferrites, such as magnetite (Fe_3O_4), can cause hyperthermia. This characteristic is used to trigger cellular reactions, accelerating tissue repair and regeneration [33]. By carefully applying magnetic hyperthermia with spinel ferrites, one can alter the local temperature and affect cellular processes that are vital to healing, including proliferation, migration, and differentiation.

5.6.1.2 Drug Delivery Methods

Therapeutic drugs can be released in a controlled and localized manner by utilizing spinel ferrites in drug delivery systems. This ability to distribute growth factors, cytokines, or other bioactive compounds directly to the target region is especially useful for stimulating tissue regeneration. Magnetic characteristics allow for external control over the kinetics of drug release, resulting in a steady and controlled flow of bioactive substances that promote tissue regeneration over time.

5.6.1.3 Imaging and Monitoring

For non-invasive visualization and tissue regeneration monitoring, MRI uses spinel ferrites as contrast agents. This helps medical professionals to evaluate the effectiveness of regeneration therapies and make well-informed choices for patient care.

5.6.2 Use of Magnetic Scaffolds in Tissue Engineering

5.6.2.1 Magnetization of Scaffold

Three-dimensional scaffolds frequently utilized in tissue engineering can be equipped with spinel ferrites. These scaffolds facilitate cell proliferation by offering a structural model that resembles the extracellular matrix. The use of spinel ferrites enhances the mechanical characteristics, providing increased stability and structural integrity—a vital component for the effective implementation of tissue engineering.

5.6.2.2 Guidance by Magnetic Field

The application of external magnetic fields can be used to direct and instruct cell migration because of the magnetic properties of spinel ferrites in scaffolds [46]. Regarding the purpose of optimizing tissue regeneration and managing the spatial arrangement of cells within the scaffold, this magnetic field guidance is especially helpful.

5.6.2.3 Modulation of Cellular Response

Scaffolds containing spinel ferrite can be magnetized to affect the behaviour of cells. The mechanical stresses experienced by the cells within the scaffold when subjected to a magnetic field cause changes in the shape, orientation, and gene expression of the cells. By promoting certain cell differentiation inside the scaffold, the magnetic regulation of cellular responses can aid in the formation of tissue with the required properties.

5.6.2.4 Elevated Body Temperature for Tissue Repair

For causing localized hyperthermia, spinel ferrites embedded in scaffolds may be used [47]. Tissue regeneration is aided by this regulated heat stimulation, which increases cellular activity such as proliferation and the production of extracellular matrix.

5.6.2.5 Osteogenesis and Angiogenesis

In the field of bone tissue engineering, the use of magnetic hyperthermia in conjunction with scaffolds containing spinel ferrite has demonstrated potential for fostering these two crucial processes for bone regrowth.

5.7 DISCUSSION ON CHALLENGES AND CONSIDERATIONS IN THE USE OF SPINEL FERRITES IN THE HEALTH SECTOR

- *Challenge of Biocompatibility and Cytotoxicity*: For spinel ferrites to be used safely in biological systems, they must be biocompatible. When these materials encounter living cells and tissues, there may be worries about possible cytotoxicity, inflammation, or immunological reactions. Thorough biocompatibility investigations, encompassing both in vitro and in vivo evaluations, are imperative for assessing the influence of spinel ferrites on cellular viability, functionality, and enduring consequences. Coatings and surface alterations can be used to improve biocompatibility.
- *Challenge of Long-Term Biostability*: One factor to consider is the long-term stability of spinel ferrites in biological settings, particularly in physiological settings. The performance and safety may be affected over time by degradation or modification. The goal of research should be to create spinel ferrite formulations that are more stable, resistant to degradation, and leak less potentially hazardous ions. To evaluate these materials' biostability, long-term research is essential.
- *Delivery that is Specific and Targeted*: It can be difficult to distribute spinel ferrite-based nanoparticles precisely and specifically to certain cells or tissues. Reduced treatment efficacy and off-target consequences can result from non-specific interactions. To increase specificity, targeted ligands can be added to the surface of spinel ferrite nanoparticles [8]. Additional developments, in targeting techniques, include active targeting with ligands that are required to bind precisely to target cell receptors.
- Obtaining regulatory approval is a challenge in transferring spinel ferrites from research labs to clinical settings. It takes work to meet regulatory requirements for efficacy, safety, and quality control. It is crucial for researchers, industry partners, and regulatory agencies to work together. The development process should incorporate a methodical approach to meeting compliance standards and regulatory requirements.
- *Reproducibility and Standardization*: The synthesis and functionalization of spinel ferrites present a difficulty in terms of achieving standardization and reproducibility. The characteristics and functionality of these materials can be affected by changes in the manufacturing process. Standardized procedures must be established for the synthesis and functionalization of spinel ferrites. To guarantee consistency, stringent characterization methods and quality control procedures should be used.
- *Possible Reactions with Biological Substances*: Proteins and nucleic acids are examples of biological macromolecules with which spinel ferrites may interact, thereby compromising their intended applications and altering their functionality. For safe and efficient use, it is essential to comprehend how these interactions may impact the biological components as well as the material.

- *Social and Ethical Issues to Consider*: The application of sophisticated materials, such as spinel ferrites, in the medical field brings up moral questions about patient permission, privacy, and the effects of new technology on society [48]. The creation and application of spinel ferrite-based technologies ought to be accompanied by clear and open communication with all parties involved, including patients and the public, as well as ethical principles.
- *Environmental Effect*: It is necessary to consider the possible environmental effects of spinel ferrites, particularly regarding disposal and waste management. It is recommended that researchers and industry implement eco-friendly procedures for the production and disposal of spinel ferrites. To reduce the influence on the environment, biodegradable formulations and recycling techniques can be investigated.
- *Price and Availability*: In resource-constrained environments, especially, the cost of synthesizing spinel ferrites and integrating them into medical technology may be a barrier to their general accessibility [49]. Investigations ought to concentrate on creating affordable synthesis techniques, and endeavours should be undertaken to enhance the worldwide availability of spinel ferrite-based technologies.

Researchers, physicians, regulatory agencies, and business partners must work together to address these issues and concerns. Persistent progress in comprehending the biochemical relationships and refining the structure and production of spinel ferrites will open the door for their secure and efficient application in the medical field.

5.8 EXPLORATION OF EMERGING TRENDS AND FUTURE DIRECTIONS IN THE APPLICATION OF SPINEL FERRITES IN HEALTHCARE

As technology develops, scientists investigate new uses and enhance current techniques, opening the door to creative approaches in medical imaging, therapy, and diagnostics. The following are some new developments and potential paths for use in medical applications of spinel ferrites (Figure 5.5):

- *Platforms with Multiple Functions and Theragnostic*: Theragnostic applications involving spinel ferrites, which combine therapeutic and diagnostic functions on a single platform, are being investigated more and more [50]. Spinel ferrite nanoparticles with multiple functions can function as carriers of therapeutic payloads and imaging agents, allowing for targeted therapy and real-time monitoring.

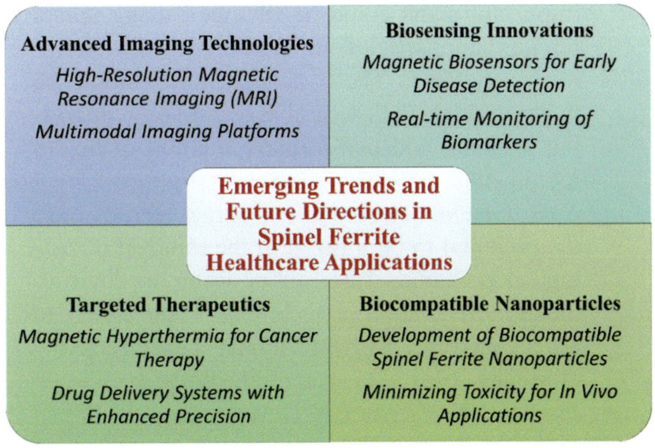

FIGURE 5.5 Trends and future directions in the application of spinel ferrites in healthcare.

- *Wireless Monitoring Devices*: An emerging trend is the use of spinel ferrites in wireless monitoring systems intended for in vivo applications. These gadgets might use the spinel ferrites' magnetic characteristics to monitor and sense physiological factors remotely, giving real-time data without requiring invasive treatments.
- *Therapeutic Use of Spinel Ferrites in Magnetic Stimulation for Neurological Illnesses*: The potential of spinel ferrites in magnetic stimulation for neurological illnesses is being investigated. The deliberate use of magnetic fields to excite brain cells may pave the way for new non-invasive therapies for diseases, including depression and Parkinson's disease.
- *Biosensing and Point-of-Care Diagnostics*: A developing trend in point-of-care diagnostics is the development of highly sensitive and portable magnetic biosensors using spinel ferrites [51]. The easy and on-site identification of pathogens, disease-related chemicals, or biomarkers may be made possible using these biosensors.
- *3D Printing of Structures that are Biocompatible*: The field of 3D bioprinting using spinel ferrites is developing. It may be possible to create biocompatible scaffolds with improved magnetic properties using bioinks containing spinel ferrite, which would enable tissue engineering and regenerative medicine applications.
- *Personalized Therapies and Precision Medicine*: With a deeper comprehension of the ways in which spinel ferrites interact with biological systems, customized treatment strategies may be developed. Customizing spinel ferrite-based therapies to unique profile of each patient could increase therapeutic effectiveness and decrease adverse effects.
- *Environmental and Clothing Health Monitoring*: Continuous data on health parameters and

important signs could be obtained using spinel ferrite nanoparticles in clothing health monitoring systems. These nanoparticles might be made to work with wearable technology, providing a handy and non-invasive method of health monitoring.

- *Environmental Remediation*: Spinel ferrites have the potential to be utilized in the removal of biological pollutants from air or water in environmental remediation projects [52]. Their ability to attract and repel magnetically may help effectively eliminate contaminants and infections.
- *Medical Bioelectronics*: Developments in spinel ferrite-based materials could have an impact on bioelectronic medicine by enabling the integration of these materials into biological system interfaces for diagnostic or therapeutic applications.

To connect the space between laboratory discoveries and clinical applications, examining these new patterns requires interdisciplinary collaboration with a focus on translational research. Research initiatives will probably uncover new opportunities as spinel ferrites keep showing potential in a variety of healthcare applications, advancing patient care and medical technology.

5.9 CONCLUSION

In conclusion, the use of spinel ferrites in the medical field offers a promising new area of application where the special magnetic and physicochemical qualities of these materials enable creative uses that have a big impact on treatments and diagnosis. With their unique crystal structure and magnetic properties, spinel ferrites have shown great promise in biomedical technology, providing solutions that close the gap between state-of-the-art research and traditional medical procedures. A prominent use for spinel ferrite nanoparticles is in MRI, where they act as contrast agents to improve the visibility of pathological alterations and anatomical structures. Their biocompatibility and adjustable magnetic characteristics make them excellent options for enhancing imaging accuracy and minimizing possible adverse effects related to conventional contrast agents. When exposure to an alternating magnetic field, spinel ferrite nanoparticles in magnetic hyperthermia show thermal generation capability. Using this feature, tailored cancer therapy can increase the efficacy of other therapeutic interventions or cause apoptosis in cancer cells by localized heating. The ability of these materials to respond magnetically makes it possible to deliver drugs precisely to tissues or cells, increasing therapy precision and reducing systemic side effects. Although spinel ferrites have great potential for use in the medical field, more research and cooperation between chemists, materials scientists, and physicians are necessary. To guarantee the safe and efficient integration of spinel ferrites into clinical settings, issues with biocompatibility, toxicity, and scalability need to be resolved.

In simple terms, application in the medical field of spinel ferrite is a prime example of how materials science and innovative healthcare intersect. Spinel ferrites have the potential to transform therapeutic interventions, imaging, and diagnostics as research explores their entire range of potential applications. They can provide new levels of precision and efficacy in the pursuit of better patient outcomes and overall well-being.

REFERENCES

1. Srivastava, R., & Yadav, B. C. (2012). Ferrite materials: introduction, synthesis techniques, and applications as sensors. *International Journal of Green Nanotechnology*, 4(2), 141–154.
2. Thakur, A., Chahar, D., & Thakur, P. (2022). Synthesis of Nanomaterials by Biological Route. In *Synthesis and Applications of Nanoparticles* (pp. 77–119). Singapore: Springer Nature Singapore.
3. Amiri, S., & Shokrollahi, H. (2013). The role of cobalt ferrite magnetic nanoparticles in medical science. *Materials Science and Engineering: C*, 33(1), 1–8.
4. Thakur, P., & Thakur, A. (2022). Nanomaterials, Their Types and Properties. In *Synthesis and Applications of Nanoparticles* (pp. 19–44). Singapore: Springer Nature Singapore.
5. Riporto, J., Demierre, A., Kilin, V., Balciunas, T., Schmidt, C., Campargue, G., ... & Bonacina, L. (2018). Bismuth ferrite dielectric nanoparticles excited at telecom wavelengths as multicolor sources by second, third, and fourth harmonic generation. *Nanoscale*, 10(17), 8146–8152.
6. Thakur, A., Thakur, P., & Khurana, S. P.. (2022). *Synthesis and Applications of Nanoparticles* (pp. 1–17). Singapore: Springer.
7. Thakur, P., Taneja, S., Sindhu, D., Lüders, U., Sharma, A., Ravelo, B., & Thakur, A. (2020). Manganese zinc ferrites: a short review on synthesis and characterization. *Journal of Superconductivity and Novel Magnetism*, 33, 1569–1584.
8. Kefeni, K. K., Msagati, T. A., Nkambule, T. T., & Mamba, B. B. (2020). Spinel ferrite nanoparticles and nanocomposites for biomedical applications and their toxicity. *Materials Science and Engineering: C*, 107, 110314.
9. Martínez-Rodríguez, N. L., Tavárez, S., & González-Sánchez, Z. I. (2019). In vitro toxicity assessment of zinc and nickel ferrite nanoparticles in human erythrocytes and peripheral blood mononuclear cell. *Toxicology in Vitro*, 57, 54–61.
10. Akhtar, K., Javed, Y., Jamil, Y., & Muhammad, F. (2020). Functionalized cobalt ferrite cubes: toxicity, interactions and mineralization into ferritin proteins. *Applied Nanoscience*, 10, 3659–3674.
11. Punia, P., Bharti, M. K., Dhar, R., Thakur, P., & Thakur, A. (2022). Recent advances in detection and removal of heavy metals from contaminated water. *ChemBioEng Reviews*, 9(4), 351–369.
12. Taguchi, T., & Suzuki, S. (1983). Health effects of ferrite and the raw materials dust. *Industrial Health*, 21(1), 1–9.
13. Magdolenova, Z., Drlickova, M., Henjum, K., Rundén-Pran, E., Tulinska, J., Bilanicova, D., ... & Dusinska, M. (2015). Coating-dependent induction of cytotoxicity and genotoxicity of iron oxide nanoparticles. *Nanotoxicology*, 9(supl), 44–56.

14. Pfaller, T., Colognato, R., Nelissen, I., Favilli, F., Casals, E., Ooms, D., ... & Oostingh, G. J. (2010). The suitability of different cellular in vitro immunotoxicity and genotoxicity methods for the analysis of nanoparticle-induced events. *Nanotoxicology*, 4(1), 52–72.
15. Tomitaka, A., Hirukawa, A., Yamada, T., Morishita, S., & Takemura, Y. (2009). Biocompatibility of various ferrite nanoparticles evaluated by in vitro cytotoxicity assays using HeLa cells. *Journal of Magnetism and Magnetic Materials*, 321(10), 1482–1484.
16. Sanpo, N., Tharajak, J., Li, Y., Berndt, C. C., Wen, C., & Wang, J. (2014). Biocompatibility of transition metal-substituted cobalt ferrite nanoparticles. *Journal of Nanoparticle Research*, 16, 1–13.
17. Klein, S., Kızaloğlu, M., Portilla, L., Park, H., Rejek, T., Hümmer, J., ... & Kryschi, C. (2018). Enhanced in vitro biocompatibility and water dispersibility of magnetite and cobalt ferrite nanoparticles employed as ROS formation enhancer in radiation cancer therapy. *Small*, 14(21), 1704111.
18. Staedler, D., Passemard, S., Magouroux, T., Rogov, A., Maguire, C. M., Mohamed, B. M., ... & Wolf, J. P. (2015). Cellular uptake and biocompatibility of bismuth ferrite harmonic advanced nanoparticles. *Nanomedicine: Nanotechnology, Biology and Medicine*, 11(4), 815–824.
19. Lickmichand, M., Shaji, C. S., Valarmathi, N., Benjamin, A. S., Kumar, R. A., Nayak, S., ... & Raj, N. A. N. (2019). In vitro biocompatibility and hyperthermia studies on synthesized cobalt ferrite nanoparticles encapsulated with polyethylene glycol for biomedical applications. *Materials Today: Proceedings*, 15, 252–261.
20. Shultz, M. D., Calvin, S., Fatouros, P. P., Morrison, S. A., & Carpenter, E. E. (2007). Enhanced ferrite nanoparticles as MRI contrast agents. *Journal of Magnetism and Magnetic Materials*, 311(1), 464–468.
21. Hong, S., Chang, Y., & Rhee, I. (2010). Chitosan-coated ferrite (Fe_3O_4) nanoparticles as a T2 contrast agent for magnetic resonance imaging. *Journal of the Korean Physical Society*, 56(3), 868–873.
22. Bárcena, C., Sra, A. K., Chaubey, G. S., Khemtong, C., Liu, J. P., & Gao, J.. (2008). Zinc ferrite nanoparticles as MRI contrast agents. *Chemical Communications*, (19), 2224–2226.
23. Du, Y., Liu, X., Liang, Q., Liang, X. J., & Tian, J. (2019). Optimization and design of magnetic ferrite nanoparticles with uniform tumor distribution for highly sensitive MRI/MPI performance and improved magnetic hyperthermia therapy. *Nano Letters*, 19(6), 3618–3626.
24. Bauer, L. M., Situ, S. F., Griswold, M. A., & Samia, A. C. S. (2016). High-performance iron oxide nanoparticles for magnetic particle imaging–guided hyperthermia (hMPI). *Nanoscale*, 8(24), 12162–12169.
25. Du, H., Akakuru, O. U., Yao, C., Yang, F., & Wu, A. (2022). Transition metal ion-doped ferrites nanoparticles for bioimaging and cancer therapy *Translational Oncology*, 15(1), 101264.
26. Wang, G., Ma, Y., Wei, Z., & Qi, M. (2016). Development of multifunctional cobalt ferrite/graphene oxide nanocomposites for magnetic resonance imaging and controlled drug delivery. *Chemical Engineering Journal*, 289, 150–160.
27. Wu, H., Liu, G., Wang, X., Zhang, J., Chen, Y., Shi, J., ... & Yang, S. (2011). Solvothermal synthesis of cobalt ferrite nanoparticles loaded on multiwalled carbon nanotubes for magnetic resonance imaging and drug delivery. *Acta Biomaterialia*, 7(9), 3496–3504.
28. Dey, C., Baishya, K., Ghosh, A., Goswami, M. M., Ghosh, A., & Mandal, K. (2017). Improvement of drug delivery by hyperthermia treatment using magnetic cubic cobalt ferrite nanoparticles. *Journal of Magnetism and Magnetic Materials*, 427, 168–174.
29. Dahiya, M. S., Tomer, V. K., & Duhan, S. (2018). Metal–ferrite Nanocomposites for Targeted Drug Delivery. In *Applications of Nanocomposite Materials in Drug Delivery* (pp. 737–760). Woodhead Publishing.
30. Rana, S., Gallo, A., Srivastava, R. S., & Misra, R. D. K. (2007). On the suitability of nanocrystalline ferrites as a magnetic carrier for drug delivery: functionalization, conjugation and drug release kinetics. *Acta Biomaterialia*, 3(2), 233–242.
31. Xiao, S., Fakhri, A., & Janani, B. J. (2021). Synthesis of spinel tin ferrite decorated on Bismuth ferrite nanostructures for synergetic photocatalytic, superior drug delivery, and antibacterial efficiencies. *Surfaces and Interfaces*, 27, 101490.
32. Sulaiman, N. H., Ghazali, M. J., Majlis, B. Y., Yunas, J., & Razali, M. (2015). Superparamagnetic calcium ferrite nanoparticles synthesized using a simple sol-gel method for targeted drug delivery. *Bio-Medical Materials and Engineering*, 26(s1), S103–S110.
33. Amiri, M., Salavati-Niasari, M., & Akbari, A. (2019). Magnetic nanocarriers: evolution of spinel ferrites for medical applications. *Advances in Colloid and Interface Science*, 265, 29–44.
34. Abdel Maksoud, M. I. A., Ghobashy, M. M., Kodous, A. S., Fahim, R. A., Osman, A. I., Al-Muhtaseb, A. A. H., ... & Ashour, A. H. (2022). Insights on magnetic spinel ferrites for targeted drug delivery and hyperthermia applications. *Nanotechnology Reviews*, 11(1), 372–413.
35. Yang, Y., Shi, H., Wang, Y., Shi, B., Guo, L., Wu, D., ... & Wu, H. (2016). Graphene oxide/manganese ferrite nanohybrids for magnetic resonance imaging, photothermal therapy and drug delivery. *Journal of Biomaterials Applications*, 30(6), 810–822.
36. Yang, K., Yu, G., Tian, R., Zhou, Z., Deng, H., Li, L., ... & Chen, X. (2021). Oxygen-Evolving manganese ferrite nanovesicles for hypoxia-responsive drug delivery and enhanced cancer chemoimmunotherapy. *Advanced Functional Materials*, 31(11), 2008078.
37. Haribabu, V., Sharmiladevi, P., Akhtar, N., Farook, A. S., Girigoswami, K., & Girigoswami, A. (2019). Label free ultrasmall fluoromagnetic ferrite-clusters for targeted cancer imaging and drug delivery. *Current Drug Delivery*, 16(3), 233–241.
38. Chen, P., Cui, B., Cui, X., Zhao, W., Bu, Y., & Wang, Y. (2017). A microwave-triggered controllable drug delivery system based on hollow-mesoporous cobalt ferrite magnetic nanoparticles. *Journal of Alloys and Compounds*, 699, 526–533.
39. Nigam, A., & Pawar, S. J. (2020). Structural, magnetic, and antimicrobial properties of zinc doped magnesium ferrite for drug delivery applications. *Ceramics International*, 46(4), 4058–4064.
40. Sakthivel, K., Mani, G., Chen, S. M., Lin, S. H., Muthumariappan, A., & Mani, V. (2018). A novel synthesis of non-aggregated spinel nickel ferrite nanosheets for developing non-enzymatic reactive oxygen species sensor in biological samples. *Journal of Electroanalytical Chemistry*, 820, 161–167.
41. Yardımcı, F. S., Şenel, M., & Baykal, A. (2012). Amperometric hydrogen peroxide biosensor based on cobalt ferrite–chitosan nanocomposite. *Materials Science and Engineering: C*, 32(2), 269–275.

42. Sanko, V., Şenocak, A., Tümay, S. O., & Demirbas, E. (2023). A novel comparative study for electrochemical urea biosensor design: effect of different ferrite nanoparticles (MFe_2O_4, M: Cu, Co, Ni, Zn) in urease immobilized Composite system. *Bioelectrochemistry*, 149, 108324.
43. Shobana, M. K. (2021). Nanoferrites in biosensors—a review. *Materials Science and Engineering: B*, 272, 115344.
44. Pham, T. N., Huy, T. Q., & Le, A. T. (2020). Spinel ferrite (AFe_2O_4)-based heterostructured designs for lithium-ion battery, environmental monitoring, and biomedical applications. *RSC Advances*, 10(52), 31622–31661.
45. Ansari, M., Bigham, A., & Ahangar, H. A. (2019). Superparamagnetic nanostructured CuZnMg mixed spinel ferrite for bone tissue regeneration. *Materials Science and Engineering: C*, 105, 110084.
46. Bigham, A., Aghajanian, A. H., Behzadzadeh, S., Sokhani, Z., Shojaei, S., Kaviani, Y., & Hassanzadeh-Tabrizi, S. A. (2019). Nanostructured magnetic Mg_2SiO_4-$CoFe_2O_4$ composite scaffold with multiple capabilities for bone tissue regeneration. *Materials Science and Engineering: C*, 99, 83–95.
47. Rabbani, A., Haghniaz, R., Khan, T., Khan, R., Khalid, A., Naz, S. S., ... & Wahid, F. (2021). Development of bactericidal spinel ferrite nanoparticles with effective biocompatibility for potential wound healing applications. *RSC Advances*, 11(3), 1773–1782.
48. Katoch, G., Rana, G., Singh, M., García-Peñas, A., Bhardwaj, S., Sharma, I., ... & Kumar, G. (2021). Recent advances in processing, characterizations and biomedical applications of spinel ferrite nanoparticles. *Ferrite: Nanostructures with Tunable Properties and Diverse Applications*, 112, 62–120.
49. Bhaduri, B., Dikshit, A. K., Kim, T., & Tripathi, K. M. (2022). Research progress and prospects of spinel ferrite nanostructures for the removal of nitroaromatics from wastewater. *ACS Applied Nano Materials*, 5(11), 16000–16026.
50. Reddy, D. H. K., & Yun, Y. S. (2016). Spinel ferrite magnetic adsorbents: alternative future materials for water purification? *Coordination Chemistry Reviews*, 315, 90–111.
51. Tripathy, A., Nine, M. J., & Silva, F. S. (2021). Biosensing platform on ferrite magnetic nanoparticles: synthesis, functionalization, mechanism and applications. *Advances in Colloid and Interface Science*, 290, 102380.
52. Skliri, E., Miao, J., Xie, J., Liu, G., Salim, T., Liu, B., ... & Armatas, G. S. (2018). Assembly and photochemical properties of mesoporous networks of spinel ferrite nanoparticles for environmental photocatalytic remediation. *Applied Catalysis B: Environmental*, 227, 330–339.

6 Spinel Nano-Ferrites as Antibacterial Agents

Kavita Yadav
G.C.W. Gurawara, Rewari, Haryana, India

Atul Thakur, Preeti Thakur and Amitender Singh*
Amity Institute of Nanotechnology, Amity University Haryana, Gurugram, India

Ravi Kant Singh
Amity Institute of Biotechnology, Amity University Uttar Pradesh, Noida, India
and
Amity Institute of Nanotechnology, Amity University Haryana, Gurugram, India

6.1 INTRODUCTION OF FERRITES

According to Singh et al. nanomaterials are particles with one or more surface or interior structure diameters between 1 and 100 nm (Singh et al. 2023). When materials are reduced to nanoscale particle sizes, their characteristics change greatly from those observed at the macro scale (Thakur et al. 2022). As the surface area of the nanoparticles is large, their efficacy is likewise increased. Nanomaterials are significant for a wide range of applications in catalysis, energy, electronics, cosmetics, and medicine in addition to fundamental research (Pathania et al. 2019; Thakur et al. 2021).

6.1.1 History of Ferrites

The discovery of stones with iron-attracting properties marked the beginning of the history of ferrites. Magnetite (Fe_3O_4) is the name given to this mineral since it was discovered in great quantities in the Magnesia province of Asia Minor. The "Lodestones," which early navigators used to find magnetic poles, were the first usage of magnetite (Mills 2004). Due to its high iron oxide content, it exhibits persistent magnetism. The Latin word "Ferrum," meaning "iron," is when the term ferrites get its derivation (Garcia-Muñoz et al. 2020). For almost 50 years, ferrites, a broad family of oxides with Fe^{3+} and other metal cations, have been studied and used in several disciplines. The crystal structure and oxygen anion arrangement around metal cations of spinel, magnetoplumbite, garnet, and ortho-ferrite are used to classify them. Spinel ferrites are among the several forms of ferrites that are widely employed in magnetic, electrical, and medicinal applications. At the American Ceramic Society's annual conference in Dallas, Texas, in 1990, Prof. Takei received recognition for his prior contributions to the field of ferrite and was given the title "The father of ferrite." Hilpert et al. were the first to report on a comprehensive investigation into the link between the chemical makeup and magnetic characteristics of different ferrites. Hilpert had at that point successfully made spinel ferrites of zinc, magnesium, cobalt, manganese, and copper. Among his work's noteworthy features were his attempts to manufacture calcium, lead, and barium for which a German patent application is filed based on his original conception of ferrite core.

Magnetite, or $Fe^{2+}Fe^{3+}{}_2O_4$ (commonly referred to as Fe_3O_4), is an important ferrite material. It is a naturally occurring spinel oxide and is arguably the oldest magnetic material. It is currently the subject of intense research because of its intriguing characteristics which have practical applications. Large amounts of naturally occurring magnetite are ecologically resistant, chemically stable, and readily available. Another significant substance based on its composition and uses is maghemite, also known as γ-Fe_2O_3, which is essentially a spinel with defects (Valenzuela 2012).

6.1.2 Definition of Ferrites

A type of ceramic compounds known as ferrites is mostly made up of metallic components such as iron oxide (Fe_2O_3 or Fe_3O_4) (Chahar et al. 2022). They are useful materials for a variety of applications, especially in electronics and telecommunications, because of their distinctive magnetic characteristics. Ferrites are recognized for their high electrical resistance and crystalline structure, two characteristics that are essential to many of their uses. To get certain magnetic and electrical characteristics, ferrites are frequently created by combining iron oxide with other metal oxides, such as zinc, nickel, manganese, and others. Because of its high electrical resistance, ferrites are often used in situations where the least amount of electrical conductivity is required. By modifying their chemical composition and production methods, ferrites' unique magnetic and electrical characteristics

* Corresponding author.

DOI: 10.1201/9781003568940-6

may be customized. Because of their adaptability, they may be tailored to fit the needs of different applications. Ferrites are also used in a variety of electronic components. For example, ferrite cores and beads are used in transformers and inductors to improve magnetic characteristics and lower energy losses, respectively, and to minimize electromagnetic interference (EMI) in electronic circuits.

6.1.3 Types of Ferrites

There are many different kinds of ferrites, and each has a special composition and magnetic characteristics (Saini et al. 2016). The main types of ferrites are illustrated in Figure 6.1 and described below:

- *Soft Ferrites*: Soft ferrites are readily magnetized and demagnetized due to their low coercivity. They are utilized in transformers, inductors, and EMI suppression devices, among other applications where you wish to regulate and modify magnetic fields. Examples of soft ferrites include the following:
    - *Manganese-Zinc (Mn-Zn) Ferrites*: Among the most typical soft ferrites are Mn-Zn ferrites. They are utilized in transformers and inductors and have a high initial permeability. The low coercivity and high electrical resistance of Mn-Zn ferrites are well-known.
    - *Nickel-Zinc (Ni-Zn) Ferrites*: Ni-Zn ferrites have remarkably greater initial permeability, when compared to Mn-Zn ferrites (Punia et al. 2021). They are employed in high-frequency devices including inductors and radiofrequency (RF) transformers (Taneja et al. 2021).
- *Hard Ferrites*: Conversely, hard ferrites are difficult to magnetize and demagnetize due to their strong coercivity. They are frequently used in loudspeakers, electric motors, refrigerator magnets, and other applications where permanent magnets are required. Examples of Hard ferrites include the following:
    - *Barium Ferrite* ($BaFe_{12}O_{19}$): Permanent magnets are made from barium ferrites, a kind of hard ferrite with a high coercivity. They are frequently used in gadgets like microwaves, hard drives, and loudspeakers, among other magnetic data storage applications.
    - *Strontium Ferrite* ($SrFe_{12}O_{19}$): Strontium ferrite is utilized in permanent magnets, much as barium ferrite. It is used in applications that need a strong and steady magnetic field because to its high coercivity.
- *Spinel Ferrites*
    - *Magnetite* (Fe_3O_4): Sometimes used unprocessed or processed into a synthetic form, magnetite is a naturally occurring ferrite. In many different applications, such as magnetic recording media, pigments, and magnetic contrast agents in medical imaging, it is frequently utilized as a magnetic pigment.
- *Hexagonal Ferrites*
    - *Yttrium Iron Garnet* (*YIG*): YIG is a hexagonal ferrite that finds usage in phase shifters, circulators, and isolators in microwave and radiofrequency applications.
- *Garnet Ferrites*
    - *Terbium-Doped Garnet* ($Tb_3Ga_5O_{12}$): In magneto-optical devices like Faraday rotators and isolators, which regulate the polarization of light in optical communications, terbium-doped garnet ferrites are used.

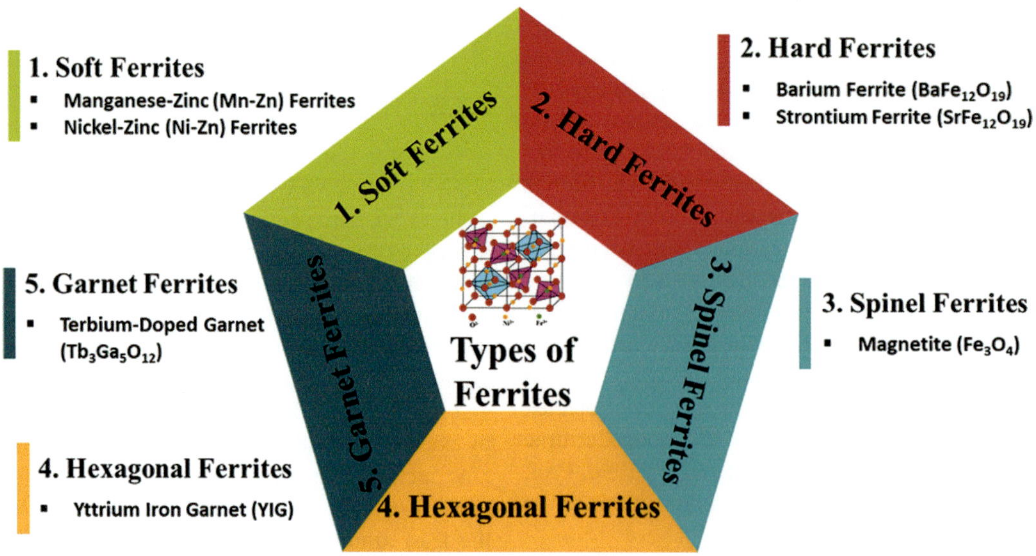

FIGURE 6.1 Different types of ferrites.

These are a few types of ferrites; each is designed with particular uses in mind due to its unique magnetic properties. The selection of ferrite type is contingent upon the specific needs of the application, including but not limited to the intended magnetic strength, coercivity, and frequency range.

6.1.4 Applications of Ferrites

Ferrites, also known as ferrimagnetic oxides, have found countless applications in the past few decades, including thin films and nanoparticles. Ferrites find most intriguing uses in high-frequency systems, power delivery devices, EMI reduction, electronic circuits as inductors, and biotechnology. Ferrites are used in many different sectors, as illustrated in Figure 6.2.

i. *Electronics*: Ferrites are utilized in parts such as ferrite cores, which improve the magnetic characteristics of inductors and transformers, and ferrite beads, which are used to reduce EMI in electronic circuits.
ii. *Telecommunications*: They are utilized to direct and regulate magnetic fields in microwave devices and antenna cores.
iii. *Permanent Magnets*: Permanent magnets for a range of uses, such as refrigerator magnets and magnetic closures, are frequently made from hard ferrites.
iv. *Automotive Industry*: Ferrites are utilized in a variety of automobile electrical components, electric vehicle motors, and sensors.
v. *Medical Devices*: They are present in MRI scanners and other medical imaging devices.
vi. *Energy Generation and Transmission*: Ferrites are used in transformers and inductors, among other power-generating and transmission devices.

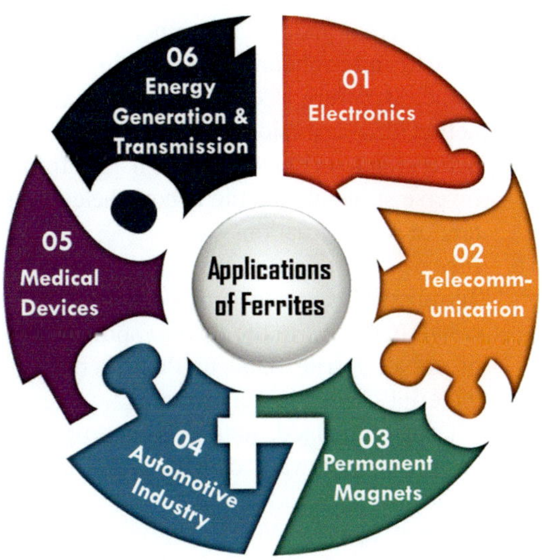

FIGURE 6.2 Applications of ferrites.

To sum up, ferrites are adaptable ceramic materials that primarily consist of iron oxide. Depending on their composition and structure, they may display certain magnetic characteristics, which make them vital parts of many contemporary technologies. They are used in a variety of applications across industries due to their strong electrical resistance and capacity to regulate and modify magnetic fields.

6.2 SPINEL FERRITES

6.2.1 Definition and Types of Spinel Ferrites

The ceramic ferrite known as "spinel ferrites," or just "spinels," have a distinct crystal structure. The name of this class of materials is derived from the mineral spinel ($MgAl_2O_4$), which has a crystal structure that is similar to theirs.

The qualities of spinel ferrites may be optimized by varying their composition and synthesis circumstances, which adds to their value. Because of this, they are still the subject of study and development in a number of disciplines, including as electronics, materials science, and nanotechnology. Because of their intriguing magnetic, electrical, and structural characteristics, spinel ferrites are beneficial for a variety of applications. The crystal structure of spinel ferrites is referred to as the "spinel structure." In this configuration, oxygen ions fill the spaces left by metal cations occupying particular positions in a face-centered cubic lattice. The spinel structure can be represented by the generic formula AB_2O_4, where A and B stand for distinct metal cations. Magnetite (Fe_3O_4), a mixed ferrite with both ferrous (Fe^{2+}) and ferric (Fe^{3+}) ions, is the most prevalent kind of spinel ferrite. By changing the metal ions in the crystal structure, other spinel ferrites may be created, giving materials customized characteristics. Strong magnetic characteristics are well-known for spinel ferrites. The precise composition and arrangement of metal ions determines whether they behave in a ferromagnetic or ferrimagnetic manner. Because of this, they may be used in applications like magnetic recording medium that call for magnetic materials.

6.2.1.1 Types of Spinel Ferrites

A range of materials having distinct metal ion combinations in the spinel crystal structure are referred to as spinel ferrites. Different magnetic, electrical, and physical characteristics result from these mixtures (Naagar et al. 2023). The following are a few typical forms of spinel ferrites as explained below and illustrated in Figure 6.3.

i. *Normal Spinel Ferrites*: A class of ferrite materials called normal spinel ferrites is defined by the particular configuration of metal ions in its crystal structure, which is referred to as the "normal spinel structure" (Mattei et al. 2011). A few examples of typical spinel ferrites are as follows:

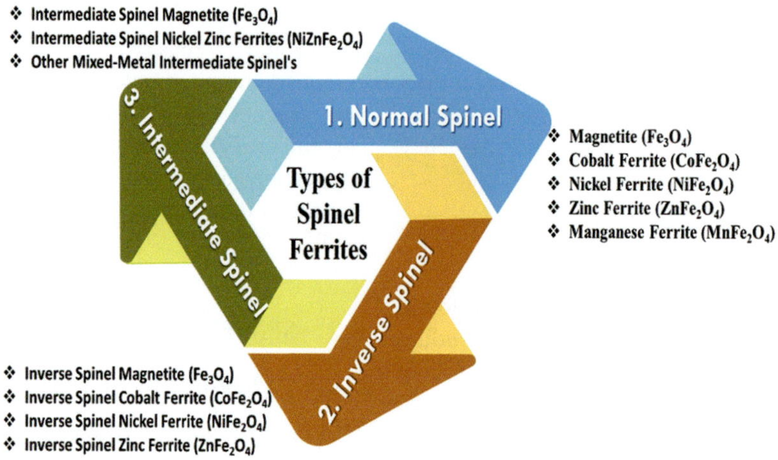

FIGURE 6.3 Types of spinel ferrites.

a. *Magnetite* (Fe_3O_4): Iron (Fe) cations in both the 2+ and 3+ oxidation states make up magnetite, a well-known typical spinel ferrite. It is both a manufactured and a natural mineral. Because of its straightforward, well-characterized structure, magnetite is frequently employed as a reference material in magnetic research.
b. *Cobalt Ferrite* ($CoFe_2O_4$): With cobalt (Co) and iron (Fe) ions, cobalt ferrite is a typical spinel ferrite. It is renowned for having a high magnetic coercivity and strong magnetic characteristics. Magnetic recording medium, high-frequency transformers, and microwave devices are just a few of the uses for this material.
c. *Nickel Ferrite* ($NiFe_2O_4$): Another common spinel ferrite that has iron (Fe) and nickel (Ni) ions is called nickel ferrite. It is utilized as a pigment in ceramics and paintings, as well as in electronics and microwave components, due to its high magnetic characteristics.
d. *Zinc Ferrite* ($ZnFe_2O_4$): Zinc ferrite, consisting of zinc (Zn) and iron (Fe) ions, is a typical spinel ferrite. Due to its high electrical resistivity, it is utilized in high-frequency and microwave applications, including as circulators and isolators.
e. *Manganese Ferrite* ($MnFe_2O_4$): Another common spinel ferrite that has iron (Fe) and manganese (Mn) ions is called manganese ferrite. It is utilized in high-frequency transformers, magnetic recording media, and some biological applications. It is well-known for its electrical insulating qualities.

Because of their electrical and magnetic qualities, these normal spinel ferrites are widely employed in many different technological applications. They are useful for several applications like high-frequency electronics, magnetic devices, data storage, and many more because of their distinct crystal structure and metal ion arrangement. The characteristics and possible uses of normal spinel ferrites are still being investigated by researchers.

ii. *Inverse Spinel Ferrites*: A kind of ferrite material called inverse spinel ferrites has a different crystal structure than regular spinel ferrites. The distribution of metal ions inside the crystal lattice of inverse spinel ferrites differs from that of conventional spinel ferrites. To be more precise, in an inverse spinel structure, certain divalent metal cations (often represented by the letter "B" in the AB_2O_4 formula) are found in the tetrahedral sites of the crystal structure, while other cations are found in the octahedral sites. The site occupancy is flipped, producing unique magnetic and electrical characteristics. Inverse spinel ferrites include the following examples:

a. *Inverse Spinel Magnetite* (Fe_3O_4): Some of the iron (Fe) cations in magnetite, which may have both normal and inverse spinel structures, are located in tetrahedral positions in the inverse spinel structure. Compared to regular magnetite, this inversion produces distinct magnetic characteristics. Although less frequent, inverse spinel magnetite is nonetheless important in some situations.
b. *Inverse Spinel Cobalt Ferrite* ($CoFe_2O_4$): Tetrahedral positions are occupied by some cobalt (Co) ions in the inverse spinel structure of cobalt ferrite. Due to this alteration, the material's electrical and magnetic characteristics are changed, making it appropriate for usage in microwave devices and other high-frequency applications.
c. *Inverse Spinel Nickel Ferrite* ($NiFe_2O_4$): Similar to cobalt ferrite, nickel ferrite in the inverse spinel structure has some of the nickel

(Ni) ions in tetrahedral sites. This inversion leads to different magnetic characteristics and is useful in high-frequency electronics.

d. *Inverse Spinel Zinc Ferrite* ($ZnFe_2O_4$): Some of zinc (Zn) ions are located in tetrahedral positions in inverse spinel structure of zinc ferrite. This configuration, which is employed in some microwave applications, alters the material's electrical characteristics.

When certain magnetic and electrical characteristics are needed, inverse spinel ferrites are very advantageous. They are employed in high-frequency electronics, where their altered crystal structure can enhance their performance at particular frequencies, such as microwave components and telecommunications equipment. Inverse spinel ferrites are useful in many different technological applications because of the changed magnetic and electrical characteristics caused by the inversion of site occupancy inside the crystal lattice.

iii. *Intermediate Spinel Ferrites*: A type of ferrite materials known as intermediate spinel ferrites is made up of those with a crystal structure halfway between that of regular (cubic) and inverse (tetrahedral-octahedral inversion) spinel structures. Intermediate spinel ferrites have particular magnetic and electrical characteristics because the distribution of metal cations in the crystal lattice is more equally balanced between octahedral and tetrahedral sites. Although they are less frequent than inverse or normal spinel ferrites, intermediate spinel ferrites have a variety of uses. Few examples of intermediate spinel ferrites are as follows:

a. *Intermediate Spinel Magnetite* (Fe_3O_4): Comparing intermediate spinel magnetite to typical spinel magnetite, the distribution of iron (Fe) cations between octahedral and tetrahedral sites is more equal. The magnetic characteristics produced by this configuration fall somewhere between those of inverse and conventional spinel magnetite. Applications for intermediate spinel magnetite include geological research and magnetic recording.

b. *Intermediate Spinel Nickel Zinc Ferrites* ($NiZnFe_2O_4$): The cations of iron (Fe), zinc (Zn), and nickel (Ni) are mixed together in these ferrites. This cation's distribution inside the intermediate spinel structure results in special magnetic and electrical characteristics. Applications for them include microwave devices, inductors, and transformers.

c. *Other Mixed-Metal Intermediate Spinel's*: Different metal cations combinations can be used in intermediate spinel ferrites to create materials with specific characteristics. These materials are used in some scientific investigations, telecommunications, and high-frequency electronics.

When the required qualities depend on the balance between octahedral and tetrahedral site occupancy, intermediate spinel ferrites may be of interest. Because of their adaptability, magnetic and electrical properties may be tailored, which makes them appropriate for some scientific and technical applications where regular and inverse spinel structures would not offer the best qualities. These are only a handful of the several spinel ferrites that exist in different compositions. The needed performance qualities and the desirable magnetic, electrical, and physical properties determine which spinel ferrite is best for a certain application. Spinel ferrites with specific characteristics are still being investigated and developed by researchers for a variety of scientific and technical uses.

6.2.2 Applications of Spinel Ferrites

i. *Magnetic Recording Media*: Floppy disks, magnetic recording tapes, and other magnetic storage media have all included spinel ferrites. Their tiny magnetic particles make them particularly suitable for high-density recording.

ii. *Pigments*: Certain spinel ferrites are utilized as pigments in paintings and ceramics, among other uses. They are used for coloring because of their stable qualities and brilliant hues.

iii. *Biomedical Applications*: Some spinel ferrites are being investigated for usage in biomedical applications, like in MRI as contrast agents and magnetic nanoparticles for medication administration.

iv. *Catalysts*: Because of their catalytic qualities, spinel ferrites are employed in heterogeneous catalysis, especially in procedures pertaining to the chemical sector.

v. *Magnetic Fluids*: In order to create magnetic fluids, which are used in a variety of settings, including as lubricants and seals in hard drives, spinel ferrites are utilized.

6.3 ANTIBACTERIAL MECHANISM OF SPINEL FERRITE NANOPARTICLES

Because of their special antibacterial qualities, spinel ferrite nanoparticles (SF NPs) have drawn interest in the field of antibacterial research. SF NPs principal antibacterial mechanism consists of their capacity to upset bacterial cell structures and obstruct cellular functions. The following are some of the main processes (as shown in Figure 6.4) by which SF NPs work against bacteria:

i. *Generation of Reactive Oxygen Species (ROS)*: ROS can be induced in bacterial cells using metallic nanoparticles. Extremely reactive substances

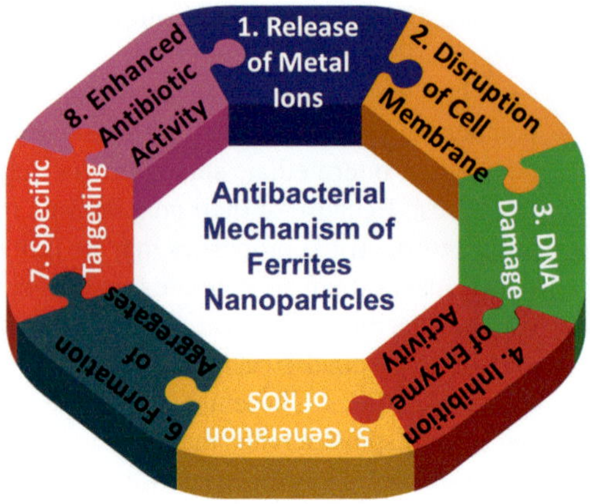

FIGURE 6.4 Antibacterial mechanism of spinel ferrites.

known as ROS, such as H_2O_2 and superoxide radicals, can damage lipids, proteins, and DNA in cells. The result of this oxidative stress may be cell death.

ii. *Release of Metal Ions*: Metal ions are released into the surroundings by SF NPs. Bacterial cells are poisonous to these metal ions, which include zinc (Zn^{2+}), copper (Cu^{2+}), and silver (Ag^{+}). They have the ability to compromise the integrity of the bacterial cell membrane and obstruct vital biological functions.

iii. *Disruption of Cell Membranes*: Bacterial cell membranes can interact with metallic nanoparticles. The negatively charged bacterial cell membrane may be attached to by positively charged nanoparticles, which can cause membrane breakdown and internal cell leakage. This can eventually result in cell death by upsetting structural integrity of cell.

iv. *Inhibition of Enzyme Activity*: Certain SF NPs have the ability to directly stop bacterial enzymes from working. For instance, the operation of the bacterial cell can be disrupted by the inhibition of enzymes involved in energy generation and metabolism by silver nanoparticles.

v. *DNA Damage*: Bacterial DNA can interact with metallic nanoparticles. Damage to DNA from this interaction can result in genetic alterations or even cell death. Damage to DNA can impede the development and replication of bacteria.

vi. *Formation of Aggregates*: Bacterial cells can be physically obstructed by SF NPs clusters that can collect and form. These clusters have the ability to stop bacterial growth and division.

vii. *Specific Targeting*: SF NPs can be functionalized by researchers using chemicals that specifically target different types of bacteria. Because of their specialization, certain infections can be destroyed while protecting beneficial microorganisms.

viii. *Enhanced Antibiotic Activity*: Antibiotics can be utilized more effectively by combining them with metallic nanoparticles. Fighting antibiotic-resistant microorganisms is one area in which this synergistic impact is very helpful. The disruption of bacterial defenses and increased susceptibility to antibiotic action can be caused by the nanoparticles.

It is noteworthy that the effectiveness of SF NPs as antibacterial agents might differ based on several aspects, including the type of metal utilized, surface charge, size, and form of the particles. When employing metallic nanoparticles for antibacterial purposes, it's also important to thoroughly evaluate their possible cytotoxicity and environmental effects. However, considering antibiotic resistance, SF NPs are a promising new weapon in the fight against bacterial diseases.

6.3.1 Reactive Oxygen Species (ROS)

ROS are chemically reactive molecules that contain oxygen. These comprise a range of peroxides and oxygen radicals, including singlet oxygen, H_2O_2, hydroxyl radicals (•OH), and superoxide anion ($O_{2\bullet-}$). ROS are naturally occurring byproducts of cellular metabolism and are crucial to many physiological functions. The various mechanisms of production of ROS are represented in Figure 6.5. On the other hand, excessive ROS production can be detrimental, resulting in oxidative stress and injury to lipids, proteins, DNA, and cells.

i. *Mitochondrial Electron Transport Chain*: The electron transport chain in the mitochondria is one of the main generators of ROS in cells. Superoxide anions ($O_{2\bullet-}$) are created when electrons escape

FIGURE 6.5 Mechanism of ROS production for antibacterial effect.

from the electron transport chain during oxidative phosphorylation. The first ROS produced in mitochondria are these superoxide anions.

ii. *NADPH Oxidase (NOX) Enzymes*: Membrane-bound NADPH oxidase enzymes are essential for producing ROS during the immunological response. They exhibit heightened activity in phagocytic cells, such as macrophages and neutrophils. Superoxide anions are created by NOX enzymes as a component of the host's defensive strategy against infections.

iii. *Enzymatic Reactions*: ROS can be produced by specific enzymatic processes within cells. For example, superoxide anions can be produced by xanthine oxidase while purine bases are being metabolized. ROS are produced by cytochrome P450 enzymes when medicines and foreign substances are detoxified. The enzymes known as nitric oxide synthase (NOS) produce nitric oxide (NO), which when combined with superoxide anions can produce peroxynitrite ($ONOO^-$), a powerful ROS.

iv. *Metabolism of Oxygen*: During aerobic respiration, oxygen is typically absorbed by cells. Superoxide anions can be formed when a little amount of molecular oxygen (O_2) is incompletely reduced during metabolic activities.

v. *Metal-Catalyzed Reactions*: Iron (Fe) and copper (Cu) are two examples of transition metals that can take part in Fenton and Haber-Weiss reactions. The process of these reactions involves the transfer of electrons, which produces one of the most powerful ROS: hydroxyl radicals (•OH).

vi. *Ultraviolet (UV) Radiation*: The production of ROS in the skin can be brought on by exposure to UV radiation from the sun. Singlet oxygen is formed when oxygen molecules are excited by UV light, and this can harm DNA and skin cells.

The production of ROS can also be triggered by exposure to poisons and contaminants. For instance, oxidative stress and ROS generation in cells can be caused by certain substances and radiation. For cells to remain healthy, the body's antioxidant defense systems and ROS generation must be in equilibrium. Oxidative stress can happen when the body's capacity to neutralize ROS is exceeded. This can cause harm to biomolecules and play a role in a number of illnesses, such as cancer, neurological conditions, and cardiovascular diseases. Vitamins C and E are examples of antioxidants that help lessen the negative effects of ROS by either neutralizing them or preventing their production.

6.3.2 Release of Metal Ions

One significant way that SF NPs, in particular, deliver their antibacterial, cytotoxic, or therapeutic effects is through the release of metal ions. This procedure might be advantageous or disadvantageous based on the situation. Figure 6.6 illustrates how the SF NPs discharge metal ions:

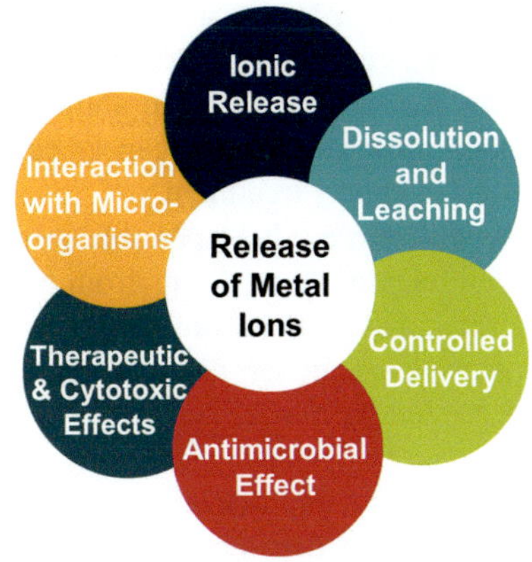

FIGURE 6.6 Discharge of metal ions in bacterial cell for antibacterial effect.

i. *Dissolution and Leaching*: The metal-metal connections between these nanoparticles may weaken or break when they come into touch with a liquid (such as water or biological fluids) or an acidic environment, which will cause the NPs to dissolve. Metal ions are released into the surrounding media as a result of this disintegration. The size, content, surface coating, and ambient conditions of the nanoparticles, among other parameters, influence the rate of dissolution and ion release.

ii. *Ionic Release*: Cu^{2+}, Zn^{2+}, Ag^+, and other positively charged metal cations are examples of the metal ions that the SF NPs can release. These metal ions can interact with biological tissues and microorganisms in the surrounding environment, as they are usually extremely reactive.

iii. *Interaction with Microorganisms*: Metal ions released by metallic nanoparticles can interact with internal components and cell membranes of microorganisms, such as bacteria. For instance, the bacterial cell membrane can be damaged by metal ions, which can cause membrane permeabilization and internal cell leakage. Essential biological functions including DNA replication, enzyme activity, and energy generation can all be hampered by metal ions.

iv. *Antimicrobial Effect*: Antimicrobial effects may arise from the release of metal ions like Zn^{2+}, Cu^{2+}, or Ag^+. These metal ions are poisonous to many different types of microbes, such as bacteria, viruses, and fungus. They cause these infections' essential processes and structural integrity to be disrupted, which results in their inactivation or death.

v. *Therapeutic and Cytotoxic Effects*: SF NPs that emit metal ions have potential medicinal uses, including the treatment of certain illnesses.

For example, silver NPs that release Ag^+ ions have been applied as antimicrobials and in wound healing. Metal-based NPs have the ability to release metal ions, which can harm cancer cells locally when used in cancer treatment.

vi. *Controlled Delivery*: It is possible to create the controlled release of metal ions from NPs for certain medicinal uses. To ensure a long-lasting therapeutic impact, researchers can create nanoparticles that release ions at a regulated rate.

It is noteworthy that the discharge of metal ions may have both beneficial and detrimental consequences, thus it is crucial to carefully evaluate the characteristics and dose of the nanoparticles. Furthermore, it is necessary to assess the possible cytotoxicity, environmental effect, and safety issues related to the usage of metal ions in diverse applications. Researchers persistently investigate and cultivate SF NPs for an extensive array of uses, encompassing their antibacterial, medicinal, and diagnostic capabilities.

6.3.3 Disruption of Cell Membrane

Different chemicals, such antimicrobial agents and NPs, can affect cells, both microbial cells (like bacteria) and human cells, through the disruption of cell membranes. Cell damage or even death may result from the breakdown of cell membranes. The possible reasons for the bacterial cell disruption are illustrated in Figure 6.7:

i. *Mechanical Disruption*: Mechanical disruption of cell membranes can be caused by substances such as NPs or physical pressures. Sharp nanoparticles, for instance, have the ability to physically pierce or harm cell membranes, resulting in cell lysis and death.

ii. *Lipid Bilayer Interaction*: The fundamental structural element of cell membranes is the lipid bilayer. The lipid bilayer can be impacted by a number of chemicals, including as surfactants, antimicrobial compounds, and specific antimicrobial peptides. They have the ability to penetrate into the lipid bilayer, destabilizing the membrane and allowing holes or pores to develop, which lets cellular contents seep out and eventually causes cell death.

iii. *Osmotic Imbalance*: Osmotic imbalance can occur when there is a disruption in the cell membrane. The membrane's capacity to regulate the flow of solutes and water into and out of the cell may be compromised by damage or permeabilization. This may lead to an inflow of ions and water, which might expand the cell to the point where it bursts from the increasing internal osmotic pressure.

iv. *Electric Field Disruption*: Electric fields are used in certain procedures, such as electroporation, to damage cell membranes. Short electrical pulses can be used to temporarily produce holes or gaps in cells' membranes, enabling materials that would not ordinarily be able to pass through to enter.

v. *Cationic Peptides*: Defensins and magainins are examples of cationic peptides with antibacterial qualities that can damage cell membranes through electrostatic interactions. These peptides are drawn to the negatively charged membranes of microbes, where they can break membranes and induce lysis of cells when they bind.

vi. *Antimicrobial Agents*: Certain antimicrobial drugs, such as detergents and polymyxins, function by rupturing the bacterial cell membrane. For example, polymyxins cause membrane disruption when they interact with the lipopolysaccharides of Gram-negative bacteria's outer membrane.

vii. *Nanoparticles*: Bacterial cell membranes can interact with certain NPs, such silver NPs. The membrane may be damaged structurally by the NPs, which might disrupt the barrier and kill cells.

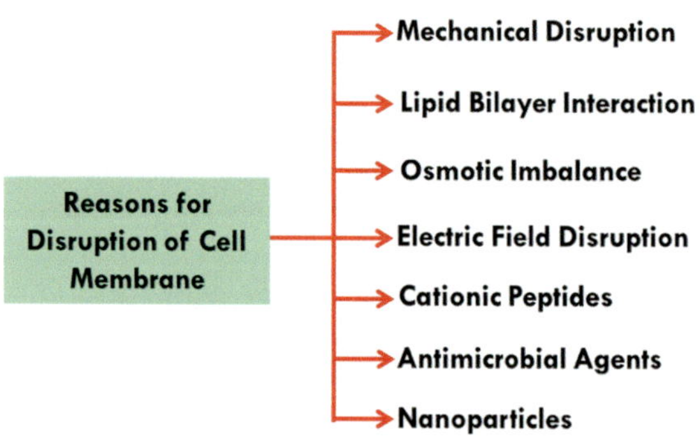

FIGURE 6.7 Different reasons for disruption of bacterial cell membrane.

One prevalent mechanism seen in antibacterial and antimicrobial techniques is the disruption of cell membranes. By undermining the structural integrity of bacteria and preventing them from carrying out essential tasks, it is an efficient method of neutralizing them. To reduce the possibility of injury to mammalian cells or tissues, it is crucial to take the selectivity of the disruption mechanism into account when targeting microbial cells.

6.3.4 Inhibition of Enzyme Activity

One way that different compounds might affect biological systems, including microbial cells, is by inhibiting the action of enzymes. In living things, biochemical processes are catalyzed by crucial proteins called enzymes. Inhibiting enzyme activity can cause biological processes to become disturbed, which can result in cell damage or death. Enzyme inhibition works by following mechanisms:

i. *Competitive Inhibition*: When a molecule—often referred to as the inhibitor—closely mimics the substrate that the enzyme typically works upon, competitive inhibition takes place. In an attempt to attach to enzyme's active site, the inhibitor and substrate compete. Enzyme's activity is essentially stopped if the inhibitor attaches itself to the active site and stops the substrate from binding. Because the inhibitor can separate from the enzyme and resume its action after the inhibitor is removed, this kind of inhibition may be reversible.

ii. *Non-competitive Inhibition*: The inhibitor and substrate do not compete with one another for binding to the active site. Rather, inhibitor attaches to an alternative regulatory site on enzyme, changing its conformation in a way that prevents the catalytic activity of the enzyme. Reversible or irreversible non-competitive inhibition is possible.
iii. *Mixed Inhibition*: Competitive and non-competitive inhibitions are combined to create mixed inhibition. Depending on the inhibitor's respective affinities for the two sites, it can attach to both the active site of enzyme and a different regulatory site, producing a mixed pattern of inhibition in mixed inhibition.
iv. *Uncompetitive Inhibition*: One less prevalent type of inhibition is uncompetitive inhibition. In this instance, the substrate is effectively trapped in the active site since inhibitor exclusively binds to complex of the enzyme and substrate. Usually, this kind of inhibition causes the activity of the enzyme and the substrate's affinity for the active site to decline.
v. *Irreversible Inhibition*: When an inhibitor creates a strong covalent link with the enzyme, rendering it permanently inactive, irreversible inhibition takes place. Typically, this kind of inhibition is non-competitive. Irreversible inhibitors include several medications and poisons.

Inhibiting certain enzymes is one tactic used to interfere with vital metabolic processes in microbial cells, causing inhibition or death in the context of antibacterial activity. Antibiotics such as penicillin, for example, block the enzymes that bacteria need to manufacture their cell walls. In a similar way, certain antimicrobial peptides can specifically target enzymes that are essential for the creation of proteins, DNA, and other parts of microorganisms. In order to control metabolic pathways or target certain enzymes linked to certain disorders, enzyme inhibition is also utilized in research and medicinal therapies. In biochemistry, pharmacology, and medical research, an understanding of and utilization of enzyme inhibition is a basic concept.

6.3.5 DNA Damage

One way by which certain metallic nanoparticles might achieve their antimicrobial properties is DNA damage caused by the NPs. These NPs have the ability to interact with bacterial cells, causing DNA damage that eventually affects vital cellular functions and may even cause cell death. When using NPs to combat germs, DNA damage usually takes the following forms as represented in Figure 6.8 and explained below:

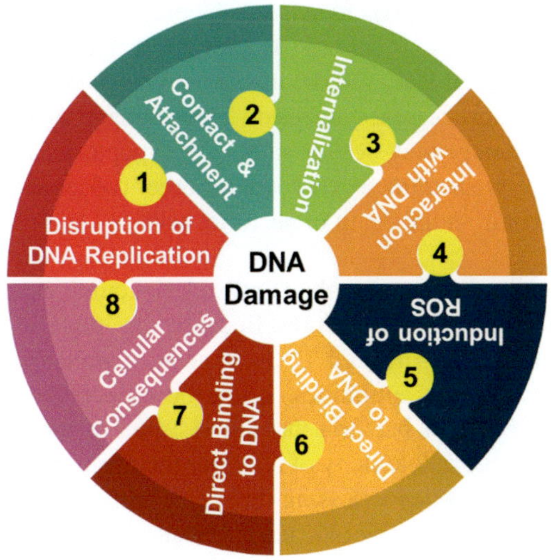

FIGURE 6.8 Different ways of bacterial cell DNA damage.

i. *Contact and Attachment*: Bacterial cells come into contact with NPs, such as copper ferrite nanoparticles (CuNPs) and silver ferrite nanoparticles (AgNPs). The cell membrane can be adhered to by the NPs.
ii. *Internalization*: The bacterial cells absorb portion of the NPs. Numerous processes, including as endocytosis, direct cell membrane penetration, or other interactions between the nanoparticle and the cell that are unique to each one, might cause this.
iii. *Interaction with DNA*: NPs have the ability to interact with DNA once they are within the bacterial cell. Damage to DNA may result from this interaction.
iv. *Induction of ROS*: Certain metallic nanoparticles react with the bacterial cell's surroundings to produce ROS, such as •OH or $O_{2}^{\bullet-}$. Because ROS are so reactive, they have the ability to oxidatively damage DNA, causing breaks and mutations.
v. *Direct Binding to DNA*: A few NPs have the ability to bond with DNA directly. DNA strands may break, cross-link, or undergo structural changes as a result of this contact.
vi. *Disruption of DNA Replication and Repair*: The bacterial cell's capacity to copy and repair its DNA is hampered by damage to its DNA. The bacteria might not be able to proliferate and preserve its genetic integrity if the DNA replication process is interfered with if there is significant DNA damage.

vii. *Cellular Consequences*: Numerous biological outcomes, including as cell cycle halt, genetic material loss, and the inability to carry out essential processes, can result from damage to DNA. Damage may occasionally cause apoptosis, a type of programmed cell death, or other processes that result in the death of bacterial cells.

SF NPs demonstrate their antibacterial qualities through these mechanisms of action. Because the mode of action of nanoparticle-induced DNA damage differs from that of conventional antibiotics, it can be especially efficient against types of bacteria that are resistant to these drugs. The possible cytotoxicity of NPs and their impact on mammalian cells when assessing their suitability for antibacterial purposes is crucial to check. Because of their potential to fight bacterial infections, researchers are still studying and developing nanoparticles.

6.4 MEASUREMENT AND DETECTION OF ANTIBACTERIAL ACTIVITY OF SPINEL FERRITES

Antimicrobial studies depend heavily on the detection of antibacterial activity. To evaluate the efficacy of antibacterial agents, antimicrobial chemicals, or new materials like nanocomposites, researchers and doctors require dependable methodologies. The following assays represented in Figure 6.9 are frequently employed to find antibacterial activity:

i. *Agar Disk Diffusion Assay*: This is a straightforward and popular technique for determining a substance's antibacterial activity. An agar plate that has been infected with bacteria is covered with a paper disk that contains the test agent. Bacterial growth is suppressed in a zone of inhibition (ZOI) created by the antibacterial agent diffusing into the agar over time. The antibacterial activity is directly correlated with the size of the ZOI.

ii. *Broth Dilution Assay (MIC/MBC)*: In liquid broth, bacteria are exposed to varying quantities of an antibiotic agent in the minimum inhibitory concentration (MIC) and minimum bactericidal concentration (MBC) experiments. The minimum concentration at which observable bacterial growth is inhibited is known as the MIC, while the minimum concentration at which bacterial death occurs is known as the MBC. The antibacterial activity of these tests may be quantified.

iii. *Time-Kill Assay*: The kinetics of antibacterial activity over time is measured by this test. To assess the bacterial vitality, samples are collected at different times after the bacteria are exposed to a specific concentration of the antibacterial agent. Time-kill curves display an agent's effectiveness in stopping or eliminating bacterial growth over a certain amount of time.

iv. *Micro dilution Assay*: This assay uses microtiter plates to evaluate antibacterial drugs on a smaller scale than the broth dilution method. It makes it possible to screen several drugs at high-throughput against various bacterial strains.

v. *Checkerboard Assay*: This test evaluates whether two or more antibacterial agents together have antagonistic or synergistic effects. The interaction between two compounds is ascertained by testing them in combination at varying doses and calculating the fractional inhibitory concentration index (FICI).

vi. *Biofilm Assays*: Bacteria that produce biofilms may be more resilient to antibacterial treatments. Specialized tests, such the crystal violet assay, measure how antibacterial drugs disrupt or impede the production of biofilms.

vii. *Fluorescent or Luminescent Assays*: Fluorescent or luminous probes that show bacterial viability can be used to measure antibacterial activity. Real-time data on the reaction of bacteria to antibacterial drugs can be obtained using fluorescence-based

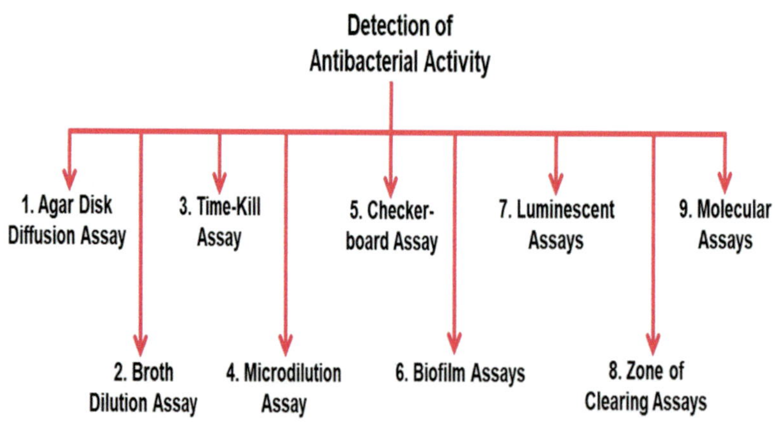

FIGURE 6.9 Different types of assays for measurement of antibacterial activity.

approaches, bioluminescence tests, or live/dead staining kits.

viii. *Zone of Clearing Assays*: These are employed in the evaluation of lytic agents, such as bacteriophages. Bacteriophages cause clearing zones (plaques) in bacterial lawns on agar plates by infecting and killing bacterial cells.

ix. *Molecular Assays*: Detecting alterations in bacterial gene expression or DNA damage in response to antibacterial drugs can be done by molecular methods like quantitative polymerase chain reaction.

Flow cytometry can also be used to quantify how antibacterial drugs affect a cell's size, membrane integrity, or other properties. The particular antibacterial drug, the kind of bacteria being investigated, and the goals of the study all influence the test selection. These techniques offer useful data for assessing a substance's antibacterial qualities and locating possible novel antibiotics or antimicrobial compounds.

6.4.1 Agar Disk Diffusion Assay

The Kirby-Bauer test, sometimes called the agar disk diffusion assay, is a commonly used, standardized technique for determining how susceptible bacteria are to different antimicrobial agents, such antibiotics or other antibacterial compounds. The effectiveness of an antibacterial agent to stop the development of a particular bacterial strain is measured using a qualitative approach. The schematic representation of the procedure for agar disk diffusion method is shown in Figure 6.10 and described below:

Materials and Reagents

- Nutrient agar plates (or another suitable agar medium)
- Sterile paper disks (also called filter paper disks)
- Antibacterial agents (e.g., antibiotic disks)
- Bacterial culture to be tested
- Sterile forceps
- Incubator (for bacterial growth)
- Ruler or calipers

Procedure

- *Preparation of Nutrient Agar Plates*
  - Use autoclaving or other appropriate techniques to sterilize the nutrition agar plates.
  - To make agar plates, pour sterile, melted agar onto sterile petri dishes and let it harden.
- *Inoculation of Bacterial Culture*
  - Using a sterile inoculating loop or swab, blot the bacterial culture to be tested onto the agar plate's surface.
  - Verify that the bacterial culture is dispersed equally throughout the surface of the agar.
- *Placement of Antibacterial Disks*
  - Place sterile paper disks onto surface of the inoculated agar plates by sterile forceps.
  - Sterilize the paper disks with antibacterial agents. Antibiotics, antiseptics, and other drugs under test could be among these agents.
- *Incubation*
  - For a certain amount of time, generally between 18 and 24 hours, incubate the infected agar plates at the proper temperature (normally 37°C for bacteria).
- *Measurement of Zones of Inhibition*
  - Check the agar plates for zones of inhibition surrounding the paper disks after incubation.
  - Using calipers or a ruler, determine each ZOI's diameter. The zone shows the location of the antibacterial agent's inhibition of bacterial growth.

Interpretation

- Examine the zones of inhibition's diameter compared to accepted interpretation standards.

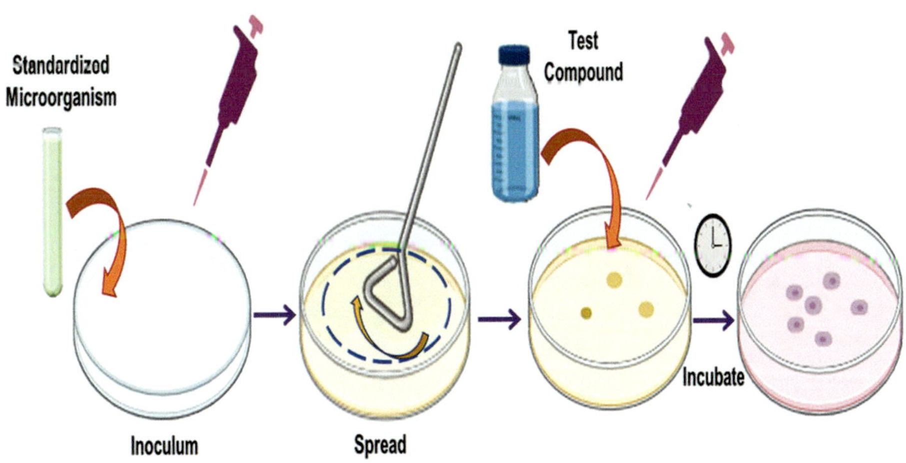

FIGURE 6.10 Schematic representation of agar disk diffusion method.

- Smaller or nonexistent zones represent resistance to the antibacterial agent; zones that meet or surpass the prescribed diameter indicate sensitivity of the bacteria to the antibacterial agent.

Recording and Reporting

- Note the outcomes, including the diameter of each ZOI, the antibacterial agent's name, and the particular type of bacteria that was tested.
- Provide the findings in a way that is consistent.

An easy and trustworthy way to determine a bacterium's initial sensitivity to different antibiotic drugs is to use the agar disk diffusion assay. It is frequently used in clinical microbiology to inform treatment choices and assess how well antibiotics work against clinical isolates of bacteria. It should be mentioned, nonetheless, that this test yields qualitative data and does not calculate the MIC. Typically, broth dilution tests or other quantitative techniques are employed for MIC determination.

6.4.2 Broth Dilution Method (MIC/MBC)

A popular technique for figuring out an antimicrobial agent's MIC and MBC is broth dilution assay. This test measures how well antibiotics or other antibacterial agents work against a particular strain of bacteria. Figure 6.11 represents the schematic illustration of broth dilution method for MIC and MBC determination and explained as:

Materials and Reagents

- Broth culture medium (commonly Mueller-Hinton broth for bacteria)
- Test tubes or microtiter plates
- Antibacterial agents (e.g., antibiotics)
- Bacterial culture to be tested
- Incubator (for bacterial growth)
- Spectrophotometer or turbidimeter (for measuring bacterial growth)
- Sterile pipettes and tips

Procedure

- *Preparation of Broth Culture*
  - Inoculate a broth culture medium with the bacterial culture that will be evaluated. Generally speaking, the bacterial concentration needs to be changed to meet a predetermined turbidity, which is frequently equal to a 0.5 McFarland standard. By doing this, it is guaranteed that the bacterial suspension is growing in the logarithmic phase.
- *Preparation of Antibacterial Agents*
  - Get ready to test a variety of dilutions of the antibacterial agent. The concentration range that the dilutions should span is high to low.
- *Inoculation of Broth Tubes or Microtiter Plates*
  - Dispense the prepared bacterial culture into test tubes or microtiter plate wells, with each tube or well containing the same initial bacterial concentration.
- *Addition of Antibacterial Agents*
  - Add the various dilutions of the antibacterial agent to the test tubes or microtiter plate wells. Each tube or well receives a different concentration of the antibacterial agent.

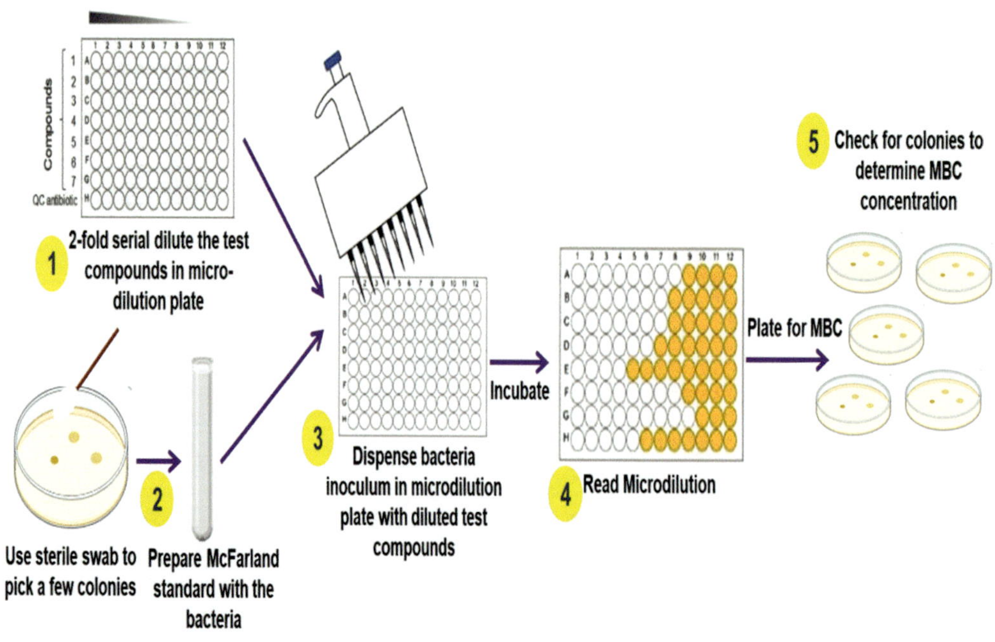

FIGURE 6.11 Schematic illustration of Broth Dilution Method for MIC/MBC estimation.

- *Incubation*
  - For a certain amount of time, generally between 18 and 24 hours, incubate the test tubes or microtiter plates at the proper temperature (normally 37°C for bacteria).
- *Measurement of Growth*
  - Using a spectrophotometer or turbidimeter, determine the optical density (OD) or turbidity of the bacterial cultures in each tube or well after incubation.

Interpretation

- *Determining the MIC*: MIC is antibacterial agent concentration at which no discernible growth occurs, as demonstrated by the lack of a rise in turbidity or OD relative to the growth control, which contains no antibacterial agent.
- *MBC determination*: Samples from tubes or wells that, based on MIC determination, exhibited no growth should be taken and streaked onto agar plates in order to measure the MBC. For a certain amount of time, incubate the agar plates at the proper temperature (usually 18–24 hours). This suggests that the bacteria are killed instead than only inhibited from growing.

Recording and Reporting

- Note the antibacterial agent's MIC and MBC values for that particular bacterial strain.

The broth dilution assay yields quantitative information on the tested agent's antibacterial activity as well as the lowest concentration needed to suppress or eradicate the bacteria. This data is useful for making judgments about the use of antibiotics, figuring out which bacteria are susceptible to them, and evaluating the potency of novel antibacterial substances.

6.4.3 Time-Kill Assay

A microbiological technique called a time-kill assay is used to evaluate an antibacterial agent's antimicrobial effectiveness over a predetermined amount of time. This test gauges how well an agent works overtime to lower or eradicate the number of viable microorganisms, most often bacteria. It offers crucial details on the kinetics of the antibacterial agent's action, such as whether it has bactericidal or bacteriostatic effects. The following describes the time-kill assay procedure:

Materials and Reagents

- Bacterial culture to be tested
- Antibacterial agent (e.g., antibiotic)
- Sterile broth culture medium (commonly Mueller-Hinton broth for bacteria)
- Sterile test tubes or microtiter plates
- Incubator (set at the appropriate temperature for bacterial growth)
- Sterile pipettes and tips
- Agar plates for plating bacterial samples (if determining bactericidal activity)
- A spectrophotometer or turbidimeter for measuring bacterial growth

Procedure

- *Preparation of Bacterial Culture*
  - Transfer the bacterial culture into a culture medium made of sterile broth. To make sure the bacterial suspension is in the logarithmic phase of growth, adjust the bacterial concentration to match a predetermined turbidity, usually equal to a 0.5 McFarland standard.
- *Preparation of Antibacterial Agent*
  - Assemble an antibacterial agent stock solution at the appropriate concentration.
- *Inoculation of Test Tubes or Microtiter Plates*
  - Transfer the ready-made bacterial culture onto sterile microtiter plate wells or test tubes. There should be an equal starting concentration of germs in each tube or well.
- *Addition of Antibacterial Agent*
  - To build a time course, add the antibacterial agent to the test tubes or wells at the correct concentration. Samples should be taken at various time intervals.
- *Incubation*
  - For the designated amount of time, incubate the test tubes or microtiter plates at the proper temperature (typically 37°C for bacteria).
- *Sampling*
  - To gauge the bacterial growth, take samples from the wells or test tubes at each time point. Usually, a spectrophotometer or turbidimeter is used to measure OD or turbidity.
- *Plating for Bactericidal Activity (Optional)*
  - Take samples from the test tubes at the designated intervals and streak them over agar plates to find the bactericidal activity. The agar plates should be incubated at the proper temperature for 18–24 hours.

Interpretation

- Examine the findings to ascertain how the bacterial population has changed over time.
- Determine the proportion of the starting population of bacteria that has been reduced at each time point.
- Ascertain if the antibacterial agent has bactericidal or bacteriostatic properties.

Recording and Reporting

- Summarize the findings, mentioning when bactericidal action was seen or when bacterial growth was reduced.

Important details on the kinetics of the antibacterial agent's activity are provided by the time-kill assay. In order to choose the best treatment plans and assess the effectiveness of antimicrobial drugs, it is essential for researchers and physicians to comprehend how quickly and efficiently the agent can suppress bacterial growth.

6.5 ANTIBACTERIAL ACTIVITY OF SPINEL FERRITES

Because iron sequestration is a common defense strategy used by many species, iron plays a major role in development of microbial diseases. Unexpected resistance to a number of antibiotics is also shown by microorganism's resistant to Fe^{+2} and Fe^{+3}. It has been demonstrated that MFe_2O_4 NPs (where M denotes metal) have outstanding antibacterial capabilities against a variety of bacteria and fungus. It has been discovered that *Escherichia coli* largely develops resistance to excess iron by altering its iron absorption mechanism. Because $CoFe_2O_4$ NPs can fight a variety of human illnesses, they are often employed in biological applications. Their efficaciousness against detrimental bacteria stems from their nano size and greater surface area-to-volume ratio. When coupled with streptomycin, a nanocomposite known as $CoFe_2O_4$@SiO_2@Ag has shown enhanced antibacterial activity against bacteria. Because the composite has a magnetic core, Ag NPs on $CoFe_2O_4$@SiO_2 inhibit aggregation and facilitate simple removal from the solution by an external magnetic field during disinfection. By adhering to the membranes of microorganisms, $CoFe_2O_4$ increases bacterial cell division, lengthens lag stage of bacterial development, and increases duration of microorganism production. The formation of antibacterial characteristics by superoxide and H_2O_2 is the reason behind antibacterial activity of $CoFe_2O_4$ NPs. The ferrite NPs' antibacterial action is caused by these efficient oxides migrating across bacterial cell membranes. It has been discovered that $ZnFe_2O_4$ NPs have substantial antibacterial action against a variety of bacterial species, including the Gram-negative *Aeromonas hydrophila* and *Vibrio harveyi* as well as the Gram-positive *Lactobacillus*, *E. coli*, and *Bacillus cereus*. It was found that the NPs' tiny crystallite size and surface area were related to their antibacterial activity (Ewunkem et al. 2021) (Ashour et al. 2018). Table 6.1 tabulate antibacterial activity of some of the ferrites along with the tested pathogen strains.

$CoFe_2O_4$ NPs' antibacterial efficacy against several bacterial species, including *B. cereus, Staphylococcus aureus, and Pseudomonas aeruginosa*, was investigated by D. Gheidari et al. The observations demonstrated that antibacterial activity of produced $CoFe_2O_4$ NPs supported the encouraging antibacterial potential of the nanoparticles at various doses. *P. aeruginosa* was shown to be the most resistant bacterium in the growth suppression zone of 12 mm in diameter at 0.48 mg/mL concentration of NPs, as demonstrated by the disk diffusion test, which yielded similar results to other antibacterial tests. Additionally, *B. cereus* proved to be the most sensitive bacterium, exhibiting a growth inhibition zone with a diameter of 12 mm at 0.06 mg/mL concentration of NPs (Gheidari et al. 2020).

6.5.1 Cobalt Ferrites

Thakur et al. investigated various properties of Cu^{2+} ions substituted Co-nano-ferrites (Thakur et al. 2012; Thakur et al. 2023). El-Khawaga *et al.* studied the antibacterial activities of $CoFe_2O_4$ taking Erythromycin as control antibiotic using disk agar distribution method. ZOI values for *S. aureus* and *E. coli* were discovered to be 7.5 and 9.7 mm, respectively, in comparison to 17.4 and 11.6 mm for the control antibiotic. The synthesized capsaicin-coated $CoFe_2O_4$ nanocomposite showed enhanced antibacterial activity against *S. aureus* (ZOI: 23.5 mm) and *E. coli* (ZOI: 17.0 mm), as per the in-vitro ZOI result. It is noteworthy

TABLE 6.1
Reported Antibacterial Activity of Different Spinel Ferrites Nanocomposites

S. No	Nano-Ferrites	Pathogens	Control (Standard Antibiotic)	ZOI Control (mm)	ZOI test NPs (mm)	Reference
1	$CoFe_2O_4$	*Escherichia coli*	Amoxicillin/clavulanic acid (10µg/mL)	17±0.57	9±0	Ashour et al. (2018)
2	$CoFe_2O_4$	*E. coli*	Erythromycin (10µg/mL)	11.6	9.7	El-Khawaga et al. (2023)
3	$CoFe_2O_4$ 0.06mg/dL	*Bacillus cereus*	–	–	12	Gheidari et al. (2020)
4	$Cu_xCo_{1-x}Fe_2O_4$	*Bacillus subtilis*	Amoxicillin/clavulanic acid (10µg/mL)	14±0.57	9±0 .57	Ashour et al. (2018)
5	$CoFe_2O_4$	*Staphylococcus aureus*	Erythromycin (10µg/mL)	17.4	7.5	El-Khawaga et al. (2023)
6	$ZnFe_2O_4$ (100µg/mL)	*S. aureus*	1% silver sulfadiazine cream	20±1.2	13±0.9	Morais et al. (2021)
7	$Zn_xCo_{1-x}Fe_2O_4$	*S. aureus*	Amoxicillin/clavulanic acid (10µg/mL)	16±1	13±0.57	Ashour et al. (2018)
8	$ZnFe_2O_4$ at 800°C	*E. coli*	Chloramphenicol	20	7.8	Morais et al. (2021)
9	$Mn_xCo_{1-x}Fe_2O_4$	*Pseudomonas aeruginosa*	Amoxicillin/clavulanic acid (10µg/mL)	21±0.57	10±0.5	Ashour et al. (2018)

that the $CoFe_2O_4$ nanocomposite coated with capsaicin exhibited notably more antibacterial activity than $CoFe_2O_4$ NPs. The idea that the $CoFe_2O_4$ nanocomposite coated with capsaicin was highly effective against Gram +ve bacteria than Gram −ve bacteria is noteworthy. The cell walls of Gram −ve bacteria are composed of lipid layers, lipopolysaccharide, and peptidoglycan, in contrast to Gram-positive bacteria, which combine very compact peptidoglycan forms (El-Khawaga et al. 2023).

Cu-substituted cobalt ferrite ($Cu_xCo_{1-x}Fe_2O_4$) was researched by Alireza Samavati *et al.* They found that, as shown by the particle size studies, NPs with greater Cu contents had smaller crystalline sizes and bigger surface-to-volume ratios. As a result, larger surface area increases their bactericidal effectiveness, which raises the rate at which *E. coli* bacteria are killed (Samavati & Ismail 2017). In their investigation of antimicrobial activity of metal-substituted Co ferrite NPs against *E. coli, Bacillus subtilis, P. aeruginosa, S. aureus, and Candida albicans*, A.H. Ashour et al. found that Zn-Co ferrite NPs were the most effective against all pathogens they studied, and Mn ferrite NPs coming at second place. Zn-Co ferrite NPs are more effective against Gram −ve than Gram +ve bacteria's. After 50, 100, and 150 kGy of gamma irradiation, Zn-Co ferrite NPs activity increased to 150.0 kGy. Zn-Co ferrite NPs that had been exposed to radiation exhibited greater efficacy against *P. aeruginosa* (15.0 mm ZOI) and *S. aureus* (16.0 mm ZOI). The decrease in Zn-Co ferrite NPs' crystallite size (from 14.64 to 12.33 nm) was the cause of this. The enhanced and promising activity of Zn-Co ferrite NPs against evaluated harmful microorganisms was also significantly and critically influenced by their surface area (Ashour et al. 2018).

6.5.2 Zinc and Copper Ferrites

Morais et al. synthesized $ZnFe_2O_4$ for studying the antibacterial activity of nanocomposite against strains of *S. aureus* and *E. coli* using agar disk diffusion method and concluded greater inhibition efficiency of $ZnFe_2O_4$ (Morais et al. 2021). In order to explore if damage of the bacterial cell membrane might result in the leaking of cytoplasmic chemicals, researchers produced $CuFe_2O_4$ and $ZnFe_2O_4$ NPs using the co-precipitation process. The results showed that $ZnFe_2O_4$ and $CuFe_2O_4$ NPs could kill *S. aureus* and *E. coli*, respectively. Gram −ve bacteria *E. coli* was more susceptible to the antibacterial effects of $ZnFe_2O_4$ NPs, while Gram +ve bacteria *S. aureus* was more vulnerable to $CuFe_2O_4$ NPs. The structures of cell wall in Gram −ve bacteria have a thickness of less than 15 nm. According to Haghniaz et al., Gram +ve bacteria are primarily composed of very thick peptidoglycans with cell wall that act as protective boundary. Antibacterial and wound healing-promoting effects of zinc ferrite NPs are greater due to their effectiveness in cell rupturing as shown in Figure 6.12 (Haghniaz et al. 2021). The findings demonstrated that the application of $CuFe_2O_4$ and $ZnFe_2O_4$ NPs may damage bacterial cell membranes,

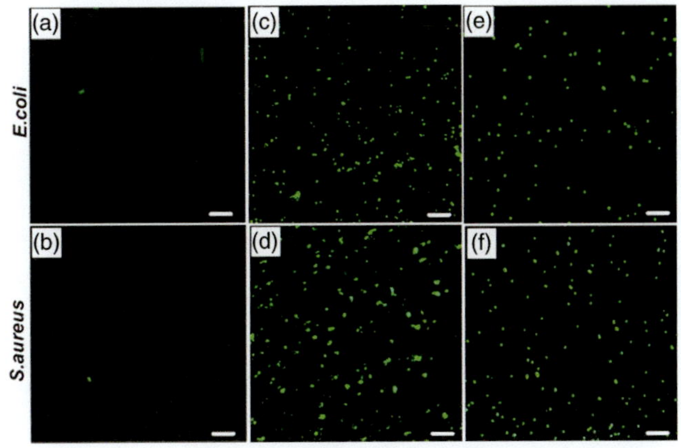

FIGURE 6.12 Membrane permeability test. Fluorescence pictures of *Staphylococcus aureus* and *Escherichia coli* demonstrating the inflow of FITC following treatment with tetracycline (positive control) and zinc ferrite ($ZnFe_2O_4$) NPs. (a) *E. coli* left untreated (negative control), (b) *S. aureus* left untreated (negative control), (c) Tetracycline-treated *E. coli* (positive control), (d) Tetracycline-treated *S. aureus* (positive control), (e) *E. coli* given 100 µg/mL $ZnFe_2O_4$ NPs treatment, (f) $ZnFe_2O_4$ NPs treated at 100 µg/mL with *S. aureus*. Scale bars indicate 100 µm (Haghniaz et al. 2021).

resulting in membrane rupture in bacteria. Figure 6.13 demonstrates membrane permeability test with and without $ZnFe_2O_4$ treatment. Bacteria's experienced membrane degradation from tetracycline and $ZnFe_2O_4$ NPs, which caused the membranes to become permeable to green-fluorescent FITC dye.

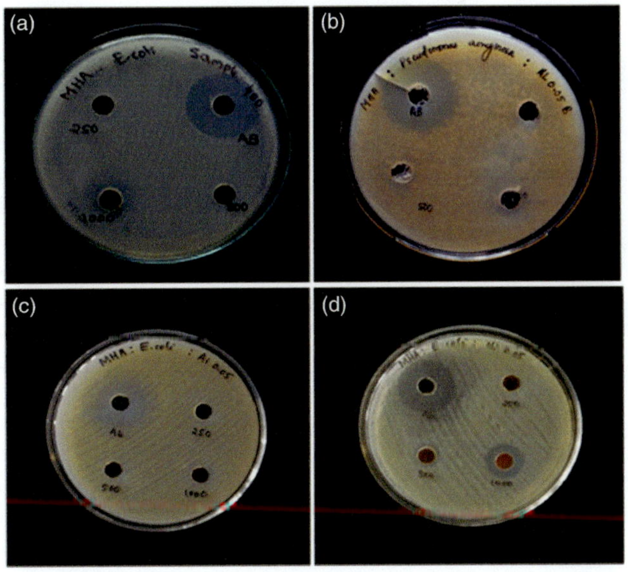

FIGURE 6.13 The agar well-cut diffusion technique to measure the antibacterial activity against *Escherichia coli* of (a–d) ZFO at 400°C, *Pseudomonas aeruginosa* of ZAFO, Antibacterial effect against *E. coli* of ZAFO (0.05), Antibacterial effect against *E. coli* of ZNFO, respectively. (Reprinted Jyothish and Jacob (2023), with permission from Elsevier License No. 5875881210455).

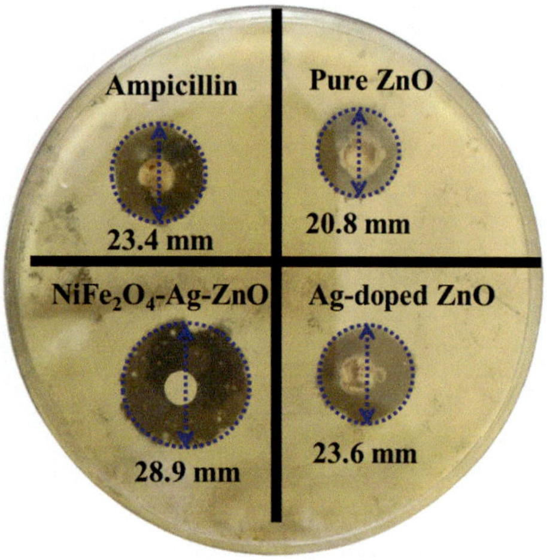

FIGURE 6.14 Comparison of antibacterial activity of ZnO, Ag-doped ZnO, and $NiFe_2O_4$-Ag-ZnO against *Escherichia coli* by agar well diffusion assay. (Reprinted from Singh et al. (2024) with permission from Elsevier vide License number 5876440121203).

6.5.3 Silver Ferrites

Singh et al. prepared a hybrid $NiFe_2O_4$-Ag-ZnO nanocomposite and reported 123.5% antibacterial activity compared to ampicillin antibiotic against bacterial strains of *E. coli* as shown in Figure 6.14 (Singh et al. 2024). Hani Nasser Abdelhamid et al. observed how *P. aeruginosa* and *S. aureus* changed morphologically in relation to concentration (0, 25, 50, 75, and 100 μg/mL) (Abdelhamid et al. 2015). TEM images of *P. aeruginosa* with bare $AgFeO_2$ at high concentrations (75 and 100 μg/mL) demonstrate total rupture of the cell structure. The bacteria had asymmetrical forms and the cell membrane was seriously damaged. According to Erika Fantozzi et al., as long as the powder grains were nanometric, the dopant ion Ag added to the spinel lattice may assist in boosting the antibacterial activity. The researchers studied $MgFe_2O_4$ and Ag-doped $MgFe_{2-x}Ag_xO_4$ NPs and studied antibacterial activity toward *E. coli* and *S. aureus* (Fantozzi et al. 2021). Incorporating silver into ferrite materials can be a method to take use of the well-established strong antibacterial characteristics of silver ions for particular applications. When employing materials containing silver, certain considerations about safety and regulations must be made, particularly when the materials come into contact with people or the environment.

6.6 CHALLENGES, FUTURE PERSPECTIVES, AND CONCLUSIONS

The last several decades have seen a notable increase in interest in nanoscale spinel ferrites because of their unique properties, which include strong antibacterial activity and stability in both chemical and biological settings. The right characteristics of magnetic spinel ferrites, along with common functionalizations and modifications, open up new possibilities for their application as antibacterial agents, biomedical devices, and water treatment technologies in the future. The future trend in applications for magnetic spinel ferrites will center on manipulating necessary magnetic characteristics by adjusting synthesis techniques, leading to the optimization of particle size, shape, and crystallinity. In addition to synthesis techniques and modification, doping is another useful technique for producing non-aggregated and mono-disperse nano-sized spinel ferrites, which affect their physical and chemical characteristics. Magnetic spinel ferrites are effective magnetic materials that can be used to treat magnetic hyperthermia. Even though spinel ferrite-based materials have been used in several studies to treat hyperthermia, the field is still in its infancy, and more research is needed to address some of the problems, such as adjusting the shape, size, and magnetic characteristics of NPs. In order to develop the photocatalytic properties of SF NPs for the organic pollutants degradation with simultaneous protection against pathogens, some challenges must be addressed in future studies.

It's crucial to remember that iron ferrites direct antibacterial action frequently stems from their conjunction with other tactics, including medication delivery. Even though iron ferrites are usually thought to be biocompatible, using them in environmental or medical applications requires regulatory permits as well as safety concerns. The goal of continuing research in this area is to maximize the usage of iron ferrites for antibacterial and related applications. More often than not, antibacterial action is linked to other minerals like copper, silver, or certain chemical compounds that have been shown to stop the development of bacteria. These substances have the potential to harm bacteria by releasing ions or interfering with the membranes of bacterial cells. Regulatory clearances can also be necessary if they are to be used in consumer or medical goods. Particularly when employing spinel ferrite NPs, safety concerns and possible toxicity should be thoroughly discussed. For these materials to be used in industrial, medicinal, or environmental settings, they must be biocompatible and long-lastingly safe.

REFERENCES

Abdelhamid, Hani Nasser, Talib, Abou, and Wu, Hui Fen. 2015. "Facile Synthesis of Water Soluble Silver Ferrite ($AgFeO_2$) Nanoparticles and Their Biological Application as Antibacterial Agents." *RSC Advances* 5 (44): 34594–602. https://doi.org/10.1039/c4ra14461a.

Ashour, A. H., El-Batal, Ahmed I., Abdel Maksoud, M. I.A., El-Sayyad, Gharieb S., Labib, Sh, Abdeltwab, E., and El-Okr, M. M.. 2018. "Antimicrobial Activity of Metal-Substituted Cobalt Ferrite Nanoparticles Synthesized by Sol–Gel Technique." *Particuology* 40 (October): 141–51. https://doi.org/10.1016/j.partic.2017.12.001.

Chahar, Deepika, Thakur, Preeti, Kumar, Rakesh, Ravelo, Blaise, and Thakur, Atul. 2022. "Nickel Substituted Co–Zn Nanoferrites: An Investigation of Structural, Electrical and Magnetic Properties." *SSRN Electronic Journal* (October). https://doi.org/10.2139/SSRN.4236202.

El-Khawaga, Ahmed M., Elsayed, Mohamed A., Fahim, Yosri A., and Shalaby, Rasha E. 2023. "Promising Photocatalytic and Antimicrobial Activity of Novel Capsaicin Coated Cobalt Ferrite Nanocatalyst." *Scientific Reports* 13 (1): 1–15. https://doi.org/10.1038/s41598-023-32323-y.

Ewunkem, Akamu J., Rodgers, Lashunta, Campbell, Daisha, Staley, Constance, Subedi, Kiran, Boyd, Sada, and Graves, Joseph L. 2021. "Experimental Evolution of Magnetite Nanoparticle Resistance in *Escherichia coli*." *Nanomaterials* 11 (3): 1–22. https://doi.org/10.3390/nano11030790.

Fantozzi, Erika, Rama, Erlinda, Calvio, Cinzia, Albini, Benedetta, Galinetto, Pietro, and Bini, Marcella. 2021. "Silver Doped Magnesium Ferrite Nanoparticles: Physico-Chemical Characterization and Antibacterial Activity." *Materials* 14 (11): 2859. https://doi.org/10.3390/ma14112859.

Garcia-Muñoz, Patricia, Fresno, Fernando, de la Peña O'Shea, Víctor A., and Keller, Nicolas. 2020. *Ferrite Materials for Photoassisted Environmental and Solar Fuels Applications*, 107–62. Springer, Cham. https://doi.org/10.1007/978-3-030-49492-6_4.

Gheidari, Davood, Mehrdad, Morteza, Maleki, Saloomeh, and Hosseini, Samanesadat. 2020. "Synthesis and Potent Antimicrobial Activity of $CoFe_2O_4$ Nanoparticles under Visible Light." *Heliyon* 6 (10). https://doi.org/10.1016/j.heliyon.2020.e05058.

Haghniaz, Reihaneh, Rabbani, Atiya, Vajhadin, Fereshteh, Khan, Taous, Kousar, Rozina, Khan, Abdul Rehman, Montazerian, Hossein, et al. 2021. "Anti-Bacterial and Wound Healing-Promoting Effects of Zinc Ferrite Nanoparticles." *Journal of Nanobiotechnology* 19 (1): 1–15. https://doi.org/10.1186/s12951-021-00776-w.

Jyothish, B., and Jacob, John. 2023. "Synthesis and Characterization of Ni^{2+} and Al^{3+} Doped Zinc Ferrite Nanoparticles for Antibacterial, Antioxidant, and Anticancer (MCF-7) Analysis." *Chemical Physics Impact* 6 (June): 100209. https://doi.org/10.1016/j.chphi.2023.100209.

Mattei, Jean Luc, Huitema, Laure, Queffelec, Patrick, Pintos, Jean François, Minard, Philippe, Sharahia, Ala, Jamnier, Benjamin, et al. 2011. "Suitability of Ni-Zn Ferrites Ceramics with Controlled Porosity as Granular Substrates for Mobile Handset Miniaturized Antennas." *IEEE Transactions on Magnetics* 47 (10): 3720–23. https://doi.org/10.1109/TMAG.2011.2148109.

Mills, Allan A. 2004. "The Lodestone: History, Physics, and Formation." *Annals of Science* 61 (3): 273–319. https://doi.org/10.1080/00033790310001642812.

Morais, Dielly Oliveira, Pancotti, Alexandre, de Souza, Guilherme Sastre, Saivish, Marielena Vogel, Braoios, Alexandre, Moreli, Marcos Lázaro, Souza, Mauro Vinícius de B., da Costa, Vivaldo G., and Wang, Jiale. 2021. "Synthesis, Characterization, and Evaluation of Antibacterial Activity of Transition Metal Oxyde Nanoparticles." *Journal of Materials Science: Materials in Medicine* 32 (9): 1–15. https://doi.org/10.1007/s10856-021-06578-8.

Naagar, Manish, Chalia, Sonia, Kharbanda, Saarthak, Thakur, Preeti, and Thakur, Atul. 2023. "Correlating Tailored Saturation Magnetization of Normal-Inverse Spinels with Noncollinear Magnetic Ordering." *Journal of Materials Science: Materials in Electronics* 34 (6): 1–13. https://doi.org/10.1007/S10854-023-09828-3.

Pathania, Abhilash, Thakur, Preeti, Trukhanov, Alex V., Trukhanov, Sergei V., Panina, Larissa V., Lüders, Ulrike, and Thakur, Atul. 2019. "Development of Tungsten Doped Ni-Zn Nano-Ferrites with Fast Response and Recovery Time for Hydrogen Gas Sensing Application." *Results in Physics* 15 (December): 102531. https://doi.org/10.1016/J.RINP.2019.102531.

Punia, Pinki, Dhar, Rakesh, Ravelo, Blaise, Trukhanov, Alex V., Panina, Larissa V., Thakur, Preeti, and Thakur, Atul. 2021. "Microstructural, Optical and Magnetic Study of Ni–Zn Nanoferrites." *Journal of Superconductivity and Novel Magnetism* 34 (8): 2131–2140. https://doi.org/10.1007/s10948-021-05967-y.

Saini, Ashish, Thakur, Atul, and Thakur, Preeti. 2016. "Matching Permeability and Permittivity of $Ni_{0.5}Zn_{0.3}Co_{0.2}In_{0.1}Fe_{1.9}O_4$ Ferrite for Substrate of Large Bandwidth Miniaturized Antenna." *Journal of Materials Science: Materials in Electronics* 27 (3): 2816–23. https://doi.org/10.1007/S10854-015-4095-8.

Samavati, Alireza, and Ismail, A. F. 2017. "Antibacterial Properties of Copper-Substituted Cobalt Ferrite Nanoparticles Synthesized by Co-Precipitation Method." *Particuology* 30 (February): 158–63. https://doi.org/10.1016/j.partic.2016.06.003.

Singh, Amitender, Wan, Fayu, Yadav, Kavita, Kharbanda, Saarthak, Thakur, Preeti, and Thakur, Atul. 2024. "A Novel Magnetic $NiFe_2O_4$-Ag-ZnO Hybrid Nanocomposite for the Escalated Photocatalytic Dye Degradation and Antibacterial Activities." *Materials Science and Engineering: B* 299 (January): 116935. https://doi.org/10.1016/j.mseb.2023.116935.

Singh, Amitender, Wan, Fayu, Yadav, Kavita, Salvi, Anand, Thakur, Preeti, and Thakur, Atul. 2023. "Synergistic Effect of ZnO Nanoparticles with Cu^{2+} Doping on Antibacterial and Photocatalytic Activity." *Inorganic Chemistry Communications* 157 (November): 111425. https://doi.org/10.1016/j.inoche.2023.111425.

Taneja, Shilpa, Chahar, Deepika, Thakur, Preeti, and Thakur, Atul. 2021. "Influence of Bismuth Doping on Structural, Electrical and Dielectric Properties of Ni–Zn Nanoferrites." *Journal of Alloys and Compounds* 859 (April): 157760. https://doi.org/10.1016/J.JALLCOM.2020.157760.

Thakur, Atul, Thakur, Preeti, and Hsu, Jen Hwa. 2012. "Magnetic Behaviour of $Ni_{0.4}Zn_{0.6}Co_{0.1}Fe_{1.9}O_4$ Spinel Nano-Ferrite." *Journal of Applied Physics* 111 (7): 1–4. https://doi.org/10.1063/1.3670606.

Thakur, Atul, Thakur, Preeti, and Paul Khurana, S. M. 2022. *Synthesis and Applications of Nanoparticles*. Springer Nature. https://doi.org/10.1007/978-981-16-6819-7.

Thakur, Preeti, Taneja, Shilpa, Chahar, Deepika, Ravelo, Blaise, and Thakur, Atul. 2021. "Recent Advances on Synthesis, Characterization and High Frequency Applications of Ni–Zn Ferrite Nanoparticles." *Journal of Magnetism and Magnetic Materials* 530 (July): 1–20. https://doi.org/10.1016/j.jmmm.2021.167925.

Thakur, Atul, Verma, Ritesh, Wan, Fayu, Ravelo, Blaise, Edelman, Irina, Ovchinnikov, Sergei, and Thakur, Preeti. 2023. "Investigation of Structural, Elastic and Magnetic Properties of Cu^{2+} Ions Substituted Cobalt Nano Ferrites." *Journal of Magnetism and Magnetic Materials* 581 (September): 170980. https://doi.org/10.1016/j.jmmm.2023.170980.

Valenzuela, Raúl. 2012. "Novel Applications of Ferrites." *Physics Research International*. https://doi.org/10.1155/2012/591839.

7 Antifungal Activity of Spinel Nano-Ferrites

Atul Thakur and Arti Rao*
Amity Institute of Nanotechnology, Amity University Haryana, Gurugram, India

Muhammad Ehsan
CY Cergy-Paris Université, École CY-Tech, Pau, France

Ritesh Verma
Graphic Era (Deemed to be) University, Dehradun, Uttarakhand, India

Preeti Thakur
Nanotechnology Wing, Innovative Science Research Society, Shimla, India
nanoLatticeX LLP, Manesar, Gurugram, India
Amity Institute of Nanotechnology, Amity University Haryana, Gurugram, India

7.1 INTRODUCTION

Fungal infections, also known as mycoses, pose a significant threat to human health, agriculture, and the environment. The rise in fungal pathogens and their growing resistance to conventional antifungal drugs have led to a pressing need for innovative solutions in the field of antifungal research. In recent years, nanotechnology has emerged as a promising avenue for combating fungal infections. Among the various nanomaterials investigated for their antifungal properties, spinel nano-ferrites have gained considerable attention due to their unique characteristics and potential applications. The choice to focus on spinel nano-ferrites in this chapter is driven by their exceptional properties. Spinel ferrites, such as zinc ferrite ($ZnFe_2O_4$), cobalt ferrite ($CoFe_2O_4$), and manganese ferrite ($MnFe_2O_4$), are a class of materials known for their remarkable magnetic and structural attributes. These attributes include high surface area, ease of synthesis, chemical stability, and biocompatibility, making them a versatile platform for various applications, including biomedical ones.[1]

The primary objective of this chapter is to explore the antifungal activity of spinel nano-ferrites, shedding light on their potential as novel antifungal agents. As conventional antifungal drugs face limitations in terms of efficacy and resistance development, nano-ferrites offer a fresh perspective. By elucidating their antifungal mechanisms and assessing their performance against a range of fungal species, we aim to contribute to the development of new, more effective antifungal strategies. We will begin by delving into the broader context of fungal infections, by first discussing the prevalence of fungal diseases in both clinical and agricultural settings, emphasizing the need for alternative treatments. Subsequently, we will introduce the fundamental properties of spinel nano-ferrites, highlighting why they are a promising candidate for antifungal research. Their magnetic properties, high surface area, and tunable surface chemistry make them an intriguing platform for interaction with fungal cells. The following sections will outline the methodology employed in this study, encompassing the synthesis of spinel nano-ferrites, characterization techniques, and antifungal testing protocols. We will then present the results of our experiments, detailing the antifungal activity of different spinel nano-ferrites against a variety of fungal strains. This will be followed by a discussion of the underlying mechanisms through which these nanomaterials exert their antifungal effects.

Ultimately, this chapter aims to provide a comprehensive understanding of the potential of spinel nano-ferrites as antifungal agents, paving the way for further research and development in this exciting field. By addressing the pressing need for novel antifungal strategies, we hope to contribute to the improvement of fungal infection management and the reduction of associated healthcare and agricultural challenges.

7.2 CATALYZING ANTIFUNGAL SOLUTIONS: SPINEL NANO-FERRITES

Spinel nano-ferrites are a fascinating class of nanomaterials that have garnered significant attention in recent years due to their diverse range of applications, including their remarkable antifungal properties. These nanoferrites are typically composed of transition metal ions,

* Corresponding author.

DOI: 10.1201/9781003568940-7

such as iron, in a spinel crystal structure, which imparts unique magnetic and chemical properties to them. The synthesis of spinel nano-ferrites involves various techniques, including sol-gel, co-precipitation, and hydrothermal methods, allowing for precise control over their size, composition, and morphology.[2] This tunability is essential in tailoring their antifungal activity. One of the key advantages of spinel nano-ferrites is their high surface area-to-volume ratio, which enhances their interactions with fungal cells.

Research has shown that spinel nano-ferrites, particularly those doped with different metal ions like zinc, cobalt, or manganese, exhibit remarkable antifungal properties. These nano-ferrites can inhibit fungal growth through multiple mechanisms, including oxidative stress, disruption of cell membranes, and interference with vital metabolic processes. Their ability to generate reactive oxygen species (ROS) plays a crucial role in damaging fungal cell components, making them a potent antifungal agent.

Spinel nano-ferrites' unique magnetic properties also contribute to their antifungal effectiveness. When exposed to an external magnetic field, these nano-ferrites can be guided to specific fungal infection sites, enhancing their localized antifungal activity. This targeted approach minimizes collateral damage to healthy tissues and reduces the risk of systemic side effects. Furthermore, the biocompatibility of spinel nano-ferrites is a significant advantage for their potential medical applications. Studies have demonstrated that these nano-ferrites exhibit minimal toxicity to mammalian cells, making them suitable for various biomedical applications, including antifungal therapies. Incorporating spinel nano-ferrites into different drug delivery systems, such as nanoparticles or coatings on medical devices, holds promise for combating fungal infections. These materials can release their antifungal agents gradually, ensuring sustained therapeutic effects while minimizing the risk of drug resistance development.

The development of spinel nano-ferrites with enhanced antifungal properties continues to be an active area of research. Scientists are exploring novel doping strategies, surface modifications, and functionalization techniques to further improve their efficacy against a wide range of fungal pathogens.[3] Additionally, the potential synergistic effects of combining spinel nano-ferrites with existing antifungal drugs are being investigated to develop more effective treatments. Spinel nano-ferrites are a promising class of nanomaterials with remarkable antifungal properties. Their unique chemical, magnetic, and biocompatible characteristics make them suitable candidates for various biomedical applications aimed at combating fungal infections. As research in this field advances, we can anticipate the development of innovative antifungal therapies that harness the full potential of spinel nano-ferrites while prioritizing patient safety and treatment efficacy.

7.3 UNIQUE PROPERTIES OF SPINEL NANO-FERRITES WHICH UNDERPIN THEIR ANTIFUNGAL ACTIVITY

Spinel nano-ferrites possess a range of properties; these properties stem from their crystal structure, composition, and surface characteristics as shown in Figure 7.1. In this discussion, we will explore nine key properties of spinel nano-ferrites that contribute to their antifungal capabilities.

- *Chemical Composition*: Spinel nano-ferrites consist of transition metal ions, typically iron, combined with other metal ions like zinc, cobalt, or manganese. This diverse composition enables the fine-tuning of their antifungal properties through doping, where different metal ions can impart specific attributes that enhance their efficacy against fungi.
- *High Surface Area*: Spinel nano-ferrites possess a high surface area-to-volume ratio due to their nanoscale size. This feature increases their contact with fungal cells, facilitating greater interactions and improving their antifungal effectiveness.
- *ROS Generation*: Spinel nano-ferrites are known to generate ROS when exposed to external stimuli or in the presence of fungal cells. These ROS, including superoxide radicals and hydrogen peroxide, induce oxidative stress in fungal cells, leading to cell damage and death.
- *Disruption of Fungal Membranes*: The interaction between spinel nano-ferrites and fungal cells can lead to the disruption of fungal cell membranes. This disruption compromises the integrity of the cell, allowing for the leakage of vital cellular components and ultimately leading to fungal cell death.
- *Interference with Metabolic Pathways*: Spinel nano-ferrites can interfere with crucial metabolic pathways within fungal cells. By disrupting these pathways, they inhibit the growth and proliferation of fungi, further enhancing their antifungal activity.
- *Targeted Magnetic Delivery*: The inherent magnetic properties of spinel nano-ferrites allow for targeted delivery to specific fungal infection sites when subjected to an external magnetic field. This targeted approach ensures that the antifungal treatment is concentrated where it is needed most, minimizing collateral damage to healthy tissues.
- *Biocompatibility*: Spinel nano-ferrites are generally biocompatible and exhibit minimal toxicity to mammalian cells. This characteristic is crucial for their potential medical applications, ensuring that they do not harm healthy tissues while effectively combating fungal infections.

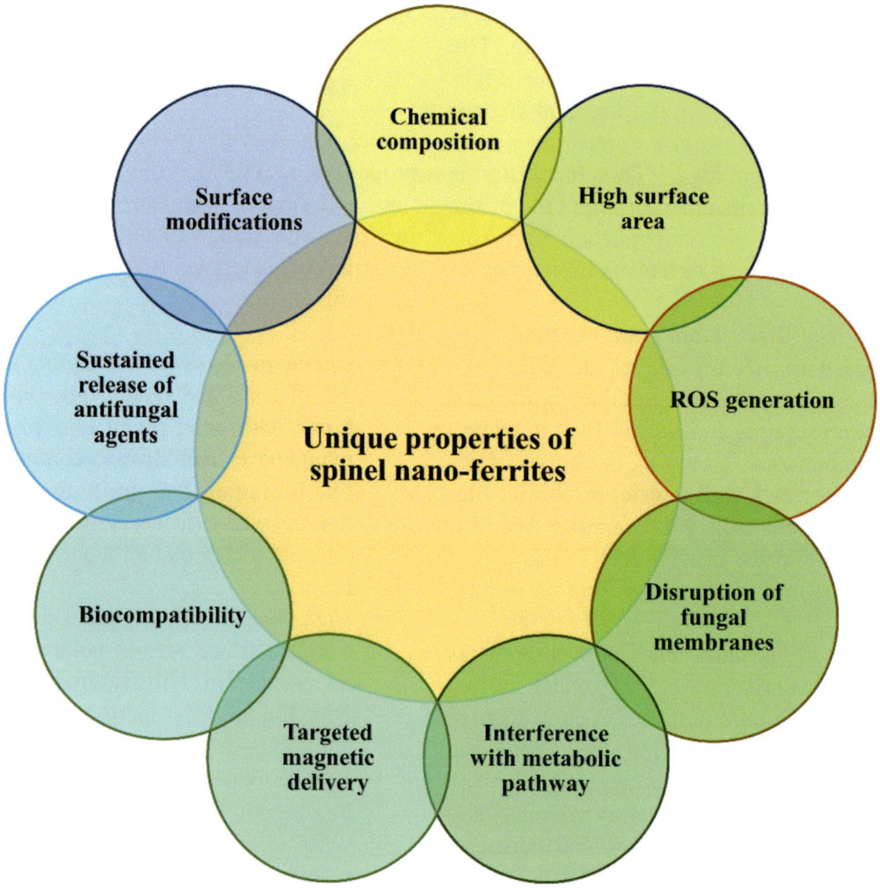

FIGURE 7.1 Unique properties of spinel nano-ferrites.

- *Sustained Release of Antifungal Agents*: Spinel nano-ferrites can be incorporated into drug delivery systems, such as nanoparticles or coatings on medical devices. These materials can gradually release their antifungal agents, providing sustained therapeutic effects over an extended period while reducing the risk of drug resistance.
- *Surface Modifications*: Researchers can modify the surface of spinel nano-ferrites to enhance their antifungal properties. Surface functionalization with specific molecules or ligands can increase their specificity for fungal cells and improve their overall performance as antifungal agents.

The multifaceted antifungal activity of spinel nano-ferrites arises from their unique combination of chemical composition, high surface area, ROS generation, membrane disruption, interference with metabolic pathways, magnetic targeting, biocompatibility, sustained release capabilities, and surface modifications.[4] These properties make spinel nano-ferrites a promising candidate for the development of advanced antifungal therapies, with the potential to address fungal infections more effectively while minimizing adverse effects on the patient.

7.4 COMMON CHALLENGES WITH ANTIFUNGAL AGENTS

Antifungal agents play a crucial role in combating fungal infections, but they often come with challenges that spinel nano-ferrites have the potential to overcome. Below, we'll discuss several common challenges associated with antifungal agents and how spinel nano-ferrites can address them; one significant challenge in antifungal therapy is the development of drug resistance by fungal pathogens.[5] Over time, fungi can adapt and become less responsive to conventional antifungal drugs. Spinel nano-ferrites, with their multiple mechanisms of action, including ROS generation and disruption of fungal membranes, can provide a new approach to combat drug-resistant fungi effectively. Limited drug solubility: Many antifungal drugs suffer from limited solubility, which can lead to poor bioavailability and therapeutic effectiveness. Spinel nano-ferrites can be functionalized to serve as nanocarriers for these drugs, improving their solubility and ensuring their controlled release at the infection site.

Some antifungal drugs can have toxic effects on healthy tissues when administered systemically. By using spinel nano-ferrites as drug carriers, it is possible to achieve targeted delivery to the fungal infection site, minimizing exposure to healthy cells and reducing systemic toxicity. Fungal biofilms are complex communities of fungi that are notoriously resistant to antifungal agents. Spinel nano-ferrites can be engineered to disrupt fungal biofilms through their membrane-disrupting properties and oxidative stress-inducing capabilities, making them effective against biofilm-associated infections. Also, some fungi can form protective structures that hinder the penetration of antifungal drugs. Spinel nano-ferrites, especially when functionalized with targeting ligands, can improve fungal penetration and access hard-to-reach infection sites, enhancing their antifungal activity. There is a limited repertoire of antifungal agents available, and new drug development is slow. Spinel nano-ferrites can serve as a platform for the delivery of various antifungal agents, allowing for a more diverse range of treatment options and potentially overcoming the limitations of current drugs.[6] Traditional antifungal drugs may have adverse side effects due to their systemic distribution. The use of spinel nano-ferrites for targeted drug delivery can minimize these side effects by concentrating the therapeutic agent at the site of infection, reducing the impact on healthy tissues.

Developing and manufacturing new antifungal drugs can be expensive. Spinel nano-ferrites, once functionalized and optimized, can potentially reduce the required drug dosage, making treatment more cost-effective in the long run.

Antifungal agents face several challenges, including drug resistance, limited solubility, systemic toxicity, biofilm resistance, and limited therapeutic options. Spinel nano-ferrites offer a promising solution to these challenges by providing targeted drug delivery, improving drug solubility, overcoming biofilm resistance, and reducing systemic toxicity. Their versatility and ability to be tailored for specific applications make them a valuable tool in the fight against fungal infections, potentially improving the efficacy and safety of antifungal therapies.[7]

7.5 FUNCTIONALIZATION OF SPINEL NANO-FERRITES

Functionalization of spinel nano-ferrites is a critical aspect of enhancing their antifungal properties, as it allows researchers to tailor these nanomaterials for specific applications (Figure 7.2). This process involves modifying the surface of spinel nano-ferrites with various functional groups or molecules to improve their efficacy against fungal infections.[8] Here, we delve into eight key aspects of functionalization in the context of spinel nano-ferrites and their antifungal potential.

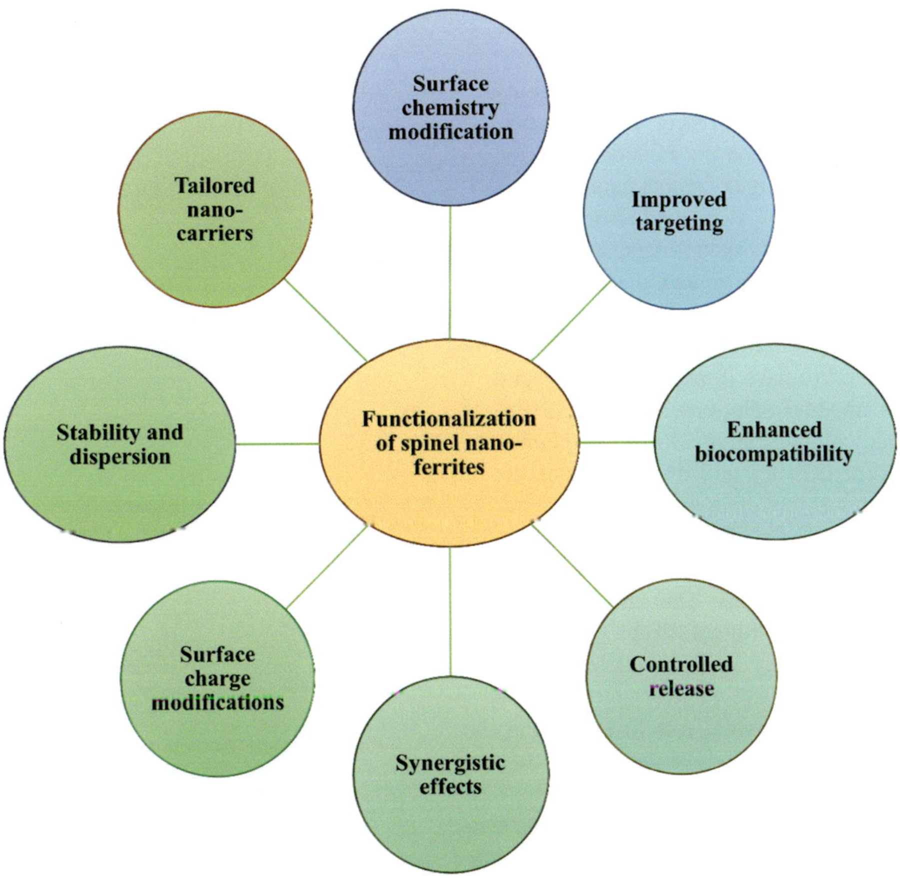

FIGURE 7.2 Key aspects of functionalization of spinel nano-ferrites and their antifungal potential.

- *Surface Chemistry Modification*: Functionalization typically starts with altering the surface chemistry of spinel nano-ferrites. This can be achieved by introducing functional groups such as amino ($-NH_2$), carboxyl ($-COOH$), or thiol ($-SH$) groups. These groups can facilitate further modification and enhance interactions with fungal cells.
- *Improved Targeting*: Functionalization allows for the incorporation of ligands or molecules that specifically target fungal cells. Ligands that recognize fungal cell receptors can increase the selectivity of spinel nano-ferrites toward pathogens while reducing their impact on healthy cells.
- *Enhanced Biocompatibility*: To ensure safe medical applications, functionalization can include biocompatible coatings that minimize cytotoxicity. Polymer coatings or surface modifications with biocompatible molecules make spinel nano-ferrites suitable for use in antifungal drug delivery systems.
- *Controlled Release*: Functionalization can also enable controlled release of antifungal agents. By attaching molecules that respond to external stimuli or enzymatic activity prevalent in fungal infections, spinel nano-ferrites can release antifungal drugs in a controlled and targeted manner, optimizing therapeutic outcomes.
- *Synergistic Effects*: Combining functionalization with other antifungal agents, such as antibiotics or natural compounds, can create synergistic effects. The functionalized nano-ferrites can act in concert with these agents to increase antifungal activity while potentially reducing the risk of drug resistance.
- *Surface Charge Modification*: Functionalization can alter the surface charge of spinel nano-ferrites. Positively charged groups can enhance interactions with negatively charged fungal cell membranes, promoting adhesion and disrupting the integrity of fungal cells.
- *Stability and Dispersion*: Functionalized spinel nano-ferrites tend to have improved stability and dispersion in aqueous solutions, which is essential for biomedical applications. Proper dispersion ensures uniform delivery and distribution of the antifungal agents at the infection site.
- *Tailored Nanocarriers*: Functionalization can transform spinel nano-ferrites into effective nanocarriers for antifungal drugs. The surface can be engineered to load and release drugs efficiently, improving drug solubility and bioavailability.

Functionalization plays a pivotal role in enhancing the antifungal properties of spinel nano-ferrites. By modifying their surface chemistry, improving targeting, ensuring biocompatibility, enabling controlled drug release, creating synergistic effects, adjusting surface charge, enhancing stability, and tailoring them as nanocarriers, researchers are able to develop highly effective antifungal therapies. These tailored nanomaterials hold significant promise in addressing fungal infections more efficiently and with fewer side effects, bringing us closer to improved treatments for fungal diseases.[9]

7.6 SYNTHESIS TECHNIQUES FOR SPINEL NANO-FERRITES AS ANTIFUNGAL AGENTS

The synthesis techniques for spinel nano-ferrites as antifungal agents are a critical aspect of research in combating fungal infections (Figure 7.3). These techniques are tailored to produce nano-ferrites with specific characteristics that enhance their antifungal properties.[10]

- *Sol-Gel Method*: The sol-gel method involves the dissolution of metal precursors in a solvent, followed by the formation of a gel. This gel is then subjected to controlled drying and heating processes to obtain nano-ferrite powders. This method allows precise control over particle size and composition, which can influence the nano-ferrites' antifungal activity.
- *Co-Precipitation*: Co-precipitation involves the simultaneous precipitation of metal ions from a solution by the addition of a precipitating agent. This method can produce nano-ferrites with a high degree of homogeneity and fine particle sizes. The choice of precipitating agent and reaction conditions can be adjusted to tailor the properties of the nano-ferrites.
- *Hydrothermal Synthesis*: Hydrothermal synthesis utilizes high-temperature and high-pressure conditions to promote the formation of nano-ferrites. This method often results in nanocrystals with unique morphologies and excellent crystallinity.

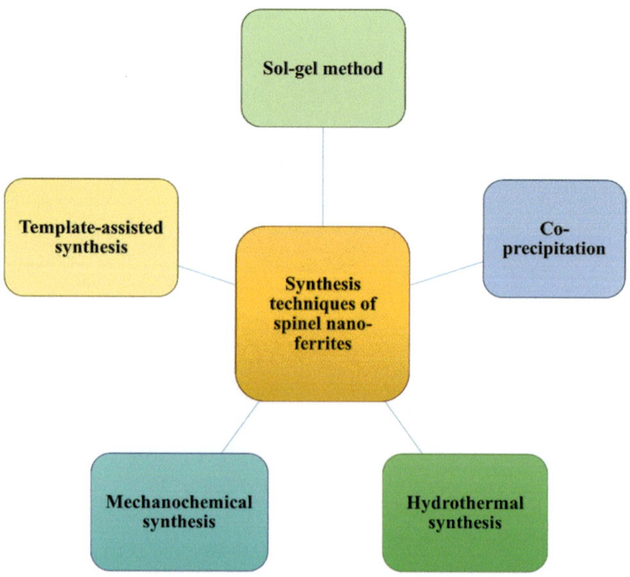

FIGURE 7.3 Synthesis techniques for spinel nano-ferrites as antifungal agents.

The controlled hydrothermal parameters play a crucial role in determining the nano-ferrites' structural properties and, consequently, their antifungal performance.

- *Mechanochemical Synthesis*: In this approach, mechanical energy is applied to precursor materials to induce chemical reactions. It offers a more environmentally friendly and energy-efficient route to produce nano-ferrites. The nanoscale mechanical forces applied during this process can lead to unique structural features that may enhance antifungal activity.
- *Template-Assisted Synthesis*: Template-assisted methods involve using templates or matrices to guide the formation of nano-ferrites with specific structures. These templates can be removed afterward, leaving behind well-defined nano-ferrite structures. This technique enables precise control over the size, shape, and porosity of the nano-ferrites.

Each of these synthesis techniques offers distinct advantages and challenges in tailoring spinel nano-ferrites for antifungal applications. Researchers often choose the most suitable method based on the desired nano-ferrite properties and intended use, underscoring the versatility of these techniques in advancing antifungal research.

7.7 MECHANISM OF ANTIFUNGAL ACTION

The mechanism of antifungal action involving spinel nano-ferrites primarily relies on their unique physical and chemical properties that hinder fungal growth, as shown in Figure 7.4.

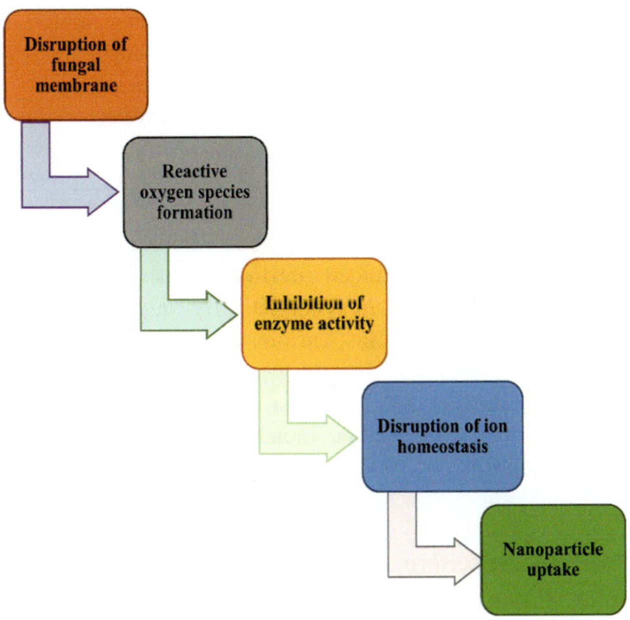

FIGURE 7.4 Mechanism of antifungal action involving spinel nano-ferrites.

7.7.1 Disruption of Fungal Membrane

The disruption of fungal membranes is a key mechanism involved in the antifungal action of spinel nano-ferrites. Spinel nano-ferrites exert their antifungal activity by interacting with the fungal cell membrane. The nano-ferrite nanoparticles, due to their unique physical and chemical properties, can adhere to the fungal cell membrane. This interaction disrupts the structural integrity of the lipid bilayer that surrounds the fungal cell.[11]

The disruption of the fungal membrane can occur through several mechanisms:

- *Physical Disruption*: The nanoparticles can physically pierce or puncture the fungal cell membrane. This physical damage compromises the membrane's ability to maintain its selective permeability and can result in the leakage of cellular contents.
- *Electrostatic Interactions*: Spinel nano-ferrites may carry a net charge, which allows them to interact electrostatically with the negatively charged components of the fungal cell membrane. This can lead to changes in membrane potential and permeability.
- *Lipid Interaction*: The nanoparticles can embed themselves within the lipid bilayer, causing alterations in the membrane's fluidity and stability. This disturbs the arrangement of lipids and proteins in the membrane, further compromising its function.
- *Formation of Pores*: In some cases, the nanoparticles can induce the formation of membrane pores or channels. These pores disrupt the membrane's barrier function and can lead to the uncontrolled movement of ions and molecules in and out of the fungal cell.

The disruption of the fungal cell membrane ultimately compromises the integrity of the cell, leading to cell death. This is a crucial step in the antifungal action of spinel nano-ferrites, as it prevents the fungus from maintaining its structural and functional integrity, ultimately inhibiting its growth and propagation.[12]

7.7.2 Reactive Oxygen Species

Creating ROS for the antifungal action of spinel nano-ferrites involves a series of chemical reactions.

Materials required

- Spinel nano-ferrite nanoparticles
- Oxygen source (e.g., atmospheric oxygen)
- Solvent (e.g., water or buffer)
- Stirring apparatus
- Reaction vessel
- pH meter
- Chemical reagents (if necessary)

Procedure

- *Preparation of Nano-Ferrite Suspension*: Weigh the desired amount of spinel nano-ferrite nanoparticles and add them to a suitable solvent (e.g., water or buffer) in a reaction vessel. The concentration can vary depending on the experimental requirements.

- *Adjusting pH (If Necessary)*: Measure the pH of the nano-ferrite suspension using a pH meter. Ensure that the pH is within the desired range for ROS production. In many cases, a near-neutral pH is suitable.
- *Oxygen Exposure*: Place the reaction vessel containing the nano-ferrite suspension under ambient conditions with access to atmospheric oxygen. Oxygen serves as the source of oxygen molecules (O_2) required for ROS generation.
- *Stirring*: Initiate stirring of the suspension to ensure uniform exposure of nano-ferrite nanoparticles to oxygen. Stirring helps in maximizing the contact between the nanoparticles and oxygen.
- *Monitoring ROS Generation*: Over time, monitor ROS generation. This can be done using various methods such as electron paramagnetic resonance (EPR) spectroscopy or by assessing the formation of specific ROS like superoxide radicals or hydroxyl radicals using suitable chemical probes.
- *Antifungal Application*: Once a sufficient amount of ROS is generated, the nano-ferrite suspension can be applied to the fungal culture. The ROS generated by the nanoparticles will interact with the fungal cells, causing oxidative stress and damage.
- *Incubation and Observation*: Incubate the fungal culture with the nano-ferrite suspension under appropriate conditions (temperature, humidity, etc.) and observe the antifungal effects. These effects may include inhibition of fungal growth, cellular damage, or cell death.
- *Analysis and Further Studies*: Analyze the results of the antifungal assay, and if necessary, conduct additional experiments to understand the specific ROS species involved and their concentrations.

The exact procedure may vary depending on the specific nano-ferrite formulation and experimental goals. Additionally, safety precautions should be taken when working with nanoparticles and ROS-producing systems in a laboratory setting.

7.7.3 Inhibition of Enzyme Activity

Some spinel nano-ferrites can inhibit the activity of key fungal enzymes required for vital metabolic processes. By binding to or altering the conformation of these enzymes, nano-ferrites disrupt the fungal cell's ability to carry out essential biochemical reactions.[13]

7.7.4 Disruption of Ion Homeostasis

Spinel nano-ferrites can also interfere with ion transport mechanisms in fungal cells. This disruption of ion homeostasis can lead to imbalances in intracellular ion concentrations, which are critical for various cellular functions. This disruption further impairs fungal cell growth and viability.

7.7.5 Nanoparticle Uptake

Fungal cells may actively take up spinel nano-ferrite nanoparticles, leading to their accumulation within the cells. This intracellular presence of nano-ferrites exacerbates their inhibitory effects on fungal growth and metabolism.

7.8 DIVERSITY OF SPINEL NANO-FERRITES USED IN ANTIFUNGAL APPLICATIONS

Spinel nano-ferrites have shown promise for their antifungal activity in various research studies. While specific formulations and applications may vary, here are some examples of spinel nano-ferrites that have been generally used for their antifungal properties:

- *Zinc Ferrite Nanoparticles*: Zinc ferrite nanoparticles, often doped with other metal ions, have been extensively studied for their antifungal potential. Their ability to generate ROS contributes to their effectiveness in damaging fungal cells and inhibiting growth.
- *Cobalt Ferrite Nanoparticles*: Cobalt ferrite nanoferrites have exhibited notable antifungal activity. Their magnetic properties make them suitable for targeted drug delivery to fungal infection sites, improving localized treatment.
- *Manganese Ferrite Nanoparticles*: Manganese ferrite nano-ferrites have been investigated for their antifungal properties due to their ability to disrupt fungal cell membranes and interfere with fungal metabolic pathways.
- *Nickel Ferrite Nanoparticles*: Nickel ferrite nano-ferrites have shown potential in combating fungal infections. They can be tailored to release antifungal agents gradually, ensuring sustained therapeutic effects.
- *Multi-Metal Doped Ferrite Nanoparticles:* Researchers have explored the doping of spinel nano-ferrites with multiple metal ions like zinc, cobalt, and manganese to create hybrid nanoparticles with enhanced antifungal activity. These combinations can synergistically improve their effectiveness against various fungal pathogens.
- *Functionalized Ferrite Nanoparticles*: Surface modification and functionalization of spinel nano-ferrites with specific molecules or ligands have been employed to enhance their targeting and specificity toward fungal cells. These functionalized nanoparticles can improve their overall antifungal performance.
- *Magnetite-Spinel Hybrid Nanoparticles*: Hybrid nanoparticles combining magnetite and spinel nano-ferrites have been developed for their magnetic targeting capabilities. They can be guided to

fungal infection sites using external magnetic fields, ensuring precise delivery of antifungal agents.

- *Polymer-Coated Ferrite Nanoparticles:* Polymer-coated spinel nano-ferrites have been designed to improve their biocompatibility and control the release of antifungal agents. These coated nanoparticles offer a versatile platform for antifungal drug delivery.
- *Nanostructured Ferrite Thin Films*: In addition to nanoparticles, nanostructured ferrite thin films have been explored as antifungal coatings for medical devices. These thin films can provide long-lasting protection against fungal colonization.

These examples represent the diversity of spinel nano-ferrites used in antifungal applications. The choice of nano-ferrite formulation depends on factors such as the specific fungal infection being targeted, the desired drug delivery method, and the extent of surface functionalization required for optimal performance. Researchers continue to investigate and innovate in this field, aiming to develop more effective and targeted antifungal therapies using spinel nano-ferrites.[14,15]

7.9 IN VITRO ANTIFUNGAL ASSAYS

In vitro antifungal assays are crucial for evaluating the antifungal activity of spinel nano-ferrites. These assays involve controlled laboratory conditions and offer a preliminary understanding of how these nanomaterials interact with fungal pathogens. Common in vitro assays include broth dilution, agar diffusion, and microdilution methods, which assess the minimum inhibitory concentration (MIC) or minimum fungicidal concentration (MFC) of spinel nano-ferrites against specific fungal strains. Additionally, growth curve analyses monitor the effect of nano-ferrites on fungal proliferation over time. Techniques such as flow cytometry and fluorescence microscopy can provide insights into nano-ferrite-induced fungal cell damage, including membrane disruption and oxidative stress. Moreover, cell viability assays, such as MTT or XTT assays, help determine the impact of nano-ferrites on fungal cell viability.[16] Collectively, these in vitro assays allow researchers to comprehensively evaluate the antifungal potential of spinel nano-ferrites and their mechanisms of action while ensuring the safety and efficacy of these nanomaterials for future biomedical applications.

7.10 SPINEL NANO-FERRITES ANTIFUNGAL VS TRADITIONAL ANTIFUNGALS

A comparative analysis of spinel nano-ferrites' antifungal properties with traditional antifungal agents reveals several key distinctions and potential advantages are provided in Table 7.1.

TABLE 7.1
Comparative Analysis of Spinel Nano-Ferrites Antifungals with Traditional Antifungal Agents

Comparative Analysis	Spinel Nano-Ferrites	Traditional Antifungals
Mechanism of action	Spinel nano-ferrites often employ multifaceted mechanisms of action, including ROS generation, membrane disruption, and interference with fungal metabolic pathways.	Conventional antifungals usually target specific components within fungal cells, such as cell walls or ergosterol synthesis. This singular mechanism can make them susceptible to drug resistance development.
Broad-spectrum vs narrow-spectrum	These nanomaterials exhibit a broader spectrum of activity, making them effective against a wide range of fungal pathogens, including drug-resistant strains.	Many traditional antifungals have a more limited spectrum and may be less effective against certain fungal species or strains.
Targeted drug delivery	Functionalized spinel nano-ferrites can be employed as nanocarriers for antifungal drugs, enabling targeted drug delivery to the infection site, reducing systemic toxicity, and enhancing drug solubility.	Systemic administration of traditional antifungals may lead to off-target effects and systemic side effects.
Resistance development	Their multifaceted mechanisms of action make it challenging for fungi to develop resistance against spinel nano-ferrites, reducing the risk of drug-resistant strains.	Fungi can develop resistance to traditional antifungals more readily due to their singular mode of action, necessitating the continuous development of new drugs.
Biofilm disruption	These nanomaterials have shown promise in disrupting fungal biofilms, making them effective against infections where biofilm formation is a challenge.	Many conventional antifungals struggle to penetrate and disrupt established biofilms.
Cost and development	The functionalization of spinel nano-ferrites for specific applications may require initial research and development efforts, but their versatility and potential to improve drug efficacy can be cost-effective in the long run.	Developing and bringing traditional antifungals to market can be costly and time-consuming, with limited therapeutic options available.
Future innovations	Ongoing research is exploring novel applications and modifications of spinel nano-ferrites, offering the potential for innovative antifungal therapies.	While effective, the development of new traditional antifungals faces challenges such as drug resistance and limited therapeutic options.

7.11 BIOMEDICAL APPLICATIONS OF SPINEL NANO-FERRITES

Spinel nano-ferrites have shown promising biomedical applications as antifungals due to their unique properties, as shown in Figure 7.5. Some examples of these applications are described in the following sections.

7.11.1 Drug Delivery Systems

Spinel nano-ferrites are synthesized with controlled sizes and surface properties, allowing for precise drug loading and release characteristics. Commonly used spinel ferrites include zinc ferrite ($ZnFe_2O_4$) and manganese ferrite ($MnFe_2O_4$). Antifungal drugs, such as amphotericin B or fluconazole, can be encapsulated within the nano-ferrite structure during or after synthesis. This encapsulation protects the drug from degradation and enhances its stability. Spinel nano-ferrites exhibit controlled drug release properties. Drug molecules trapped within the nano-ferrite lattice can be released gradually over time. This controlled release ensures sustained therapeutic drug concentrations at the infection site, optimizing treatment efficacy. Functionalization of the nano-ferrite surface with ligands or antibodies enables targeted drug delivery.[17] These functionalized nano-ferrites can recognize specific fungal markers or receptors, allowing for selective drug release at the fungal infection site while sparing healthy tissues. Controlled and targeted drug delivery minimizes systemic exposure to antifungal drugs, reducing the risk of side effects commonly associated with these medications. Spinel nano-ferrites possess inherent magnetic properties, making them suitable for magnetic resonance imaging (MRI). This dual functionality allows for simultaneous drug delivery and non-invasive monitoring of treatment progress through imaging.[18] Combining the antifungal activity of spinel nano-ferrites with the encapsulated drug's therapeutic effect can result in synergistic antifungal action, enhancing the overall treatment efficacy.

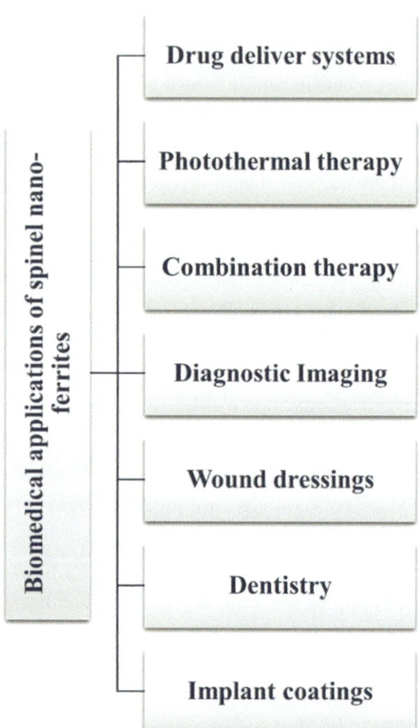

FIGURE 7.5 Promising biomedical applications of spinel nano-ferrites as antifungals.

The drug delivery system of spinel nano-ferrites for antifungal therapy offers controlled drug release, targeted delivery, reduced side effects, and the potential for imaging guidance. These features make spinel nano-ferrites a promising platform for improving the treatment of fungal infections.[18]

7.11.2 Photothermal Therapy

Spinel nano-ferrites, such as manganese ferrite ($MnFe_2O_4$) or zinc ferrite ($ZnFe_2O_4$), exhibit excellent photothermal properties. When exposed to near-infrared (NIR) light, these nano-ferrites efficiently absorb and convert light energy into heat. Functionalized spinel nano-ferrites can be designed to selectively bind to fungal cells. This targeting can be achieved through surface modifications, such as attaching antibodies or ligands that recognize fungal-specific markers. Once the functionalized nano-ferrites have accumulated at the fungal infection site, they can be activated using NIR laser irradiation. The NIR light penetrates deep into tissues without causing significant damage to healthy cells.

Under NIR irradiation, the nano-ferrites rapidly generate localized heat due to their photothermal properties. This heat is focused on the fungal cells, causing thermal damage to their cellular structures. The elevated temperature induced by the nano-ferrites disrupts fungal cell membranes, proteins, and other vital components. This leads to fungal cell death and the destruction of the infection. Photothermal therapy with spinel nano-ferrites offers high selectivity for fungal cells, minimizing damage to surrounding healthy tissues. This targeted approach reduces the risk of side effects compared to traditional antifungal treatments.

Photothermal therapy using nano-ferrites can be combined with other antifungal agents, creating a synergistic effect that enhances the overall therapeutic outcome. The magnetic properties of spinel nano-ferrites also make them suitable for MRI. This allows for real-time monitoring of the treatment's progress and effectiveness.

In conclusion, photothermal therapy using spinel nano-ferrites as antifungal agents offers a selective and minimally invasive approach for treating fungal infections. The ability to generate localized heat with NIR light makes it a promising strategy to target and destroy fungal cells while sparing healthy tissues.

7.11.3 Combination Therapy

Combination therapy using spinel nano-ferrites as antifungal agents is an effective strategy to enhance the treatment of fungal infections. Spinel nano-ferrites, such as zinc ferrite ($ZnFe_2O_4$) or manganese ferrite ($MnFe_2O_4$), possess unique properties that can be leveraged for combination therapy.[19]

When used in combination with traditional antifungal drugs, spinel nano-ferrites can enhance the overall antifungal activity. The nano-ferrites can disrupt fungal cell membranes, making it easier for the drug to enter and exert its effects. Combination therapy reduces the likelihood of drug resistance development by targeting fungi through multiple mechanisms. This makes it more challenging for fungi to adapt and become resistant to treatment. Spinel nano-ferrites can provide controlled drug release, ensuring a sustained therapeutic drug concentration at the infection site while minimizing systemic exposure. Functionalized nano-ferrites can be designed to target specific fungal species or infection sites, improving treatment precision and reducing off-target effects.[20] Combination therapy using nano-ferrites can be tailored to individual patient needs by adjusting the drug type, dosage, and release kinetics based on the specific characteristics of the fungal infection.

Combination therapy with spinel nano-ferrites as antifungal agents offers a multifaceted approach to combat fungal infections. It combines the strengths of nano-ferrite-mediated antifungal mechanisms with traditional drug therapy, resulting in enhanced efficacy and reduced drug resistance.

7.11.4 Diagnostic Imaging

Utilizing spinel nano-ferrites for diagnostic imaging in the context of antifungal applications is an innovative approach. Spinel nano-ferrites, such as manganese ferrite ($MnFe_2O_4$) and zinc ferrite ($ZnFe_2O_4$), possess inherent magnetic properties.[21] These properties make them valuable contrast agents for MRI in the diagnosis and monitoring of fungal infections. Functionalization of spinel nano-ferrites with specific ligands or antibodies can enable targeted imaging.[22] These functionalized nano-ferrites can recognize fungal-specific markers, allowing for the selective visualization of fungal infections.

The use of spinel nano-ferrites as contrast agents enhances the accuracy of MRI by providing clear and distinct imaging of fungal lesions. This aids healthcare professionals in identifying the precise location, size, and extent of the infection. MRI with spinel nano-ferrite contrast agents allows for real-time monitoring of the progression of fungal infections during treatment. This helps healthcare providers assess the effectiveness of antifungal therapies and make timely adjustments if necessary. Unlike some other imaging techniques, MRI does not involve ionizing radiation. This means that patients can undergo repeated MRI scans for monitoring purposes without significant health risks. Spinel nano-ferrites can be combined with other imaging agents or modalities, such as positron emission tomography (PET) or computed tomography (CT), to provide comprehensive information about fungal infections.

MRI with spinel nano-ferrite contrast agents can aid in tailoring antifungal treatment plans to individual patients by offering insights into the infection's response to therapy. In addition to clinical use, spinel nano-ferrite-enhanced MRI can be valuable in research settings for studying the pathogenesis of fungal infections and evaluating the efficacy of experimental antifungal treatments.[23] Diagnostic imaging using spinel nano-ferrites as contrast agents in MRI is a powerful tool for the accurate and targeted visualization of fungal infections. It offers real-time monitoring, precision, and the potential for personalized antifungal therapy.

7.11.5 Wound Dressings

Using spinel nano-ferrites in wound dressings as antifungal agents is an innovative approach. Spinel nano-ferrites, such as zinc ferrite ($ZnFe_2O_4$) or manganese ferrite ($MnFe_2O_4$), can be incorporated into wound dressing materials, creating nanocomposite dressings. Spinel nano-ferrites possess inherent antifungal properties due to their ability to generate oxidative stress and disrupt fungal cell membranes.[24] When integrated into wound dressings, they provide an additional layer of protection against fungal infections. These nanocomposite dressings serve a dual purpose—they can both prevent fungal infections in wounds and treat existing infections. This is especially crucial for patients with compromised immune systems or chronic wounds susceptible to fungal colonization.

The nano-ferrites within the wound dressing can be engineered to release antifungal agents, such as amphotericin B or fluconazole, in a controlled manner. This ensures a sustained therapeutic concentration at the wound site. In addition to their antifungal properties, spinel nano-ferrites may have a positive impact on wound healing. Their ability to modulate oxidative stress and inflammation can promote tissue regeneration and reduce the risk of secondary infections. By delivering antifungal agents directly to the wound site, nanocomposite dressings reduce systemic exposure, minimizing the risk of systemic side effects associated with oral or intravenous antifungal medications.

Nano-ferrite-embedded wound dressings can provide prolonged protection against fungal infections, reducing the need for frequent dressing changes and healthcare interventions. Healthcare providers can tailor the choice of nano-ferrite-embedded dressings and antifungal agents based on the specific needs and conditions of the patient, allowing for personalized wound care. This application also opens up research opportunities to study the effectiveness of nano-composite wound dressings in clinical settings and optimize their design for various wound types. Wound dressings incorporating spinel nano-ferrites as antifungal agents offer a multifaceted approach to wound care. They provide both preventive and therapeutic benefits against fungal infections while potentially enhancing the wound healing process.[25]

7.11.6 Dentistry

The use of spinel nano-ferrites in dentistry as antifungal agents represents a novel and promising approach to combating fungal infections in the oral cavity. Fungal infections in the oral cavity, such as oral candidiasis (thrush), can be common, especially among immunocompromised individuals or those with denture-related issues. Spinel nano-ferrites, such as zinc ferrite ($ZnFe_2O_4$) or manganese ferrite ($MnFe_2O_4$), can be incorporated into various dental materials, including composites, cements, and denture bases. These nano-ferrites exhibit inherent antifungal properties due to their ability to induce oxidative stress and disrupt fungal cell membranes. When integrated into dental materials, they provide a continuous source of protection against fungal colonization. Denture bases made with nano-ferrite-infused polymers can help prevent fungal biofilm formation on denture surfaces, reducing the risk of oral fungal infections in denture wearers.

Dental composites containing spinel nano-ferrites can be used in restorative procedures. These composites not only restore teeth but also inhibit fungal growth, contributing to the maintenance of oral health. In patients at risk of oral thrush, such as those undergoing chemotherapy or immunosuppressive therapies, denture base materials and restorations with nano-ferrites can serve as a preventive measure against fungal infections. By locally delivering antifungal protection through dental materials, the potential for systemic side effects associated with oral antifungal medications is reduced.

Patients with dentures or dental restorations made from nano-ferrite-infused materials may experience improved comfort and oral health by avoiding the discomfort and inconvenience of fungal infections. The application of spinel nano-ferrites in dentistry offers research opportunities to investigate the long-term efficacy and safety of these materials in clinical settings. The integration of spinel nano-ferrites into dental materials represents an innovative approach to preventing and managing fungal infections in the oral cavity. This application has the potential to improve oral health outcomes, particularly for vulnerable patient populations.

7.11.7 Implant Coatings

The use of spinel nano-ferrites as antifungal coatings for medical implants is an emerging and promising strategy to prevent fungal infections associated with implant placement. Medical implants, such as catheters or orthopedic devices, can be susceptible to microbial colonization, including fungal infections. These infections can lead to serious complications. Spinel nano-ferrites, like manganese ferrite ($MnFe_2O_4$) or zinc ferrite ($ZnFe_2O_4$), can be applied as coatings on the surface of medical implants.[26] These coatings provide a protective barrier against fungal adhesion and proliferation.

Spinel nano-ferrites possess intrinsic antifungal properties, primarily through the generation of oxidative stress and disruption of fungal cell membranes.[27] When used as coatings, they deter fungal growth on the implant surface. Fungal biofilms on implant surfaces can be particularly challenging to treat. Nano-ferrite coatings inhibit the formation of biofilms, reducing the risk of persistent infections. Nano-ferrite coatings are biocompatible and do not negatively impact the host tissue's response to the implant. This ensures that the implant functions effectively while providing antifungal protection. By preventing fungal infections, nano-ferrite coatings can help extend the lifespan of medical implants, reducing the need for costly replacements and minimizing patient discomfort.

Coated implants can lower the risk of systemic fungal infections, as localized fungal growth on the implant surface is controlled. Researchers can tailor the composition and thickness of nano-ferrite coatings to suit specific implant types and infection risks, allowing for a customized approach to implant protection. This application opens up research opportunities to explore the long-term efficacy and safety of nano-ferrite coatings on various types of medical implants.

The utilization of spinel nano-ferrites as antifungal coatings for medical implants represents a valuable approach to preventing implant-associated fungal infections.[28] This innovation has the potential to improve patient outcomes and reduce the burden of implant-related complications.

These examples highlight the versatility of spinel nano-ferrites in various biomedical applications for combating fungal infections. Their unique properties, including controlled drug delivery, photothermal therapy, and MRI, make them valuable tools in the fight against fungal diseases.[29]

7.12 SUMMARY

The chapter dedicated to the "Antifungal Activity of Spinel Nano-ferrites" provides valuable insights into the potential applications of nano-ferrites as effective antifungal agents. In the face of growing fungal resistance to traditional treatments, the exploration of alternative solutions is of paramount importance.[30] The chapter commences by discussing the fundamental significance of developing novel antifungal strategies and introduces the synthesis techniques employed to create spinel nano-ferrites. These techniques, such as sol-gel, co-precipitation, hydrothermal synthesis, mechanochemical synthesis, and template-assisted synthesis, enable researchers to tailor nano-ferrite properties for optimal antifungal performance.[31–33] A pivotal aspect of this chapter is the in-depth analysis of the antifungal activity of spinel nano-ferrites. The mechanisms behind this antifungal action are explored, shedding light on the interactions between nano-ferrites and fungal cells, which is crucial for understanding their efficacy.

Furthermore, the chapter addresses the structural characteristics of spinel nano-ferrites, including their crystal structure, morphology, and surface properties. These properties play an important role in influencing the antifungal behavior of the nano-ferrites, and their detailed examination contributes to a comprehensive understanding of their action. The chapter also touches upon potential practical applications of spinel nano-ferrites in fields such as pharmaceuticals, agriculture, and environmental remediation. It emphasizes the need for further research to optimize the nano-ferrites for real-world antifungal solutions, highlighting the promising yet evolving nature of this research field.

This chapter offers a comprehensive overview of the synthesis, mechanisms, and potential applications of these nanomaterials as antifungal agents. It underscores their significance in addressing the challenge of fungal infections while acknowledging the ongoing efforts required to harness their full potential.

REFERENCES

1. Singh, A., Suhag, D. (2023). Polymers and Polymeric Composites in Nano/Bio-Medicine. In: Suhag, D., Thakur, A., Thakur, P. (eds) *Integrated Nanomaterials and Their Applications*. Springer, Singapore. https://doi.org/10.1007/978-981-99-6105-4_8
2. Abdelhamid, H. N., Talib, A., Wu, H.-F. (2015). Facile synthesis of water-soluble silver ferrite ($AgFeO_2$) nanoparticles and their biological application as antibacterial agents. *RSC Advances*, 5, 34594–34602.
3. Abdelsalam, H. K. (2018). Enhancing the structural and spectroscopic properties of Cr^{3+} ion-doped Ni/Cd/Zn nanoferrite to be applied to industrial applications. Journal of Superconductivity and Novel Magnetism, 31, 4063–4077.
4. Srivastava, N., Mishra, G., Saxena, S. K. (2022). Environmental Nanomedicine. In: Thakur, A., Thakur, P., Khurana, S.P. (eds) *Synthesis and Applications of Nanoparticles*. Springer, Singapore. https://doi.org/10.1007/978-981-16-6819-7_22
5. Bharamagoudar, R. C., Patil, A. S., Vijay, S. N. M., Laxmi, M. K., Kankanawadi, B. (2018). Magnetic and antibacterial studies of nanoferrites prepared by self-propagating high-temperature synthesis route. *Acta Chemica IASI*, 26(2), 249–262.
6. Cavassin, E. D., de Figueiredo, L. F., Otoch, J. P., Seckler, M. M., de Oliveira, R. A., Franco, F. F., Marangoni, V., Zucolloto, V., Levin, A. S. S., Costa, S. F. (2015). Comparison of methods to detect the in vitro activity of silver nanoparticles (AgNP) against multidrug resistant bacteria. *Journal of Nanobiotechnology*, 13(1), 64.
7. El-Bassuony, A. a. H., Abdelsalam, H. K. (2020b). Synthesis, characterization, magnetic and antimicrobial properties of silver chromite nanoparticles. *Journal of Materials Science: Materials in Electronics*, 31(4), 3662–3673.
8. Perween, N., Khan, H. M., Fatima, N. (2019). Silver nanoparticles: An upcoming therapeutic agent for the resistant Candida infections. *Journal of Microbiology & Experimentation*, 7(1).
9. El-Bassuony, A. a. H., Abdelsalam, H. K. (2020a). Impacts of hematite, bunsenite and maghemite impurities on the physical and antimicrobial properties of silver nanoparticles. *European Physical Journal Plus*, 135(1).
10. Abdelsalam, H. K. (2018). Enhancing the structural and spectroscopic properties of Cr^{3+} ion-doped Ni/Cd/Zn Nanoferrite to be applied to industrial applications. *Journal of Superconductivity and Novel Magnetism*, 31(12), 4063–4077.
11. El-adly, A., Shabana, I. (2018). Antimicrobial Activity of Green Silver Nanoparticles against Fluconazole-resistant Candida albicans in Animal Model. *Egyptian Journal of Botany*, 58(1), 119–132. doi: 10.21608/ejbo.2018.1292.1110
12. Sharma, P., Sharma, A., Sharma, M., Bhalla, N., Estrela, P., Jain, A., Thakur, P., Thakur, A. (2017). Nanomaterial fungicides: in vitro and in vivo antimycotic activity of cobalt and nickel nanoferrites on phytopathogenic fungi. *Global Challenges*, 1(9), 1700041.
13. Villamizar-Gallardo, R. A., Cruz, J. F. O., Ortíz-Rodriguez, O. O. (2016). Fungicidal effect of silver nanoparticles on toxigenic fungi in cocoa. *Pesquisa Agropecuaria Brasileira*, 51(12), 1929–1936.
14. El-Bassuony, A. a. H. (2017b). Enhancement of structural and electrical properties of novelty nanoferrite materials. *Journal of Materials Science: Materials in Electronics*, 28(19), 14489–14498.
15. Ahmed, D. I. R. Y. a. I. S. (2016b). Applications of nickel nanoparticles for control of fusarium wilt on lettuce and tomato. *International Journal of Innovative Research in Science, Engineering and Technology*, 5(5), 7378–7385.
16. He, L., Liu, Y., Mustapha, A., Lin, M. (2011). Antifungal activity of zinc oxide nanoparticles against *Botrytis cinerea* and *Penicillium expansum*. *Microbiological Research*, 166(3), 207–215.
17. Pariona, N., Mtz-Enriquez, A., Sánchez-Rangel, D., Carrión, G., Paraguay-Delgado, F., Rosas-Saito, G. (2019b). Green-synthesized copper nanoparticles as a potential antifungal against plant pathogens. *RSC Advances*, 9(33), 18835–18843.
18. Abdel-Aziz, M. M., Emam, T. M., Elsherbiny, E. A. (2020b). Bioactivity of magnesium oxide nanoparticles synthesized from cell filtrate of endobacterium *Burkholderia rinojensis* against *Fusarium oxysporum*. *Materials Science and Engineering: C*, 109, 110617.
19. Dhanda, N., Thakur, P., Sun, A. C. A., Thakur, A. Structural, optical and magnetic properties along with antifungal activity of Ag-doped Ni-Co nanoferrites synthesized by eco-Friendly route. *Journal of Magnetism and Magnetic Materials*. 572 (2023) 170598. https://doi.org/10.1016/j.jmmm.2023.170598
20. Mandal, S., Natarajan, S., Tamilselvi, A., Mayadevi, S. (2016). Photocatalytic and antimicrobial activities of zinc ferrite nanoparticles synthesized through soft chemical route: a magnetically recyclable catalyst for water/wastewater treatment. *Journal of Environmental Chemical Engineering*, 4(3), 2706–2712.
21. Thakur, P., Kumari, S., Singh, S., Ravelo, B., Wan, F., Thakur, A. (2023). Mg-Zn ferrite nanoparticles: a brief review on synthesis, characterizations and applications. *ChemBioEng Reviews*, 10(5), 711–736.
22. Kariim, I., Abdulkareem, A. S., Kamardeen, A. O., Aliyu, A., Mohammed, I. A., Abdullahi, H. (2017). Characterization and antibacterial activity of nickel ferrite doped α-alumina nanoparticle. *Engineering Science and Technology, an International Journal*, 20(2), 563–569.
23. Rehman, Ansari, Alzohairy, Alomary, Jermy, Shahzad, Tashkandi, et al. (2019). Antibacterial and antifungal activity of novel synthesized neodymium-substituted cobalt ferrite nanoparticles for biomedical application. *Processes*, 7(10), 714. MDPI AG. Retrieved from
24. Singh, A., Wan, F., Yadav, K., Kharbanda, S., Thakur, P., Thakur, A. (2024). A novel magnetic $NiFe_2O_4$-Ag-ZnO hybrid nanocomposite for the escalated photocatalytic dye degradation and antibacterial activities. *Materials Science and Engineering: B*. 299, 116935. https://doi.org/10.1016/j.mseb.2023.116935

25. González-Merino, A. M., Hernández-Juárez, A., Betancourt-Galindo, R., Fuentes, Y. M. O., Valdez-Aguilar, L. A., Limon-Corona, M. L. (2021). Antifungal activity of zinc oxide nanoparticles in *Fusarium oxysporum-Solanum lycopersicum* pathosystem under controlled conditions. *Journal of Phytopathology*, 169(9), 533–544.
26. Nami, S., Aghebati-Maleki, A., Aghebati-Maleki, L. (2021). Current applications and prospects of nanoparticles for antifungal drug delivery. *EXCLI Journal*, 20, 562–584.
27. Cruz-Luna, A. R., Cruz-Martínez, H., Vásquez-López, A., Medina, D. I. (2021). Metal nanoparticles as novel antifungal agents for sustainable agriculture: Current advances and future directions. Journal of Fungi (Basel, Switzerland), 7(12), 1033.
28. Parveen, S., Wani, A. H., Shah, M. A., Devi, H. S., Bhat, M. Y., Koka, J. A. (2018). Preparation, characterization and antifungal activity of iron oxide nanoparticles. *Microbial Pathogenesis*, 115, 287–292.
29. Gole, D. A., Kapatkar, S. B., Mathad, S. N., Chavan, R. R. (2020). In vitro antimicrobial activity of cobalt ferrite nanoparticles synthesized by co-precipitation method. *Acta Chemica IASI*, 28(2), 225–236.
30. El-adly, A., Shabana, I. (2018). Antimicrobial activity of green silver nanoparticles against fluconazole-resistant *Candida albicans* in animal model. *Egyptian Journal of Botany*, 58(1), 119–132.
31. Kumari, S., Wan, F., Thakur, A., Singh, S., Thakur, P. (2023). Retrievable nano-sized Co–Mg–Zn ferrite as an active photocatalyst and adsorbent for removing organic and biological contaminants for water purification. *Ceramics International*, 50(2), 3210–3221.
32. Chahar, D., Kumar, D., Thakur, P., Thakur, A. (2023). Visible light induced photocatalytic degradation of methylene blue dye by using Mg doped Co-Zn nanoferrites. *Materials Research Bulletin*, 162, 112205.
33. Thakur, P., Taneja, S., Sindhu, D., Lüders, U., Sharma, A., Ravelo, B., Thakur, A. (2020). Manganese zinc ferrites: A short review on synthesis and characterization. *Journal of Superconductivity and Novel Magnetism*, 33, 1569–1584.

8 Cancer Detection and Diagnosis with Spinel Nano-Ferrites

Atul Thakur and Preeti Thakur*
Nanotechnology Wing, Innovative Science Research Society, Shimla, India
nanoLatticeX LLP, Manesar, Gurugram, India
Amity Institute of Nanotechnology, Amity University Haryana, Gurugram, India

Anand Salvi, Manish Shandilya, and Renuka Kanojia
Amity School of Applied Sciences, Amity University Haryana, Gurugram, India

Hui-Min David Wang
Graduate Institute of Biomedical Engineering, National Chung Hsing University, Taichung, Taiwan

8.1 INTRODUCTION

As per the World Health Organization (WHO), cancer stands as one of the primary contributors to global mortality, resulting in approximately 10 million fatalities in the year 2020 [1]. The projected figure is anticipated to experience a rise, reaching approximately 12 million by the year 2030 [1, 2]. In the year 2020, the number of deaths attributed to lung cancer reached around 1.80 million, making it the leading cause of mortality. This was followed by colon and rectum cancer, which accounted for approximately 935,000 deaths, liver cancer with approximately 830,000 deaths, stomach cancer with approximately 769,000 deaths, and breast cancer with approximately 685,000 deaths. Timely diagnosis and prompt treatment have been shown to increase the likelihood of survival and reduce morbidity rates [3]. The advancements in the field of nanotechnology have resulted in notable research breakthroughs in the domains of medicine and pharmaceuticals. The utilisation of nanomaterials exhibits significant promise in the field of cancer therapy [4–6]. The advancement of cancer diagnostic and therapeutic technologies has facilitated their use in various medical procedures, including magnetic resonance imaging (MRI), gene therapy, immunotherapy, and computed tomography [6]. Radiotherapy, chemotherapy, and surgery continue to be the preferred modalities for efficacious cancer therapies [7, 8]. The efficacy of conventional approaches is frequently compromised due to many factors, encompassing post-administration difficulties and the adaptive characteristics exhibited by cancer cells. Surgical procedures are characterised by their intrusive nature, which might lead to the occurrence of unfavourable side effects. In addition, the exposure of malignant tumours to chemical agents creates a selection pressure that enables cancer cells to adapt, persist, and proliferate, leading to the development of treatment resistance [9]. Viral vectors have emerged as the preferred means of gene delivery due to their notable transfection efficiency and inherent transducing properties. Regrettably, the utilisation of viral proteins has been hindered by the presence of associated immunogenic reactions, potential incorporation into the host genome, production-related obstacles, and issues pertaining to toxicity [1, 9]. Therefore, it is imperative to develop non-viral technologies that are more resilient, effective, and tailored to specific targets in order to address the aforementioned requirement in gene delivery and cancer treatment. Nanoparticles (NPs) possess various advantages in comparison to alternative delivery systems, including their straightforward production and convenient functionalisation capabilities and Furthermore, the versatility of this technology allows for customisation to suit specific applications, such as therapy and diagnostics. In addition, these entities have the ability to sequester hydrophobic compounds, such as pharmaceutical agents used in cancer treatment, thereby enhancing their solubility, compatibility with biological systems, and selectivity. Over the course of time, several NPs have been developed and utilised in the field of biomedicine.

The aforementioned entities encompass organic linear polymers, hyperbranched polymers, dendrimers, liposomes, micelles, and inorganic NPs. Several inorganic NPs have been documented in both pre-clinical and clinical investigations for their potential applications in the treatment, diagnosis, and detection of specific disorders. The types of magnetic NPs encompass gold, silica, carbon-based, and iron oxide-based varieties [10–12]. Figure 8.1 depicts the inorganic and organic NPs that are frequently employed in the field of nanomedicine.

Magnetic NPs consist of three main types: (i) metallic NPs, such as iron (Fe) and cobalt (Co); (ii) alloys, including gold/iron (Au/Fe); and (iii) iron oxide-based NPs, including magnetite (Fe_3O_4), maghemite (γ-Fe_2O_3), and ferrites (such as $CoFe_2O_4$, $MnFe_2O_4$, and $ZnFe_2O_4$) [13, 14]. The removal of NPs that contain heavy metals is frequently a complex task. Nevertheless, a number of iron-oxide NPs have been

* Corresponding author.

DOI: 10.1201/9781003568940-8

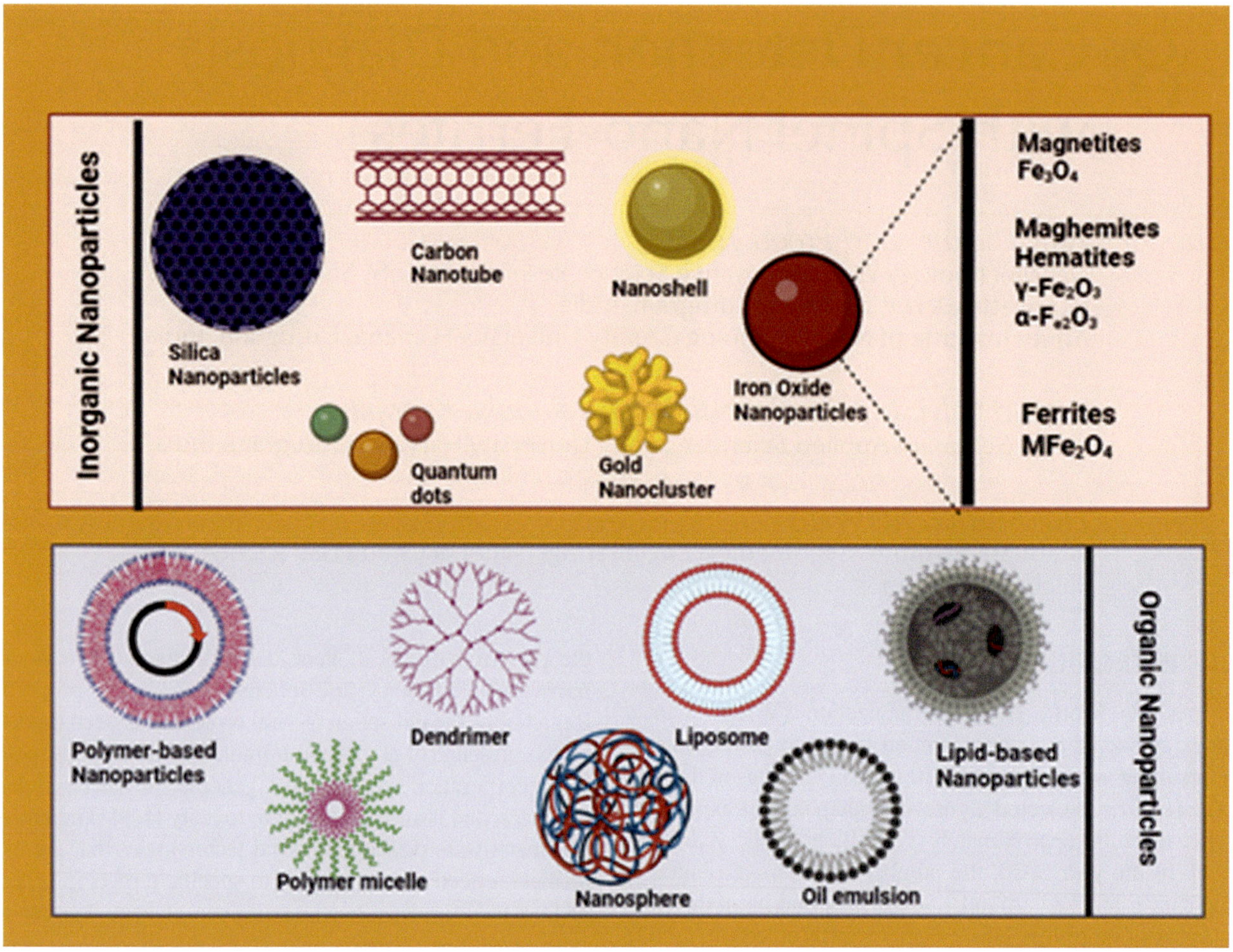

FIGURE 8.1 Several categories of organic and inorganic nanoparticles that are frequently utilised in the field of nanomedicine.

authorised for therapeutic application, demonstrating negligible harmful effects [14, 15]. Cobalt ferrite NPs that have been coated with polymers, including polyvinyl alcohol (PVA), polyvinylpyrrolidone (PVP), and polyethylene glycol (PEG), have demonstrated no harmful effects when tested at doses of up to 150 μg/mL. According to a study conducted in a laboratory setting, it was shown that the use of magnesium ferrite NPs ($MgFe_2O_4$ NPs) coated with chitosan, PEG, and PVA effectively facilitated the delivery of doxorubicin to cells affected by colorectal and breast cancer. This delivery method demonstrated low levels of toxicity when tested at doses of up to 100 μg/mL. A study further documented the reported phenomenon of chitosan-coated $Mg_{0.5}Co_{0.5}Fe_2O_4$ NPs effectively delivering the anticancer medication 5-fluorouracil (5-FU) specifically to cancer cells, while sparing healthy cells [15, 16]. Zinc ferrite NPs ($ZnFe_2O_4$ NPs) have garnered significant interest in some applications owing to their biocompatible nature and minimal toxicity when utilised at concentrations lower than 125 μg/mL [16]. The little toxicity seen can be primarily ascribed to the elemental makeup of the NPs, which enhances their biocompatibility and degradability. In addition, it is imperative for magnetic NPs to possess desirable magnetic characteristics, including a high saturation magnetisation (MS) and a low coercivity (HC), alongside consistent nanoscale dimensions below 100 nm [17]. The ensuing sections provide a summary of the main aspects pertaining to the use of iron-oxide NPs in the context of cancer therapy.

8.2 IRON OXIDE NANOPARTICLES IN CANCER DIAGNOSTICS AND THERAPY

Iron oxide NPs with distinctive features such as large surface areas, small particle sizes, and superparamagnetism have been mentioned in literature for their potential use in diagnostic techniques and targeted drug delivery [17–19]. When superparamagnetic iron oxide NPs (SPIONs) are subjected to an external magnetic field, the magnetic domains within the NPs align themselves with the direction of the field. Upon removal, the magnetisation of these NPs reverts back to zero. Ferromagnetic NPs have also been investigated as a distinct category of magnetic NPs within the field of nanomedicine. Nevertheless, it should be noted that these NPs exhibit a state of permanent magnetisation, allowing them to maintain their magnetic properties even in the absence of an external magnetic field. In the case of both superparamagnetic and ferromagnetic NPs, the application of an external magnetic field causes their magnetic moments to align with the field direction. This alignment process results in the generation of heat, which serves as the fundamental mechanism for hyperthermia [19–21]. Originally, magnetite (Fe_3O_4), hematite (α-Fe_2O_3), and maghemite (γ-Fe_2O_3) have been extensively employed. Magnetite NPs exhibit significant prevalence and possess intriguing characteristics attributed to the presence of two valence states, namely Fe^{2+} and Fe^{3+} [20]. The materials under consideration feature inherent ferromagnetic characteristics and have a distinctive reverse cubic spinel structure. The composition of their structure consists of tetrahedral (A) sites, octahedral (B) sites, and layers that are a combination of tetrahedral and octahedral sites, including iron ions Fe^{2+} and Fe^{3+}. Approximately 50% of the Fe^{3+} ions are located at A sites, which are coordinated by four oxygen atoms. Conversely, the combination of Fe^{2+}/Fe^{3+} ions is situated at site B, where they are surrounded by six oxygen atoms [20]. The utilisation of polymer-based magnetic microspheres containing magnetite NPs was initially demonstrated for the purpose of magnetic medication delivery and targeting by Wu *et al.* [21]. Multiple studies have documented the effective production of magnetite NPs coated with various polymers such as chitosan, PEG, polyethyleneimine (PEI), and dextran. These synthesised NPs have exhibited favourable characteristics suitable for their application as contrast agents in clinical MRI [22, 23]. The application of dextran coating has been documented to enhance the physical characteristics of SPIONs for their utilisation as contrast agents in MRI and hyperthermia therapy. In a separate investigation, the utilisation of magnetite NPs that were modified with PEI and galactose was explored for the purpose of delivering small interfering RNA (siRNA) specifically targeting hepatocellular cancer. Encouraging outcomes were observed in this study. Maghemite exhibits a spinel structure that is akin to magnetite, however with the presence of vacant divalent iron ions. The size of the unit cell is slightly reduced in comparison to that of magnetite due to the presence of cationic vacancies and the smaller ionic radius of Fe^{3+} ions relative to Fe^{2+} ions [24]. In this study, the researchers examined the application of citrate-coated maghemite NPs in vitro for the purpose of conducting combined magnetic hyperthermia and photodynamic therapy.

The objective of this investigation was to explore the potential of this treatment approach for the targeted treatment of specific types of cancer, such as glioblastoma. Polystyrene-coated iron oxide NPs were employed in an initial clinical experiment to augment MRI of the gastrointestinal tract (GIT). Common examples of coating agents found in many substances are dextran/carboxydextran in Ferumoxtran, Ferumoxide, and Ferucarbotran; PEG in Feruglose; and aminosilane in Nanotherm™. The GastroMARK® product employs iron oxide NPs coated with siloxane for MRI of the GIT. Ferumoxytol, γ-Fe_2O_3 compound coated with polyglucose-sorbitol-carboxymethyl-ether, was created and received approval in the year 2000 as a contrast agent for MRI in the diagnosis of several types of cancer. The fate of NPs within a biological system is often influenced by the characteristics of their coating and their hydrodynamic size. This encompasses the various interactions between the coated iron oxide NPs and their surrounding cellular environment, as well as the processes of uptake, accumulation, circulation, and subsequent clearance of these NPs from the human body [24, 25]. Moreover, Figure 8.2 elucidates the investigations employing iron oxides in the context of

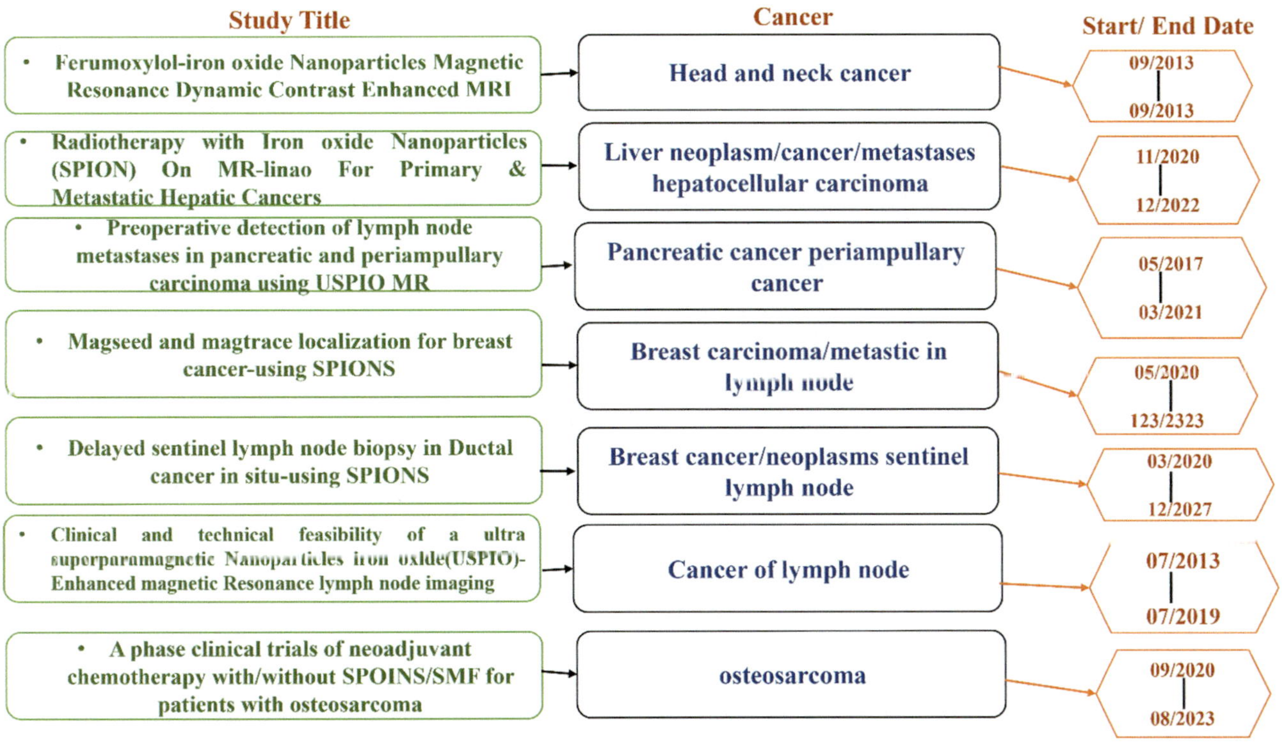

FIGURE 8.2 Characteristics of the iron oxide nanoparticles employed in cancer clinical trials that have been concluded or halted since 2019, as well as those that are still ongoing or actively recruiting participants.

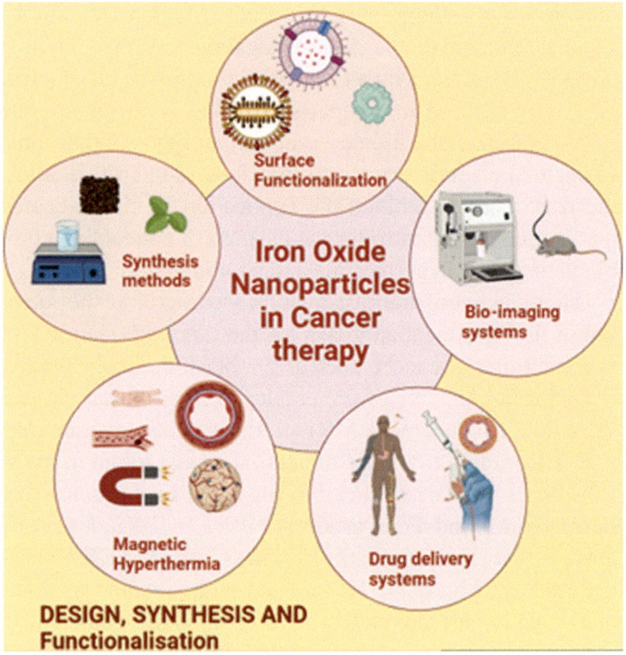

FIGURE 8.3 The utilisation of iron oxide nanoparticles in the field of cancer diagnosis and therapy.

cancer research that have progressed to the stage of clinical trials, including those that have already been terminated or concluded subsequent to the year 2019.

The clinical trial stage has been reached by iron oxide NPs, indicating the need of surface functionalisation as a necessary preliminary step for medical purposes such as MRI, hyperthermia, or drug delivery. Figure 8.3. shows This functionalisation often involves the use of surfactants or polymers [25].

The Phenomenon of Magnetic Hyperthermia During hyperthermia therapy, the application of an alternating magnetic field (AMF) to the NPs leads to an increase in temperature ranging from 42 to 45°C [25]. The generation of heat can be attributed to three main mechanisms: magnetic hysteresis loss, Néel relaxation, and Brown relaxation. Hysteresis losses are reported in iron oxide NPs that exhibit the presence of several magnetic domains. The constant alignment that ensues and the variations in energy release might give rise to the development of heat [25, 26]. Conversely, Néel relaxation involves the fast fluctuations in the alignment of magnetic moments with respect to the crystal lattice. The relaxation process is hindered by the presence of anisotropic energy, which promotes the alignment of the magnetic domain in a specific direction with respect to the crystal lattice [26]. On the other hand, the phenomenon of Brownian relaxation arises from the rotational motion of particles within a given medium. The cause of the delay in this particular instance might be attributed to the adhesive properties that impede the motion of particles within the medium [26, 27]. The creation of heat is influenced by various elements, including the dimensions, morphology, and magnetic characteristics of the NPs. The introduction of an external magnetic field of various sizes addresses several issues encountered in the distribution and delivery of therapeutic agents. These challenges include non-specificity, toxicity in healthy cells, and reduced therapeutic efficacy.

Upon arrival at the designated location, NPs frequently employ controlled drug release methods that rely on several parameters including pH, light, heat stimuli, and redox stimuli [27]. Due to the diminutive dimensions of NPs, a proclivity exists for their passive accumulation within tumour tissues as a result of the increased enhanced permeability and retention (EPR) effect [28]. Passive targeting is employed in order to exploit the weakened vascular and micro-tumour environment characterised by variations in pH levels, temperature, and inadequate lymphatic drainage [29]. Figure 8.4 depicts the altered vascular network in cancer cells, which enhances the EPR effect. Solid tumours, on the other hand, pose additional obstacles, including tumour heterogeneity and matrix barriers, such as fibrosis collagen. Active targeting involves the functionalisation or customisation of NPs to facilitate their specific recognition by ligand receptors or interactions with antigens and antibodies [29, 30].

The phenomenon of receptor-mediated transcytosis, which involves the internalisation of NPs delivering therapeutic agents, is selectively induced in cancer cells through the overexpression of certain receptors [25, 26]. Polymer-coated iron oxide NPs enable the encapsulation of drugs within their matrix, facilitating trigger-based release upon arrival at a tumour site [30]. Numerous papers have demonstrated the suitability of functionalised magnetites as carriers for anticancer medicines, including doxorubicin, 5-FU, morin, and ciprofloxacin [30, 31].

In recent times, ferrites have emerged as viable biocompatible NPs for drug delivery in the context of anticancer treatment. There has been a notable surge in research focused on utilising ferrites for the purpose of targeted drug delivery. The study revealed that the utilisation of chitosan and PEG as coatings for manganese ferrite NPs resulted in a significant enhancement in the encapsulation effectiveness of methotrexate [31], The in vitro study demonstrated that the utilisation of chitosan-functionalised NPs resulted in an improved delivery of 5-FU in MCF-7 cells [31]. In both studies, the activation of drug release was observed in response to acidic conditions. The study observed a comparable drug delivery system that responds to stimuli employing $ZnFe_2O_4$ and $Ag_{1-x}Zn_xFe_2O_4$ NPs, while varying the pH conditions [30]. The investigation of $Zn_xMg_{(1-x)}Fe_2O_4$ NPs for drug administration also yielded confirmation of their favourable drug-loading capacity and drug release patterns [32]. In addition, the investigation of cobalt ferrite NPs doped with gadolinium ions (Gd^{3+}) has demonstrated their potential application in targeted drug administration and contrast enhancement in MRI. These NPs exhibited a sustained release of drugs ranging from 90.6% to 95% over a period of 24 hours, specifically at a pH value of 7.4 [24, 29]. The utilisation of lime peel extract was employed in the synthesis of $NiFe_2O_4$ NPs, which were subsequently

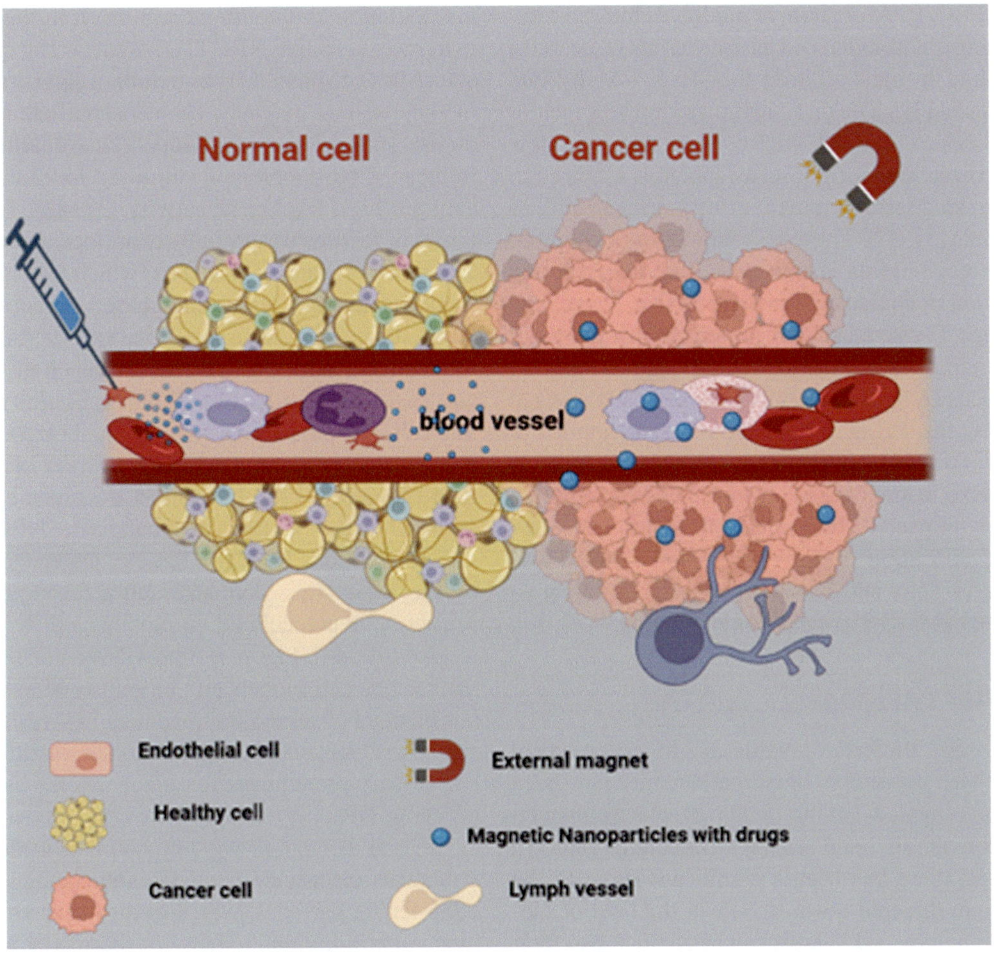

FIGURE 8.4 An illustration of the improved permeability and retention effect in cancer cells enables passive targeting.

examined for their potential as antioxidants, anticancer agents, and antibiotics [31].

In a separate investigation, the synthesis of bismuth-doped nickel ferrite NPs ($NiFe_{2-x}Bi_xO_4$ NPs) was conducted using the co-precipitation technique. These NPs were suggested to possess potential utility in the context of magnetic carrier targeting [33, 34].

Imaging systems refer to technologies and methodologies employed for capturing, processing, and displaying visual information in various fields such as medicine, astronomy MRI is an indispensable modality for visualising anatomical structures and pathological conditions in clinical diagnostics, particularly in the identification and characterisation of neoplastic growths. The efficacy of MRI is contingent upon the magnetic resonance (MR) signal emitted by the tissues under examination, which can be modified by the presence of contrast agents. There exist two primary categories of MR contrast agents, namely positive contrast agents (also known as T1-weighted agents) and negative contrast agents (sometimes referred to as T2/T2-weighted agents). Positive contrast chemicals have the ability to reduce the longitudinal relaxation time (T1) of protons, leading to the production of a vivid image. Negative contrast agents have the effect of reducing the transverse relaxation time (T2) of protons and exhibit a tendency to decay quickly in the transverse direction, resulting in the production of a dark image [34–36]. Gadolinium (Gd)-based contrast agents of type 1 (T1) have been widely utilised in clinical settings due to their ability to enhance picture brightness and improve resolution [35].

However, these displays exhibit certain drawbacks including limited longevity, inadequate cellular absorption, and potential hazards for individuals with renal and liver disorders. The unbound gadolinium ions have little efficacy in terms of post-administration clearance, hence leading to toxicity within natural environments. Superparamagnetic iron-oxide NPs have been extensively utilised and documented to demonstrate superior MRI signal contrast compared to gadolinium-based counterparts, mostly attributed to their elevated MS [33, 36].

Moreover, due to the inherent biomineral properties of Fe, it is plausible that these agents exhibit biodegradability and biocompatibility in living organisms, in contrast to contrast agents based on metals or metal alloys [31, 32]. The application of PEG, PVA, dextran, or modified chitosan coatings has demonstrated enhanced properties in these NPs, mostly attributed to their reduced long-term toxicity and extended shelf life [31]. The investigation of ferrites, including $CoFe_2O_4$, as potential T2 MRI contrast agents is now underway. Nevertheless, further investigation is required to ascertain their long-term cytotoxic effects prior to initiating clinical trials [31, 37]. The exploration of their potential as T1 MRI contrast agents is now underway, necessitating the consideration of characteristics such as size, shape, and surface coatings. For instance, it is preferable that these NP sizes are less than 5 nm in order to provide a high-quality T1 image resolution [37, 38]. Furthermore, the relaxation period is influenced by the sort of surface modification applied to the NPs. Previous studies have noted that hydrophilic coatings, which serve to stabilise water molecules surrounding NPs, negatively impact T2 by reducing its value. The study described the utilisation of $CoFe_2O_4$ @$MnFe_2O_4$ NPs to improve the specificity and sensitivity of MRI and fluorescence labelling in MGC-803 cell lines and tumours [37, 38].

8.3 FERRITES (MFe_2O_4)

Nanocarriers as Preferred Vehicles in Biomedical Applications The utilisation of spinel ferrites has been extensively investigated in the fields of electromagnetics, information storage, and sensor technologies [39–41]. In recent times, there has been a significant surge in the level of attention directed towards NPs in the field of biomedicine. This heightened interest can be attributed to their unique optical and magnetic characteristics, as well as the potential for customisation through alterations in their original composition or cation replacement [40, 41].

The materials under consideration exhibit a cubic spinel structure in a single phase, characterised by the formula MFe_2O_4, where M represents one of the following elements: magnesium (Mg), manganese (Mn), nickel (Ni), copper (Cu), zinc (Zn), cobalt (Co), or iron (Fe) [42]. Mixed ferrites consist of combinations of different oxidation states that exist at both locations. This is observed in the compound $MnFe_2O_4$. On the other hand, the compound $CoFe_2O_4$ can adopt either the inverse or mixed spinel structure, depending on several synthesis parameters, including the duration of the process [39, 40]. The cubic structure exhibits a greater abundance of octahedral sites compared to tetrahedral sites, with both types of sites possessing magnetic moments that are aligned in antiparallel orientations. This configuration leads to an increased level of magnetisation. The concept involves the integration of distinct characteristics provided by the substituted molecule, often a divalent cation metal like Mn, Co, Ni, or Ca, with those exhibited by the iron oxide NPs. Ferrites of this sort finally demonstrate exceptional magnetic properties, significant surface area, diminutive dimensions, and the necessary biocompatibility for utilisation in biomedical applications [39].

Ferrites have a high degree of versatility due to their nanoscale dimensions, which enable precise surface manipulation and consequently result in the production of extremely selective NPs. The magnetic and structural characteristics of spinel ferrites exhibit a significant reliance on factors such as magnetic moment, particle size and distribution, shape, and crystallinity. These qualities are notably influenced by the process employed for manufacture, indicating a high level of sensitivity [39–42]. The biocompatibility of ferrites was initially examined and documented in a study conducted by Zhu et al., which focused on magnetic fluid and magneto liposomes based on cobalt-ferrite [41]. After subjecting the tissue to various doses and durations of treatment with the NPs, no evidence of tissue damage was detected. Furthermore, the capillaries and parenchymal cells remained unharmed by the therapy. Subsequently, intermittent accounts of the application of ferrites in the field of biomedicine have been documented. Kudr et al. reported on the synthesis of cobalt ferrite NPs for hyperthermic applications [43, 44]. Toxicity experiments conducted by the researchers demonstrated a favourable therapeutic impact on hepatocellular carcinoma cells both in vitro and in vivo [40, 44]. Figure 8.5 provides a concise overview of the various cell models and in vivo systems in which there has been an observed improvement in the dispersion of ferrite NPs. Over time, the application of a coating has become a necessary prerequisite for enhancing the biocompatibility of ferrite NPs.

Several elements have a crucial role in determining the structural, electrical, magnetic, and chemical properties of nano-ferrite particles, which in turn have an impact on the biocompatibility and toxicity of these NPs [40]. The selection of synthesis processes frequently determines several characteristics of particles, including size, crystallinity, content, and site occupancy. Various synthesis processes can be employed in the production of materials, such as sonochemical, mechanochemical, hydrothermal, co-precipitation, and sol-gel approaches [33, 34]. These are discussed in depth in the below sections.

8.3.1 Synthesis Methods of Ferrites

In recent decades, there has been a significant focus on the development of a broad range of methodologies for synthesising nanomaterials, employing both bottom-up and top-down approaches [33].

Various approaches are employed in the production of ferrite NPs in order to cater to the intricate functionality demanded by the field of human healthcare and medicine [35]. The manufacture of magnetic NPs can be broadly classified into three primary categories: biological, physical, and chemical approaches, as seen in Figure 8.6 [35]. The selection of a synthesis method is of paramount importance in determining the characteristics of NPs, particularly in relation to their intended uses. The requirement of uniformity in NPs is crucial due to the significant influence of their size on their structural, electrical, optical, and magnetic characteristics. The synthesis methods enumerated in Figure 8.5 are widely employed in the manufacture of cubic

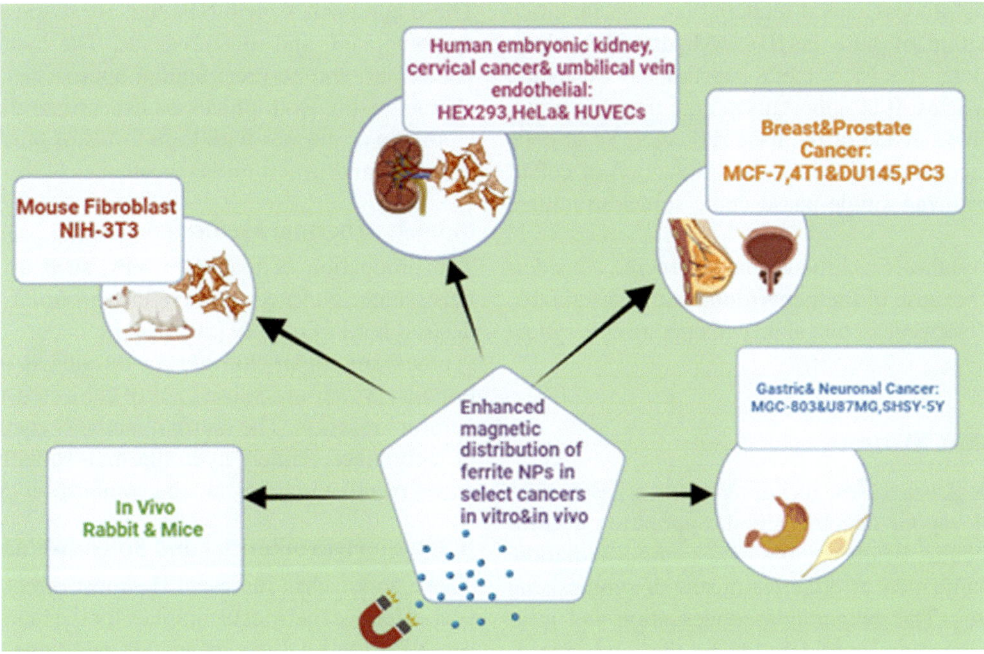

FIGURE 8.5 The utilisation of cancer cell models and in vivo systems to study the magnetic distribution of ferrite nanoparticles.

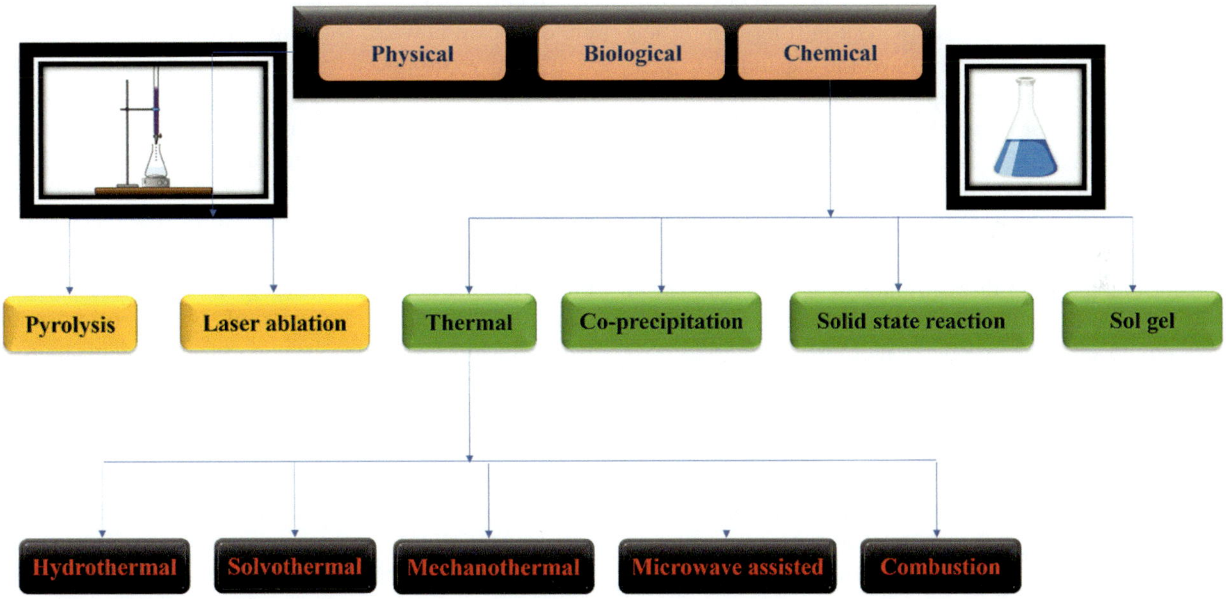

FIGURE 8.6 Synthesis techniques for cubic spinel ferrite materials.

spinel ferrite materials, since they are known for their reliability, simplicity, and effectiveness in achieving the requisite characteristics [37].

8.3.2 Biological Methods

The utilisation of biological approaches in green synthesis is widely regarded as a straightforward and cost-effective approach. These alternatives have garnered significant interest as viable substitutes for physical and chemical methodologies. Various plant components, including roots, leaves, stems, latex, fruit pericarp, fruit juice, and seeds, have been utilised in the synthesis of ferrite NPs. These plant extracts serve as both stabilising and reducing agents in the process [44]. Fungi, algae, and bacteria have been documented as agents that aid in the synthesis of ferrites NPs, so offering a useful and new biotechnological resource for the production of very stable NPs [45].

The synthesis of ferrite NPs with a single-phase cubic spinel structure can be achieved through a bacterial-mediated process. In this process, metal-reducing bacteria facilitate the transformation of pure Fe(III) oxyhydroxides, along with soluble metal species or metal-substituted akaganeite, into ferrite particles. It is important to note that this transformation occurs outside of the bacterial cells. Yeast cells have been effectively utilised for the production of cobalt ferrite NPs featuring single-phase cubic spinel structures [44, 45].

Biological synthesis techniques have garnered considerable attention because of their environmentally sustainable nature, cost-effectiveness, and ability to operate at very low temperatures.

8.3.3 Physical Methods

Various physical processes, including plasma, laser ablation, chemical vapour decomposition/evaporation decomposition, molecular beam epitaxy, and gamma radiation, have been documented as effective means of synthesising ferrite NPs [46]. The evaporation-condensation and laser ablation techniques are considered to be the most significant procedures among the options available, primarily due to their reduced potential for contamination. The utilisation of these approaches yields NPs with particle sizes that are more evenly dispersed compared to methods involving the use of wet chemicals [46, 47]. Furthermore, it has been observed that these NPs have the capability to generate nanoscale-sized NPs, approximately 100 nm in size, rendering them appropriate for utilisation in the field of biomedicine [35, 38, 47]. The structural and magnetic characteristics of nanocrystalline NiZn and Zn ferrite thin films generated through laser ablation deposition were investigated by Sorescu *et al.* The researchers achieved particle sizes of 30 and 50 nm through the use of laser ablation in distilled water. Figure 8.6 illustrates various synthesis processes employed for the production of ferrite materials possessing a cubic spinel structure.

8.3.4 Wet Chemistry Methods

Wet chemistry techniques employ chemical reactions in a liquid medium to produce NPs. The methods employed exhibit distinct characteristics in terms of reliability and uniqueness, offering a significant degree of control and repeatability in the production of magnetic ferrite NPs. Every chemical synthesis technique possesses distinct characteristics, presenting a range of benefits and drawbacks. The manufacture of ferrite NPs is typically conducted using low-temperature processes. The co-precipitation approach has been widely employed for the manufacture of magnetic NPs in many biological applications. Additional techniques encompass the glycol-thermal/solvothermal approach [48–50], The synthesis techniques employed in this study include hydrothermal methods, auto-combustion and co-precipitation techniques [48, 49], sol-gel method, forced hydrolysis, CTAB-assisted hydrothermal method, microwave-assisted hydrothermal method, and the citrate precursor method. These approaches yield NPs that are shape-controlled, unagglomerated, and monodisperse. The hydrothermal, solvothermal, and co-precipitation approaches have a shared characteristic. Both processes necessitate the use of metal chlorides or nitrates with high levels of purity, in stoichiometric quantities, as initial substances [49].

8.3.4.1 Thermal Methods

The production of magnetic NPs, such as ferrites, often necessitates elevated temperatures in order to achieve the desired level of quality [50]. The procedure commences by synthesising metal chlorides or nitrates as precursor compounds, which are subsequently subjected to a high-temperature reaction. The thermal methods encompass a range of techniques, namely hydrothermal, solvothermal, microwave, mechano-chemical, and combustion [51].

8.3.4.2 Hydrothermal and Solvothermal Methods

These approaches represent the most direct, efficient, and cost-effective thermal techniques for the manufacture of ferrite NPs. Optimal experimental conditions, encompassing factors such as temperature, pressure, and reaction duration, can yield samples of exceptional quality [51]. Both of these processes utilize Fe^{3+} and divalent metal M^{2+} compounds, specifically chlorides or nitrates, as precursors. The salts are dissolved in water with vigorous agitation, while a base is gradually introduced to induce the formation of a precipitate. After the pH regulation process, the mixture undergoes filtration to eliminate the presence of chlorides or nitrate ions. Subsequently, the filtered mixture is transferred to an autoclave or pressure reactor and allowed to settle down to ambient temperature. In the glycol-thermal technique, the substitution of water with ethylene glycol is observed. Following the process of heating, the mixes that have been cooled are subjected to a series of subsequent procedures, which include washing, drying, and homogenisation [52]. One notable benefit of the hydrothermal approach is in its capacity for integration with several complementary processes, including microwave, electrochemistry, ultrasound, and optical radiations. This integration enhances the overall efficacy of material synthesis, facilitating the production of novel materials. By optimising several parameters such as reaction time, temperature, reactant concentration, solvent type, and precursors, it is possible to monitor the growth and form of particles using this particular technique [53].

8.3.4.3 Microwave Method

The microwave synthesis technique is a recently developed approach for synthesising ferrite NPs, in addition to other existing synthesis techniques [54]. In recent years, there has been significant interest in the use of microwave-assisted chemical reactions [53]. The utilisation of this approach offers numerous benefits, including enhanced heating efficiency, accelerated reaction kinetics, increased crystallisation extent, improved product yield, and reduced particle sizes characterised by a narrower distribution. The expeditious development of structure under microwave conditions may be attributed to the phenomenon of restricted superheating in the solutions. This methodology has the potential to produce ferrite NPs on a larger scale, contingent upon their intended application [51, 53].

8.3.4.4 Mechano-Chemical Method

The approach utilised in this methodology is founded around the manipulation of solid materials, wherein mechanical and chemical interactions are interconnected at the molecular level. The phenomenon under consideration exhibits a recurring process of distortion, breakage, and fusion of a chemical amalgamation, employing interactions involving solid acids derived from hydrated substances, crystal hydrates, as well as basic and acidic salts. Typically, the chemical reaction involved in this method is conducted at elevated temperatures, followed by subsequent reduction in temperature throughout the ball milling process. The careful determination of appropriate synthesis conditions, including stoichiometry, chemical reaction route of starting materials, and milling settings, is of most importance [50–52].

8.3.4.5 The Combustion Method

This method presents a straightforward and cost-effective approach to the manufacture of ferrite NPs. Additionally, it necessitates a decrease in the amount of time and energy expended during the synthesis procedure. This process involves the combination of stoichiometrically balanced quantities of metal chlorides or nitrates with high levels of purity, in the presence of oxidising agents such as urea, glycine, and hydrazides. Glycerine, a fuel, serves as a reducing agent that facilitates the process of combustion [51, 52]. The individual batches are introduced onto a silica dish characterised by its glassy composition, where they are subjected to homogenisation to generate a slurry. This slurry formation is attributed to the inherent properties of the metal nitrates. Subsequently, the substance is subjected to heat, whereupon the process of combustion occurs over a duration of approximately 20 minutes, but the actual ignition period may be less than 10 seconds. During the process of combustion, the generation of foams occurs, resulting in the production of a voluminous and fluffy substance [50]. The aforementioned phenomenon frequently exhibits characteristics of purity and homogeneity [50].

8.3.4.6 The Sol-Gel Method

The multistep process encompasses hydrolysis, condensation, and polymerisation reactions of metal precursors, leading to the subsequent creation of a gel [51]. The utilisation of this approach presents the benefit of achieving precise stoichiometric regulation, leading to the production of ultrafine NPs with excellent purity and a restricted size distribution. Furthermore, this procedure may be carried out at low temperatures, allowing for a very short processing time. The sol-gel process is characterised by its simplicity and cost-effectiveness. In summary, the process involves the co-precipitation of metal salts in an aqueous solution using a metal hydroxide or base. This mixture is then processed to create a colloidal sol or an inorganic or metallo-organic precursor. The resulting solution can be condensed into a gel, which is then subjected to firing to generate fine-grained polycrystalline ferrite NPs [51]. In this study, the synthesis of pure spinel nickel ferrite NPs was conducted utilising the sol-gel method, with polyacrylic acid (PAA) serving as the chelating agent. The manipulation of the molar ratios of PAA to total metal ions and the adjustment of the calcination temperature was utilised to regulate the size distribution, specific surface area, and crystallinity of ferrite NPs [51].

8.3.4.7 Solid-State Method

This particular methodology exhibits characteristics such as operational simplicity, cost efficiency, and applicability to large-scale synthesis of NPs. The process basically entails the precise combination of high-purity raw metal oxides or metal salts with a few drops of methanol, followed by grinding using a high-energy ball machine for a specific duration [52]. One significant advantage of employing this technique is the ability to synthesise samples without the need for any solvent [54]. The production of $Co_{1-x}Zn_xFe_2O_4$ ($x = 0$ and 0.5) spinel ferrite NPs by various approaches demonstrated that the solid-state approach effectively yielded particles with dimensions in the nanoscale range [55].

8.4 POTENTIAL OF FERRITES IN BIOMEDICAL APPLICATIONS

The burgeoning body of research on ferrites in the context of drug administration has exhibited promising results, indicating their considerable potential for wider application in the field of biomedicine. In order to meet the requirements for utilisation in biomedical applications, nanocarriers must exhibit the essential characteristics of biodegradability, biocompatibility, and in vivo stability, as emphasised in Figure 8.7.

The application of these NPs is influenced by various physical criteria, including magnetic properties, size, capacity for drug/gene binding, and physiological parameters such as internal trafficking to target vascular supply and body weight. The optimisation of these elements has the potential to yield a therapy that is both desired and exact [55]. Additional metrics, including form, stoichiometry, and surface structure, were also documented [55]. In order to exert an impact on the physical and chemical characteristics of magnetic NPs. The qualities of these features are largely contingent upon the specific synthesis method employed and the chemical composition involved [55]. The subsequent discussion delves into the aforementioned criteria and their impact on the utilisation of ferrite NPs in the field of biomedicine.

8.4.1 Particle Size and Shape

The dimension of magnetic NPs is a significant physical characteristic that is manipulated to customise their magnetic characteristics and surface area. Numerous scholars

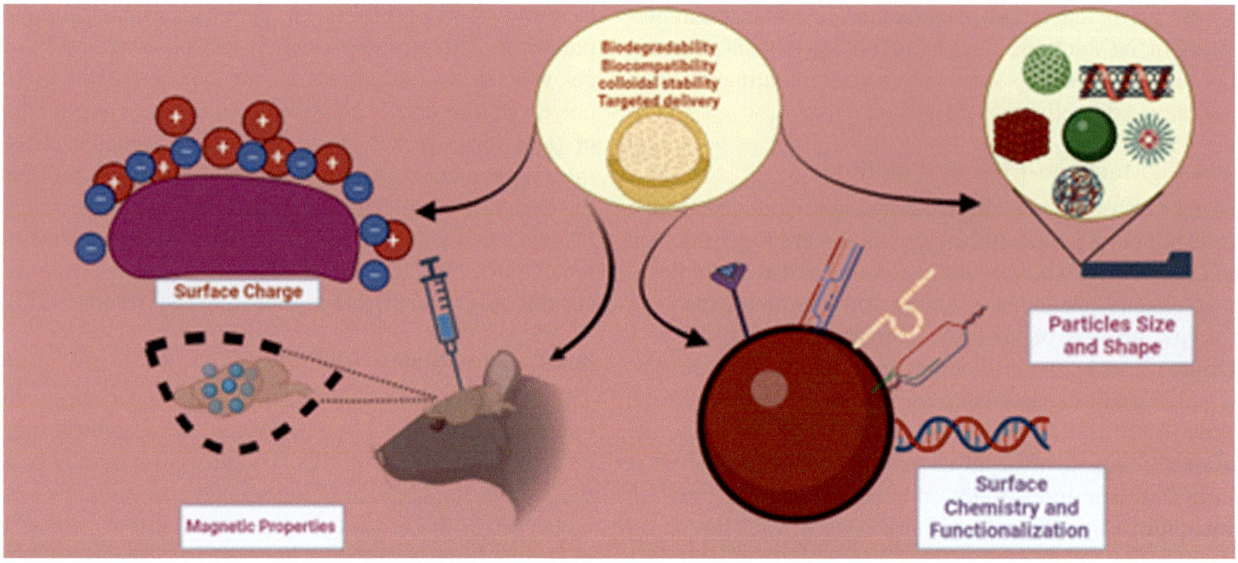

FIGURE 8.7 When considering the application of magnetic nanoparticles (NPs) in biomedical settings, certain parameters need to be taken into account.

have conducted investigations into the controlled size synthesis of iron oxide NPs and have observed that particles with dimensions less than 100 nm exhibit the phenomenon of superparamagnetism [56]. Superparamagnetic NPs exhibit magnetisation only when subjected to an external magnetic field and return to a non-magnetic condition upon removal of the external magnetic field. The aforementioned behaviour hinders particle activity in the absence of an applied field, resulting in reduced particle aggregation [45]. This characteristic is beneficial for in vivo applications [40]. The determination of NP size is influenced by two important factors: the boiling point of the solvent and the reaction duration during synthesis. This conclusion is supported by the observed variety in sizes of superparamagnetic NPs that have been reported [40]. Moreover, it has been discovered that the hydrodynamic diameters of NPs play a significant role in affecting the injectability, biodistribution, and clearance of hyperthermia or MRI agents. The determination of NP size is influenced by two key factors: the boiling point of the solvent and the reaction duration during synthesis [41].

Nevertheless, the production of magnetic NPs with desirable dimensions has posed a significant scientific and practical obstacle thus far. The magnetic characteristics of the NPs are significantly influenced by their morphology. The alteration in morphology can result in distinct crystal facets and atomic configurations, hence influencing the collective characteristics of the NPs [45]. The manipulation of these physical attributes is contingent upon the synthesis process or design selected by the NP. According to a theoretical investigation, it has been postulated that the morphology of NPs has an impact on the duration of circulation in the bloodstream, commonly referred to as the particle's half-life [53, 54]. This finding was substantiated by a study which shown that filamentous polymer micelles exhibit extended circulation durations of over one week following administration, in contrast to their spherical counterparts which only persisted for approximately two to three days. The observed outcomes were mostly ascribed to the inclination of the particles to align themselves with the direction of blood flow. Additional research has indicated that magnetic metal oxide NPs with spherical and cubic shapes exhibit distinct magnetic characteristics [56].

Numerous studies have demonstrated the manifestation of superparamagnetic characteristics and decreased aggregation in spherical NPs [56]. One of the primary obstacles encountered in the biomedical use of magnetic NPs is to their potential toxicity. According to a recent paper, it has been proposed that rod-shaped magnetic NPs exhibit lower toxicity compared to spherical-shaped NPs [56]. This further emphasises the significant influence that the size and form of NPs have on their characteristics. Physical synthesis technologies, including as gas-phase deposition and electron beam lithography, are complex operations that face limitations in effectively controlling the size and shape of particles within the nanometer scale. Nevertheless, the wet chemical synthesis methods exhibit characteristics of simplicity, manageability, and efficiency, hence enabling significant control over the size, content, and occasionally the morphology of the NPs. Numerous techniques have been documented in the literature, which are capable of regulating the dimensions and morphology of targeted NPs [57]. The consideration of procedures for achieving desired morphology, size, and shape is ongoing due to the variations in attributes seen in the NPs synthesised through diverse pathways. According to Serrano García *et al.*, the manipulation of iron oxide precursor to base concentrations, solvents, surfactants, the nature of surfactants, initial reactant concentrations, and reaction temperature and time can provide control over the size and dynamics of synthesised NPs [58].

8.4.2 Surface Chemistry and Functionalisation

NPs surfaces are specifically engineered to enhance their biocompatibility, solubility, and stability, while concurrently mitigating aggregation, optimising size distribution, and minimising toxicity towards healthy cells and tissues [58]. Furthermore, the interplay between NPs and the immune system, plasma proteins, and extracellular matrix is a crucial factor that is affected by both charge and hydrophobicity. Therefore, the surface chemistry of NPs plays a pivotal role in determining their level of biocompatibility and biodegradability. Numerous research have documented the utilisation of coated ferrite NPs exhibiting favourable biocompatibility in the fields of medication administration and hyperthermia [24, 27, 30, 34, 38]. The application of the coating can be accomplished by the utilisation of organic biopolymers or inorganic layers [24]. The characteristics of the coating and the hydrodynamic size have the potential to influence the behaviour of NPs inside a biological system, including processes such as cellular uptake and accumulation, circulation, and clearance from the body. In addition, it is imperative to employ suitable surface functionalisation techniques in order to facilitate the conjugation of biomolecules to NPs, hence enabling the achievement of desired applications [30, 31].

8.4.2.1 Organic Polymers

The utilisation of biopolymers for surface functionalisation of NPs is a prevalent practice, using both natural and manufactured polymers. Magnetic NPs are frequently enveloped with organic polymers, including chitosan, dextran, PVA, PEG, and PVP [32, 33]. Polymers play a significant role in the prevention of oxidation, hence facilitating enhanced absorption by bodily tissues and ensuring improved stability throughout therapeutic interventions. Dextran is an inherently biocompatible, neutral, and branching polysaccharide composed of glucose subunits, making it a natural polymer. Dextran and its derivatives, namely carboxydextran and carboxymethyl dextran, have been widely employed as polymers in numerous iron oxide-based NPs licenced for clinical use in MRI and hyperthermia therapy. These polymers have a long-standing history of being utilised in such applications [32, 33]. In their study, Ramnandan *et al.* conducted a comparative analysis of various coatings, including chitosan, PEG, and PVA, on $MgFe_2O_4$ NPs for the purpose of in vitro doxorubicin delivery. The findings of their investigation revealed that chitosan exhibited the most beneficial outcomes. The cytotoxicity of the same polymers was examined on $Mg_{0.5}Co_{0.5}Fe_2O_4$ NPs in certain cell types [59].

Chitosan is a linear polysaccharide derived from chitin and possesses significant reactive functional groups, notably amines. Chitosan is recognised for its capacity to enhance biocompatibility and biodegradability, while also offering a means to covalently attach medicines, imaging agents, and targeting ligands. PVA and PEG are examples of synthetic organic polymers that possess certain characteristics, such as improved particle uniformity, which effectively prevent particle coagulation [15, 16]. PVA is a substance that is capable of dissolving in water and exhibits characteristics such as emulsification and adhesion. These features contribute to the prevention of particle aggregation. PEG is widely acknowledged for its function in maintaining prolonged circulation within biological systems. The polymer's steric hindrance and stabilising effects [19, 22, 23] facilitate the evasion of reticuloendothelial system (RES) by the NPs.

8.4.2.2 Inorganic Compounds

Inorganic coating materials, such as silica, carbon, metals, and oxides, have been extensively investigated for their application in iron-oxide NPs within the field of biomedicine [30, 35]. It is worth noting that there is a limited selection of inorganic materials that can be utilised for surface coatings of magnetic NPs. Gold (Au) and silica (SiO_2) have historically been widely utilised in various applications, mostly owing to their biocompatibility and distinctive characteristics [35].

Silica exhibits a high degree of stability and demonstrates stability throughout a broad spectrum of pH values. The substance exhibits inert properties and enhances the dispersion of NPs by the formation of a protective layer that mitigates the dipolar attraction among the NPs. The reduction in the aggregation of noun phrases (NPs) results in enhanced stability [55]. The selection of suitable coating methods in the process of silica-coating is of utmost importance, as it has an indirect influence on the therapeutic effectiveness of NPs. The use of a silica coating facilitates the conjugation of additional biomolecules, ligands, fluorophores, dyes, and quantum dots [56].

The Stöber method, microemulsion, and aerosol pyrolysis are the three most often utilised techniques. The Stöber process involves the synthesis of silica through the hydrolysis and condensation of a sol-gel precursor, leading to the formation of iron oxide@SiO_2 NPs. It is imperative to generate a uniform silica layer devoid of silica particles lacking a core, as the presence of such particles may result in a diminished therapeutic dosage of NPs in MRI and hyperthermia treatment. The achievement of enhanced efficiency of the NPs as contrast agents can be attained by changing the thickness of silica shell [56].

Numerous studies have been conducted to examine the utilisation of hybrid NPs consisting of iron oxide coated with gold, aiming to harness the distinctive characteristics offered by these two materials [58]. Autonomous gold NPs (AuNPs) provide favourable characteristics including straightforward synthesis, robust stability, and distinctive optical features. When applied as a coating on iron oxide NPs, they have the ability to inhibit corrosion that may occur under acidic conditions. Furthermore, these NPs exhibit localised surface plasmon resonance, rendering them very suitable for applications such as cancer therapy by imaging or photothermal ablation [60]. The presence of a gold shell serves to effectively hinder the oxidation process of the iron oxide core, hence offering a highly favourable chemistry for surface functionalisation [60]. Just like the silica layer, it is crucial to meticulously adjust the thickness and geometry of the shell, since these factors might influence the magnetic properties of the NPs. Consequently, this may lead to a decrease in their effectiveness as contrast agents in MRI [58]. One of the significant obstacles frequently faced during the development of a drug delivery system for cancer therapy is the inherent difficulty in achieving therapeutic concentrations of the medicine at the specific areas of the disease. One of the contributing factors is the indiscriminate internalisation of NPs by organs in a healthy state.

Additionally, the interactions between NPs and opsonins, as well as their attachment to plasma proteins, lead to the creation of a corona [61]. The process of opsonisation is mostly regulated by the hydrophobicity and charge of the NPs. Therefore, it is vital to possess knowledge regarding the surface chemistry and the consequential effects of the coating. NPs that possess hydrophobic properties and carry an electric charge have been seen to exhibit reduced circulation durations because of opsonisation. This phenomenon triggers detection by the RES and consequently leads to the prompt removal of these NPs from the bloodstream [61]. Furthermore, the presence of hydrophobic groups on the surface of NPs can result in the aggregation of NPs and their rapid elimination by the RES. Therefore, the surface characteristics of NPs, including their charge, play a crucial role in restricting interactions between NPs and their host systems [62]. Numerous investigations have been conducted to explore the physical and magnetic characteristics of both uncoated and coated magnetic NPs with the aim of devising coating techniques that improve the NPs' properties. The objective is to achieve enhanced cellular uptake and efficient therapeutic outcomes on a consistent basis.

8.5 CONCLUSIONS

The field of cancer theranostics continues to be a pressing and crucial area of research, as evidenced by the increasing mortality rates associated with different types of cancer worldwide. Significant advancements have been made in the field of nanotherapeutics, demonstrating exponential growth in efforts to tackle the limitations associated with conventional therapeutic approaches. One of the primary obstacles in the utilisation of NPs for cancer treatment pertains to their restricted cellular absorption, a factor that is contingent upon their size and structure. In order to minimise the elimination by the RES prior to reaching the intended place, it is imperative to meticulously regulate these parameters. Due to the intrinsic tendency of magnetic NPs to aggregate and agglomerate, it is crucial to optimise their alterations in order to address the resulting implications on NP stability and toxicity. Although it is crucial to make adjustments, such as in the polymer coating, it is important to ensure that these modifications do not undermine the therapeutic effectiveness of the delivery system. Ideally, it is desirable for magnetic NPs to demonstrate superparamagnetic characteristics, characterised by a high MS and an HC. Ferrites have been identified as a promising subset of iron-oxide NPs in the field of nanomedicine. These properties can be attributed to their distinct and beneficial ultrastructural and magnetic characteristics. Moreover, due to its status as a vital mineral within the human body, iron is anticipated to undergo inherent degradation mechanisms, hence mitigating any potential harmful effects. The production methods for spinel ferrites have been emphasised due to their significant influence on the nanostructural properties of NPs. The utilisation of biopolymers, such as PEG, PVA, and chitosan, has become a customary approach in the development of NPs intended for nanomedicine applications, mostly due to their inherent ability to enhance the biocompatibility of the NPs. The existing research has made significant progress in enhancing the capabilities of ferrites in cancer theranostics. However, further endeavours are necessary to refine the size, shape, surface properties, and drug-loading capacities of these materials. These factors play a crucial role in determining the effectiveness of magnetic NPs as vehicles for therapeutic delivery.

REFERENCES

1. S. Senapati, A. K. Mahanta, S. Kumar, and P. Maiti, "Controlled Drug Delivery Vehicles for Cancer Treatment and Their Performance," *Signal Transduct Target Ther*, vol. 3, no. 1, p. 7, March 2018, doi: 10.1038/s41392-017-0004-3.
2. J. Peng, X. Liang, and L. Calderon, "Progress in Research on Gold Nanoparticles in Cancer Management," *Medicine*, vol. 98, no. 18, May 2019, doi: 10.1097/MD.0000000000015311.
3. J. Ferlay *et al.*, "Cancer Statistics for the Year 2020: An Overview," *Int J Cancer*, vol. 149, no. 4, pp. 778–789, August 2021, doi: 10.1002/IJC.33588.
4. C. Auría-Soro *et al.*, "Interactions of Nanoparticles and Biosystems: Microenvironment of Nanoparticles and Biomolecules in Nanomedicine," *Nanomaterials (Basel)*, vol. 9, no. 10, October 2019, doi: 10.3390/NANO9101365.
5. C. Fornaguera, and M. J. García-Celma, "Personalized Nanomedicine: A Revolution at the Nanoscale," *J Pers Med*, vol. 7, no. 4, December 2017, doi: 10.3390/JPM7040012.
6. A. Shakeel, and S. Rawat (2023). Nanomaterials in Bioimaging and Diagnostics. In *Integrated Nanomaterials and Their Applications* (pp. 311–327), doi: 10.1007/978-981-99-6105-4_15.
7. H. Hermawan, D. Ramdan, J. R. P. Djuansjah, H. Hermawan, D. Ramdan, and J. R. P. Djuansjah, "Metals for Biomedical Applications," *Biomedical Engineering – From Theory to Applications*, August 2011, doi: 10.5772/19033.
8. Y. Gao *et al.*, "Nanotechnology-Based Intelligent Drug Design for Cancer Metastasis Treatment," *Biotechnol Adv*, vol. 32, no. 4, pp. 761–777, 2014, doi: 10.1016/J.BIOTECHADV.2013.10.013.
9. C. Martinelli, C. Pucci, and G. Ciofani, "Nanostructured Carriers as Innovative Tools for Cancer Diagnosis and Therapy," *APL Bioeng*, vol. 3, no. 1, p. 11502, March 2019, doi: 10.1063/1.5079943.

10. R. Singh, P. Thakur, A. Thakur, H. Kumar, P. Chawla, J. V. Rohit, R. Kaushik, and N. Kumar, Colorimetric Sensing Approaches of Surface-Modified Gold and Silver Nanoparticles for Detection of Residual Pesticides: a Review. *Int J Environ Anal Chem*, vol. 101, no. 15, pp. 3006–3022, 2020, doi: 10.1080/03067319.2020.1715382.
11. A. Thakur, P. Thakur, and S. P. Khurana. (2022). *Synthesis and Applications of Nanoparticles* (pp. 1–17). Singapore: Springer.
12. L. Krishnia, P. Thakur, and A. Thakur. (2022). Synthesis of Nanoparticles by Physical Route. In *Synthesis and Applications of Nanoparticles* (pp. 45–59). Singapore: Springer Nature Singapore
13. S. M. Dizaj, S. Jafari, and A. Y. Khosroushahi, "A Sight on the Current Nanoparticle-Based Gene Delivery Vectors," *Nanoscale Res Lett*, vol. 9, no. 1, pp. 1–9, 2014, doi: 10.1186/1556-276X-9-252.
14. J. Dobson, "Gene Therapy Progress and Prospects: Magnetic Nanoparticle-Based Gene Delivery," *Gene Ther*, vol. 13, no. 4, pp. 283–287, February 2006, doi: 10.1038/SJ.GT.3302720.
15. M. Singh, and M. Ariatti, "A Cationic Cytofectin with Long Spacer Mediates Favourable Transfection in Transformed Human Epithelial Cells," *Int J Pharm*, vol. 309, no. 1–2, pp. 189–198, February 2006, doi: 10.1016/J.IJPHARM.2005.11.023.
16. M. Masserini, "Nanoparticles for Brain Drug Delivery," *ISRN Biochem*, vol. 2013, pp. 1–18, May 2013, doi: 10.1155/2013/238428.
17. Y. Bertrand *et al.*, "Transport Characteristics of a Novel Peptide Platform for CNS Therapeutics," *J Cell Mol Med*, vol. 14, no. 12, pp. 2827–2839, December 2010, doi: 10.1111/J.1582-4934.2009.00930.X.
18. A. C. Anselmo, and S. Mitragotri, "A Review of Clinical Translation of Inorganic Nanoparticles," *AAPS J*, vol. 17, no. 5, pp. 1041–1054, May 2015, doi: 10.1208/S12248-015-9780-2.
19. G. Kandasamy, and D. Maity, "Recent Advances in Superparamagnetic Iron Oxide Nanoparticles (SPIONs) for in Vitro and in Vivo Cancer Nanotheranostics," *Int J Pharm*, vol. 496, no. 2, pp. 191–218, December 2015, doi: 10.1016/J.IJPHARM.2015.10.058.
20. S. M. Dadfar *et al.*, "Iron Oxide Nanoparticles: Diagnostic, Therapeutic and Theranostic Applications," *Adv Drug Deliv Rev*, vol. 138, pp. 302–325, January 2019, doi: 10.1016/J.ADDR.2019.01.005.
21. W. Wu, Z. Wu, T. Yu, C. Jiang, and W. S. Kim, "Recent Progress on Magnetic Iron Oxide Nanoparticles: Synthesis, Surface Functional Strategies and Biomedical Applications," *Sci Technol Adv Mater*, vol. 16, no. 2, April 2015, doi: 10.1088/1468-6996/16/2/023501.
22. D. Ramnandan, S. Mokhosi, A. Daniels, and M. Singh, "Chitosan, Polyethylene Glycol and Polyvinyl Alcohol Modified $MgFe_2O_4$ Ferrite Magnetic Nanoparticles in Doxorubicin Delivery: A Comparative Study In Vitro," *Molecules*, vol. 26, no. 13, July 2021, doi: 10.3390/MOLECULES26133893.
23. S. Mngadi, S. Mokhosi, M. Singh, and W. Mdlalose, "Chitosan-Functionalized $Mg0.5Co0.5Fe_2O_4$ Magnetic Nanoparticles Enhance Delivery of 5-Fluorouracil In Vitro," *Coatings*, vol. 10, no. 5, p. 446, May 2020, doi: 10.3390/COATINGS10050446.
24. S. Moise, E. Céspedes, D. Soukup, J. M. Byrne, A. J. El Haj, and N. D. Telling, "The Cellular Magnetic Response and Biocompatibility of Biogenic Zinc- and Cobalt-Doped Magnetite Nanoparticles," *Sci Rep*, vol. 7, no. 1, pp. 1–11, January 2017, doi: 10.1038/srep39922.
25. H. W. Yang, M. Y. Hua, H. L. Liu, C. Y. Huang, and K. C. Wei, "Potential of Magnetic Nanoparticles for Targeted Drug Delivery," *Nanotechnol Sci Appl*, vol. 5, no. 1, pp. 73–86, August 2012, doi: 10.2147/NSA.S35506.
26. Y. Wang *et al.*, "A Magnetic Nanoparticle-Based Multiple-Gene Delivery System for Transfection of Porcine Kidney Cells," *PLOS ONE*, vol. 9, no. 7, p. e102886, July 2014, doi: 10.1371/JOURNAL.PONE.0102886.
27. L. S. Arias, J. P. Pessan, A. P. M. Vieira, T. M. T. De Lima, A. C. B. Delbem, and D. R. Monteiro, "Iron Oxide Nanoparticles for Biomedical Applications: A Perspective on Synthesis, Drugs, Antimicrobial Activity, and Toxicity," *Antibiotics*, vol. 7, no. 2, p. 46, June 2018, doi: 10.3390/ANTIBIOTICS7020046.
28. E. Tombácz, R. Turcu, V. Socoliuc, and L. Vékás, "Magnetic Iron Oxide Nanoparticles: Recent Trends in Design and Synthesis of Magnetoresponsive Nanosystems," *Biochem Biophys Res Commun*, vol. 468, no. 3, pp. 442–453, December 2015, doi: 10.1016/J.BBRC.2015.08.030.
29. R. A. Revia, and M. Zhang, "Magnetite Nanoparticles for Cancer Diagnosis, Treatment, and Treatment Monitoring: Recent Advances," *Mater Today*, vol. 19, no. 3, pp. 157–168, April 2016, doi: 10.1016/J.MATTOD.2015.08.022.
30. S. Kumari, F. Wan, A. Thakur, S. Singh, and P. Thakur (2023). Retrievable Nano-Sized Co–Mg–Zn Ferrite as an Active Photocatalyst and Adsorbent for Removing Organic and Biological Contaminants for Water Purification. *Ceram Int*, doi: 10.1016/j.ceramint.2023.11.071.
31. H. D. Wang, P. Thakur, and A. Thakur (2023). Biocompatibility, Bio-Clearance, and Toxicology. In *Integrated Nanomaterials and Their Applications* (pp. 201–221), doi: 10.1007/978-981-99-6105-4_10.
32. M. Khalkhali, K. Rostamizadeh, S. Sadighian, F. Khoeini, M. Naghibi, and M. Hamidi, "The Impact of Polymer Coatings on Magnetite Nanoparticles Performance as MRI Contrast Agents: A Comparative Study," *DARU, J Pharm Sci*, vol. 23, no. 1, pp. 1–12, September 2015, doi: 10.1186/S40199-015-0124-7.
33. Y. Zhang *et al.*, "Synthesis of Superparamagnetic Iron Oxide Nanoparticles Modified With MPEG-PEI via Photochemistry as New MRI Contrast Agent," *J Nanomater*, vol. 2015, 2015, doi: 10.1155/2015/417389.
34. H. Unterweger *et al.*, "Dextran-Coated Superparamagnetic Iron Oxide Nanoparticles for Magnetic Resonance Imaging: Evaluation of Size-Dependent Imaging Properties, Storage Stability and Safety," *Int J Nanomedicine*, vol. 13, pp. 1899–1915, March 2018, doi: 10.2147/IJN.S156528.
35. P. Thakur, S. Kumari, S. Singh, B. Ravelo, F. Wan, and A. Thakur. (2023). MG-ZN Ferrite Nanoparticles: A Brief Review on Synthesis, Characterizations, and Applications. *ChemBioEng Rev*, vol. 10, no. 5, pp. 711–736, doi: 10.1002/cben.202200063.
36. L. B. De Paula, F. L. Primo, M. R. Pinto, P. C. Morais, and A. C. Tedesco, "Evaluation of a Chloroaluminium Phthalocyanine-Loaded Magnetic Nanoemulsion as a Drug Delivery Device to Treat Glioblastoma Using Hyperthermia and Photodynamic Therapy," *RSC Adv*, vol. 7, no. 15, pp. 9115–9122, January 2017, doi: 10.1039/C6RA26105A.
37. X. Liu *et al.*, "Magnetic Nanomaterials-Mediated Cancer Diagnosis and Therapy," *Prog Biomed Eng*, vol. 4, no. 1, p. 012005, November 2021, doi: 10.1088/2516-1091/AC3111.
38. C. Liang, X. Zhang, Z. Cheng, M. Yang, W. Huang, and X. Dong, "Magnetic Iron Oxide Nanomaterials: A Key Player in Cancer Nanomedicine," *View*, vol. 1, no. 3, p. 20200046, September 2020, doi: 10.1002/VIW.20200046.

39. H. Ching-Chien *et al.*, "Effect of Particle Size of as-Milled Powders on Microstructural and Magnetic Properties of Y3MnxAl0.8-xFe4.2O12 Ferrites," *J Am Ceram Soc*, vol. 102, no. 6, pp. 3525–3534, June 2019, doi: 10.1111/JACE.16211.
40. M. Mahmoudi, H. Hofmann, B. Rothen-Rutishauser, and A. Petri-Fink, "Assessing the in Vitro and in Vivo Toxicity of Superparamagnetic Iron Oxide Nanoparticles," *Chem Rev*, vol. 112, no. 4, pp. 2323–2338, April 2012, doi: 10.1021/CR2002596.
41. N. Zhu *et al.*, "Surface Modification of Magnetic Iron Oxide Nanoparticles," *Nanomaterials (Basel)*, vol. 8, no. 10, October 2018, doi: 10.3390/NANO8100810.
42. M. Wu, and S. Huang, "Magnetic Nanoparticles in Cancer Diagnosis, Drug Delivery and Treatment," *Mol Clin Oncol*, vol. 7, no. 5, August 2017, doi: 10.3892/MCO.2017.1399.
43. J. Kudr *et al.*, "Magnetic Nanoparticles: From Design and Synthesis to Real World Applications," *Nanomaterials*, vol. 7, no. 9, p. 243, August 2017, doi: 10.3390/NANO7090243.
44. D. Chang *et al.*, "Biologically Targeted Magnetic Hyperthermia: Potential and Limitations," *Front Pharmacol*, vol. 9, no. AUGUST, p. 386237, August 2018, doi: 10.3389/FPHAR.2018.00831.
45. B. Jeyadevan, "Present Status and Prospects of Magnetite Nanoparticles-Based Hyperthermia," *J Ceram Soc Jpn*, vol. 118, no. 1378, pp. 391–401, 2010, doi: 10.2109/jcersj2.118.391.
46. S. Gul, S. B. Khan, I. U. Rehman, M. A. Khan, and M. I. Khan, "A Comprehensive Review of Magnetic Nanomaterials Modern Day Theranostics," *Front Mater*, vol. 6, July 2019, doi: 10.3389/fmats.2019.00179.
47. H. M. Williams, "The Application of Magnetic Nanoparticles in the Treatment and Monitoring of Cancer and Infectious Diseases," *Biosci Horiz: Int J Stud Res*, vol. 10, August 2017, doi: 10.1093/biohorizons/hzx009.
48. H. Shirazi, M. Daneshpour, S. Kashanian, and K. Omidfar, "Synthesis, Characterization and in Vitro Biocompatibility Study of Au/TMC/Fe_3O_4 Nanocomposites as a Promising, Nontoxic System for Biomedical Applications," *Beilstein J Nanotechnol*, vol. 6, pp. 1677–1689, August 2015, doi: 10.3762/bjnano.6.170.
49. R. M. Patil *et al.*, "Superparamagnetic Iron Oxide/Chitosan Core/Shells for Hyperthermia Application: Improved Colloidal Stability and Biocompatibility," *J Magn Magn Mater*, vol. 355, pp. 22–30, April 2014, doi: 10.1016/j.jmmm.2013.11.033.
50. M. Suciu *et al.*, "Applications of Superparamagnetic Iron Oxide Nanoparticles in Drug and Therapeutic Delivery, and Biotechnological Advancements," *Beilstein J Nanotechnol*, vol. 11, pp. 1092–1109, July 2020, doi: 10.3762/bjnano.11.94.
51. H. Zhang *et al.*, "Leucine-Coated Cobalt Ferrite Nanoparticles: Synthesis, Characterization and Potential Biomedical Applications for Drug Delivery," *Phys Lett A*, vol. 384, no. 24, p. 126600, August 2020, doi: 10.1016/j.physleta.2020.126600.
52. S. Lal, J. Verma, and C. J. F. Van Noorden, "Nanoparticles for Hyperthermic Therapy: Synthesis Strategies and Applications in Glioblastoma," *Int J Nanomedicine*, p. 2863, June 2014, doi: 10.2147/IJN.S57501.
53. K. Ulbrich, K. Holá, V. Šubr, A. Bakandritsos, J. Tuček, and R. Zbořil, "Targeted Drug Delivery with Polymers and Magnetic Nanoparticles: Covalent and Noncovalent Approaches, Release Control, and Clinical Studies," *Chem Rev*, vol. 116, no. 9, pp. 5338–5431, May 2016, doi: 10.1021/acs.chemrev.5b00589.
54. Z. Li, S. Tan, S. Li, Q. Shen, and K. Wang, "Cancer Drug Delivery in the Nano Era: An Overview and Perspectives," *Oncol Rep*, vol. 38, no. 2, pp. 611–624, February 2017, doi: 10.3892/or.2017.5718.
55. Y. I. Golovin *et al.*, "Non-Heating Alternating Magnetic Field Nanomechanical Stimulation of Biomolecule Structures via Magnetic Nanoparticles as the Basis for Future Low-Toxic Biomedical Applications," *Nanomaterials*, vol. 11, no. 9, p. 2255, August 2021, doi: 10.3390/nano11092255.
56. S. L. Gribanovsky, A. O. Zhigachev, D. Y. Golovin, Y. I. Golovin, and N. L. Klyachko, "Mechanisms and Conditions for Mechanical Activation of Magnetic Nanoparticles by External Magnetic Field for Biomedical Applications," *J Magn Magn Mater*, vol. 553, p. 169278, July 2022, doi: 10.1016/j.jmmm.2022.169278.
57. F. Broders-Bondon, T. H. Nguyen Ho-Bouldoires, M.-E. Fernandez-Sanchez, and E. Farge, "Mechanotransduction in Tumor Progression: The Dark Side of the Force," *J Cell Biol*, vol. 217, no. 5, pp. 1571–1587, May 2018, doi: 10.1083/jcb.201701039.
58. R. Serrano García, S. Stafford, and Y. Gun'ko, "Recent Progress in Synthesis and Functionalization of Multimodal Fluorescent-Magnetic Nanoparticles for Biological Applications," *Appl Sci*, vol. 8, no. 2, p. 172, January 2018, doi: 10.3390/app8020172.
59. N. Dhanda, P. Thakur, A. Sun, and A. Thakur. (2023). Structural, Optical and Magnetic Properties Along with Antifungal Activity of Ag-Doped Ni-Co Nanoferrites Synthesized by Eco-Friendly Route. *J Magn Magn Mater*, 572, 170598, doi: 10.1016/j.jmmm.2023.170598.
60. N. Bertrand, J. Wu, X. Xu, N. Kamaly, and O. C. Farokhzad, "Cancer Nanotechnology: The Impact of Passive and Active Targeting in the Era of Modern Cancer Biology," *Adv Drug Deliv Rev*, vol. 66, pp. 2–25, February 2014, doi: 10.1016/j.addr.2013.11.009.
61. A. Zeeshan, M. Farhan, and A. Siddiqui, "Nanomedicine and Drug Delivery: A Mini Review," *Int Nano Lett*, vol. 4, no. 1, pp. 1–7, February 2014, doi: 10.1007/S40089-014-0094-7.
62. J. Huang *et al.*, "Magnetic Nanoparticle Facilitated Drug Delivery for Cancer Therapy with Targeted and Image-Guided Approaches," *Adv Funct Mater*, vol. 26, no. 22, pp. 3818–3836, June 2016, doi: 10.1002/adfm.201504185.

9 Spinel Nano-Ferrites for Targeted Drug Delivery

Manish Shandilya, Renuka Kanojia, and Anand Salvi*
Amity School of Applied Sciences, Amity University Haryana, Gurugram, India

Larissa V. Panina
National University of Science and Technology, MISIS, Moscow, Russia

*Preeti Thakur and Atul Thakur**
Nanotechnology Wing, Innovative Science Research Society, Shimla, India
nanoLatticeX LLP, Manesar, Gurugram, India
Amity Institute of Nanotechnology, Amity University Haryana, Gurugram, India

9.1 INTRODUCTION

The most important technology of the twenty-first century is thought to be nanotechnology (Amiri et al., 2019; Lanzalaco & Armelin, 2017). Nanotechnology refers to the examination of nanoparticles (NPs) that possess a diameter ranging from around 1 to 100 nm. NPs are commonly differentiated from microparticles, fine particles, and coarse particles based on their size range. Microparticles typically range from 1 to 1000 μm, while fine particles fall within the 100–2500-nm size range (Ouvinha de Oliveira et al., 2014). Coarse particles, on the other hand, span from 2500 to 10,000 nm. The smaller size of NPs gives rise to distinct physical and chemical properties, such as colloidal, optical, and electrical properties (Singh et al., 2020). Due to their increased susceptibility to Brownian motion, colloidal particles often exhibit limited sedimentation tendencies (Singh et al., 2020). NPs possess distinct characteristics that differentiate them from bulk materials due to their reduced size, including enhanced chemical reactivity, increased energy absorption capacity, and improved biological mobility. Many scientific, industrial, ecological, and technical disciplines are currently developing because of the study and engineering of nanoscale materials. Agriculture, the food business, chemistry, biology, medicine, and the primary disciplines that have profited from the significant technical advancements made in nanotechnology (Singh et al., 2020). Recent years have seen a notable increase in the biological application of biomaterials with nanostructures. Because of their small size, magnetic NPs can be used in biomedicine (Mendonça et al., 2002). They can be used for a number of purposes, including medication administration, diagnostics, treatment, and magnetic resonance imaging (MRI) as well as in noninvasive tumours, because of their optical and magnetic characteristics, imaging, and the early detection of cancerous development (Ouvinha de Oliveira et al., 2014).

The constant search for novel drug delivery methods is causing a revolutionary paradigm change in the field of modern medicine (Abdel Maksoud et al., 2022). The pursuit of accuracy, effectiveness, and fewer adverse effects has prompted scientists to investigate nanotechnology as a potentially useful field. Spinel nano-ferrites are among the many types of nanomaterials that have shown great promise for transforming tailored medication delivery (Abdel Maksoud et al., 2022; Rivas et al., 2012). The synergistic interaction between spinel nano-ferrites and drug delivery (Figure 9.1) is thoroughly explored in this chapter, which also explains the specific features of these nano-ferrites and how they might be used to develop customised therapies (Kefeni et al., 2020a; Kumar & Mohammad, 2011).

The drawbacks of conventional methods, which range from systemic toxicity concerns to insufficient targeting, highlight the necessity for more accurate and effective drug delivery systems (Abdel Maksoud et al., 2022; Kefeni et al., 2020a). With its capacity to regulate materials at the nanoscale, nanotechnology emerged as a ray of hope in response to this necessity, providing previously unheard-of control over therapeutic interventions. As we go deeper into the field of nanotechnology, we observe that spinel nano-ferrites are a promising class of nanomaterials that may be able to overcome the drawbacks of traditional drug delivery (Srivastava & Yadav, 2012). The readers are introduced to the basic structural and magnetic properties that distinguish spinel nano-ferrites in this section. Their utility in targeted drug delivery is based on their distinct crystal structure, content, and magnetic properties (Jadhav et al., 2020), which offer a level of precision and control that is critical to achieving therapeutic efficacy (Kefeni et al., 2020a). The ideal drug-targeting delivery systems are those that hold their loaded pharmaceuticals in the circulation systems and only release them in response to internal or external triggers such as pH or enzymes. Perfect drug-targeting delivery systems release loaded drugs only after exposure to internal or external triggers, such as enzymes, pH, or magnetic

* Corresponding authors.

DOI: 10.1201/9781003568940-9

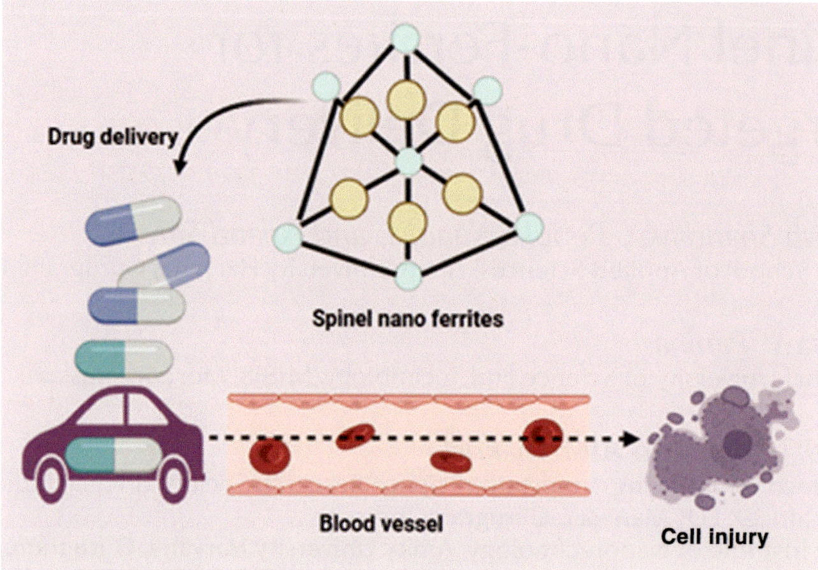

FIGURE 9.1 Synergistic interaction between spinel nano-ferrites and drug delivery.

fields. Spinel ferrites (SFs), chemically and thermally stable magnetic materials, have been used in various applications, including MRI, electronic devices, information storage, and drug delivery (Jang et al., 2009; Srivastava & Yadav, 2012). Recent interest has shifted towards the application of SFs and their derivative composites, as conventional delivery methods have limitations in successful treatment (Jermy et al., 2020).

In addition, the section highlights the gap that spinel nano-ferrites seek to fill in the field of medication delivery and briefly addresses the growing interest in nanotechnology as a whole. The details of creating spinel nano-ferrites, adding medicinal chemicals to their surfaces, and using their magnetic characteristics for precise delivery will become clear as we move through the next portions of this chapter. Through this investigation, we hope to gain a comprehensive understanding of the potential these nanomaterials have to transform drug delivery and welcome a new era of tailored and focused treatments.

9.2 FERRITES

Ferrite is a ceramic substance that possesses magnetic characteristics (Silva et al., 2019), rendering it highly advantageous for a wide range of electronic applications (Dahiya et al., 2018; Srivastava & Yadav, 2012). These chemicals represent a distinct category of chemical compounds that consist of lanthanides and rapid transition metals (Melagiriyappa & Jayanna, 2009). Ferrites are characterised by their hardness, brittleness, iron content, and typically exhibit a grey or black colouration (Gadkari et al., 2009; Jadhav et al., 2020). Additionally, they possess a polycrystalline structure. These objects exhibit electrical non-conductivity. These entities consist of iron oxide and one or more other metals in their chemical composition (Gadkari et al., 2009; Liu et al., 2008). A ferrite can be produced through the chemical reaction between ferric oxide and other metals, such as magnesium, aluminium, barium, manganese, copper, nickel, cobalt, or iron. A ferrite can often be denoted by the chemical formula $M(Fe_xO_y)$, where M indicates a metal capable of forming divalent bonds, including the above-mentioned components (Jacobo et al., 2004). Ferrites are a category of magnetic materials that are widely utilised in numerous applications, including microwave devices, computer memory chips, magnetic recording media, transformer cores, rod antennas, and various fields within telecommunication and electrical engineering (Kefeni et al., 2020a; Verma et al., 1999).

Ferrites composed of iron oxide (Fe_2) Beads and cores are commonly employed in the design of equipment to mitigate and disperse elevated amounts of high-frequency noise generated by electromagnetic devices (Deraz, 2010).

9.3 STRUCTURE AND PROPERTIES OF SPINEL NANO-FERRITES

The determination of the structure of SFs in the context of $MgAl_2O_4$ was initially conducted by Bragg and Nishikawa in the year 1915. The general structural formula for a spinel ferrite is commonly denoted as MFe_2O_4, where M represents a divalent metal ion such as Fe, Co, Ni, Mn, Mg, Cu, Zn, etc., or a combination thereof (Abdel Maksoud et al., 2020; Gao et al., 2019). The spinel structure is classified under the *Fd3m* space group (Abdel Maksoud et al., 2022). Ferrite is a material that resembles ceramic and has magnetic qualities that are helpful in a variety of electronic devices (Ashour et al., 2018). These are a special class of chemicals that include fast-transition metals and lanthanides. Ferrites are polycrystalline, hard, brittle, and contain iron. They are often grey or black in colour. They don't conduct electricity.

In its chemical combination, iron oxide and one or more additional metals make up their composition. Spinel nano-ferrites are members of a group of materials called spinel structures, which are distinguished by a particular crystal structure (Gao et al., 2019). Two different kinds of metal ions are organised in a certain way inside the crystal lattice to form this structure (Gao et al., 2019). Iron and other transition metals are commonly present in spinel nano-ferrites, which are oxides made of metal cations (Ashour et al., 2018). These nano-ferrites' electrical and magnetic characteristics are largely dependent on their composition (Reyes-Rodríguez et al., 2017).

The SF MFe_2O_4 is made up of eight cubic cells that hold 56 ions and has a cubic unit cell. The distribution of these ions was as follows: 32 anions of oxygen (O^{2-}), 8 cations of M^{2+}, and 16 ferric cations as well (Abdel Maksoud et al., 2020, 2022). The oxygen ions have a substantial radius and established an almost crowded cubic face-centred structure among smaller metal cations that make up the sub-locations states that are part of the space group *Fd*3*m* (Gao et al., 2019). SFs are typically characterised by two vibrational bands, υ1 and υ2, which stem from the stretching vibration of A-site groups (tetrahedral) and B-site groups (octahedral), respectively (Reyes-Rodríguez et al., 2017). The crystal lattice's magnetic ion arrangement causes spinel nano-ferrites to behave ferromagnetically (Ghodake et al., 2017). This characteristic, which makes it possible to manipulate them with external magnetic fields, is essential for their use in targeted drug administration.

9.4 PROPERTIES OF SPINEL NANO-FERRITES

SFs have attracted significant attention from researchers due to their distinct electrical, dielectric, electronic, mechanical, magnetic, optical, and catalytic characteristics (Silva et al., 2019). The SFs can be characterised as magnetic semiconductors due to their distinctive combination of ferrimagnetic and semiconductor features (Jadhav et al., 2020). These materials exhibit distinct physicochemical characteristics, such as a high Curie temperature, significant magnetic crystalline anisotropy, favourable thermal stability, and moderate chemical composition (Gul et al., 2020). Furthermore, these materials possess exceptional magnetic properties, adjustable dimensions and morphology, a large specific surface area, active surface sites, notable chemical stability, and the convenience of facile modification or functionalisation (Balakrishnan et al., 2009). The qualities discussed are influenced by various factors, including the specific divalent cation present in the spinel structure (in addition to Fe^{3+}), as well as its distribution between tetrahedral and octahedral sites (Guardia et al., 2012). Other factors that can impact these properties include the synthesis method employed, the grain size of the material, its morphology, and the temperature at which annealing occurs (Silva et al., 2019). Moreover, there has been a significant amount of research conducted on SFs in the field of water treatment (Takkar et al., 2022). This is mostly due to their cost-effectiveness, strong catalytic activity, intriguing band-gap properties, selectivity, capacity to be recovered and reused in subsequent cycles without significant loss of activity, and low toxicity (Silva et al., 2019). Due to their high saturation magnetisation (Ms), spinel nano-ferrites are frequently sensitive to external magnetic fields. Drug delivery systems use this feature for magnetic targeting. Due to their remarkable magnetic and structural properties, magnetic NPs are widely used in a variety of applications, including MRI, high-density magnetic recording media, magnetically guided drug delivery (Kumar & Mohammad, 2011), and hyperthermia (Jang et al., 2009). Ms, the greatest value of magnetisation, approaches as the magnetic field (H) strength increases (Massoudi et al., 2020). The magnetisation variations show distinct performance when the magnetic field is cancelled (Silva et al., 2019). Low magnetisation remains in the SF, which is known as residual magnetisation (Mr). The impacted magnetisation in SF constantly decreases when the magnetic field intensity gets closer to a negative value by decreasing it to negative states (Sasikala et al., 2016). The spinal nano-ferrites' surface can be modified to improve biocompatibility and make it easier to load medicinal chemicals. Coatings or functional groups to interact with medications or target ligands are examples of surface modifications. In order to use spinel nano-ferrites for medication administration, it is imperative to evaluate their biocompatibility (Kumar & Mohammad, 2011). In order to guarantee safety and reduce any potential negative consequences, it is crucial to understand (Kumar & Mohammad, 2011) how these materials interact with biological systems. For spinel nano-ferrites to be used as medication carriers, their physiological stability is essential. The usefulness of the nanomaterials depends on ensuring that they do not change in an undesirable way and stay intact.

The section "Structure and Properties of Spinel Nano-ferrites" offers a basis for understanding the various characteristics that make these NPs promising for focused medication delivery. Their potential to revolutionise drug delivery methods is attributed to a combination of factors, including their spinel structure, magnetic capabilities, nanoscale size, and surface features.

9.5 METHOD OF SYNTHESIS OF SPINEL NANO-FERRITES

In today's world, several methods are employed for the synthesis of SFs (Figure 9.2), encompassing solid state, co-precipitation, sol-gel, hydrothermal, microemulsion, solvothermal, template method, sono-chemical, electrochemical, and other techniques (Kefeni et al., 2015, 2020b). However, so far, a universally applicable method for synthesising SFs NPs has not been established, with each synthesis approach possessing its own distinct set of merits and drawbacks (Kumar & Mohammad, 2011). Moreover, the form, structure, and characteristics of magnetic ferrites

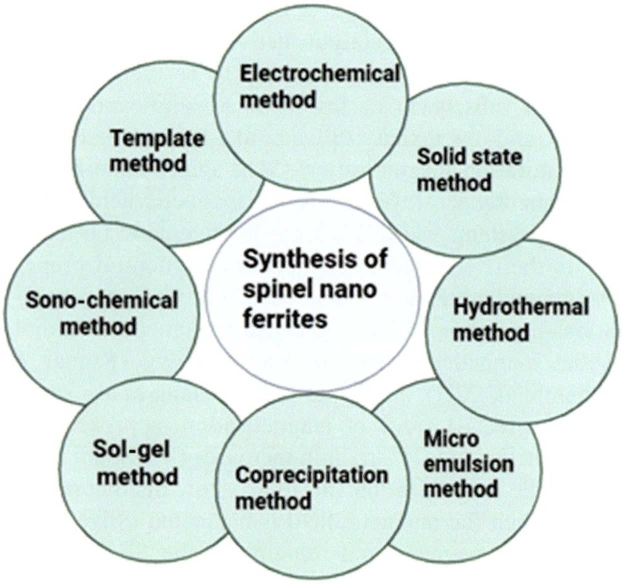

FIGURE 9.2 Different methods for the synthesis of spinel nano-ferrites.

are significantly affected by the chosen synthesis methods, which in turn might impact their catalytic activity when employed as catalysts.

9.5.1 Sol-Gel Method

It involves making a colloidal solution (sol), which then gels to create a network (gel). Spinel nano-ferrite is then obtained by calcining and drying the gel (Jalili et al., 2016). A product containing a highly chemically active component can be solidified through a solution, sol, or gel using the sol-gel process (Lanzalaco & Armelin, 2017). An oxide or other compound can then be heat-treated to further solidify it under mild conditions. Sol is a colloidal system that possesses liquid properties. Small or large molecules make up the dispersed particles. While the gel is made up of colloidal particles or polymers in a sol or solution that, under specific circumstances, join to one another to form a spatial network structure, the particles in the gel are uniformly dispersed and range in size from 1 to 100 nm (Rahimi et al., 2014). The hydrolysis and polycondensation, gelation, ageing, drying, densification, and crystallisation are the sequential steps in the sol-gel process (Jalili et al., 2016).

It takes a bottom-up strategy. By using a compound with a high chemical activity as a precursor and uniformly mixing it with these raw ingredients in the liquid phase, the sol-gel technique creates a stable, clear sol system in solution through hydrolysis and condensation chemical processes (Kefeni et al., 2017). The ageing colloidal particles and the sol gradually polymerise to produce a gel with a three-dimensional network structure. In order to prepare molecular and even nano-substructure materials, the gel is dried, sintered, and hardened (Jalili et al., 2016). A sol is a suspension of microscopic particles dispersed in a solution medium, whereas a solution is a single-phase system. This is the main distinction between a sol and a solution. Precursor materials are hydrolysed and condensed in solution media during a conventional sol-gel process. Precursor materials that are typically selected are inorganic and inorganic salts, as well as metal alkoxides. The sol-gel process uses regulated hydrolysis and condensation reactions to control the size of nanoscale clusters as well as the final product's morphology and microstructure (Jalili et al., 2016). The concentration of the reactive species and the amount of ageing influence the size of these particles as well. The sol-gel method is an easy, straightforward, and affordable technique (Kefeni et al., 2017). The resulting end products have a high degree of purity and are generally homogenous in character.

9.5.2 Solid State Method

The process entails the direct chemical reaction between oxide or carbonate precursors under elevated temperatures, resulting in the formation of nano-ferrites (Lanzalaco & Armelin, 2017). The solid-state approach is a widely employed technique for synthesising single crystal and polycrystalline powders of phosphates and arsenates including transition metals and monovalent cations. The synthesis described is a widely employed technique for producing polycrystalline material using solid reagents. This methodology is employed to induce a chemical transformation from solid precursor materials, resulting in the formation of a novel solid product characterised by a precisely defined crystalline arrangement (Rahimi et al., 2014). The process involves the combination, palletisation, and controlled temperature heating of fine-grained metal compounds for a specified duration. Certain metal compounds, such as metal oxides or salts, necessitate the application of extreme conditions, such as elevated temperature and pressure, in order to induce reactions within a molten flux or a quickly condensing vapour phase (Sharifi et al., 2012). This particular procedure is commonly known as "shake and bake" or "heat and beat" chemistry. Characterising the reaction rate in solid-state synthesis holds significant importance. Solid-state reactions must reach full completion due to the limited purification options available for the resulting solids (Lanzalaco & Armelin, 2017). The kinetics of the solid-state reaction are influenced by several reaction conditions, encompassing the structural characteristics, morphology, and surface areas of the reactants, as well as the diffusion rate and thermodynamic parameters connected with the reaction (Rahimi et al., 2014). The chemical and physical characteristics of the end products are influenced by the chemical precursors and methods of manufacture. The present methodology involves the initiation of reactions involving metal compounds through the utilisation of an energy source, followed by the propagation of these processes through the heat generated during the production of products and byproducts (Lanzalaco & Armelin, 2017).

9.5.3 Hydrothermal Method

The hydrothermal process is considered a non-conventional approach for the synthesis of nano-crystalline inorganic materials. A hydrothermal reaction is characterised as a chemical process that occurs in the presence of aqueous solvents at elevated temperature and pressure (Lanzalaco & Armelin, 2017). This reaction leads to the creation of solid products that exhibit very low solubility under typical conditions. The technology is both economically efficient and environmentally sustainable. Temperature, water content, pressure, and reaction duration are the four essential physical factors in hydrothermal processing. The synthesis method commonly referred to as the solvothermal process involves the utilisation of solvents other than water (Sharifi et al., 2012). In a standard experimental procedure, the precursor substance is dissolved in an appropriate solvent and subsequently put into a stainless-steel container known as an autoclave. Typically, water is the recommended solvent in this experiment. After the completion of the transfer process, the vessel is securely sealed and exposed to thermal treatments. The temperature is maintained within the range of 80–250°C during the synthesis of basic nanostructures (Zhang & Zhu, 2012). At elevated temperatures, the supercritical solvent has a reaction with the precursor material, leading to the creation of products. Surfactants play a significant role in these reactions, in conjunction with the precursor materials, in order to obtain a nanostructure that exhibits the desired morphology (Ding et al., 2015). Furthermore, there is a diminishing presence of the uncomplicated and refined procedure. Aluminium phosphate nanosheets were obtained by the hydrothermal synthesis process.

9.5.4 Co-Precipitation Method

The concurrent precipitation of metal cations from their individual salts results in the creation of nano-ferrite particles. Co-precipitation is a frequently employed method in the production of NPs, which results in the achievement of a uniform distribution (Kefeni et al., 2015).

In the context of spinel ferrite NPs (SFNPs), the procedure involves the combination of divalent transition cations and ferric ions in aqueous solutions, with a volume-to-volume ratio of 1:2. In order to achieve superior characteristics in the production of SFNPs, the synthesis technique necessitates precise calibration and pH verification. The pH of the solution is often regulated by the addition of either NH_4OH or NaOH solutions (Nabiyouni et al., 2015). Subsequently, the solution will undergo a vigorous agitation, with or without subsequent drying. Several scientific investigations have been conducted using the co-precipitation technique to manufacture NPs. In a recent study conducted by Gul et al., the synthesis of $Al_xZnFe_{2-x}O_4$ NPs was documented. The co-precipitation technique was employed, wherein solutions containing the initial ingredients were dissolved in deionised water and agitated at a stoichiometric ratio. The pH of all the solutions was maintained at a constant value of 10 by gradually adding the NH_4OH solution. Subsequently, the solution underwent a noticeable colour transformation, transitioning from an orange hue to a significantly darker shade of brown, accompanied by the formation of precipitates resembling the colour of coffee, which settled at the bottom of the beaker (Amiri & Shokrollahi, 2013). The precipitates undergo a drying process and are thereafter subjected to crushing and annealing at a temperature of 600°C for a duration of 8 hours. The examination of magnetic properties was based on the assumption that the NPs that were created had exceptional magnetic qualities. Furthermore, the optical examinations conducted on the NPs revealed that these surface functionalities exhibit additional properties related to photocatalysis (Li et al., 2015). The present study examines the photocatalytic degradation of methylene blue using spinel samples (Gul et al., 2020). The separation of ferrite particles from contaminated water can be achieved with ease by utilising a magnetic field. Therefore, ferrite NPs can be effectively utilised and conveniently recycled for various photocatalytic applications (Gul et al., 2020). This proposed method is characterised by its relative simplicity, cost-effectiveness, and ability to facilitate the synthesis of NPs with precise control over their composition.

9.5.5 Micro Emulsion Method

Microemulsions are thermodynamically stable systems consisting of two immiscible liquids and a surfactant, exhibiting isotropic properties (Ali et al., 2014). The utilisation of the microemulsion approach results in the production of NPs that exhibit a limited range of sizes, hence achieving a narrow size distribution. Furthermore, the size of these NPs may be easily regulated by making modest modifications to the synthesis conditions (Rodríguez-Rodríguez et al., 2019). This methodology has three essential constituents, namely water, oil, and surfactant, which provide the manipulation and regulation of many characteristics, including particle size, geometry, morphology, homogeneity, and surface area. The methods can be classified as regular micelle methods, namely oil-in-water (o/w), and reverse micelle methods, specifically water-in-oil (w/o). Both approaches employ surfactants at concentrations exceeding the critical micelle concentration (CMC) (Ali et al., 2014). The aforementioned technique is recognised for its environmentally sustainable nature, as well as its efficacy at lower temperatures and the ability to reuse surfactants for several synthesis cycles (Ali et al., 2014). Furthermore, it holds potential for commercial-scale manufacturing (Ali et al., 2014). However, a significant drawback arises from the suboptimal crystalline structure of synthesised SFs, as well as the necessity for higher quantities of solvents and the resulting increased polydisperse caused by the slower nucleation rate at lower temperatures. Several spinel ferrites have been produced using the microemulsion process, including MFe_2O_4 where M represents Co, Ni, and Zn, as well as Mn-Zn ferrite, among others (Rodríguez-Rodríguez et al., 2019).

9.5.6 Template Method

Template synthesis is a straightforward and accessible process that has facilitated the production of highly advanced nanomaterials. The methodology is founded upon the straightforward yet efficacious concept that the pores of a host material can serve as a blueprint for guiding the development of novel materials (Ali et al., 2014). The template has a dual purpose. Firstly, it facilitates the replication of the framework with optimal reproducibility and serves as a foundational structure for organising the various functionalities of a device, including the active components and diverse interfaces. The typical procedure for the synthesis of nanostructured materials employing templates involves the sequential execution of three main steps: (1) preparation of the template, (2) guided synthesis of the desired materials utilising the template, and (3) potential removal of the template, if deemed essential. The approach employed in this study has successfully synthesised various SFs, such as MFe_2O_4 (where M represents Ni, Mg, or Co), $ZnFe_2O_4$, $NiFe_2O_4$, $MgFe_2O_4$, and $CoFe_2O_4$.

9.5.7 Sono-Chemical Methods

The sono-chemical process is a recently developed, efficient, and relatively simple approach for synthesising ferrite NPs, which eliminates the need for the utilisation of poisonous, dangerous, and costly chemicals (Ghobashy, 2017b). This approach is useful for the production of NPs due to its ease and simplicity. The ability to manipulate reaction conditions, achieve precise control over particle size distribution, produce NPs with strong monodispersed, and generate NPs of extremely small dimensions. Hence, the temperature and strength of ultrasonic waves are significant variables that can exert a direct influence on the particle size of SFs (Ghobashy, 2017a). During sono-chemical synthesis, ultrasound irradiation causes sound waves to travel through the liquid, resulting in the formation of microbubbles. Simultaneously, the solid particles undergo in-situ calcination as a result of the high-energy collisions caused by the ultrasound (Bang & Suslick, 2010). Through the utilisation of ultrasonication, it is feasible to attain atomic-level mixing, hence enabling the production of crystalline phases at comparatively low temperatures (Bang & Suslick, 2010).

9.6 ELECTROCHEMICAL METHOD

Electrochemical synthesis involves the application of an electric current between multiple electrodes, namely the anode and cathode, which are immersed in an electrolyte solution (Ouvinha de Oliveira et al., 2014). In this particular methodology, the anode undergoes oxidation, resulting in the formation of metal ion species within the electrolyte. Subsequently, the metal ion species are reduced back to their metallic form by the cathode, facilitated by the presence of stabilising agents. The current density, pH, electrolyte composition, electrode selection, applied cell potential, temperature, stirring, and electrode spacing are critical variables that can exert a substantial impact on the crystallinity and particle size of the resulting product (Ouvinha de Oliveira et al., 2014). The electrochemical synthesis process is characterised by its simplicity, affordability, environmental friendliness, and cost-effectiveness.

9.7 IMPORTANCE OF SPINEL FERRITES IN DRUG DELIVERY

The utilisation of SFNPs as a drug delivery agent, regulated by an external magnetic field, has garnered significant attention due to their simplicity, effectiveness, ease of manufacture, and capacity to tailor their properties for specific biological applications (Mody et al., 2014). Approximately 1 year ago, the German scientist Paul Ehrlich (1854–1915) introduced the notion of a "magic bullet" capable of selectively targeting and eliminating sick tissue. During the 1950s, significant advancements were made in the field of drug delivery, particularly in the area of microencapsulated drug particles. Subsequently, there has been a significant surge in the number of papers pertaining to this particular field of biomedicine (Guo et al., 2015). Numerous analyses and research papers pertaining to NPs generated from silica, gold, polymers, and other materials are presently accessible in the academic literature. Consequently, it is evident that magnetism does not constitute an indispensable attribute in the design of a drug delivery nanosystem, as a diverse array of nanomaterials can be employed for this purpose (Chan et al., 1993). In recent decades, there has been a significant increase in the interest in the utilisation of magnetic NPs for medication delivery purposes. The conventional approach to drug administration involves indiscriminate distribution among cells and tissues, which can lead to metabolic instability, causing toxicity throughout the body and diminishing the effectiveness of therapy (Jordan et al., 2009). One further noteworthy characteristic of SFNPs is their capacity to encapsulate cytotoxic medicines inside the matrix of polymers and subsequently transport them to cells (Karimi et al., 2014). The utilisation of SFNPs enables the encapsulation and controlled release of medications, allowing for their efficient circulation within the body without leakage (Veiseh et al., 2010). By harnessing the influence of an external magnetic field, these NPs may be directed towards the specific tumour site, facilitating targeted delivery of therapeutic agents and minimising impact on healthy cells. This approach has promise for enhancing the efficacy of cancer treatment. The medicine will undergo release and then exhibit therapeutic effects upon reaching its intended areas of action (Mattingly et al., 2015). Multifunctional NPs composed of SFNPs, anticancer medicines, semiconductors for cell imaging, and biocompatible coating agents have demonstrated significant potential in the field of cancer research. Previous studies have provided evidence that NPs of $CoFe_2O_4$, with a lauric acid capping agent, have potential as a drug delivery system due to their ability to release drugs in a pH-sensitive manner (Yang et al., 2014). In addition, the application of biocompatible materials to encapsulate or cover SFNPs may potentially enhance their stability and mitigate their cytotoxicity within cellular environments. In order to be suitable for magnetically driven drug delivery applications, SFNPs should exhibit a lack of Mr subsequent to the removal of an external magnetic field (Teng & Pong, 2018). Additionally, this feature enables the preservation of colloidal stability and the prevention of aggregation, hence facilitating their utilisation in biomedical applications. The aggregation of SFNPs is hypothesised to be influenced by the magnetic attraction between the particles (Maravajhala et al., 2012).

9.8 DRUG LOADING ONTO SPINEL NANO-FERRITES

The loading of therapeutic compounds onto spinel nano-ferrites constitutes a pivotal stage in the advancement of drug delivery systems. The impact of drug loading efficiency on the therapeutic effectiveness, release kinetics, and overall performance of carriers based on nano-ferrites is significant (Chevalier et al., 2015). Various techniques can be utilised to facilitate the loading of pharmaceuticals onto spinel nano-ferrites (Figure 9.3), with the selection of a specific approach contingent upon the characteristics of the drug and the features inherent to the nano-ferrite material. The following are often employed strategies:

9.8.1 Adsorption

The drug molecules exhibit affinity towards the surface of spinel nano-ferrites via physical or chemical interactions. The proposed method exhibits simplicity and versatility, rendering it ideal for a diverse array of pharmaceutical compounds (Amiri et al., 2019).

9.8.2 Encapsulation

The medication is incorporated into the nano-ferrite structure either during the production process or through additional procedures. The purpose of drug encapsulation is to safeguard the drug from various environmental influences and facilitate prolonged release (Amiri et al., 2019; Lanzalaco & Armelin, 2017).

9.8.3 Co-Precipitation with Drug

The co-precipitation of drug and nano-ferrite precursors results in the concurrent synthesis of the drug-loaded nano-ferrite (Kumar & Mohammad, 2011; Natarajan & Tomich, 2020). The nano-ferrite facilitates the homogeneous dispersion of the medication.

9.8.4 Layer by Layer Assembly

A multilayered structure is formed by assembling alternating layers of polyelectrolytes and nano-ferrites loaded with drugs (Mohapatra et al., 2011). This technology offers fine manipulation of drug release kinetics, making it well-suited for sustained-release applications.

9.8.5 Chemical Conjugation

The drug molecules are chemically bonded to the surface of nano-ferrites. The formulation provides stability and

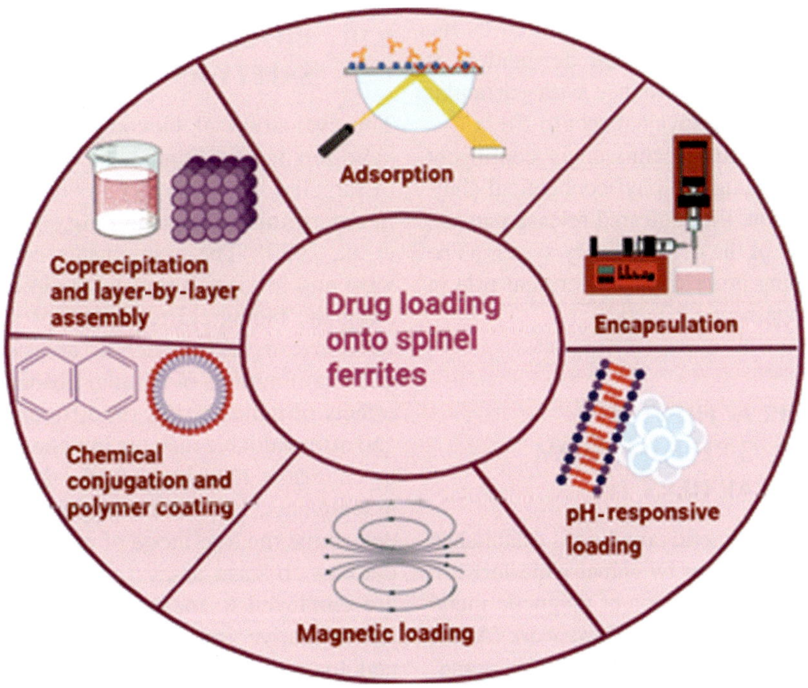

FIGURE 9.3 Different methods of loading drugs through spinel nano-ferrites.

facilitates regulated release, preventing the untimely release of the medicine (Ouvinha de Oliveira et al., 2014).

9.8.6 Polymer-Coating

A biocompatible polymer is utilised to coat the surface of nano-ferrites, enabling the transportation of medication molecules (Dahiya et al., 2018). The utilisation of this technology improves the biocompatibility of the medicine and establishes a protective barrier.

9.8.7 Magnetic Loading

The utilisation of magnetic NPs for drug delivery involves the loading of pharmaceuticals onto these particles, which are subsequently directed towards the desired location through the application of external magnetic fields (Amiri et al., 2019; Guo et al., 2015). This technology facilitates the precise administration of therapeutic agents to designated anatomical sites.

9.8.8 Hydrophobic/Hydrophilic Interactions

The loading process is facilitated by leveraging the hydrophobic or hydrophilic properties of the medication and the nano-ferrite surface (Veiseh et al., 2010). The substance exhibits versatility, making it ideal for a diverse range of medications characterised by varying features.

9.8.9 pH-Responsive Loading

pH-responsive linkers or materials are employed for the purpose of loading pharmaceutical agents onto nano-ferrites, wherein the subsequent release is initiated by the pH conditions prevailing at the intended destination. It facilitates regulated release in accordance with particular physiological circumstances (Maravajhala et al., 2012).

The selection of the loading technique is contingent upon various criteria, including the physicochemical characteristics of the medication, the preferred release pattern, and the intended purpose of the drug delivery system. The optimisation of the loading process plays a critical role in attaining the intended therapeutic effects.

9.9 STRATEGIES FOR ACHIEVING SUSTAINED DRUG RELEASE FROM NANO-FERRITE CARRIERS

The fundamental basis for magnetic targeting is established by the intrinsic magnetism shown by spinel nano-ferrites, which can be attributed to the inclusion of magnetic metal ions such as iron within their crystalline framework (Amiri et al., 2019). The exact targeting of drug-loaded nano-ferrites to specific anatomical regions within the body is achieved by carefully applying external magnetic fields. The mobility of the magnetic nanocarriers is influenced by the direction and strength of the field, enabling them to be guided to the desired locations. The application of induced magnetic forces and torques results in the alignment and propulsion of nanocarriers, hence improving their localisation at the intended target location (Mattingly et al., 2015). The use of controlled navigation inside the bloodstream or tissues serves to enhance the targeted delivery of carriers, hence increasing the accuracy and precision of drug administration. The utilisation of magnetic targeting enables the precise accumulation of drug-loaded nano-ferrites at the desired tissue or organ, hence reducing the occurrence of off-target effects and augmenting the therapeutic concentration (Dahiya et al., 2018). The utilisation of imaging techniques, such as MRI, for real-time monitoring, offers significant insights into the efficacy of magnetic targeting, allowing for necessary modifications. The efficiency of magnetic targeting can be enhanced through many approaches, such as ensuring the biocompatibility of magnetic nanocarriers, refining their magnetic characteristics, and investigating potential synergies with other targeting technologies (Amiri et al., 2019). Furthermore, it is imperative to take into account the practical considerations of clinical translation, including the viability of utilising external magnetic fields within a clinical environment (Mattingly et al., 2015). This is essential in order to effectively incorporate magnetic targeting into practical medical applications in real-world scenarios. By effectively addressing these translational issues, it is possible to facilitate the smooth integration of magnetic targeting technologies into clinical practices, hence enhancing treatment effects (Khezerlou et al., 2021).

9.10 BIOCOMPATIBILITY AND SAFETY ASSESSMENT

The evaluation of biocompatibility and safety of spinel nano-ferrites is a fundamental aspect in the advancement of targeted drug delivery systems, as it plays a vital role in determining their suitability for clinical use (Al-Rawi et al., 2020). The assessments encompass a range of in vitro and in vivo experiments aimed at thoroughly evaluating the biological responses of spinel nano-ferrites. In vitro investigations, such as assessments of cell viability and examinations of cellular absorption, shed light on the effects of nano-ferrites on different types of cell cultures and offer valuable insights into their cellular internalisation mechanisms (Mushtaq et al., 2021). Hemocompatibility assessments, such as hemolysis assays, are conducted to determine the likelihood of red blood cell harm. Similarly, cytokine release assays and oxidative stress evaluations are employed to measure immune responses and potential oxidative impacts, respectively. The assessment of genotoxicity, through the utilisation of assays such as the micronucleus assay, provides valuable information regarding the possible chromosomal harm caused by exposure to spinel nano-ferrites. In order to get insights into the systemic effects, the evaluation of biodistribution, organ toxicity, and immunological responses is conducted through the utilisation of animal models in in vivo investigations (Amiri et al., 2019). The study of tissue biopsies obtained from animals that have been exposed to a certain stimulus shows possible histopathological alterations. Additionally, the assessment of immune cell infiltration at the site of injection or in the specific tissue of interest yields valuable insights into the immunological response. The investigation of long-term impacts is conducted through the examination of chronic exposure studies, which provide valuable knowledge regarding the outcomes of extended periods of exposure, such as chronic toxicity and the buildup of substances in organs. The assessment of cellular responses and immunological reactions in relation to adjustments made to the surface of nano-ferrite is of utmost importance in determining the biocompatibility of such modifications (Srinivasan et al., 2018). The assessment of clinical significance entails replicating pertinent physiological parameters in both in vitro and in vivo investigations to more accurately represent the circumstances in which spinel nano-ferrites will be employed. Finally, the strict adherence to regulatory standards, such as the compliance with Good Laboratory Practice (GLP) recommendations, plays a crucial role in maintaining the integrity and dependability of safety assessments. This, in turn, facilitates the smooth transition of research findings into the regulatory approval processes. This integrated methodology for evaluating biocompatibility and safety offers a comprehensive comprehension of the potential hazards linked to spinel nano-ferrites, hence facilitating the design of secure and efficient targeted drug delivery systems for therapeutic purposes.

9.11 FUTURE PROSPECTS AND CHALLENGES

Spinel nano-ferrites hold great promise for targeted drug delivery, but they also present challenges. These include enhanced targeting precision, multifunctional nanoplatforms, responsive and dynamic delivery systems, personalised medicine, and biomimetic design. However, challenges include ensuring long-term biocompatibility and safety, addressing large-scale production, understanding clearance mechanisms and biodistribution, ensuring regulatory approval, and ensuring cost-effectiveness. Research and development can refine targeting strategies using advanced magnetic guidance, external stimuli, and combination approaches. Additionally, designing nano-ferrites that respond to dynamic physiological changes can improve therapeutic efficacy. Personalised medicine can also be developed by tailoring delivery systems to individual patient profiles. Additionally, addressing regulatory approval and cost-effectiveness is crucial for successful integration into clinical practice. Overcoming these challenges will enable the realisation of the full potential of spinel nano-ferrites in revolutionising targeted drug delivery systems.

9.12 CONCLUSION

In conclusion, this book chapter has presented a thorough examination of the application of spinel nano-ferrites in targeted drug delivery systems. The distinctive characteristics of spinel nano-ferrites, such as their intrinsic magnetism, compatibility with biological systems, and adaptable production techniques, render them very prospective for the advancement of sophisticated pharmaceutical delivery systems. Researchers have a wide range of options at their disposal when it comes to tailoring drug delivery systems to meet specific therapeutic requirements. These options include various mechanisms for controlled drug release, such as diffusion-controlled and magnetic-responsive release. This diversity allows for a spectrum of possibilities in the field of drug delivery. The comprehensive investigation of in vitro and in vivo experiments, evaluations of hemocompatibility, analysis of immunological reactions, and examination of long-term consequences have provided valuable insights into the safety characteristics and biocompatibility of spinel nano-ferrites. Comprehending the complexities of these biological interactions is of utmost importance for the effective transition of spinel nano-ferrite-based drug delivery systems from laboratory experimentation to clinical use.

REFERENCES

Abdel Maksoud, M. I. A., Elgarahy, A. M., Farrell, C., Al-Muhtaseb, A. H., Rooney, D. W., & Osman, A. I. (2020). Insight on water remediation application using magnetic nanomaterials and biosorbents. *Coordination Chemistry Reviews*, *403*, 213096. https://doi.org/10.1016/j.ccr.2019.213096

Abdel Maksoud, M. I. A., Ghobashy, M. M., Kodous, A. S., Fahim, R. A., Osman, A. I., Al-Muhtaseb, A. H., Rooney, D. W., Mamdouh, M. A., Nady, N., & Ashour, A. H. (2022). Insights on magnetic spinel ferrites for targeted drug delivery and hyperthermia applications. *Nanotechnology Reviews*, *11*(1), 372–413. https://doi.org/10.1515/ntrev-2022-0027

Ali, R., Khan, M. A., Mahmood, A., Chughtai, A. H., Sultan, A., Shahid, M., Ishaq, M., & Warsi, M. F. (2014). Structural, magnetic and dielectric behavior of $Mg_{1-x}Ca_xNi_yFe_{2-y}O_4$ nano-ferrites synthesized by the micro-emulsion method. *Ceramics International*, *40*(3), 3841–3846. https://doi.org/10.1016/j.ceramint.2013.08.024

Al-Rawi, N. N., Anwer, B. A., Al-Rawi, N. H., Uthman, A. T., & Ahmed, I. S. (2020). Magnetism in drug delivery: The marvels of iron oxides and substituted ferrites nanoparticles. *Saudi Pharmaceutical Journal*, *28*(7), 876–887. https://doi.org/10.1016/j.jsps.2020.05.003

Amiri, M., Salavati-Niasari, M., & Akbari, A. (2019). Magnetic nanocarriers: Evolution of spinel ferrites for medical applications. *Advances in Colloid and Interface Science*, *265*, 29–44. https://doi.org/10.1016/j.cis.2019.01.003

Amiri, S., & Shokrollahi, H. (2013). Magnetic and structural properties of RE doped Co-ferrite (RE=Nd, Eu, and Gd) nano-particles synthesized by Co-precipitation. *Journal of Magnetism and Magnetic Materials*, *345*, 18–23. https://doi.org/10.1016/j.jmmm.2013.05.030

Ashour, A. H., El-Batal, A. I., Maksoud, M. I. A. A., El-Sayyad, G. S., Labib, S., Abdeltwab, E., & El-Okr, M. M. (2018). Antimicrobial activity of metal-substituted cobalt ferrite nanoparticles synthesized by sol–gel technique. *Particuology, 40*, 141–151. https://doi.org/10.1016/j.partic.2017.12.001

Balakrishnan, S., Bonder, M. J., & Hadjipanayis, G. C. (2009). Particle size effect on phase and magnetic properties of polymer-coated magnetic nanoparticles. *Journal of Magnetism and Magnetic Materials, 321*(2), 117–122. https://doi.org/10.1016/j.jmmm.2008.08.055

Bang, J. H., & Suslick, K. S. (2010). Applications of ultrasound to the synthesis of nanostructured materials. *Advanced Materials, 22*(10), 1039–1059. https://doi.org/10.1002/adma.200904093

C.F. Chan, D., B. Kirpotin, D., & A. Bunn, P. Jr. (1993). Synthesis and evaluation of colloidal magnetic iron oxides for the site-specific radiofrequency-induced hyperthermia of cancer. *Journal of Magnetism and Magnetic Materials, 122*(1–3), 374–378. https://doi.org/10.1016/0304-8853(93)91113-L

Chevalier, M. T., Gonzalez, J., & Alvarez, V. (2015). Biodegradable polymeric microparticles as drug delivery devices. *IFMBE Proceedings, 49*, 187–190. https://doi.org/10.1007/978-3-319-13117-7_49

Dahiya, M. S., Tomer, V. K., & Duhan, S. (2018). Metal-ferrite nanocomposites for targeted drug delivery. In *Applications of Nanocomposite Materials in Drug Delivery*. Elsevier Inc. https://doi.org/10.1016/B978-0-12-813741-3.00032-7

Deraz, N. M. (2010). Glycine-assisted fabrication of nanocrystalline cobalt ferrite system. *Journal of Analytical and Applied Pyrolysis, 88*(2), 103–109. https://doi.org/10.1016/j.jaap.2010.03.002

Ding, Z., Wang, W., Zhang, Y., Li, F., & Liu, J. P. (2015). Synthesis, characterization and adsorption capability for Congo red of $CoFe_2O_4$ ferrite nanoparticles. *Journal of Alloys and Compounds, 640*, 362–370. https://doi.org/10.1016/j.jallcom.2015.04.020

Gadkari, A. B., Shinde, T. J., & Vasambekar, P. N. (2009). Structural analysis of Y3+-doped Mg-Cd ferrites prepared by oxalate co-precipitation method. *Materials Chemistry and Physics, 114*(2–3), 505–510. https://doi.org/10.1016/j.matchemphys.2008.11.011

Gao, Y., Wang, Z., Pei, J., & Zhang, H. (2019). Structural, elastic, thermal and soft magnetic properties of Ni-Zn-Li ferrites. *Journal of Alloys and Compounds, 774*, 1233–1242. https://doi.org/10.1016/j.jallcom.2018.09.394

Ghobashy, M. M. (2017a). Combined ultrasonic and gamma-irradiation to prepare TiO_2@PET-g-PAAc fabric composite for self-cleaning application. *Ultrasonics Sonochemistry, 37*, 529–535. https://doi.org/10.1016/j.ultsonch.2017.02.014

Ghobashy, M. M. (2017b). In-situ core-shell polymerization of magnetic polymer nanocomposite (PAAc/Fe_3O_4) particles via gamma radiation. *Nanocomposites, 3*(1), 42–46. https://doi.org/10.1080/20550324.2017.1316600

Ghodake, U. R., Kambale, R. C., & Suryavanshi, S. S. (2017). Effect of Mn^{2+} substitution on structural, electrical transport and dielectric properties of Mg-Zn ferrites. *Ceramics International, 43*(1), 1129–1134. https://doi.org/10.1016/j.ceramint.2016.10.053

Guardia, P., Corato, R., Di, Lartigue, L., Wilhelm, C., Espinosa, A., Garcia-hernandez, M., Gazeau, F., Manna, L., & Pellegrino, T. (2012). *Guardia2012.Pdf. ACS Nano, 6*(4), 3080–3091.

Gul, S., Yousuf, M. A., Anwar, A., Warsi, M. F., Agboola, P. O., Shakir, I., & Shahid, M. (2020). Al-substituted zinc spinel ferrite nanoparticles: Preparation and evaluation of structural, electrical, magnetic and photocatalytic properties. *Ceramics International, 46*(9), 14195–14205. https://doi.org/10.1016/j.ceramint.2020.02.228

Guo, H., Chen, W., Sun, X., Liu, Y. N., Li, J., & Wang, J. (2015). Theranostic magnetoliposomes coated by carboxymethyl dextran with controlled release by low-frequency alternating magnetic field. *Carbohydrate Polymers, 118*, 209–217. https://doi.org/10.1016/j.carbpol.2014.10.076

Jacobo, S. E., Duhalde, S., & Bertorello, H. R. (2004). Rare earth influence on the structural and magnetic properties of NiZn ferrites. *Journal of Magnetism and Magnetic Materials, 272–276* (III), 2253–2254. https://doi.org/10.1016/j.jmmm.2003.12.564

Jadhav, V. V., Shirsat, S. D., Tumberphale, U. B., & Mane, R. S. (2020). Properties of ferrites. In *Spinel Ferrite Nanostructures for Energy Storage Devices* (Issue M). Elsevier Inc. https://doi.org/10.1016/b978-0-12-819237-5.00003-1

Jalili, N. A., Muscarello, M., & Gaharwar, A. K. (2016). Nanoengineered thermoresponsive magnetic hydrogels for biomedical applications. *Bioengineering & Translational Medicine, 1*(3), 297–305. https://doi.org/10.1002/btm2.10034

Jang, J. T., Nah, H., Lee, J. H., Moon, S. H., Kim, M. G., & Cheon, J. (2009). Critical enhancements of MRI contrast and hyperthermic effects by dopant-controlled magnetic nanoparticles. *Angewandte Chemie – International Edition, 48*(7), 1234–1238. https://doi.org/10.1002/anie.200805149

Jermy, R., Ravinayagam, V., Alamoudi, W., Almohazey, D., Elanthikkal, S., Dafalla, H., Rehman, S., Chandrasekar, G., & Baykal, A. (2020). Tuning pH sensitive chitosan and cisplatin over spinel ferrite/silica nanocomposite for anticancer activity in MCF-7 cell line. *Journal of Drug Delivery Science and Technology, 57*(December 2019), 101711. https://doi.org/10.1016/j.jddst.2020.101711

Jordan, A., Wust, P., Fähling, H., John, W., Hinz, A., & Felix, R. (2009). Inductive heating of ferrimagnetic particles and magnetic fluids: Physical evaluation of their potential for hyperthermia. *International Journal of Hyperthermia, 25*(7), 499–511. https://doi.org/10.3109/02656730903287790

Karimi, Z., Mohammadifar, Y., Shokrollahi, H., Asl, S. K., Yousefi, G., & Karimi, L. (2014). Magnetic and structural properties of nano sized Dy-doped cobalt ferrite synthesized by co-precipitation. *Journal of Magnetism and Magnetic Materials, 361*, 150–156. https://doi.org/10.1016/j.jmmm.2014.01.016

Kefeni, K. K., Msagati, T. M., & Mamba, B. B. (2015). Synthesis and characterization of magnetic nanoparticles and study their removal capacity of metals from acid mine drainage. *Chemical Engineering Journal, 276*, 222–231. https://doi.org/10.1016/j.cej.2015.04.066

Kefeni, K. K., Msagati, T. A. M., & Mamba, B. B. (2017). Ferrite nanoparticles: Synthesis, characterisation and applications in electronic device. *Materials Science and Engineering: B, 215*, 37–55. https://doi.org/10.1016/j.mseb.2016.11.002

Kefeni, K. K., Msagati, T. A. M., Nkambule, T. T., & Mamba, B. B. (2020a). Spinel ferrite nanoparticles and nanocomposites for biomedical applications and their toxicity. *Materials Science and Engineering C, 107*(September 2019), 110314. https://doi.org/10.1016/j.msec.2019.110314

Kefeni, K. K., Msagati, T. A. M., Nkambule, T. T., & Mamba, B. B. (2020b). Spinel ferrite nanoparticles and nanocomposites for biomedical applications and their toxicity. *Materials Science and Engineering C, 107*, 110314. https://doi.org/10.1016/j.msec.2019.110314

Khezerlou, S., Babazadeh, M., Mehrizad, A., Gharbani, P., & Es'haghi, M. (2021). Preparation of hydroxyapatite-calcium ferrite composite for application in loading and sustainable release of amoxicillin: Optimization and modeling of the process by response surface methodology and artificial neural network. *Ceramics International*, *47*(17), 24287–24295. https://doi.org/10.1016/j.ceramint.2021.05.140

Kumar, C. S. S. R., & Mohammad, F. (2011). Magnetic nanomaterials for hyperthermia-based therapy and controlled drug delivery. *Advanced Drug Delivery Reviews*, *63*(9), 789–808. https://doi.org/10.1016/j.addr.2011.03.008

Lanzalaco, S., & Armelin, E. (2017). A review on recent progresses in biomedical applications. *Gels*, *3*(36), 16–25. https://doi.org/10.3390/gels3040036

Li, H., Qin, L., Feng, Y., Hu, L., & Zhou, C. (2015). Preparation and characterization of highly water-soluble magnetic Fe_3O_4 nanoparticles via surface double-layered self-assembly method of sodium alpha-olefin sulfonate. *Journal of Magnetism and Magnetic Materials*, *384*, 213–218. https://doi.org/10.1016/j.jmmm.2015.01.065

Liu, X., Cheng, B., Hu, J., Qin, H., & Jiang, M. (2008). Semiconducting gas sensor for ethanol based on $LaMg_xFe_{1-x}O_3$ nanocrystals. *Sensors and Actuators, B: Chemical*, *129*(1), 53–58. https://doi.org/10.1016/j.snb.2007.07.102

Maravajhala, V., Papishetty, S., & Bandlapalli, S. (2012). Nanotechnology in development of drug delivery system. *International Journal of Pharmaceutical Sciences*, *3*(1), 84–96.

Massoudi, J., Smari, M., Nouri, K., Dhahri, E., Khirouni, K., Bertaina, S., Bessais, L., & Hlil, E. K. (2020). Magnetic and spectroscopic properties of Ni-Zn-Al ferrite spinel: From the nanoscale to microscale. *RSC Advances*, *10*(57), 34556–34580. https://doi.org/10.1039/d0ra05522k

Mattingly, S. J., Otoole, M. G., James, K. T., Clark, G. J., & Nantz, M. H. (2015). Magnetic nanoparticle-supported lipid bilayers for drug delivery. *Langmuir*, *31*(11), 3326–3332. https://doi.org/10.1021/la504830z

Melagiriyappa, E., & Jayanna, H. S. (2009). Structural and magnetic susceptibility studies of samarium substituted magnesium-zinc ferrites. *Journal of Alloys and Compounds*, *482*(1–2), 147–150. https://doi.org/10.1016/j.jallcom.2009.03.112

Mendonça, M. H., Godinho, M. I., Catarino, M. A., Da Silva Pereira, M. I., & Costa, F. M. (2002). Preparation and characterisation of spinel oxide ferrites suitable for oxygen evolution anodes. *Solid State Sciences*, *4*(2), 175–182. https://doi.org/10.1016/S1293-2558(01)01238-9

Mody, V. V., Cox, A., Shah, S., Singh, A., Bevins, W., & Parihar, H. (2014). Magnetic nanoparticle drug delivery systems for targeting tumor. *Applied Nanoscience (Switzerland)*, *4*(4), 385–392. https://doi.org/10.1007/s13204-013-0216-y

Mohapatra, S., Rout, S. R., Maiti, S., Maiti, T. K., & Panda, A. B. (2011). Monodisperse mesoporous cobalt ferrite nanoparticles: Synthesis and application in targeted delivery of antitumor drugs. *Journal of Materials Chemistry*, *21*(25), 9185–9193. https://doi.org/10.1039/c1jm10732a

Mushtaq, S., Shahzad, K., Saeed, T., Ul-Hamid, A., Abbasi, B. H., Ahmad, N., Khalid, W., Atif, M., Ali, Z., & Abbasi, R. (2021). Biocompatibility and cytotoxicity in vitro of surface-functionalized drug-loaded spinel ferrite nanoparticles. *Beilstein Journal of Nanotechnology*, *12*, 1339–1364. https://doi.org/10.3762/bjnano.12.99

Nabiyouni, G., Julaee, M., Ghanbari, D., Aliabadi, P. C., & Safaie, N. (2015). Room temperature synthesis and magnetic property studies of Fe_3O_4 nanoparticles prepared by a simple precipitation method. *Journal of Industrial and Engineering Chemistry*, *21*, 599–603. https://doi.org/10.1016/j.jiec.2014.03.025

Natarajan, P., & Tomich, J. M. (2020). Understanding the influence of experimental factors on bio-interactions of nanoparticles: Towards improving correlation between in vitro and in vivo studies. *Archives of Biochemistry and Biophysics*, *694*(September), 108592. https://doi.org/10.1016/j.abb.2020.108592

Ouvinha de Oliveira, R., de Santa Maria, L. C., & Barratt, G. (2014). Nanomedicine and its applications to the treatment of prostate cancer. *Annales Pharmaceutiques Francaises*, *72*(5), 303–316. https://doi.org/10.1016/j.pharma.2014.04.006

Rahimi, R., Maleki, A., Maleki, S., Morsali, A., & Rahimi, M. J. (2014). Synthesis and characterization of magnetic dichromate hybrid nanomaterials with triphenylphosphine surface modified iron oxide nanoparticles ($Fe_3O_4@SiO_2@PPh_3@Cr_2O_7^{2-}$). *Solid State Sciences*, *28*, 9–13. https://doi.org/10.1016/j.solidstatesciences.2013.11.013

Reyes-Rodríguez, P. Y., Cortés-Hernández, D. A., Escobedo-Bocardo, J. C., Almanza-Robles, J. M., Sánchez-Fuentes, H. J., Jasso-Terán, A., De León-Prado, L. E., Méndez-Nonell, J., & Hurtado-López, G. F. (2017). Structural and magnetic properties of Mg-Zn ferrites ($Mg_{1-x}Zn_xFe_2O_4$) prepared by sol-gel method. *Journal of Magnetism and Magnetic Materials*, *427*(June), 268–271. https://doi.org/10.1016/j.jmmm.2016.10.078

Rivas, J., Bañobre-López, M., Piñeiro-Redondo, Y., Rivas, B., & López-Quintela, M. A. (2012). Magnetic nanoparticles for application in cancer therapy. *Journal of Magnetism and Magnetic Materials*, *324*(21), 3499–3502. https://doi.org/10.1016/j.jmmm.2012.02.075

Rodríguez-Rodríguez, A. A., Moreno-Trejo, M. B., Meléndez-Zaragoza, M. J., Collins-Martínez, V., López-Ortiz, A., Martínez-Guerra, E., & Sánchez-Domínguez, M. (2019). Spinel-type ferrite nanoparticles: Synthesis by the oil-in-water microemulsion reaction method and photocatalytic water-splitting evaluation. *International Journal of Hydrogen Energy*, 12421–12429. https://doi.org/10.1016/j.ijhydene.2018.09.183

Sasikala, A. R. K., Unnithan, A. R., Yun, Y. H., Park, C. H., & Kim, C. S. (2016). An implantable smart magnetic nanofiber device for endoscopic hyperthermia treatment and tumor-triggered controlled drug release. *Acta Biomaterialia*, *31*, 122–133. https://doi.org/10.1016/j.actbio.2015.12.015

Sharifi, I., Shokrollahi, H., & Amiri, S. (2012). Ferrite-based magnetic nanofluids used in hyperthermia applications. *Journal of Magnetism and Magnetic Materials*, *324*(6), 903–915. https://doi.org/10.1016/j.jmmm.2011.10.017

Silva, F. G. da, Depeyrot, J., Campos, A. F. C., Aquino, R., Fiorani, D., & Peddis, D. (2019). Structural and magnetic properties of spinel ferrite nanoparticles. *Journal of Nanoscience and Nanotechnology*, *19*(8), 4888–4902. https://doi.org/10.1166/jnn.2019.16877

Singh, T. A., Das, J., & Sil, P. C. (2020). Zinc oxide nanoparticles: A comprehensive review on its synthesis, anticancer and drug delivery applications as well as health risks. In *Advances in Colloid and Interface Science* (Vol. 286). Elsevier B.V. https://doi.org/10.1016/j.cis.2020.102317

Srinivasan, S. Y., Paknikar, K. M., Bodas, D., & Gajbhiye, V. (2018). Applications of cobalt ferrite nanoparticles in biomedical nanotechnology. *Nanomedicine*, *13*(10), 1221–1238. https://doi.org/10.2217/nnm-2017-0379

Srivastava, R., & Yadav, B. C. (2012). Ferrite materials: Introduction, synthesis techniques, and applications as sensors. *International Journal of Green Nanotechnology: Biomedicine*, *4*(2), 141–154. https://doi.org/10.1080/19430892.2012.676918

Takkar, R., Kaur, B., Rana, S., Kumar, P., Khanra, P., & Dhiman, M. (2022). Usage of magnetic spinel nano-ferrites in waste water treatment. *ECS Transactions*, *107*(1), 10237–10244. https://doi.org/10.1149/10701.10237ecst

Teng, Y., & Pong, P. W. T. (2018). One-pot synthesis and surface modification of lauric-acid-capped $CoFe_2O_4$ nanoparticles. *IEEE Transactions on Magnetics*, *54*(11), 2–6. https://doi.org/10.1109/TMAG.2018.2834524

Veiseh, O., Gunn, J. W., & Zhang, M. (2010). Design and fabrication of magnetic nanoparticles for targeted drug delivery and imaging. *Advanced Drug Delivery Reviews*, *62*(3), 284–304. https://doi.org/10.1016/j.addr.2009.11.002

Verma, A., Goel, T. C., Mendiratta, R. G., & Gupta, R. G. (1999). High-resistivity nickel-zinc ferrites by the citrate precursor method. *Journal of Magnetism and Magnetic Materials*, *192*(2), 271–276. https://doi.org/10.1016/S0304-8853(98)00592-7

Yang, Y., Li, Y., & Sun, Q. Y. (2014). Archaeal and bacterial communities in acid mine drainage from metal-rich abandoned tailing ponds, Tongling, China. *Transactions of Nonferrous Metals Society of China (English Edition)*, *24*(10), 3332–3342. https://doi.org/10.1016/S1003-6326(14)63474-9

Zhang, H., & Zhu, G. (2012). One-step hydrothermal synthesis of magnetic Fe_3O_4 nanoparticles immobilized on polyamide fabric. *Applied Surface Science*, *258*(11), 4952–4959. https://doi.org/10.1016/j.apsusc.2012.01.127

10 Benefits of Spinel Nano-Ferrites in Catalysis

*Pooja, Nisha Yadav, and Gyandshwar Kumar Rao**
Department of Chemistry, Biochemistry and Forensic Sciences,
Amity University Haryana, Gurugram, India

10.1 INTRODUCTION

Ferrites are a type of non-metallic compound consisting mainly of iron oxide combined with small amounts of metallic elements like Cu, Ni, Co and Zn with high metallic activity. The word ferrite is derived from the Latin word "ferrum" which means iron. The composition of iron oxide and the metal is ferrimagnetic and electrically non-conducting. The unequal opposing magnetic moment of ferrimagnetic materials assists in maintaining spontaneous magnetisation. Generally, ferrites are naturally magnetised and can be further magnetised to make permanent magnets. This property is known as ferrimagnetism. The general formula for a ferrite is MFe_2O_4, where M is an atom that is bivalent (normally a transition metal, like Cu, Zn, Mg and Co). On the basis of the crystalline structure, ferrites can be divided into three types: ortho ferrite ($MFeO_3$), where M is a metallic ion that is trivalent like a rare-earth ion [1, 2], spinel ferrite (MFe_2O_4), garnet ferrite ($M_3Fe_5O_{12}$) and hexagonal ferrite ($MFe_{12}O_{19}$) where M is a transition metal which is bivalent (Fe, Co, Cu, Ni, Zn etc.). Ferrites can be categorised into two types: hard ferrites and soft ferrites, on the basis of the properties of magnetisation and demagnetisation [3]. Hard ferrites are employed as permanent magnets because they maintain their magnetisation once they are magnetised. They have a high coercive strength and a tiny hysteresis loop. Ferrites that are easily magnetised and demagnetised are known as soft ferrites. For soft ferrites, the hysteresis loop is small. They have low coercive strength and are resistant to changes in the magnitude of the magnetic field. They show soft magnetic properties like high saturation magnetisation, low coercivity, high permeability and strong chemical and thermal stability [4]. Soft ferrites are widely used in various fields such as biomedical industries [5], in drug delivery, disease diagnosis and catalysis. Soft nano-ferrites are effective catalysts in a number of organic reactions which include oxidation, dehydrogenation, alkylation and photocatalytic dye degradation because of their huge surface area, selectivity, reduced reaction time due to their smaller size and ease of recovery, which also allows them to retain their catalytic activity for many cycles [6, 7]. By using an external magnet to effectively remove soft nano-ferrites from reaction mixtures, catalyst loss caused by centrifugation and filtration can be prevented and the catalyst can be easily recovered and recycled [8].

10.2 WHY DO WE PREFER SOFT NANO-FERRITES?

Soft ferrites are preferred in the majority due to their high permeability [9], high magnetic induction, saturation induction [10–12], low power losses [13–20] and excellent chemical stability. Because of their large surface area, these ferrites have higher catalytic activity in terms of high product yield and reduced reaction time. It is easy to recover ferrite nanocatalysts and use them without losing their catalytic activity by applying an external magnetic field (Figure 10.1).

10.3 SYNTHESIS

Soft ferrite nanoparticles can be synthesised using two different approaches, which are top-down and bottom-up (Figure 10.2). In top-down, the bulk substance is broken down into nanosized particles. The use of metal oxides in the majority of cases, the requirement for an extremely high reaction temperature, the heterogeneous nature of the products, the presence of impurities and crystal defects, a broad size variation and surface structure imperfections are just a few disadvantages of this technique. However, the bottom-up method involves the assembly of tiny atomic building blocks to form nanoparticles. Out of the two approaches, the bottom-up method is the most advantageous way

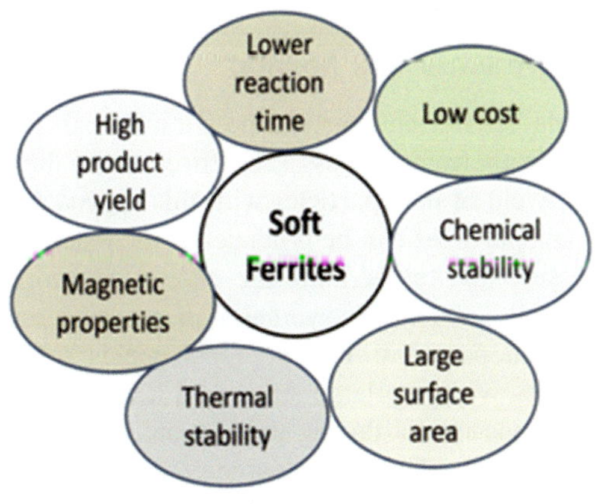

FIGURE 10.1 Characteristics of soft ferrites.

* Corresponding author.

DOI: 10.1201/9781003568940-10

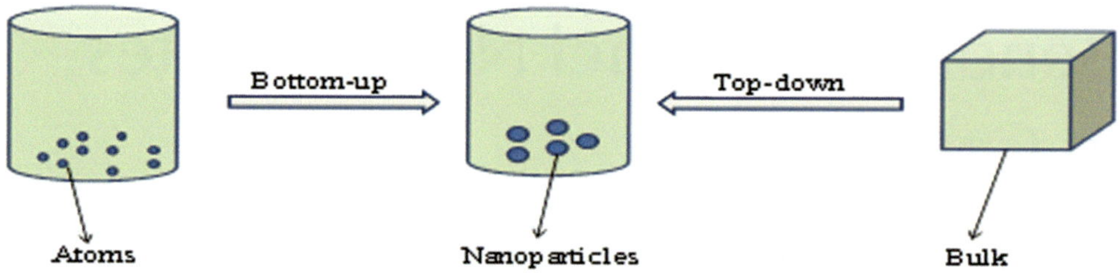

FIGURE 10.2 Bottom-up and top-down approaches for the synthesis of nanoparticles.

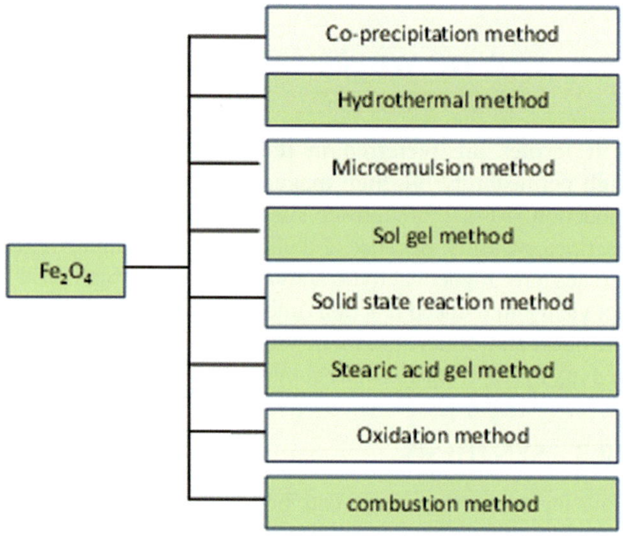

FIGURE 10.3 Various methods for synthesis of soft ferrite nanoparticles.

to synthesise nanoparticles since the end products are extremely pure, homogenous and have a small size variation of the product.

Soft ferrites can be synthesised using a variety of methods (Figure 10.3), and the method specified will depend on the particular composition and properties needed for the synthesis of soft ferrite which is as follows.

10.3.1 Hydrothermal Method

Large-scale ferrite nanoparticle preparation can be done by the hydrothermal process [21]. In essence, there is a very high yield of nanoparticles with this approach. High-quality nanoparticles can be produced if the reaction duration, pressure and temperature are chosen carefully. This technique allows for the synthesis of massive amounts of ferrite nanoparticles. This approach yields soft ferrite nanoparticles with regulated sizes, and it necessitates the use of aqueous media during production. However, there are certain drawbacks to this approach, such as the necessity for a specialised reactor and high temperatures and pressures.

10.3.2 Co-Precipitation Method

Nanomaterials can be synthesised easily and conventionally using co-precipitation [22–25]. This process yields ferrites with a homogeneous structure, regulated size and excellent purity. In order to create a homogenous solution, this process usually starts with inorganic salts (such as chloride, sulphate and nitrate) that are dissolved in water or any other suitable solvent. The solution's pH must be raised to 7–9 before the solvent is evaporated in order to produce nanoparticle precipitates. It should be noted that crystal development and particle aggregation are dependent on temperature, concentration of salt, pH and rate of pH change. The precipitation-induced solid mass is extracted and cleaned. The co-precipitation method has many benefits because it is relatively easy to use and requires an aqueous medium for synthesis. The morphology and size of the generated nanoparticles are well controlled. However, the synthesis of nano-ferrites using this approach is laborious since the resulting ferrite powder is not very crystalline which is a drawbacks of this process.

10.3.3 Sol-Gel Method

The technique has also been used to produce ferrite nanomaterials [26, 27]. As the name suggests, the sample to be synthesised in this method must pass through two phases, namely sol and gel. A colloidal or molecular suspension of solid ion particles in a solvent is called a sol. Once the solvent from the sol begins to evaporate and the remaining particles begin to form a continuous network with one another, a semi-rigid mass known as a gel is created. Colloidal powder is the final product that is generated in this process. The process is advantageous because the resulting particles are tiny and have a homogenous size distribution. This approach is cost-effective, and the nanoparticles can be synthesised at low temperatures.

10.3.4 Combustion Method

Nanomaterials can be synthesised at a low cost and with great effectiveness using this process [28]. This method is quick, flexible and easy to use for preparing nanomaterials. The advantage of this approach is that it takes less time and energy to finish the synthesis. The final nanoparticles are homogeneous and pure. Urea and glycine are typically preferred fuels in this process. Utilising these fuels homogenous nano-ferrites with controlled stoichiometry can be prepared.

10.3.5 Solid-State Reaction Method

There are several advantages of this method to prepare nanoparticles, i.e., the synthesis of soft ferrite nanoparticles can be achieved without the need for expensive and dangerous organic solvents, and all the necessary materials are easily obtained at reasonable price. The synthesis process can be carried out simply and affordably at room temperature and atmospheric pressure.

10.3.6 Oxidation Method

A chemical technique for creating ferrite nanoparticles is oxidation. The produced ferrite particles have an extended morphology and are irregular in shape. The advantage of this method is that the nanoparticles are formed with a homogeneous and narrow size distribution, however this method can also produce small-sized ferrite colloids.

10.3.7 Stearic Acid Gel Method

Common fatty acids include stearic acid, also known as octadecanoic acid, which is found in plant and animal fat as a glycerol ester. The stearic acid gel is a common method used by researchers to create nanomaterials.

10.3.8 Microemulsion Method

This approach is used to create nanoparticles with exact dimensions and shapes. Although it entails complicated processes and high temperatures [29], an additional strategy that gets around the need for the microemulsion procedure is the hydrothermal method. Soft ferrite nanoparticles can be prepared chemically by sonicating salts such as $Fe(CH_3COO)_2$ and Fe $(acac)_3$ [30–32].

10.4 STRUCTURE OF SOFT FERRITES

Soft ferrites have a crystal structure that is typically cubic or tetragonal, depending on the specific composition and processing conditions. The most common soft ferrites have a spinel structure. In a spinel structure, the cations (metal ions) are distributed between octahedral and tetrahedral sites within the crystal lattice. The most commonly used soft ferrites have the chemical formula MFe_2O_4 where M can represent various metal cations such as Cu and Mn.

The spinel structure of soft ferrites is often described as a face-centred cubic (fcc) structure with oxygen ions forming a close-packed arrangement (Figure 10.4). The cations are situated in both octahedral and tetrahedral sites, and the distribution of metal cations between these sites contributes to the unique magnetic properties of soft ferrites. The specific arrangement of metal ions into spinel structure allows soft ferrites to have high magnetic permeability and low coercivity.

10.5 CATALYTIC ACTIVITIES OF SPINEL FERRITES

The primary purpose of a catalyst is to accelerate chemical reactions, and the process is known as catalysis. Higher catalyst activity allows for the reduction of reactor size and operating conditions, including pressure and temperature, which lowers operating expenses. Because of their special qualities like superior yield, shortened reaction times and outstanding selectivity, reusable and easily separation due

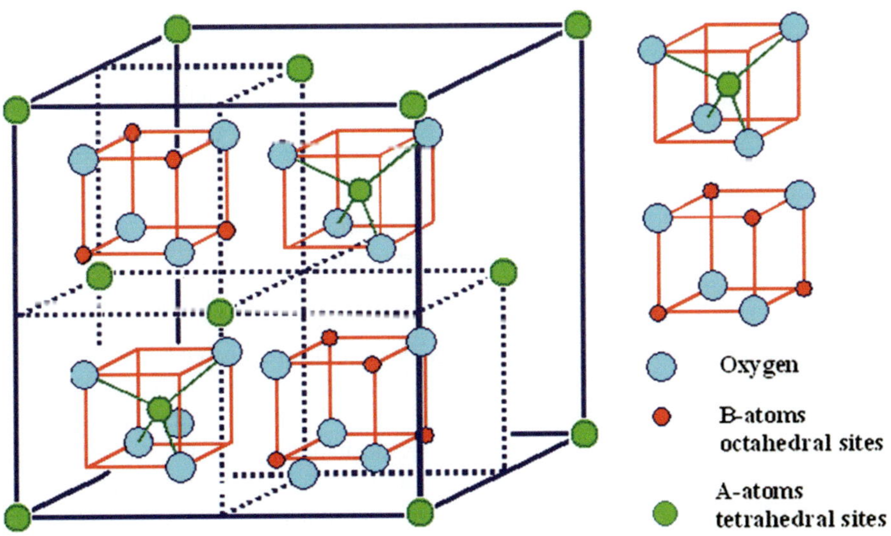

FIGURE 10.4 Structure of soft ferrites.

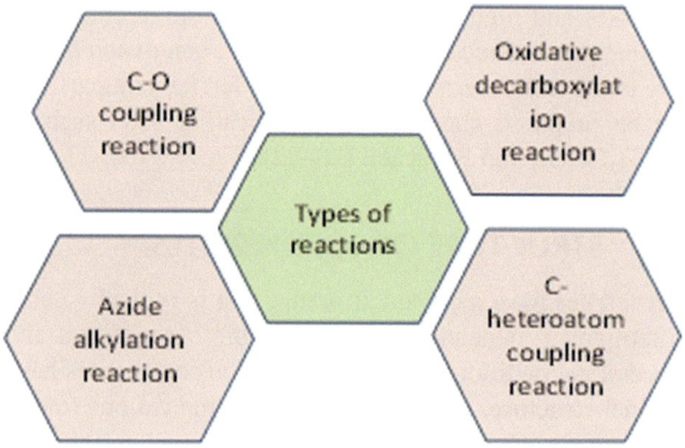

FIGURE 10.5 Different types of reaction catalysed by soft spinel ferrites.

to their magnetic nature, magnetic nanoparticles (MNPs) have drawn attention in catalysis. Catalysis can be either homogeneous or heterogeneous. Figure 10.5 depicts different types of reaction catalysed by soft ferrites. When the catalyst and the reactant used in the reaction are in the same state of matter is referred as homogenous catalysis. Typically, the solvent is used to dissolve homogeneous catalysts including substrates. Conversely, heterogeneous catalysis refers to catalysis in which the catalyst used in the reaction and the reacting chemicals are not in the same state of matter.

10.5.1 SPINEL FERRITES AS CATALYSTS FOR ORGANIC REACTIONS

Spinel ferrites play a very important role in organic synthesis as a catalyst. Because of the large size and greater ability to accommodate reactant species, the octahedral site of the soft ferrite exhibits more catalytic activity than the tetrahedral site. Ferrites are easier to handle, requires less workup and prevents metal contamination when used as a catalyst. The advantages of soft spinel ferrites are their ease of recovery and recycling, good yields under mild reaction conditions and their versatility in accommodating multifunctional groups. In recent years, there has been a significant increase in the application of MNPs as effective catalysts in chemical synthesis. When it comes to organic reactions, nanocatalysts' which have high surface area, low number of coordination sites and reactive geometries enable them more effective processes than those of conventional methods. As a result, the least amount of catalyst is used while maximising the rate of reaction.

10.5.1.1 Copper Ferrites

Copper nano-ferrites ($CuFe_2O_4$) are the most often used catalysts in a wide range of novel processes. $CuFe_2O_4$ role in the breakdown of methanol into CO and H_2 is one of its significant catalytic reactions [31]. Converting CO to CO_2 is another use of $CuFe_2O_4$ [32]. Without the use of co-catalysts, additives or ligands, Kharisov et al. reported the synthesis of β, γ-unsaturated ketones and the allylation of acid chlorides utilising $CuFe_2O_4$ nanomaterial as a reusable heterogeneous initiator [33]. Using tetrahydrofuran (THF) as a solvent and citrate as a precursor, the nanocatalyst was made at room temperature using the sol-gel method. This process produced encouraging results for three cycles in a row in very low reaction time. A wide range of acid chlorides reacted with allyl bromide and cinnamyl chloride under the specified conditions; however, the reaction with long-chain aliphatic acid chloride allyl ketone was failed to produce the desire product. Under the specified conditions, a wide variety of acid chlorides reacted with allyl bromide and cinnamyl chloride; however, allyl ketone was not produced in the reaction with long-chain aliphatic acid chloride. Both steric and electrical influences had no effect on the $CuFe_2O_4$ nanocatalyst's ability to acylate allyl halides. All acylation processes were seen to be followed by allylic rearrangement in order to remove the double bond from the conjugation. The mechanism involves the attack of $CuFe_2O_4$ nanocatalyst electrophilically at the γ carbon atom of the allylic group without undergoing prototropic rearrangement after which it combines with the acyl halides to produce stable allyl ketone.

In a different study, $CuFe_2O_4$ nanoparticles were used for the C–O coupling also known as Ullmann reaction, involving derivatives of amino, ketone, methyl, methoxy, cyano and aryl halides in addition to phenols. Sensitive functional groups including MeCO, NH_2 and CN did not require any kind of protection as the catalyst demonstrated high yields [34].

There have been reports that $CuFe_2O_4$ nanocatalyst offers an effective, easy, magnetically recoverable, ligand-free method for the oxidative decarboxylation of phenylacetic acid, which results in the synthesis of ketones and aldehydes [35]. According to the suggested process, active Cu species are produced when phenylacetic acid is decarboxylated. These active Cu species are then further oxidised by peroxy cuprate, which forms the appropriate aldehyde (Figure 10.6). When an electron-donating group rather than an electron-withdrawing group was bonded to the aromatic ring, it was discovered that the yield of the intended product was greater. Additionally, it was found that the substituent's location on the aromatic ring affected the product yield, the ortho substituents showed less yields than meta and para substituents. One contributing factor was found to be steric hindrance as it reduces the yield of ortho-substituted compounds.

In an individual study, several toluene derivatives were oxidised under mild reaction conditions with Cu nanoparticles synthesised through a green technique utilising neem leaf extract, without the need for any solvent, yielding a good yield of benzoic acid (Figure 10.7). Up to five cycles has been performed using the catalyst without losing its activity [36]. It has been found that toluene derivatives with electron-donating groups were more reactive for the oxidative reaction since they showed higher yields than those having electron-withdrawing groups. Notably, severe reaction

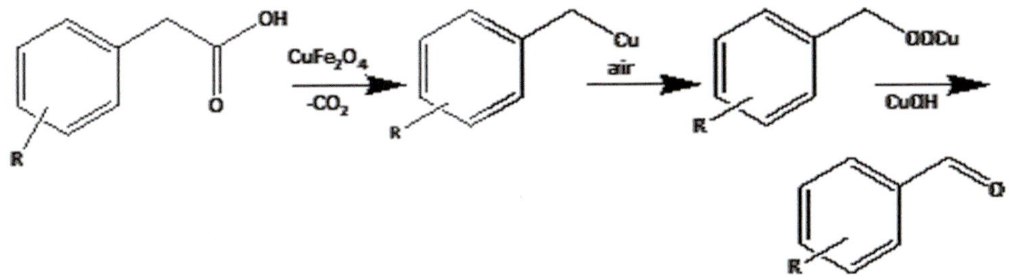

FIGURE 10.6 Proposed mechanism for the oxidative carboxylation reaction.

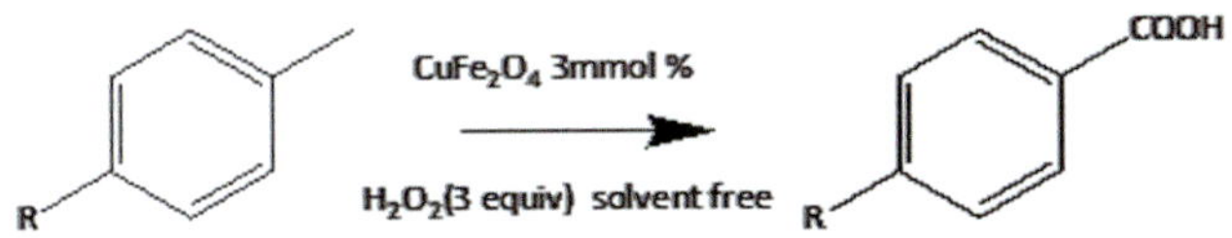

FIGURE 10.7 Benzyl alkanes oxidation to produce benzoic acid.

conditions were necessary for the successful oxidation of toluene derivatives having electron-withdrawing groups, such as 4-nitrotoluene. It was also noted that the outstanding yield was obtained when electron-rich 2-hydroxytoluene was used as substrate. Remarkably, substituted toluene compounds experienced effective oxidation irrespective of the substituent group.

After being washed with ethanol, the catalyst could be used up to four more times without losing its ability to oxidise toluene. After the fifth cycle, however, the reactivity decreased as a result of a change in the nanoparticles' shape. $CuFe_2O_4$ catalysts were also utilised to synthesise biaryls from aryl-boronic acid in a variety of solvents, including ethanol, THF and methanol (Figure 10.8). The catalyst was also employed to catalyse the oxidation of several aromatic and aryl alcohols [37]. Using methanol as the solvent the higher product yield (83%–96%) was obtained in low reaction time with maximum catalyst recovery.

10.5.1.2 Zinc Ferrite

These nano-ferrites have been used as catalysts in organic reactions. The study examined the function of $ZnFe_2O_4$

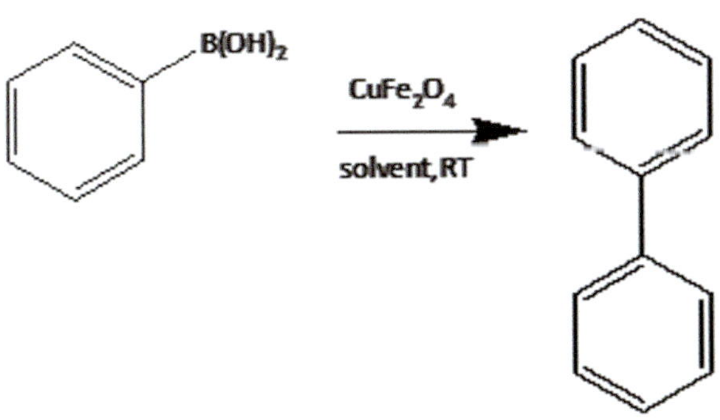

FIGURE 10.8 Formation of biaryls using $CuFe_2O_4$ nanocatalyst.

doped with neodymium as a catalyst in the oxidative conversion of methane and coupling processes [38]. Strong correlations were found between the structure of the oxide, the catalytic action and the defects resulting from substitution.

The redox character of the reaction mixture was a significant factor in the development of certain lattice defects by facilitating the removal of lattice oxygen linked to Fe^{3+}. The presence of oxygen defects countered the positive charge that resulted from the reduction of Fe^{3+} to Fe^{2+}. When octahedrally coordinated Nd^{3+} were introduced, new kinds of defects were created that made methane combustion easier. Methanol, a green fuel substitute used in fuel cells, was shown to be converted to CO and H_2 at relatively low temperatures by $ZnFe_2O_4$ nanocatalysts [39]. Mössbauer spectra supported the catalytic experiment's considerable transformation of the ferrite phase caused by the medium. Without the requirement for any ligands, Pd substituted $ZnFe_2O_4$ was studied for Heck and Suzuki cross-coupling reactions under anaerobic circumstances [40]. The $ZnFe_2O_4$ substitute showed remarkable catalytic activity. Under ideal circumstances, a large variety of boronic acids and aryl halides were investigated for the Suzuki reaction. The loading of the catalyst was low and it was easily recovered and reused. Similarly, different aryl halides with different alkenes were taken for the Heck reaction, with the exception of aryl chloride, which could not catalysed by the substituted $ZnFe_2O_4$.

Pd-doped $ZnFe_2O_4$ was synthesised as a magnet-recyclable catalyst employing polydopamine as a crosslinking agent and a covalent organic framework (COF) as a carrier. Excellent catalytic activity was shown by the catalyst in the Suzuki coupling and the reduction of p-nitrophenol [41]. By preventing Pd nanoparticle aggregation and leaching, the COF have been found to promote the catalyst's activity and stability. An additional study has been carried out for super-paramagnetic Pd-$ZnFe_2O_4$ nanoparticles as an effective heterogeneous catalyst for carbon-carbon and carbon oxygen cross-coupling reactions. The catalyst has been found suitable for the cyanation of aromatic halides without cyanide (Figure 10.9), the Heck-Mastuda reaction for a diazonium salt in an aqueous medium, the Sonogashira coupling reaction with ethanol and the Ullmann coupling reaction with phenols to obtain the corresponding coupled products (Figure 10.10) [42]. The coupling reaction could

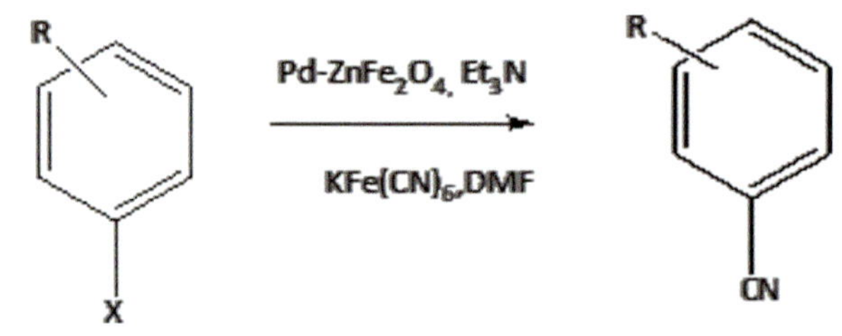

FIGURE 10.9 Aromatic halides are cyanated using $K_4[Fe(CN)_6]$ with Pd-$ZnFe_2O_4$ MNP as catalyst.

not proceed in the absence of the catalyst showing the important of the catalyst in the processes.

Zinc ferrite crystals with a spinel structure have increased ferrite stability, which increases efficiency. Numerous studies on zinc ferrites have been conducted for their application for photocatalytic degradation of organic pollutants in water. $ZnFe_2O_4$, for instance, demonstrates photocatalytic activity, but by combining it with graphene and carbon nanotubes or another semiconductor. The efficiency of $ZnFe_2O_4$ can be further increased by combining it with graphene oxide which has been found to serve as a stable and effective catalyst for the MB dye degradation in aqueous solution.

10.5.1.3 Manganese Ferrite

The catalytic activity of manganese ferrite MNPs has been reported quite good. The decarboxylative Sonogashira reaction produces diaryl acetylenes by reacting arene diazonium salts with phenyl propiolic acid, has been catalysed by Pd-doped, magnetically separable $MnFe_2O_4$ nanoparticles [43]. With the use of ultrasound assistance and a one-pot co-precipitation technique, the catalyst was synthesised without using any ligand. Both symmetrical and unsymmetric diaryl acetylenes were synthesised using the catalyst. Both electron-rich and electron-deficient arene diazonium salts results in good to excellent yield of products. A minor steric effect was noted at the arene diazonium salt's orthogonal location. The suggested mechanism involved the oxidative addition of iodobenzamide to a nanocatalyst in the presence of an alkynyl carboxylic acid, followed by in situ transmetallation. Ultimately, diarylacetylene was generated by decarboxylation followed by reductive elimination.

Another study examined the catalytic properties of a $MnFe_2O_4$-supported core-shell-structured manganese oxide for the oxidation reaction of benzyl alcohol at temperatures ranging from 60 to 90°C. Hexane was employed as the solvent and oxygen as the oxidant. The magnetic nanocatalyst showed 100% selectivity towards benzaldehyde and moderate activity under ambient conditions. It was demonstrated that the interaction between alcohol and activated Mn sites was restricted by nucleophilic solvents. The yield towards the reaction of benzaldehyde increased with longer reaction times and higher temperatures, however the reactivity towards dibenzyl ether have been found to decrease under aforementioned conditions.

$MnFe_2O_4$-supported magnetically separable palladium catalysts were recently reported to have good catalytic activity with no by-products for the conversion of hydrogenate nitrobenzene to aniline derivatives [44].

10.6 CONCLUSION

The use of spinel ferrites ($CuFe_2O_4$, $ZnFe_2O_4$ and $MnFe_2O_4$) as heterogeneous catalysts in organic reactions has been advanced significantly. The catalytic performance of these nanocatalysts in a variety of organic processes is briefly reviewed in this chapter. Soft ferrite nanocatalyst's showed superior magnetic characteristics and good product yield in short reaction time. The larger surface area allows the higher activity and their simple recovery from reaction systems with little loss in activity and reusability make them important candidates for practical applications. Various organic reactions like condensation reactions, C–C coupling, cyclisation, alkylation, oxidation of different alkenes, dehydrogenation processes and the synthesis of organic compounds such as acetylferrocene chalcones and arylamines have been effectively catalysed by MFe_2O_4. Transition metal ferrites play an important role in activating particular groups of reactants, establishing covalent bonds and act as Lewis's acids to improve electrophilicity. Future studies and research must look into the significant factor of catalyst leaching which could improve the catalyst loss during repeated reuse. As heterogeneous catalysts in organic processes, rare earth-doped metal ferrites have not received much attention till date. More investigation is needed to examine the catalytic properties of ferrites, especially those containing rare earth metals, for a variety of chemical transformations.

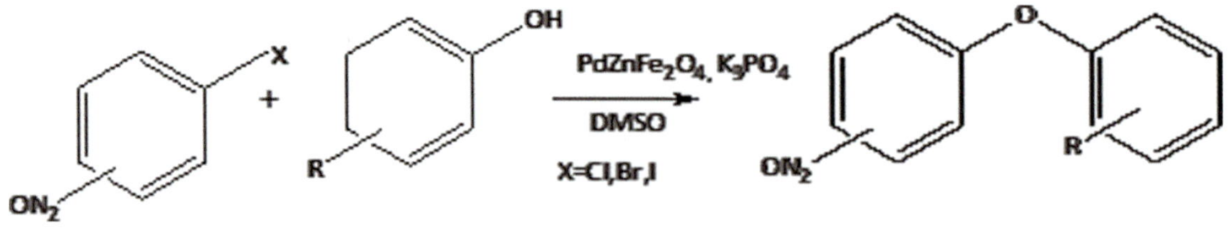

FIGURE 10.10 Reaction of carbon-oxygen cross-coupling reaction catalysed by Pd-$ZnFe_2O_4$ MNP.

REFERENCES

1. Dhiman, P.; Rana, G. C.; Goyal, D.; Goyal, A. Basics of Ferrites: Types and Structures. *Engineering Materials* 2021, 1–25. https://doi.org/10.1007/978-981-16-7454-9_1
2. Goldman, A. *Crystal Structure of Ferrites*. Springer eBooks 1999, 207–227. https://doi.org/10.1007/978-1-4615-4917-8_11
3. Chattopadhyay, S.; Bandyopadhyay, A.; Nath, M. *Physics of Ferrite Ceramics*. Elsevier eBooks 2022, 165–185. https://doi.org/10.1016/b978-0-323-89956-7.00009-7
4. Younes, A.; Kherrouba, N.; Amir, M. Magnetic, Optical, Structural and Thermal Properties of Copper Ferrite Nano-structured Synthesized by Mechanical Alloying. *Micro & Nano Letters* 2021, 16 (4), 251–256. https://doi.org/10.1049/mna2.12040
5. Pham, T. N.; Huy, T. Q.; Le, A.-T. Spinel Ferrite (AFe_2O_4)-Based Heterostructured Designs for Lithium-Ion Battery, Environmental Monitoring, and Biomedical Applications. *RSC Advances* 2020, 10 (52), 31622–31661. https://doi.org/10.1039/d0ra05133k
6. Corma, A.; Garcia, H. Supported Gold Nanoparticles as Catalysts for Organic Reactions. *Chemical Society Reviews* 2008, 37 (9), 2096–2126. https://doi.org/10.1039/B707314N
7. Davis, S. E.; Ide, M. S.; Davis, R. J. Selective Oxidation of Alcohols and Aldehydes over Supported Metal Nanoparticles. *Green Chem.* 2013, 15 (1), 17–45. https://doi.org/10.1039/c2gc36441g
8. Lu, A.-H.; Salabas, E. L.; Schüth, F. Magnetic Nanoparticles: Synthesis, Protection, Functionalization, and Application. *Angewandte Chemie International Edition* 2007, 46(8), 1222–1244. https://doi.org/10.1002/anie.200602866
9. Sanatombi, S.; Sumitra, S.; Ibetombi, S. Influence of Sintering on the Structural, Electrical, and Magnetic Properties of Li–Ni–Mn–Zn Ferrite Synthesized by Citrate Precursor Method. *Iranian Journal of Science and Technology Transaction A-Science* 2017. https://doi.org/10.1007/s40995-017-0405-8
10. Islam, R.; Hakim, M. A.; Rahman, M. O.; Narayan Das, H.; Mamun, M. A. Study of the Structural, Magnetic and Electrical Properties of Gd-Substituted Mn–Zn Mixed Ferrites. *Journal of Alloys and Compounds* 2013, 559, 174–180. https://doi.org/10.1016/j.jallcom.2012.12.080
11. Zhao, Y.; Wang, H.; Li, J.; Fang, Y.; Kang, Y.; Zhao, T.; Zhao, C. Regulating the Spin-State of Rare-Earth Ce Single Atom Catalyst for Boosted Oxygen Reduction in Neutral Medium. *Advanced Functional Materials* 2023. https://doi.org/10.1002/adfm.202305268
12. Fang, X.; Wu, R.; Peng, L.; Sin, J. K. O. A Novel Silicon-Embedded Toroidal Power Inductor with Magnetic Core. *IEEE Electron Device Letters* 2013, 34 (2), 292–294. https://doi.org/10.1109/led.2012.2234077
13. Fujita, A.; Gotoh, S. Temperature Dependence of Core Loss in Co-Substituted MnZn Ferrites. *Journal of Applied Physics* 2003, 93 (10), 7477–7479. https://doi.org/10.1063/1.1557952
14. Wang, S.-F.; Hsu, Y.; Chen, C.-H. Effects of Nb_2O_5, TiO_2, SiO_2, and CaO Additions on the Loss Characteristics of Mn-Zn Ferrite. *Journal of Electroceramics* 2014, 33 (3–4), 172–179. https://doi.org/10.1007/s10832-014-9943-z
15. Baguley, C.; Madawala, U. K.; Carsten, B. The Influence of Remanence on Magnetostrictive Vibration and Hysteresis in Mn-Zn Ferrite Cores. *IEEE Transactions on Magnetics* 2012, 48 (5), 1844–1850. https://doi.org/10.1109/tmag.2011.2174251
16. Tsakaloudi, V.; Kogias, G.; Zaspalis, V. T. Process and Material Parameters towards the Design of Fast Firing Cycles for High Permeability MnZn Ferrites. *Journal of Alloys and Compounds* 2014, 588, 222–227. https://doi.org/10.1016/j.jallcom.2013.11.047
17. Töpfer, J.; Angermann, A. Complex Additive Systems for Mn-Zn Ferrites with Low Power Loss. *Journal of Applied Physics* 2015, 117 (17). https://doi.org/10.1063/1.4918692
18. Marracci, M.; Tellini, B. Hysteresis Losses of Minor Loops Versus Temperature in MnZn Ferrite. *IEEE Transactions on Magnetics* 49 (2013) 2865–2869. https://doi.org/10.1109/TMAG.2012.2219877
19. Nien, H. H.; Huang, C. C.; Wang, M. Y.; Lin, C.; Changchien, S.-K. Estimation of Eddy-Current Loss for MnZn Ferrite Cores. *Advanced Materials Research* 2011. https://doi.org/10.4028/www.scientific.net/amr.382.204
20. Yang, W.; Wang, Y. G. Effects of TiO_2 and Co_2O_3 Combination Additions on the Elemental Distribution and Electromagnetic Properties of Mn–Zn Power Ferrites. *Journal of Magnetism and Magnetic Materials* 2015, 384, 13–17. https://doi.org/10.1016/j.jmmm.2015.02.020
21. Wang, J.; Sun, J.-H.; Sun, Q.; Chen, Q. One-Step Hydrothermal Process to Prepare Highly Crystalline Fe_3O_4 Nanoparticles with Improved Magnetic Properties. *Materials Research Bulletin* 2003, 38 (7), 1113–1118. https://doi.org/10.1016/s0025-5408(03)00129-6
22. Thakur, S.; Rai, R.; Sharma, S. Structural Characterization and Magnetic Study of $NiFe_xO_4$ Synthesized by Co-Precipitation Method. *Materials Letters* 2015, 139, 368–372. https://doi.org/10.1016/j.matlet.2014.10.014
23. Zi, Z.; Sun, Y.; Zhu, X.; Yang, Z.; Dai, J.; Song, W. Synthesis and Magnetic Properties of $CoFe_2O_4$ Ferrite Nanoparticles. *Journal of Magnetism and Magnetic Materials* 2009, 321 (9), 1251–1255. https://doi.org/10.1016/j.jmmm.2008.11.004
24. Nabiyouni, G.; Julaee, M.; Ghanbari, D.; Aliabadi, P. C.; Safaie, N. Room Temperature Synthesis and Magnetic Property Studies of Fe_3O_4 Nanoparticles Prepared by a Simple Precipitation Method. *Journal of Industrial and Engineering Chemistry* 2015, 21, 599–603.https://doi.org/10.1016/j.jiec.2014.03.025
25. Sagadevan, S.; Chowdhury, Z. Z.; Rafique, R. F. Preparation and Characterization of Nickel Ferrite Nanoparticles via Co-Precipitation Method. *Materials Research* 2018, 21 (2). https://doi.org/10.1590/1980-5373-mr-2016-0533
26. Haralkar, S. J.; Kadam, R. H.; More, S. S.; Shirsath, S. E.; Mane, M. L.; Patil, S.; Mane, D. R. Intrinsic Magnetic, Structural and Resistivity Properties of Ferromagnetic $Mn0.5Zn0.5Al_xFe_2{-}XO_4$ Nanoparticles. *Materials Research Bulletin* 2013, 48 (3), 1189–1196. https://doi.org/10.1016/j.materresbull.2012.12.018
27. Thakur, P.; Chahar, D.; Taneja, S.; Bhalla, N.; Thakur, A. A Review on MnZn Ferrites: Synthesis, Characterization and Applications. *Ceramics International* 2020, 46 (10), 15740–15763. https://doi.org/10.1016/j.ceramint.2020.03.287
28. Sivakumar, P.; Ramesh, R.; Ramanand, A.; Ponnusamy, S.; Muthamizhchelvan, C. Preparation and Properties of Nickel Ferrite ($NiFe_2O_4$) Nanoparticles via Sol–Gel Auto-Combustion Method. *Materials Research Bulletin* 2011, 46 (12), 2204–2207. https://doi.org/10.1016/j.materresbull.2011.09.010
29. Rashad, M. M.; Soltan, S.; Ramadan, A. A.; Bekheet, M. F.; Rayan, D. A. Investigation of the Structural, Optical and Magnetic Properties of $CuO/CuFe_2O_4$ Nanocomposites

Synthesized via Simple Microemulsion Method. *Ceramics International* 2015, 41 (9), 12237–12245. https://doi.org/10.1016/j.ceramint.2015.06.046
30. Pinkas, J.; Reichlová, V.; Zbořil, R.; Moravec, Z.; Bezdička, P.; Matějková, J. Sonochemical Synthesis of Amorphous Nanoscopic Iron(III) Oxide from $Fe(Acac)_3$. *Ultrasonics Sonochemistry* 2008, 15 (3), 257–264. https://doi.org/10.1016/j.ultsonch.2007.03.009
31. Vijayakumar, R.; Koltypin, Y.; Felner, I.; Gedanken, A. Sonochemical Synthesis and Characterization of Pure Nanometer-Sized Fe_3O_4 Particles. *Materials Science and Engineering: A* 2000, 286 (1), 101–105. https://doi.org/10.1016/s0921-5093(00)00647-x
32. Bhagwat, V.; Humbe, A. V.; More, S. D.; Jadhav, K. M. Sol-Gel Auto Combustion Synthesis and Characterizations of Cobalt Ferrite Nanoparticles: Different Fuels Approach. *Materials Science and Engineering: B* 2019, 248, 114388. https://doi.org/10.1016/j.mseb.2019.114388
33. Kharisov, B. I.; Dias, H. V. R.; Kharissova, O. V. Mini-Review: Ferrite Nanoparticles in the Catalysis. *Arabian Journal of Chemistry* 2019, 12 (7), 1234–1246. https://doi.org/10.1016/j.arabjc.2014.10.049
34. Velinov, N.; Koleva, K.; Tsoncheva, T.; Manova, E.; Paneva, D.; Tenchev, K.; Kunev, B.; Mitov, I. Nanosized $Cu0.5Co0.5Fe_2O_4$ Ferrite as Catalyst for Methanol Decomposition: Effect of Preparation Procedure. *Catalysis Communications* 2013, 32, 41–46. https://doi.org/10.1016/j.catcom.2012.12.002
35. Lou, J.-C.; Chang, C. Catalytic Oxidation of CO over a Catalyst Produced in the Ferrite Process. *Environmental Engineering Science* 2006, 23 (6), 1024–1032. https://doi.org/10.1089/ees.2006.23.1024
36. Yang, S.; Wu, C.; Zhou, H.; Yang, Y.; Zhao, Y.; Wang, C.; Yang, W.; Xu, J. An Ullmann C-O Coupling Reaction Catalyzed by Magnetic Copper Ferrite Nanoparticles. *Advanced Synthesis & Catalysis* 2012, 355 (1), 53–58. https://doi.org/10.1002/adsc.201200600
37. Rahman, T.; Borah, G.; Gogoi, P. K. Spinel Structured Copper Ferrite Nano Catalyst with Magnetic Recyclability for Oxidative Decarboxylation of Phenyl Acetic Acids. *Catalysis Letters* 2020, 150 (8), 2267–2272. https://doi.org/10.1007/s10562-020-03131-0
38. Al-Hunaiti, A.; Al-Said, N.; Halawani, L.; Haija, M. A.; Baqaien, R.; Taher, D. Synthesis of Magnetic $CuFe_2O_4$ Nanoparticles as Green Catalyst for Toluene Oxidation under Solvent-Free Conditions. *Arabian Journal of Chemistry* 2020, 13, 4945–4953.
39. Papa, F.; Patron, L.; Carp, O.; Paraschiv, C.; Balint, I. Catalytic Activity of Neodymium Substituted Zinc Ferrites for Oxidative Conversion of Methane. *Journal of Molecular Catalysis A-Chemical* 2009, 299 (1–2), 93–97. https://doi.org/10.1016/j.molcata.2008.10.036
40. Loan, T. T.; Huy, D. K.; Duong, N. P.; Hoan, T. D.; Soontaranon, S.; Klysubun, W. Facile Synthesis and Characterization of $ZnFe_2O_4$/ZnO Nanocomposite: An Insight into Structure and Formation. *Radiation Physics and Chemistry* 2022, 193, 109977. https://doi.org/10.1016/j.radphyschem.2022.109977
41. Singh, A. S.; Patil, U. B.; Nagarkar, J. M. Palladium Supported on Zinc Ferrite: A Highly Active, Magnetically Separable Catalyst for Ligand Free Suzuki and Heck Coupling. *Catalysis Communications* 2013, 35, 11–16. https://doi.org/10.1016/j.catcom.2013.02.003
42. Lv, K.; Yan, G.; Wang, G.; Chen, Z.; Hu, J. Palladium-Decorated Covalent Organic Framework Supported on Zinc Ferrite as Magnetic Catalyst for Suzuki Reaction and p-Nitrophenol Reduction. *Catalysis Letters* 2022. https://doi.org/10.1007/s10562-022-04113-0
43. Singh, A. S.; Shendage, S. S.; Nagarkar, J. M. Palladium Supported on Zinc Ferrite: An Efficient Catalyst for Ligand Free C–C and C–O Cross Coupling Reactions. *Tetrahedron Letters* 2013, 54 (47), 6319–6323. https://doi.org/10.1016/j.tetlet.2013.09.027

11 Water Purification Using Spinel Nano-Ferrites

*Preeti Thakur**
Nanotechnology Wing, Innovative Science Research Society, Shimla, India
nanoLatticeX LLP, Manesar, Gurugram, India
Amity Institute of Nanotechnology, Amity University Haryana, Gurugram, India

Atul Thakur
Nanjing University of Information Science & Technology, Nanjing, China
Nanotechnology Wing, Innovative Science Research Society, Shimla, India
nanoLatticeX LLP, Manesar, Gurugram, India
Amity Institute of Nanotechnology, Amity University Haryana, Gurugram, India

Neetu Dhanda
Amity University Haryana, Gurugram, India

An-Cheng Aidan Sun
Yuan Ze University, Taoyuan, Taiwan

11.1 INTRODUCTION

The most valuable resource in human society is water, and access to drinking water is a basic need. But we are far from being able to meet the demands of the world, and this issue will only get worse over time. Demand increases as a result of population growth, climate change, and declining water quality [1]. Only 2.5% of the world's seas and oceans are used to store fresh water, leaving 700 million people without access to drinking water. Developing nations and countries in sub-Saharan Africa are particularly affected by this issue. So, in these impacted areas, water treatment must be put into practice. The quality of water that can be produced with current technology to treat it is no longer sufficient to suit both human and environmental needs. Hence reuse, recycling, and repurposing are today's "needs." Organic, inorganic, and biological pollutants can all be found in water [2]. Some pollutants are harmful to people and ecosystems because they are poisonous and carcinogenic. Some heavy metals are well-known, very harmful water contaminants [3]. One of the deadliest elements, arsenic, has a long history and is well recognized. Nickel, cadmium, zinc, chromium, lead, copper, mercury, and other heavy metals are also important water contaminants with severe toxicity. Figure 11.1 shows the location of stream sampling sites, surface-water intakes, and drinking-water-treatment plants. In high quantities, sulfates, nitrates, phosphates, chlorides, fluorides, selenides, oxalates, and chromates cause harmful consequences and alter the flavor of water. Fluorosis, for instance, is brought on by excessive fluoride levels in water.

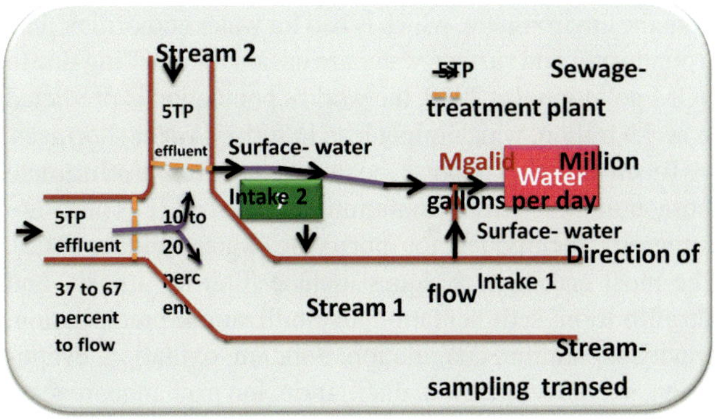

FIGURE 11.1 Schematic diagram showing location of stream sampling sites, surface-water intakes, and drinking-water-treatment plant.

Toxicities are linked to organic pollutants, including fertilizers, pesticides, hydrocarbons, plasticizers, phenols, biphenyls, detergents, greases, and oils [4]. Personal care products (PCPs) and medications are examples of emerging pollutants. PCPs typically exhibit resistance to natural biodegradation. These substances typically originate from hospital and residential water, which also contains (i) drug products, (ii) medications, (iii) packaging, both unmetabolized and metabolized, and (iv) detergent additives. In water, the content of PCP varies from ng/L to g/L. As treatment amenities are not designed to eliminate stable, low-concentrated contaminants, traditional water treatment does not produce satisfactory results. To obtain oil under high pressure, shale or rocks are fractured hydraulically. Among other volatile contaminants, aromatic

* Corresponding author.

DOI: 10.1201/9781003568940-11

hydrocarbons, like benzene, toluene, ethylbenzene, and xylenes, designated as BTEX, make up the majority of generated water [5]. When exposed to air, these volatile organic molecules readily evaporate from the water. BTEX compounds contain persistent carcinogenic pollutants and are hazardous. Ethylbenzene 700, benzene 5, xylenes 10,000, and toluene 1,000 are the current maximum contamination levels (MCL, g/L = ppb) for BTEXs in drinking water [6]. Eliminating BTEX contaminants is difficult since conventional water purification techniques (filter, sedimentation, and flocculation) don't remove organic chemicals at low quantities. The cost of the media and distribution channels employed, as well as the cost of their arrangement and dissemination, all affects working expenses. Many waterborne infections are caused by microbes.

One of the key characteristics of water quality is natural organic matter (NOM), which also influences the purification method. Total organic and dissolved carbon is sometimes used to represent NOM. NOM is pervasive and derived from plant or animal leftovers. It disproves the efficacy of procedures like filtration, flocculation, and carbon adsorption. Even worse, NOM creates unwanted oxidation products when treated with chlorine. Numerous additional microorganisms can reproduce in NOM, which promotes the development of bacteria in aquatic environments. The variety of NOM makes it impossible to eliminate all associated objects from the environment, which is bad for water authorities. The groundwater and surface water are unsafe for drinking due to NOM pollution. By 2020, the world's population is predicted to be 7.9 billion, which might lead to serious water shortages.

To ensure the public's health, it is vital to eliminate these pollutants from contaminated water [7]. There are numerous techniques for purifying water (Figure 11.2). The most crucial techniques include filtering, micro- and ultrafiltration, sedimentation, crystallization, precipitation, gravity separation, coagulation, flotation, oxidation, evaporation, solvent extraction, distillation, ion exchange, reverse osmosis (RO), electrodialysis, adsorption, electrolysis, setting-out, membrane separation, centrifugal, fluidization, demineralization and neutralization, oxidation and reduction, and others. Depending on the type of contaminated water and the intended use, any of the aforementioned techniques may be combined. Although these techniques are effective, recent well-known anthropogenic pollutants (caused by modern human lifestyles) make it difficult to filter or treat contaminated water. In addition, as contaminants are removed, treatment plant efficiency is reduced. As a result, providing potable water is not "100%" effective using current technologies [8]. Thus, it is crucial to develop new and better water filtration technologies.

11.2 DRINKING WATER PURIFICATION – CHALLENGES

Our idea of pure drinking water has changed dramatically along with human civilization. The flavor of the water was a widely used indicator of purity in the earliest civilizations. Water was acknowledged as a representation of the beginning of life and for its therapeutic benefits; it was not seen as a disease vector [9]. The invention of the microscope by Antonie van Leeuwenhoek in the 17th century began to alter how people perceived purity. It allowed us to look ahead of suspended particles, such as the small particles, and to the microorganisms. Our concept of pure drinking water has altered as a result of Louis Pasteur's discoveries (microorganism diseases-based studies) and John Snow (the connection between the cholera's development in London and the water quality). It's interesting to note that the Metropolis Water Act of 1852, the first law ever passed, was issued in 1852. Every human being now has a legal right to get clean water. The age of water filtration has arrived. Our awareness of water purity, its impact on physical condition, and purification techniques have drastically changed over

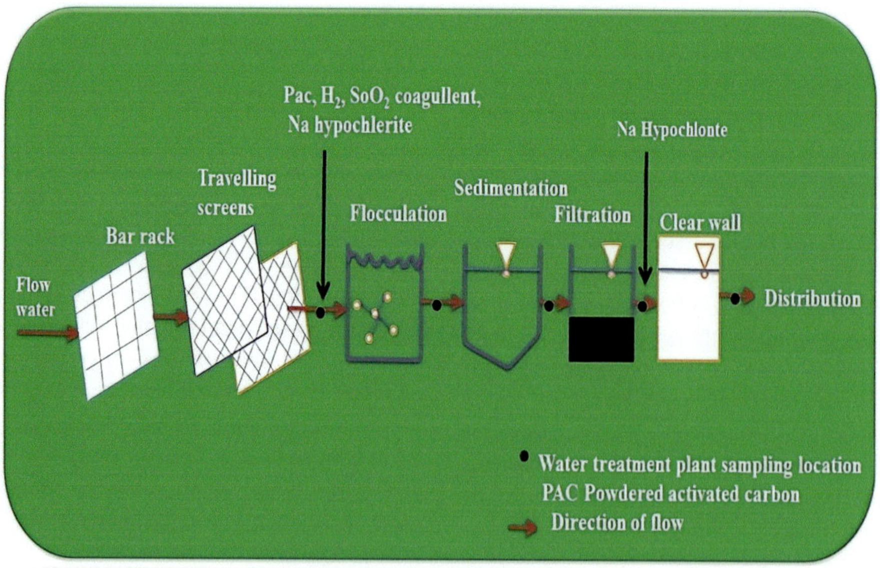

FIGURE 11.2 Chemical and physical processes used in drinking-water-treatment plants.

the past 150 years [10]. Evidence from history suggests that even in early settlements, the value of pure water was highlighted. Several techniques for purifying water, including coarse sand and charcoal filters, were described in the early scriptures. Ancient civilizations originally used aqueducts to build effective water transportation networks (Valley, Indus, Greek, American, and Roman civilizations). The founder of medicine, Hippocrates, made a connection between the value of water and general human health. Sir Francis Bacon put forth the theory of desalination using sand filter and test in the early 1600s, in 1627. The quantity of accessible water and the quality of drinking water must both be considered when determining the rigorousness of drinking water shortage [11]. We must clearly understand that the enduring accessibility of water all the way through ground and surface water assets is becoming a major concern as a result of climate change. It follows naturally that everyone must economically take water from the sea to provide everyone with access to clean drinking water. Even if the price of saltwater desalination using RO has decreased by about one-third over the past 30 years, it is still far from being cost-effective. In this regard, it is important to recall the following quote from former US President John F. Kennedy from 1961: "If we could ever competitively, at a cheap rate, get fresh water from saltwater, this would be in the long-term interests of humanity and could dwarf any other scientific accomplishments" [12]. The subsequent aspect of the issue is a result of the quick industrialization that has occurred on a worldwide scale. Economic growth has significantly improved the human's life economic situation, but it has also been accompanied by numerous environmental problems that have mostly gone unaddressed, partly because there are no practical economic remedies. Regulations for numerous harmful species found in drinking water have been developed by several organizations, including the USEPA, WHO, and EU. Halogenated organics and organochlorine pesticides are still on the USEPA's radar for potential future regulation. Other important contributors include inorganic salts, microorganisms, and metals. Since its founding (in 1974), the USEPA's regulatory coverage for safe water has improved more than four times, with changes to the maximum level of several contaminants. In terms of the number of people affected, it is evident that drinking water contamination has reached worldwide levels. Due to their extensive usage, a variety of contaminants, including lead and pesticides, are having an impact on the world's water sources. Geochemical pollutants, such as arsenic and fluoride, are already present in a few countries. Microorganisms are the main category of pollutants that harm drinking water. Overall, 90% of the 1.8 million individuals who die each year from diarrheal illnesses, primarily in underdeveloped nations, are children under the age of five [13, 14]. Increased sanitation cuts diarrhea morbidity by 32% and improved water availability cut it by 6%–25%. However, in places where there is a high risk of contamination, enhancing quality is a challenging aspect. One drop of oil, for instance, can make water unfit for human consumption. It is essential to have a more comprehensive understanding of the issue to comprehend the fine metal nanoparticles application for the filtration of consumption water. It is pretty obvious that some of these pesticides contain extremely hazardous resistant groups and are therefore very challenging to break through typical artificial degradation pathways. The pesticide's harmful effects on human health and their lifelong consequences on the food chain were only beginning to be recognized by society in the early 1960s. There is no denying the significance of the global increase in food security brought about by commercial pesticide use. Similar to this, utilization of dichlorodiphenyltrichloroethane to combat malaria resulted in the survival of a significant number of people [15]. However, the enduring effects of excessive use gradually began to show how contaminated the environment was. Over 100 individuals died in Kerala, one of India's states, in 1958 as a result of consuming wheat flour contaminated with parathion, marking the country's first instance of these effects. The US FDA had previously shown interest in looking into public concerns regarding the rising amount of pesticides in food. It's vital to remember that throughout this time, inorganic pesticides predominated, and organic pesticide use was just getting started. The Federal Economic Poisons Act's decision to designate pesticides as a split chemical additive was one of the law's crucial judgments (USA) [16]. The trade-offs associated within pesticide usage for agricultural activities were acknowledged in a study released by the Advisory Committee Panel of US President's Science. The report made clear acknowledgment of two facts: pesticides have contributed significantly to increasing agricultural output, but their toxicity, variety, and environmental perseverance may have a significant negative impact on human health, and this impact need not only to be felt by those working in agriculture [17]. Over time, it became evident that the release of pesticides into the environment was much greater in scope due to the deficiency of sufficient preparation for the controlled application of pesticides. The severity of this issue was increased when seen in the situation of the developing world. Scientific studies over time have determined how such dangerous compounds are gradually making their way into the natural food chain. It is crucial to emphasize that during this time, the ability to detect pesticides at incredibly low concentrations has improved. Following that, numerous lines of research were conducted on the microbial, photochemical, and chemical adsorption methods of pesticide degradation. In 1985, the first academic study on the use of microorganisms to break down pesticides was published [18]. The current state of affairs is not encouraging: 5.05 billion metric tons of pesticides were used globally in 2001, valued at $31.76 billion. India uses 570 grams of agrochemicals per acre as opposed to 2,500 grams in the USA, 3000 grams in Europe, and 12,000 grams in Japan. Despite low overall consumption in India relative to other industrialized nations, there are numerous regions in India where pesticide usage has been very high. Many organic insecticides have great environmental stability and high-water solubility, like endosulfan, which has 2 ppm solubility. However, the concentration of the dispersions utilized is substantially higher, causing saturation solubility. Similar properties are shared by the other most used pesticides, such as dichlorodiphenyltrichloroethane (DDT), malathion, and chlorpyrifos; applied concentrations are very high, while solubility is very low. As a result, the quantities of pesticides found in India's surface waters range from ppt to ppm [19]. Similar to this, a new problem is created by pesticide derivatives and conformational isomers, which are occasionally more stable, poisonous, and highly soluble. For instance, endosulfan is a mixture of stereoisomers known as "α-endosulfan" and "β-endosulfan" and is a mixture of stereoisomers. The more thermodynamically stable of the two, β-endosulfan, undergoes isomeric conversion slowly and irreversibly. Although there are no national statistics on the level of pesticide contamination in drinking water, regional assessments carried out across India unmistakably indicate considerable pesticide contamination [20]. Endosulfan caused previously unrecognized impacts in towns where it was aerially sprayed for 25 years, including genetic deformations. It is also crucial to understand that, to ensure the population's access to food, the usage of pesticides and other agrochemicals in many developing nations is anticipated to rise proportionately as the population grows. Due to the extraordinary persistence of some organic compounds and the continuous commercial use of several chemicals, the issue is also expected to influence developed nations. The persistence and volume of consumption of several organic compounds in the US market are covered in an excellent review study. The issue described above will worsen as water resources become scarcer and the amount of anthropogenic contamination in the water supply increases for causes other than those already addressed. Without clean drinking water, all development initiatives are reduced to theoretical concepts, and real advancement in human development becomes an unattainable goal [21]. Investigating emerging methods for more economical and pure water is necessary for this setting.

11.3 NANOMATERIAL REQUIREMENT IN WATER PURIFICATION

Standards for drinking water have been amended numerous times as our knowledge of how water quality affects health has changed. For instance, the permitted limits for well-known pollutants were periodically updated (e.g., lead), the definition of contaminant is getting more precise (e.g., a maximum concentration of organic residues vs. pesticides), and novel contaminants (e.g., RDX) are subject to regulation [22]. The contamination concentrations will likely hit molecular limits at this stage of evolution, meaning that the level at which they pose dangers may drop to a few ions, molecules, or atoms per glass of ingested water. Through several changes over the past 50 years, the recommended maximum level of arsenic in drinking water, as per the WHO International criteria, has decreased from 200 ppb to 10 ppb. The permitted level of lead in drinking water was decreased from 100 ppb to 10 ppb during this time. The same is true for adjustments to drinking water standards made by other organizations like the USEPA (1976–2001: allowed lindane concentration reduction from 4 ppb to 0.2 ppb, allowed arsenic concentration reduction from 50 ppb to 10 ppb, reduction in allowed lead concentration from 50 ppb to 15 ppb). The technology used to get rid of these impurities must thus advance to molecular levels. Highly effective molecular capturing agents must be able to capture them during single encounter occurrences. Minimum electricity use, economic viability, and environmental friendliness are also essential criteria for such a system [23]. Importantly, the chemistry ought to work even at concentrations of one part per billion. The concentrations involved in the majority of chemical reactions that we observe in our environment are so high that there are numerous interactions between the parties reacting. On the other hand, the encounters likewise decrease correspondingly at a dilution of 109. Traditional methods for removing these minute contaminants generate trash that further harms the environment. For instance, fluoride has an average adsorption capacity of 10 mg/g with traditional sorbents like activated alumina. For the construction of a residential purifier unit catering to the total 10 ppm fluoride removal, a minimum of 6.5 kg of activated alumina is needed. All of these present a number of challenging criteria. Numerous molecular technologies have been exquisitely incorporated by nature into various living organisms. Consider how effectively high-quality water is stored in watermelons. The water quality preserved in tender coconuts is amazing when compared to the feed water. All of the undesired material is rejected through a series of filtration and transportation activities before this happens. The movement of glucose across the erythrocyte membrane serves as an illustration of mediated transport. Water flowing through the well-researched aquaporins, the water channels in epithelial and renal cells, is another illustration of this kind of molecular transport. The aquaporin family members (more than 200) have been discovered in plants, microbes, vertebrates, and invertebrates [24]. There are at least ten aquaporins in humans. The first member of the aquaporin family to be found, aquaporin-1, transports water at a rate of 109 water molecules/s. Unexpectedly, the transport's activation energy through channels is almost as low as the one connected to self-diffusion in bulk water. Even while the aquaporin's porous channel, whose size range is similar to that of the water molecule, allows for transport, it is unexpected that the flow is so specifically directed toward the water that even ion movement is not possible. This has been explained in terms of how electrostatic repulsion and hydrophobic-hydrophilic interactions help water flow through numerous successive gateways. The region comprising asparagine-proline-alanine (NPA) motifs is the first major interaction site where water interacts with aquaporin. Since there are hydrophobic areas close to the NPA motifs, water molecules are forced to interact more forcefully. Because the aromatic side chains of phenylalanine, histidine, and positively charged arginine block the flow of positive ions, they create the second interaction site [25]. The hydrophobic side chain of phenylalanine causes the water molecules to align in a way that enforces strong hydrogen bonds with the amino acids arginine and histidine. Aquaporin expertise has been applied to the purification of water, and when combined with the many capabilities of nanoparticles, a multifunctional system for purifying drinking water may be created. It is also crucial to understand that nature has created several systems for purifying water as a result of the extraordinary diversity in living systems. For instance, (i) more than ten different aquaporins with specialized functionality are expressed in organs ranging from the kidney to red blood cells to the brain in humans. (ii) As a result of metal-chelation by free histidine, many flowering plants in the genus *Alyssum* demonstrate unusually high heavy metal tolerance (3% of leaf dry mass) [26]. (iii) It is known that a variety of bacteria and fungi can oxidize, hydrolyze, reduce, and immobilize hazardous chemicals. A handful of the examples found in nature have been applied to the filtration of drinking water, with RO membranes serving as iconic examples. Over 50% of the installed capacity for desalination is currently provided by RO systems. The cost of desalination has been decreased as a result of several techniques created to manage pore diameters. Desalination's specific energy has decreased from over 10 kWh/m^3 in the 1980s to less than 4 kWh/m^3. The methods still aren't the most energy-efficient ones, though; for instance, the least amount of energy used for desalination with 50% recovery is 0.97 kWh/m^3. While the currently explored methods for reducing fouling and improving pore size distribution will increase efficiency, the use of molecular transport mechanisms will assist us in filtering water in the way that uses the least amount of energy. Numerous strategies for drinking water filtration are being studied in addition to membrane research. As of now, it is acknowledged that there is no prospect of one solution fitting all circumstances due to the variations in water quality around the world. Nanotechnology will continue to be important in the development of effective molecular mechanisms for purification of water. The chemistry of fine metal particles is one of those methods that are widely used, and it is covered in depth here. The surface locations of the adsorbent particles are often used in adsorption applications (because bulk diffusion is hampered kinetically in the nonexistence of porous structure). When an adsorbent particle is shrunk from bulk to nanoscale, the surface-to-volume ratio dramatically rises [27]. It causes more atoms and molecules to be available on the surface for the adsorption of pollutants. As a result, each adsorbent particle's accessible surface energy likewise increases dramatically. This offers drinking water filtration several benefits, including the following: contaminant removal that is efficient even at low concentrations; less trash will be produced after treatment since less nanomaterial will be needed than in its bulk form. This occurs because more adsorbent atoms and molecules are present and actively used for adsorption per unit mass of the adsorbent. Novel reactions are feasible at the nanoscale but not in the corresponding bulk material when the number of surface atoms increases (i.e. surface energy) [28]. The breakdown of pesticides by noble metal nanoparticles, which cannot be accomplished by noble metals in their bulk form, is an example of one such process.

11.4 NANOTECHNOLOGY'S SIGNIFICANCE IN WATER PURIFICATION

Innovative approaches to water filtration were made possible by nanotechnology. Reactivity, high aspect ratio, electrostatic, variable pore volume, and hydrophobic and hydrophilic interactions are characteristics of nanomaterials that are beneficial for catalysis, adsorption, sensing, and optoelectronics. Nanotechnology-enabled technologies offer high-performance, cost-effective water and wastewater treatment solutions since they are extremely efficient, flexible, and multifunctional. Materials made of nano-objects are strong and have a high specific surface area. In other words, contact with contaminants and/or microorganisms is controlled by a large surface-to-volume ratio. The development of water treatment technologies using nanotechnology poses significant challenges to current practices. Nanotechnology has the potential to be economically applied to the filtration and use of non-traditional water sources. Nanomaterials and industrial wastewater have a significant and common relationship [29]. The current remediation solutions are efficient, but they are also expensive and time-consuming. Nanotechnologies are useful for treating wastewater since they get rid of impurities and aid in recycling to produce clean water. This results in the industry spending less money, time, and labor to address numerous environmental challenges. It should be highlighted that nanomaterials for water purification must be safe for the environment and non-toxic. When unsafe particles come into contact with the human body, they can seriously harm essential organs. Nanomaterials may translocate to different organs due to their dimensional properties, increasing the risk of biological harm. As a result, safety data sheets, Standard Operating Procedure (SOPs), and other relevant normative documents must precisely contain toxicity performance tests before introduction into the industry. An overview of nanotechnology's uses in water and wastewater treatment is presented in this chapter.

11.5 WATER PURIFICATION METHODS

11.5.1 Traditional Purification Methods

The tainted water was treated utilizing a variety of scientific techniques not just in ancient times but also in modern times. Some of these techniques are still used in rural regions. These less time-consuming, least expensive procedures include various processes, including solar disinfection, filtration, sedimentation, chemical disinfection, the three-pot method, boiling, and aeration. The following is a brief explanation of several conventional water purifying methods.

11.5.1.1 Boiling

Even now, especially in rural areas, this technique is extensively employed. Although boiling clean water at high temperatures does not reduce chemical pollution, it does tend to destroy the various bacteria and germs present in the water. This approach can be used to clear up pathogens or biodegradable contaminations. If the water is cloudy or filthy, cleaning it before boiling can increase the finished product's quality [30]. When water is cooked at temperatures close to 100°C, many species cannot live. The organisms present in the water are unlikely to live for longer than 30 min, even if the water is heated to 70–75°C. One of the simplest water treatment techniques that don't require any extra equipment is boiling. Additionally, boiling tends to temporarily lower the water's salinity and hardness. This technology has certain inherent advantages, but it also has some drawbacks, such as the need for fuel and the difficulty of the process when treating huge amounts of water. One of the greatest ways to get rid of bacteria, viruses, parasites, and other disease-causing organisms is to boil them.

11.5.1.2 Filtration

Filtering includes passing liquid through a thin membrane that separates the liquid's suspended particles. Bypassing the water through a thin, orderly, and clean cotton cloth or filter paper, the solid dirt or mud can be eliminated from the water. This procedure is quite effective if the water solely contains solid trash. This method does not produce findings that are satisfactory when the water is highly murky. Microorganisms cannot be strained using this approach, and dissolved gases and contaminants cannot be removed [31].

11.5.1.3 Solar Disinfection

Solar disinfection is a popular home cure that is both practical and inexpensive. The water in transparent plastic bottles/glass or clear plastic bags must be exposed to direct sunlight for many hours to eliminate any disease-causing bacteria. Due to the reduced intensity of solar irradiation under cloudy or wintery conditions, it may occasionally take several days to fully disinfect the water using this method [32]. The sun's intensity, the water's turbidity, light absorption, and the initial bacterial concentration all affect how effectively this procedure works. Although the time element makes this technology less favored, it has low operating expenses [33].

11.5.1.4 Chlorination

Chemical remediation of tainted water is both secure and economical. Chlorine is the most commonly used chemical in water treatment. The bacteria in the contaminated water are successfully destroyed by employing chlorine tablets. The chlorine disinfectant of choice is hypochlorite, including hypochlorite, calcium, and sodium hypochlorite [34]. Since this procedure can't remove any suspended sediments, it is necessary to filter the water before chlorination even though it is quite successful at killing microorganisms in the water. The chlorine concentration in the water and contact duration must be specified so that if the chlorine concentration is low, the contact time should be extended, and vice versa, for efficient disinfection [35]. However, the WHO report states that drinking chlorinated tap water with a chlorine concentration beyond the recommended levels can result in bladder cancer in people.

11.5.1.5 Three-Pot Method

The three-pot method is a traditional residential procedure that allows dirty water to settle gradually over time. If the water is murky, it is maintained in one pot for a long time until the dirt at the bottom settles, at which point it is gently transferred to the second pot. The remaining dirt settles at the bottom after being left there once again for a long enough time and then water is added to the third pot. The pots used for this need to be routinely cleaned and sterilized. Although it is affordable and simple to use, this approach cannot completely purify water [36].

11.5.1.6 Sedimentation

The sedimentation process aids in the removal of suspended particles from consumable water. The suspended particles fall to the bottom of the container due to gravity. After sedimentation, chlorine must be added because water cannot be germ-free without it. The weight of the suspended particles plays a significant role in determining how effectively the sedimentation technique works. The principles of buoyancy make it clear that a particle in water will settle downward at a constant rate when the buoyant force of the fluid acting upward is equal to the weight of the particle acting downward [37].

11.5.1.7 Aeration

The two general approaches of air diffusion and waterfall aeration are used in the aeration process. The waterfall aeration technique adds air to the water as it falls from a height, oxidizing it and removing scent and flavor. In most cases, gases like hydrogen sulfide are to blame for the taste and odor of the water [38]. In contrast, air diffusion aeration pumps air into the water through porous plates or perforated pipes, creating low-velocity bubbles that allow more water molecule surfaces to remain in touch with the air for a prolonged period [39]. The cost-effectiveness and reduced labor requirements of the aeration process sparked the creation of additional aeration technologies, including microwave desorption, pulsed vacuum post-process and heat addition, steam addition and removal, combinations of dissimilar gases, and pressure set points [40].

11.5.2 Advanced Purification Methods

Advanced purification methods provide high-quality consumable water from the cleansed water. RO, advanced oxidation, and micro-, ultra-, and nanofiltration are examples of advanced purification techniques. The effectiveness of

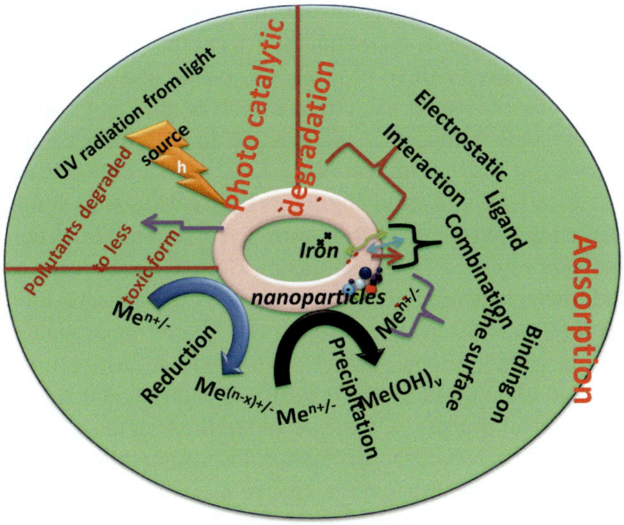

FIGURE 11.3 Photocatalytic and adsorptive processes of iron-based nanoparticles for the removal of pollutants.

various filtration systems in removing various contaminants that are often present in water is shown in Figure 11.3.

In the process of advanced water purification, all of the contaminated water is first allowed to pass through microfiltration, ultrafiltration, and nanofiltration, which uses hair-sized tubes carrying sophisticated membranes that may remove pathogens from the water [41]. RO, which involves filtration through nano-membranes while operating at high pressure, is the next step, and it tends to eliminate the remaining pollutants, including organic and metal ion impurities. The water is then further cleansed using an improved oxidation process while being exposed to UV radiation, which reduces the remaining contaminants [42]. With the use of high-purity chemical reagents, including $FeCl_3$, kaolin, H_2SO_4, and NaOH, refinery wastewater can be decontaminated using the catalytic vacuum distillation (CVD) process. Additionally, ozonation of the water followed by filtration is another efficient method for purifying water [43].

11.6 NANOPARTICLES AS ADSORBENTS

The process of adsorption only occurs on surfaces. The molecules of a dissolved solid or gaseous liquid adhere to any surface during the adsorption process to form a film or layer of molecules on that surface. The adsorbate amount adsorbed at equilibrium, which is often determined using the adsorption system's material balance, is how adsorption performance is expressed. These include chemical and electrostatic interactions between pollutants and magnetic nanomaterials since various biomolecules are present on the surfaces of magnetic nanoparticles. The presence of hydroxyl/OH functional groups is primarily responsible for diffusion between the adsorbent and adsorbate, surface precipitation, redox reactions, and ion exchange [44]. Adsorption is the method that is most frequently employed to remove contaminants from drinking water among all other methods of water treatment. This method removes the impurities from the water efficiently, is simple to use, and does not produce any secondary pollutants. The enormous specific surface area and surface multifunctionality of nanoparticles are the inherent qualities that make them effective adsorbents [45]. These characteristics enable nanoparticles to interact chemically and bond with nearby molecules with ease. The nanoparticles exhibit excellent long-term stability in addition to being effective adsorbers. Nanoadsorbents can be divided into four categories: carbon nanotubes (CNTs), activated carbon, polymeric nanoadsorbents, and nanoparticles. Activated carbon had been utilized as an adsorber for the purification of water, but after the development of CNTs, this method was superseded by the application of CNTs. CNTs are separated into two groups based on how many outer layers are present around the tubes [46]. Single-walled carbon nanotubes (SWCNTs) are CNTs with a single layer around their tubular structure, while multi-walled CNTs have more than one layer (multi-walled carbon nanotubes [MWCNT]). In general, MWCNTs are favored over SWCNTs because they are easier to make and have lower synthesis costs. Additionally, MWCNTs can be easily manufactured with acceptable purity, while SWCNTs are exceedingly difficult to purify. Carbon can be utilized as an adsorbent in several additional carbon nanomaterials (CNM) forms, including carbon fibers, carbon beads, and nanoporous carbon. CNMs exhibit a strong ability to adsorb a variety of pollutants, have quick kinetics, and also have the unique ability to select aromatics. Another class of magnetic materials being explored in the treatment of water are nanoparticles with ferrite structures [47]. $MnFe_2O_4$ nanoparticles have been tried for the removal of Cr(VI), while $CuFe_2O_4$ and $ZnFe_2O_4$ ones showed some efficiency for capturing Mo(VI). However, studies were conducted under conditions that were not appropriate for the purification of drinking water. There have also been reports of hollow $CoFe_2O_4$ nanoparticles as possible U(VI) adsorbents [35]. In contrast to high-valent heavy metals, ferrite nanoparticles performed better in the treatment of arsenic-contaminated water. More specifically, employing contaminated tap water, tiny $MgFe_2O_4$ nanoparticles were able to achieve a substantial capacity of roughly 10mg/g for both As(III) and As(V). $NiFe_2O_4$, $MnFe_2O_4$, and $CoFe_2O_4$ nanoparticles for arsenic removal were also the subject of several investigations [48]. Antibiotic degradation, mineralization, and inactivation were significantly accelerated by magnetically recoverable graphene-supported $MnFe_2O_4$ nanoparticles.

11.6.1 Adsorption of Toxic Metals

As soon as they are consumed, the hazardous metals found in water pose a serious health risk to both people and marine life. The most dangerous toxic metal ions that can be found in wastewater include Ni^{2+}, Hg^{2+}, As^{3+}, Cd^{2+}, Cr^{6+}, Mn^{2+}, and Pb^{2+} [49–51]. They frequently cause major issues, such as lung cancer, kidney failure, skin dermatitis, pulmonary impairment, liver damage, and central nervous system dyspnea. A surplus of the aforementioned metals also has a detrimental influence on agricultural productivity because they make plants more phototoxic, which harms the environment and food safety. Because these metals are dangerous, $CoFe_2O_4$ nanoparticles and their corresponding nanocomposites are used as adsorbents to remove them from water. For $CoFe_2O_4$ nanoparticles and their composite materials, the extreme absorption capacity was obtained with a pH of 3.7–8.1, respectively. $CoFe_2O_4$ nanoparticles were discovered to have a better adsorption capability when used at higher temperatures [52]. Additionally, it was found that modifying the $CoFe_2O_4$ nanoparticles' surface with other substances improved their adsorption capacity. Therefore, care must be taken while selecting the materials to create the heterojunction with $CoFe_2O_4$ nanoparticles or to modify their surface to enhance the adsorption capacity [53]. In addition, it is essential to check the $CoFe_2O_4$ nanoparticles' propensity for adsorbing additional dangerous metals. In general, it is important to use $CoFe_2O_4$ nanoparticles, which are abundant, inexpensive, and have superparamagnetic capabilities, to remove dangerous metals on a large scale, from the lab [54].

11.6.2 Dyes Adsorption

In many other industrial sectors, dyes are extensively used, particularly in the production of clothing. These dyes are released into water supplies, which might contaminate the water. Their colors maintain their vibrancy in the water because many of them have an azo group and have little decomposable substance. Given that dyes are known to be carcinogenic, mutagenic, and harmful to both humans and marine life [55], it is impossible to avoid utilizing adsorbents that are effective at removing colors from wastewater. Spinel ferrite nanoparticles (SFNPs) were dyed, and different colors were removed from them using $CoFe_2O_4$ nanoparticles and their matching nanocomposites as adsorbents [56]. While it was discovered through the kinetic investigation that the pseudo-second-order kinetic model is the dominant one and is followed by the majority of catalysts during the dye removal process, it was discovered through the adsorption isotherm investigation that dye absorption follows various models depending on the kind of dye being detached.

$CoFe_2O_4$ nanoparticles and the related nanocomposites may be used as low-cost catalysts to remove colors from water or wastewater due to their high absorption capacities [57]. As a result, $CoFe_2O_4$ nanoparticles can successfully handle the problem of future dyes in wastewater remediation as well as the sustainability problem of wastewater purification if further research is led to uncover their shortcomings and also be able to develop them for industrial water treatment. Bibi et al. [58] used the Fe_2O_3 nanoparticles depicted in Figure 11.4 to demonstrate the degradation of dye. An electron (e) and hole (h+) pair are created when the Fe_2O_3 nanoparticle is exposed to radiation, and the electron is excited from the valence band to the conduction band while the h+ remains in the VB [58]. The oxidative breakdown of dye is caused by the hydroxyl radical, which is created when

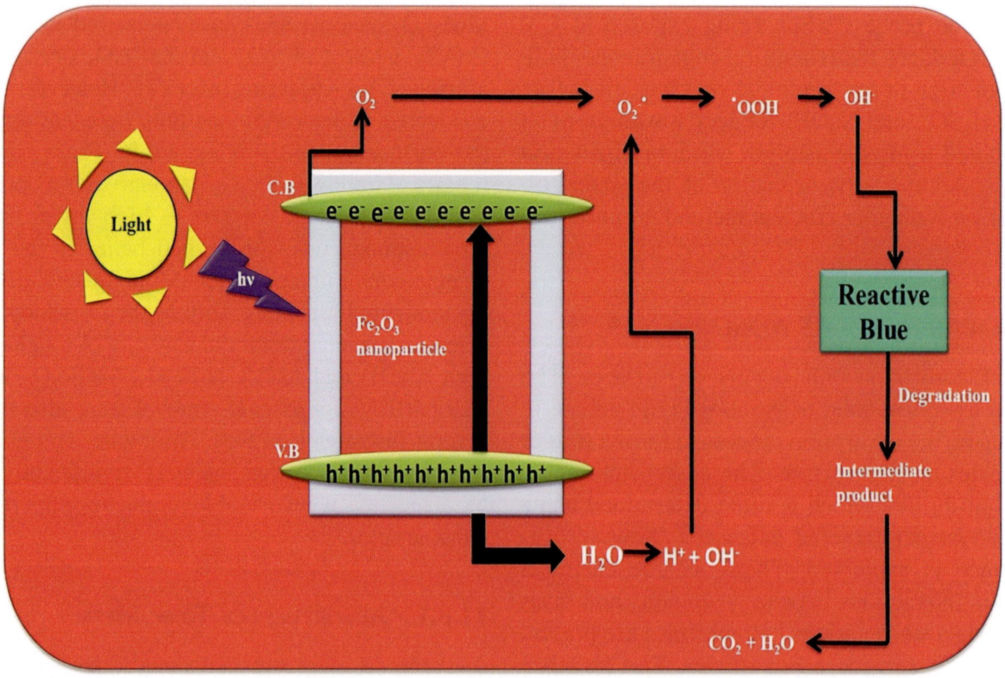

FIGURE 11.4 The mechanism of dye (pollutants) removal by Fe_2O_3 nanoparticles.

water is converted into this hole (h+). On the other hand, superoxide radicals are created when free electrons combine with molecular oxygen. Additionally, hydroxyl radicals are produced from the superoxide radical. The hydroxyl radical is a potent oxidizing agent that non-selectively breaks down organic molecules into inert byproducts [59].

11.6.3 Removal of Other Contaminants

$CoFe_2O_4$ nanoparticles and their nanocomposites were utilized as a catalyst for numerous reactions in addition to the planned implementations in water. The tetragonal $CoFe_2O_4$ manufactured with the aid of electrospinning technique was found to be effective in removing acid fuchsine from water [60], whereas the $CoFe_2O_4$ produced by the sol-gel combustion method demonstrated large catalytic action for gallic acid elimination from water in the occurrence of H_2O_2. By adding $CoFe_2O_4$ nanoparticles and their corresponding nanocomposites, additional compounds such as 4-chlorophenol, 2,4-dichlorophenoxyacetic acid, atorvastatin, phosphate, atrazine, and nitrophenol, with the help of acetaminophen scavenging, peroxymonosulfate, and phenols [61]. By introducing a composite of chitosan and $CoFe_2O_4$ nanoparticles, the pharmaceutical chemicals diltiazem hydrogen chloride and Na naproxen that disrupt the endocrine systems have been separated from the water. $CoFe_2O_4$ nanoparticles and their nanocomposites have been used in water in a variety of ways, although their use in photodegradation or as an absorbent for pollutants and pesticides that are still evolving is limited [62]. To use this magnetic catalyst on these contaminants, research is therefore needed. $CoFe_2O_4$ nanoparticles and the corresponding composites have applications outside of the purification of water. They served as various photocathode materials in the water splitting reaction, as a gas sensor for gases like hydrogen, carbon monoxide, ethane, and petroleum gas [63], as a catalyst for the water gas shift, and in bio-medication for their antibacterial activity when combined with chitosan and LDH-doxycycline. Cobalt ferrite nanoparticles and their nanocomposites have been used in the field of catalysis for several significant reactions, including the oxidation of styrene-to-styrene oxide, asymmetric hydrosilylation of ketones [64], cyanation of aryl halides, three-component formation of triazoles in water, and oxidative dehydrogenation of but-1-ene; this highly practical superparamagnetic catalyst has shown its use in numerous industries [65]. Table 11.1 represents the comparison of different nanoparticles for the removal of various contaminants.

TABLE 11.1
Various Nanoparticles Comparison for the Removal of Different Contaminants

Spinel Ferrite Nanoparticles	Size	Contaminants	pH	Reaction Time (min)	Temperature (°C)	Kinetic Model	Adsorption Capacity (mg/g)	Ref
Fe_3O_4@chitosan	15–40	AO 7	3–4	–	25	–	1215	[66]
S-$CoFe_2O_4$ (sulfur modified)	38.6	AOII	6	15	25	PSO	495	[67]
Fe_3O_4@SiO_2-NH_2	–	AO 10	3	–	25	–	61.33	[68]
Fe_3O_4@graphene	–	CR	–	–	25	PSO	33.66	[69]
Amine/Fe_3O_4-resin	9–10	AR 18	7	–	25	PSO	99.4	[70]
Fe_3O_4@SiO_2	200	CR	5.3	–	35	PSO	54.6	[71]
Fe_3O_4/MoS_2	–	CR	–	120	–	–	71	[33]
Cellulose/Fe_3O_4/activated carbon	–	CR	4	–	25	PSO	66.1	[72]
$NiFe_2O_4$	10–20	CR	7	–	18	PSO	97.1	[73]
$CoFe_2O_4$@ZnO	70	Direct blue 71	6	300	25	–	71%	[74]
$CoFe_2O_4$	30–40	CR	7	–	18	PSO	244.5	[75]
Fe_3O_4@NH_2	10	MB	7	–	20	–	6.2	[76]
Activated carbon-$NiFe_2O_4$	23.4	MO	3	30	30	PSO	33	[77]
Fe_3O_4	40–300	As(III)	7	–	22.5	–	23.8	[78]
$MnFe_2O_4$@GO	7.5	As(III)	6.5	–	25	PSO	146	[79]
$MnFe_2O_4$	30–50	As(III)	7	24	25	PSO	93.8	[80]
γ-Fe_2O_3	3–8	As(V)	5–7.6	60	–	–	45	[81]
$MnFe_2O_4$	30–50	As(V)	3	24	25	PSO	90.4	[82]
$CuFe_2O_4$	20–90	Cd (II)	6	–	–	–	17.54	[83]
Fe_3O_4-citric acid	15–27	Cd (II)	5	27	25	–	10.8	[84]
$MnFe_2O_4$-graphene	150–170	Cd (II)	7	180	37	–	76.9	[85]
Fe_3O_4	14.8	Cr (VI)	6	30	25	PSO	15.4	[86]
γ-Fe_2O_3@δ-FeOOH	15	Cr (VI)	2.5	30	25	–	25.8	[87]
$CoFe_2O_4$	12–21	Cr (VI)	6	120	55	PSO	16.73	[88]
$ZnFe_2O_4$	20	Cr (VI)	2.5	–	–	–	34.2	[89]
Fe_3O_4@chitosan	30	Cr (VI)	3	100	25	PSO	55.8	[90]

11.7 CONCLUSION

Water and wastewater treatment is one of the applications for SFNPs, which are also used in other fields. SFNPs are used to remove organic and inorganic impurities from water and wastewater by adsorbing them. Additionally, the photocatalytic activity of SFNPs contributes to the destruction of contaminants. Their use in membrane modification, as well as the sensing and detection of pollutants in wastewater, is turning into a popular study topic. They are extremely significant and readily transportable in an aqueous medium due to their strong reactivity, but their toxicity is a worry. Therefore, before commercialization and widespread application, it is crucial to take into account and thoroughly examine their toxicity. However, to avoid unintended consequences and advance sustainable applications in diverse fields, environmental health and safety research must work hand in hand with the development of new technologies for the synthesis of SFNPs. Applications of SFNPs on actual industrial wastewater samples with several pollutants present are scarce or nonexistent in the literature and call for additional studies. SFNPs have been found to have great qualities, and it is simple to functionalize their surface with harmless chemicals, opening up a promising prospect for the next generation to address issues with water quality. It is conceivable to draw the conclusion that SFNPs and SFNCs have strong adsorption and disinfection capacities toward pollutants and pathogens, respectively, based on the review. Since conventional chemicals are currently used for water and wastewater treatment, the development of better treatment technology using SFNPs/SFNCs not only resolves the current societal issues associated with poor water quality but also replaces them with improved facilities and the provision of safer water quality under sustainable circumstances. Additionally, the reuse of wastewater, reduction of waste generation and environmental contamination, ease of use, and affordability of the treatment technology make the usage of SFNPs/SFNCs an attractive choice for the treatment of water and wastewater. This is a result of the minimal dose requirement, simple external magnetic field recovery, and several cycles.

REFERENCES

1. D. Suhag, A. Thakur, P. Thakur (eds) *Integrated Nanomaterials and Their Applications*. Springer Nature Singapore, pp. 1–499, 2023. https://doi.org/10.1007/978-981-99-6105-4
2. S. Kumari, F. Wan, A. Thakur, S. Singh, P. Thakur. Retrievable nano-sized Co–Mg–Zn ferrite as an active photocatalyst and adsorbent for removing organic and biological contaminants for water purification, *Ceram. Int.* 50 (2023) 3210–3221. https://doi.org/10.1016/j.ceramint.2023.11.071
3. A. Thakur, P. Punia, R. Dhar, R. K. Aggarwal and P. Thakur. Separation of cadmium and chromium heavy metals from industrial wastewater by using Ni-Zn nanoferrites, *Adv. Nano Res.* 12 (2022) 457–465. https://doi.org/10.12989/anr.2022.12.5.457
4. P. Thakur, Y. Verma, A. Thakur. Toxicity of Nanomaterials: An Overview. In: A. Thakur, P. Thakur, S.M.P. Khurana (eds), *Synthesis and Applications of Nanoparticles*. Springer, Singapore, pp. 535–544, 2022. https://doi.org/10.1007/978-981-16-6819-7_25
5. A. Thakur, P. Thakur (eds) *Soft Nanoferrites for Biomedical and Environmental Applications*. CRC Press. 2024. https://doi.org/10.1201/9781003517139
6. H. Liang, W. Chen, R. Wang, Z. Qi, J. Mi, Z. Wang, X-shaped hollow α-FeOOH penetration twins and their conversion to α-Fe_2O_3 nanocrystals bound by high-index facets with enhanced photocatalytic activity, *Chem. Eng. J.* 274 (2015) 224–230, https://doi.org/10.1016/j.cej.2015.03.125
7. S. Nizamuddin, M.T.H. Siddiqui, N.M. Mubarak, H.A. Baloch, E.C. Abdullah, S.A. Mazari, G.J. Griffin, M.P. Srinivasan, A. Tanksale, *Iron Oxide Nanomaterials for the Removal of Heavy Metals and Dyes from Wastewater*, Elsevier Inc., 2018. 10.1016/B978-0-12-813926-4.00023-9
8. S. Andreescu, J. Njagi, C. Ispas, M.T. Ravalli, Applications of advanced nanomaterials for environmental monitoring, *J. Environ. Monit.* 11 (2009) 27–40, https://doi.org/10.1039/b811063h
9. T. Singh, S. Shukla, P. Kumar, V. Wahla, V.K. Bajpai, Application of nanotechnology in food science: Perception and overview, *Front. Microbiol.* 8 (2017) 1–7, https://doi.org/10.3389/fmicb.2017.01501
10. M. Ismail, K. Akhtar, M.I. Khan, T. Kamal, M.A. Khan, A.M. Asiri, J. Seo, S.B. Khan, Pollution, toxicity and carcinogenicity of organic dyes and their catalytic bioremediation, *Curr. Pharm. Des.* 25 (2019) 3645–3663, https://doi.org/10.2174/1381612825666191021142026
11. A. Thakur, P. Mathur, M. Singh, Controlling the properties of manganese-zinc ferrites by substituting In^{3+} and Al^{3+} ions, *Zeitschrift Fur Phys. Chemie.* 221 (2007) 837–845, https://doi.org/10.1524/zpch.2007.221.6.837
12. S. Parveen, A.H. Wani, M.A. Shah, H.S. Devi, M.Y. Bhat, J.A. Koka, Preparation, characterization and antifungal activity of iron oxide nanoparticles, *Microb. Pathog.* 115 (2018) 287–292, https://doi.org/10.1016/j.micpath.2017.12.068
13. A.A. Adegoke, T.A. Stenström, Metal oxide nanoparticles in removing residual pharmaceutical products and pathogens from water and wastewater, *Nanoparticles Pharmacother.* (2019) 561–589, https://doi.org/10.1016/b978-0-12-816504-1.00016-8
14. P. Mathur, A. Thakur, M. Singh, A study of nano-structured Zn-Mn soft spinel ferrites by the citrate precursor method, *Phys. Scripta* 77 (2008) 045701, https://doi.org/10.1088/0031-8949/77/4/045701
15. M. Sujana, S. Mohanty, Characterization and fluoride uptake studies of nano-scale iron oxide-hydroxide synthesized by microemulsion method, *Int. J. Eng. Sci. Technol.* 2 (2011) 1–12, https://doi.org/10.4314/ijest.v2i8.63775
16. N. Dhanda, P. Thakur, A. Thakur, Green synthesis of cobalt ferrite: A study of structural and optical properties, *Mater. Today: Proc.* 73 (2023) 237–240, https://doi.org/10.1016/j.matpr.2022.07.202
17. B.D. Deshpande, P.S. Agrawal, M.K.N. Yenkie, S.J. Dhoble, Prospective of nanotechnology in degradation of waste water: A new challenges, *Nano-Struct. Nano-Obj.* 22 (2020) 1–14, https://doi.org/10.1016/j.nanoso.2020.100442
18. A. Thakur, P. Thakur, J.H. Hsu, Enhancement in dielectric and magnetic properties of In^{3+} substituted Ni-Zn nanoferrites by coprecipitation method, *IEEE Trans. Magn.* 47 (2011) 4336–4339, https://doi.org/10.1109/TMAG.2011.2156394
19. A.G. Leonel, A.A.P. Mansur, H.S. Mansur, Advanced functional nanostructures based on magnetic iron oxide nanomaterials for water remediation: A review, *Water Res.* 190 (2021), https://doi.org/10.1016/j.watres.2020.116693

20. M. Anjum, R. Miandad, M. Waqas, F. Gehany, M.A. Barakat, Remediation of wastewater using various nanomaterials, *Arab. J. Chem.* 12 (2019) 4897–4919, https://doi.org/10.1016/j.arabjc.2016.10.004
21. A. Baruah, V. Chaudhary, R. Malik, V.K. Tomer, Nanotechnology based solutions for wastewater treatment, *Nanotechnol. Water Wastewater Treat.* (2019) 337–368, https://doi.org/10.1016/b978-0-12-813902-8.00017-4
22. R. Mukhopadhyay, D. Bhaduri, B. Sarkar, R. Rusmin, D. Hou, R. Khanam, S. Sarkar, J. Kumar Biswas, M. Vithanage, A. Bhatnagar, Y.S. Ok, Clay–polymer nanocomposites: Progress and challenges for use in sustainable water treatment, *J. Hazard. Mater.* 383 (2020), https://doi.org/10.1016/j.jhazmat.2019.121125
23. V.K. Chaturvedi, A. Kushwaha, S. Maurya, N. Tabassum, H. Chaurasia, M.P. Singh, Wastewater Treatment through Nanotechnology: Role and Prospects. In: *Restoration of Wetland Ecosystem: A Trajectory towards a Sustainable Environment*, pp. 227–247, (2020). https://doi.org/10.1007/978-981-13-7665-8_14
24. S. Kumari, N. Dhanda, A. Thakur, V. Gupta, S. Singh, R. Kumar, S. Hameed, P. Thakur, 'Nano Ca–Mg–Zn ferrites as tuneable photocatalyst for UV light-induced degradation of rhodamine B dye and antimicrobial behavior for water purification', *Ceram. Intern* 2022, ISSN 0272-8842, https://doi.org/10.1016/j.ceramint.2022.12.107
25. S. Manikandan, N. Karmegam, R. Subbaiya, G. Karthiga Devi, R. Arulvel, B. Ravindran, M. Kumar Awasthi, Emerging nano-structured innovative materials as adsorbents in wastewater treatment, *Bioresour. Technol.* 320 (2020), https://doi.org/10.1016/j.biortech.2020.124394
26. R. Trivedi, J. Bergi, Application of bio nanoparticles in wastewater treatment, *Adv. Oxid. Process. Effl. Treat. Plants.* (2021) 177–197, https://doi.org/10.1016/b978-0-12-821011-6.00010-4
27. N.A. Khan, S.U. Khan, S. Ahmed, I.H. Farooqi, A. Dhingra, A. Hussain, F. Changani, Applications of nanotechnology in water and wastewater treatment: A review, *Asian J. Water, Environ. Pollut.* 16 (2019) 81–86, https://doi.org/10.3233/AJW190051
28. R. Prasad, V. Kumar, M. Kumar, D. Choudhary, Nanobiotechnology in bioformulations, nanobiotechnology bioformulations, nanotechnol. *Life Sci.* (2019) 187–224, https://doi.org/10.1007/978-3-030-17061-5
29. R. Arau´jo, A.C.M. Castro, A. Fiu´za, The use of nanoparticles in soil and water remediation processes, *Mater. Today Proc.* Elsevier Ltd. (2015) 315–320, https://doi.org/10.1016/j.matpr.2015.04.055
30. L. Ling, X.Y. Huang, W.X. Zhang, Enrichment of precious metals from wastewater with core-shell nanoparticles of Iron, *Adv. Mater.* 30 (2018) 1–6, https://doi.org/10.1002/adma.201705703
31. R. Bhateria, R. Singh, A review on nanotechnological application of magnetic iron oxides for heavy metal removal, *J. Water Process Eng.* 31 (2019), https://doi.org/10.1016/j.jwpe.2019.100845
32. R.K. Ibrahim, M. Hayyan, M.A. AlSaadi, A. Hayyan, S. Ibrahim, Environmental application of nanotechnology: Air, soil, and water, environ, *Sci. Pollut. Res.* 23 (2016) 13754–13788, https://doi.org/10.1007/s11356-016-6457-z
33. H.J. Song, S. You, X.H. Jia, J. Yang, MoS_2 nanosheets decorated with magnetic Fe_3O_4 nanoparticles and their ultrafast adsorption for wastewater treatment, *Ceram. Int.* 41 (2015) 13896–13902, https://doi.org/10.1016/j.ceramint.2015.08.023
34. N. Dhanda, S. Kumari, R. Kumar, D. Kumar, A.-C. Sun, P. Thakur, A. Thakur, Influence of Ni over magnetically benign Co ferrite system and study of its structural, optical, and magnetic behaviour, *Inorg. Chem. Commun.* 151 (2023) 110569, ISSN 1387-7003, https://doi.org/10.1016/j.inoche.2023.110569
35. S. Taghipour, S.M. Hosseini, B. Ataie-Ashtiani, Engineering nanomaterials for water and wastewater treatment: Review of classifications, properties, and applications, *New J. Chem.* 43 (2019) 7902–7927, https://doi.org/10.1039/c9nj00157c
36. G. Crini, E. Lichtfouse, Advantages, and disadvantages of techniques used for wastewater treatment, *Environ. Chem. Lett.* 17 (2019) 145–155, https://doi.org/10.1007/s10311-018-0785-9
37. N. Dhanda, P. Thakur, A.-C. Sun, A. Thakur, Structural, optical and magnetic properties along with antifungal activity of Ag-doped Ni-Co nanoferrites synthesized by eco-friendly route, *J. Magn. Magn. Mater.* 572 (2023) 170598, ISSN 0304-8853, https://doi.org/10.1016/j.jmmm.2023.170598
38. E.A. Deliyanni, G.Z. Kyzas, K.S. Triantafyllidis, K.A. Matis, Activated carbons for the removal of heavy metal ions: A systematic review of recent literature focused on lead and arsenic ions, *Open Chem.* 13 (2015) 699–708, https://doi.org/10.1515/chem-2015-0087
39. S. Manna, S. Prakash, P. Das, Synthesis of graphene oxide nano-materials coated bio-char using carbonaceous industrial waste for phenol separation from water, *Colloids Surf., A Physicochem. Eng. Asp.* 581 (2019), https://doi.org/10.1016/j.colsurfa.2019.123818.
40. S.P. Goutam, G. Saxena, D. Roy, A.K. Yadav, R.N. Bharagava, Green synthesis of nanoparticles and their applications in water and wastewater treatment, *Bioremed Ind. Waste Environ. Saf.* (2020) 349–379, https://doi.org/10.1007/978-981-13-1891-7_16
41. L. Zhao, J. Deng, P. Sun, J. Liu, Y. Ji, N. Nakada, Z. Qiao, H. Tanaka, Y. Yang, Nanomaterials for treating emerging contaminants in water by adsorption and photocatalysis: Systematic review and bibliometric analysis, *Sci. Total Environ.* 627 (2018), https://doi.org/10.1016/j.scitotenv.2018.02.006.
42. N. Dhanda, P. Thakur, R. Kumar, T. Fatima, S. Hameed, Y. Slimani, A.-C.A. Sun, A. Thakur, 'Green-synthesis of Ni-Co nanoferrites using aloe vera extract: Structural, optical, magnetic, and antimicrobial studies', *Appl Organomet Chem.* 2023, e7110. https://doi.org/10.1002/aoc.7110
43. Y.V. Konyukhov, Heavy-metal extraction from wastewater by means of iron nanopowder, *Steel Transl.* 48 (2018) 135–141, https://doi.org/10.3103/S0967091218020080.
44. G.N. Hlongwane, P.T. Sekoai, M. Meyyappan, K. Moothi, Simultaneous removal of pollutants from water using nanoparticles: A shift from single pollutant control to multiple pollutant control, *Sci. Total Environ.* 656 (2019) 808–833, https://doi.org/10.1016/j.scitotenv.2018.11.257
45. Z. Cai, Y. Sun, W. Liu, F. Pan, P. Sun, J. Fu, An overview of nanomaterials applied for removing dyes from wastewater, *Environ. Sci. Pollut. Res.* 24 (2017) 15882–15904, https://doi.org/10.1007/s11356-017-9003-8
46. S.M. Dadfar, K. Roemhild, N.I. Drude, S. von Stillfried, R. Knüchel, F. Kiessling, T. Lammers, Iron oxide nanoparticles: Diagnostic, therapeutic and theranostic applications, *Adv. Drug Deliv. Rev.* 138 (2019) 302–325, https://doi.org/10.1016/j.addr.2019.01.005
47. D. Chahar, S. Taneja, S. Bisht, S. Kesarwani, P. Thakur, A. Thakur, P.B. Sharma. Photocatalytic activity of Cobalt substituted Zinc ferrite for the degradation of methylene blue dye under visible light irradiation. *J. Alloys Compd.* 851 (2021) 156878. https://doi.org/10.1016/j.jallcom.2020.156878

48. P. Punia, M.K. Bharti, R. Dhar, P. Thakur, A. Thakur. Recent advances in detection and removal of heavy metals from contaminated water. *ChemBioEng Reviews*. 9 (2022) 351–369. https://doi.org/10.1002/cben.202100053
49. M.K. Bharti, S. Gupta, S. Chalia, I. Garg, P. Thakur, A. Thakur. Potential of magnetic nanoferrites in removal of heavy metals from contaminated water: Mini review. *J. Supercond. Nov. Magn.* 33 (2020) 3651–3665. https://doi.org/10.1007/s10948-020-05657-1
50. W. Wu, C.Z. Jiang, V.A.L. Roy, Designed synthesis and surface engineering strategies of magnetic iron oxide nanoparticles for biomedical applications, *Nanoscale* 8 (2016) 19421–19474, https://doi.org/10.1039/c6nr07542h
51. P. Punia, R.K. Aggarwal, R. Kumar, R. Dhar, P. Thakur, A. Thakur. Adsorption of Cd and Cr ions from industrial wastewater using Ca doped Ni–Zn nanoferrites: Synthesis, characterization and isotherm analysis. *Ceram. Int.* 48 (2022) 18048–18056. https://doi.org/10.1016/j.ceramint.2022.02.234
52. A.P. Lagrow, M.O. Besenhard, A. Hodzic, A. Sergides, L.K. Bogart, A. Gavriilidis, N.T.K. Thanh, Unravelling the growth mechanism of the co-precipitation of iron oxide nanoparticles with the aid of synchrotron X-Ray diffraction in solution, *Nanoscale* 11 (2019) 6620–6628, https://doi.org/10.1039/c9nr00531e
53. B. Banerji, S.K. Pramanik, S. Mandal, N.C. Maiti, K. Chaudhuri, Synthesis, characterization, and cytotoxicity study of magnetic (Fe_3O_4) nanoparticles and their drug conjugate, *RSC Adv.* 2 (2012) 2493–2497, https://doi.org/10.1039/c2ra01118b
54. S. Taghizadeh, A. Berenjian, M. Zare, New perspectives on iron-based nanostructures, *Processes* 8 (2020) 1–27, https://doi.org/10.3390/pr8091128
55. P. Thakur, D. Chahar, A. Thakur. Visible light assisted photocatalytic degradation of methylene blue dye using Ni doped Co-Zn nanoferrites. *Adv. Nano. Res.* 12 (4) (2022) 415–426. https://doi.org/10.12989/anr.2022.12.4.415
56. A. Singh, F. Wan, K. Yadav, S. Kharbanda, P. Thakur, A. Thakur. A novel magnetic $NiFe_2O_4$-Ag-ZnO hybrid nanocomposite for the escalated photocatalytic dye degradation and antibacterial activities. *Mater. Sci. Eng. B.* 299 (2024) 116935. https://doi.org/10.1016/j.mseb.2023.116935
57. S.F. Hasany, I. Ahmed, J. Rajan, A. Rehman, Systematic review of the preparation techniques of iron oxide magnetic nanoparticles, *Nanosci. Nanotechnol.* 2 (2013) 148–158, https://doi.org/10.5923/j.nn.20120206.01
58. I. Bibi, N. Nazar, S. Ata, M. Sultan, A. Ali, A. Abbas, K. Jilani, S. Kamal, F.M. Sarim, M.I. Khan, F. Jalal, M. Iqbal, Green synthesis of iron oxide nanoparticles using pomegranate seeds extract and photocatalytic activity evaluation for the degradation of textile dye, *J. Mater. Res. Technol.* 8 (2019) 6115–6124, https://doi.org/10.1016/j.jmrt.2019.10.006
59. D. Maiti, P. Sujatha Devi, Selective formation of iron oxide and oxyhydroxide nanoparticles at room temperature: Critical role of concentration of ferric nitrate, *Mater. Chem. Phys.* 154 (2015) 144–151, https://doi.org/10.1016/j.matchemphys.2015.01.057
60. K. Qiao, W. Tian, J. Bai, L. Wang, J. Zhao, Z. Du, X. Gong, Application of magnetic adsorbents based on iron oxide nanoparticles for oil spill remediation: A review, *J. Taiwan Inst. Chem. Eng.* 97 (2019) 227–236, https://doi.org/10.1016/j.jtice.2019.01.029
61. S. Rahimi, R.M. Moattari, L. Rajabi, A.A. Derakhshan, M. Keyhani, Iron oxide/hydroxide (a, c-FeOOH) nanoparticles as high potential adsorbents for lead removal from polluted aquatic media, *J. Ind. Eng. Chem.* 23 (2015) 33–43, https://doi.org/10.1016/j.jiec.2014.07.039
62. N. Guo, X. Lv, Q. Li, T. Ren, H. Song, Q. Yang, Removal of hexavalent chromium from aqueous solution by mesoporous a-FeOOH nanoparticles: Performance and mechanism, *Micropor. Mesopor. Matr.* 299 (2020), https://doi.org/10.1016/j.micromeso.2020.110101
63. R. Zamiri, H.A. Ahangar, A. Zakaria, G. Zamiri, H.R. Bahari, G.P.C. Drummen, Hydrothermal synthesis of goethite (a-FeOOH) nanorods in the presence of ethylenediamine: thiourea, *J. Nanopart. Res.* 16 (2014) 1–10, https://doi.org/10.1007/s11051-014-2333-2
64. V. Rădit,oiu, L. Diamandescu, M. Cosmin Corobea, A. Rădit,oiu, N. Popescu-Pogrion, C. Andi Nicolae, A facile hydrothermal route for the synthesis of a-FeOOH with controlled morphology, *J. Cryst. Growth.* 348 (2012) 40–46, https://doi.org/10.1016/j.jcrysgro.2012.03.032
65. N. Nishida, S. Amagasa, Y. Kobayashi, Y. Yamada, Synthesis of superparamagnetic d-FeOOH nanoparticles by a chemical method, *Appl. Surf. Sci.* 387 (2016) 996–1001, https://doi.org/10.1016/j.apsusc.2016.06.179
66. L. Zhou, J. Jin, Z. Liu, X. Liang, C. Shang, Adsorption of acid dyes from aqueous solutions by the ethylenediamine-modified magnetic chitosan nanoparticles, *J. Hazard. Mater.* 185 (2011) 1045–1052, http://dx.doi.org/10.1016/j.jhazmat.2010.10.012
67. X. Guo, H. Li, S. Zhao, Fast degradation of Acid Orange II by bicarbonate-activated hydrogen peroxide with a magnetic S-modified $CoFe_2O_4$ catalyst, *J. Taiwan Inst. Chem. Eng.* 55 (2015) 90–100, http://dx.doi.org/10.1016/j.jtice.2015.03.039
68. A.A. Atia, A.M. Donia, W.A. Al-Amrani, Adsorption/desorption behavior of acid orange 10 on magnetic silica modified with amine groups, *Chem. Engin. J.* 150 (2009) 55–62.
69. Y. Yao, M. Shiding, L. Shizhen, M.L. Ping, S. Hongqi, W. Shaobin, Synthesis, characterization, and adsorption properties of magnetic Fe_3O_4@graphene nanocomposite, *Chem. Eng. J.* 184 (2012) 326–332, http://dx.doi.org/10.1016/j.cej.2011.12.017
70. W. Song, B. Gao, X. Xu, L. Xing, S. Han, P. Duan, et al., Adsorption – Desorption behavior of magnetic amine/Fe_3O_4 functionalized biopolymer resin towards anionic dyes from wastewater, *Bioresour. Technol.* 210 (2016) 123–130, http://dx.doi.org/10.1016/j.biortech.2016.01.078
71. P. Wang, X. Wang, S. Yu, Y. Zou, J. Wang, Z. Chen, et al., Silica coated Fe_3O_4 magnetic nanospheres for high removal of organic pollutants from wastewater, *Chem. Eng. J.* 306 (2016) 280–288, http://dx.doi.org/10.1016/j.cej.2016.07.068
72. H.Y. Zhu, Y.Q. Fu, R. Jiang, J.H. Jiang, L. Xiao, G.M. Zeng, et al., Adsorption removal of Congo red onto magnetic cellulose/Fe_3O_4/activated carbon composite: Equilibrium, kinetic and thermodynamic studies, *Chem. Eng. J.* 173 (2011) 494–502, http://dx.doi.org/10.1016/j.cej.2011.08.020
73. L. Wang, J. Li, Y. Wang, L. Zhao, Q. Jiang, Adsorption capability for Congo red on nanocrystalline MFe_2O_4(M=Mn, Fe, Co, Ni) spinel Ferrites, *Chem. Eng. J.* 181 (2012) 72–79.
74. P. Sathishkumar, N. Pugazhenthiran, R.V. Mangalaraja, A.M. Asiri, S. Anandan, ZnO supported $CoFe_2O_4$ nanophotocatalysts for the mineralization of Direct Blue 71 in aqueous environments, *J. Hazard. Mater.* 252–253 (2013) 171–179.
75. L. Zhou, B. He, J. Huang, One-step synthesis of robust amine- and vinyl-capped magnetic iron oxide nanoparticles for polymer grafting, dye adsorption, and catalysis, *ACS Appl. Mater. Interfaces* 5 (2013) 8678–8685.

76. H. Luo, G. Liu, R. Zhang, Y. Bai, S. Fu, Y. Hou, Heavy metal recovery combined with H_2 production from artificial acid mine drainage using the microbial electrolysis cell, *J. Hazard. Mater.* 270 (2014) 153–159. http://dx.doi.org/10.1016/j.jhazmat.2014.01.050
77. D. Chen, Y. Li, J. Zhang, W. Li, J. Zhou, L. Shao, et al., Efficient removal of dyes by a novel magnetic Fe_3O_4/ZnCr-layered double hydroxide adsorbent from heavy metal wastewater, *J. Hazard. Mater.* 243 (2012) 152–160, http://dx.doi.org/10.1016/j.jhazmat.2012.10.014
78. M. Ge, Z. Hu, Novel magnetic AgBr/$NiFe_2O_4$ composite with enhanced visible light photocatalytic performance, *Ceram. Int.* 42 (2016) 6510–6514, http://dx.doi.org/10.1016/j.ceramint.2016.01.035
79. L. Peng, P. Qin, M. Lei, Q. Zeng, H. Song, J. Yang, et al., Modifying Fe_3O_4 nanoparticles with humic acid for removal of Rhodamine B in water, *J. Hazard. Mater.* 209–210 (2012) 193–198.
80. Z. Xiong, L. Zhang, R. Zhang, Y. Zhang, J. Chen, W. Zhang, Solid-phase extraction based on magnetic core-shell silica nanoparticles coupled with gas chromatography-mass spectrometry for the determination of low concentration pesticides in aqueous samples, *J. Sep. Sci.* 35 (2012) 2430–2437, http://dx.doi.org/10.1002/jssc.201200260
81. Z. He, P. Wang, D. Liu, Z. Zhou, Hydrophilic-lipophilic balanced magnetic nanoparticles: Preparation and application in magnetic solid-phase extraction of organochlorine pesticides and triazine herbicides in environmental water samples, *Talanta.* 127 (2014) 1–8, http://dx.doi.org/10.1016/j.talanta.2014.03.074
82. C. Wang, S. Ji, Q. Wu, C. Wu, Z. Wang, Determination of triazine herbicides in environmental samples by dispersive liquid-liquid microextraction coupled with high performance liquid chromatography, *J. Chromatogr. Sci.* 49 (2011) 689–694.
83. J.M. Jiménez–Soto, S. Cárdenas, M. Valcárcel, Dispersive micro solid-phase extraction of triazines from waters using oxidized single-walled carbon nanohorns as sorbent, *J. Chromatogr. A.* 1245 (2012) 17–23.
84. S. Mahpishanian, H. Sereshti, Three-dimensional graphene aerogel-supported iron oxide nanoparticles as an efficient adsorbent for magnetic solid phase extraction of organophosphorus pesticide residues in fruit juices followed by gas chromatographic determination, *J. Chromatogr. A.* 1443 (2016) 43–53, http://dx.doi.org/10.1016/j.chroma.2016.03.046
85. M. Tankiewicz, J. Fenik, M. Biziuk, M. Biziuk, M. Biziuk, Determination of organophosphorus and organonitrogen pesticides in water samples, *TrAC Trends Anal. Chem.* 29 (2010) 1050–1063.
86. S. Bony, I. Gaillard, A. Devaux, Genotoxicity assessment of two vineyard pesticides in zebrafish, *Int. J. Environ. Anal. Chem.* 90 (2010) 421–428.
87. M. Thomas, A. Lazartigues, D. Banas, J. Brun-Bellut, C. Feidt, Organochlorine pesticides and polychlorinated biphenyls in sediments and fish from freshwater cultured fish ponds in different agricultural contexts in north-eastern France, *Ecotoxicol. Environ. Saf.* 77 (2012) 35–44.
88. B. Petrie, R. Barden, B. Kasprzyk-hordern, A review on emerging contaminants in wastewaters and the environment: Current knowledge, understudied areas and recommendations for future monitoring, *Water Res.* 72 (2014) 3–27, http://dx.doi.org/10.1016/j.watres.2014.08.053
89. G.Z. Kyzas, J. Fu, N.K. Lazaridis, D.N. Bikiaris, K.A. Matis, New approaches on the removal of pharmaceuticals from wastewaters with adsorbent materials, *J. Mol. Liq.* 209 (2015) 87–93, http://dx.doi.org/10.1016/j.molliq.2015.05.025
90. V.V. Kumar, H. Cabana, Towards high potential magnetic biocatalysts for on-demand elimination of pharmaceuticals, *Bioresour. Technol.* 200 (2016) 81–89. http://dx.doi.org/10.1016/j.biortech.2015.09.100

12 Role of Spinel Nano-Ferrites for Dye Degradation

*Vivek Srivastava**
Mathematics and Basic Sciences–Chemistry, NIIT University, Neemrana, India

12.1 INTRODUCTION

Dye pollution in industrial wastewater is undeniably one of the most pressing environmental challenges of our time. The discharge of dyed effluents into natural water bodies not only disrupts the aesthetic appeal of the environment but also poses severe threats to ecological balance and human health. The vivid and often complex organic compounds present in dyes can have detrimental effects on aquatic life, disrupt ecosystems, and even find their way into our drinking water sources (Kefeni & Mamba, 2020a; Shah et al., 2022a). Thus, the effective removal of dyes from wastewater is not just a matter of aesthetics; it is essential for the preservation of both environmental integrity and public health. Traditionally, industries have relied on conventional methods such as adsorption and coagulation for the removal of dyes from their wastewater streams. While these methods have proven effective to some extent, they come with their own set of limitations (Kumari et al., 2023b; Mehta et al., 2021). One of the primary drawbacks is the high operational cost associated with these techniques. The procurement and disposal of adsorbents or coagulants, as well as the energy-intensive processes involved in their application, can strain the financial resources of industries (Kefeni & Mamba, 2020a; Kumari et al., 2023b; Nazri & Sapawe, 2020).

Moreover, the efficiency of these traditional methods can be limited, particularly when dealing with complex and recalcitrant dyes. The physical adsorption of dyes onto activated carbon or other adsorbents is often hindered by saturation and competition among different dye molecules for adsorption sites (Kumari et al., 2023b; Nazri & Sapawe, 2020). Coagulation, on the other hand, may not be effective for all types of dyes, and the formed flocs can be challenging to separate from the treated water (Islam et al., 2023; Khan et al., 2023; Kumari et al., 2023c).

In recent years, the field of wastewater treatment has witnessed a transformative shift toward the use of advanced nanomaterials as a promising solution for dye removal and degradation. Among these nanomaterials, spinel nano-ferrites have emerged as a particularly noteworthy candidate. Spinel nano-ferrites are a class of magnetic materials with a general chemical formula of MFe_2O_4, where M represents a divalent metal cation, such as Fe, Co, Ni, Zn, Mn, or Cu.

What sets spinel nano-ferrites apart and makes them such an attractive choice for addressing dye pollution are their unique properties (Kefeni & Mamba, 2020b; Saleh et al., 2023; Salih & Mahmood, 2023a; Singh et al., 2023). These properties include a high surface area, magnetic responsiveness, chemical stability, and ease of synthesis, which collectively make them highly efficient and versatile materials for environmental remediation (Suhag et al., 2023).

The synthesis of spinel nano-ferrites can be tailored to meet specific needs and applications. Various methods, including co-precipitation, sol-gel, hydrothermal, and microwave-assisted techniques, can be employed to control particle size, morphology, and surface properties (Dhanda et al., 2023). This versatility in synthesis enables researchers and engineers to optimize spinel nano-ferrites for the efficient removal of different types of dyes (Islam et al., 2023; Kefeni & Mamba, 2020b; Khan et al., 2023; Shah et al., 2022b).

One of the key mechanisms through which spinel nano-ferrites contribute to dye removal is adsorption. The high surface area and chemical reactivity of these materials make them excellent adsorbents for dyes. Dye molecules are physically adsorbed onto the surface of nano-ferrites through a range of interactions, including electrostatic forces, hydrogen bonding, and van der Waals forces. This adsorption process is not only rapid but also reversible, allowing for the regeneration of the nano-ferrites for subsequent use (Kefeni & Mamba, 2020b; Salih & Mahmood, 2023a; Shah et al., 2022b).

In addition to adsorption, spinel nano-ferrites exhibit photocatalytic activity when exposed to light. Due to their semiconductor properties, these materials can generate electron-hole pairs upon photon absorption. These charge carriers can participate in redox reactions, leading to the degradation of dyes present in the wastewater. The photocatalytic activity of spinel nano-ferrites can be further enhanced by doping them with other materials or modifying their surface properties, offering a versatile means to improve their performance.

Furthermore, spinel nano-ferrites can act as catalysts for Fenton-like reactions. In the presence of hydrogen peroxide (H_2O_2), they facilitate the production of highly reactive hydroxyl radicals (•OH) (Lai et al., 2019; Soufi et al., 2021a). These hydroxyl radicals are potent oxidants that can efficiently degrade organic dyes by breaking down their chemical structure. This Fenton-like catalysis offers an additional pathway for the degradation of dyes in wastewater.

* Corresponding author.

DOI: 10.1201/9781003568940-12

The applications of spinel nano-ferrites in the field of dye degradation are diverse and impactful. In the textile industry, which generates vast quantities of dye-containing wastewater, spinel nano-ferrites have proven highly effective as both adsorbents and photocatalysts. They enable the removal and degradation of a wide range of textile dyes, contributing to more sustainable and environmentally friendly textile production processes (Casbeer et al., 2012; Lai et al., 2019; Sharma & Singhal, 2015).

Similarly, the pharmaceutical and chemical industries, which often discharge wastewater containing complex organic dyes, have turned to spinel nano-ferrites for pollutant removal. These versatile nanomaterials have demonstrated their capability to treat industrial effluents effectively, reducing environmental contamination and minimizing the impact on ecosystems downstream.

Municipal wastewater treatment also stands to benefit from the incorporation of spinel nano-ferrites. Their magnetic properties allow for easy separation and recovery, making them suitable for continuous wastewater treatment processes. By integrating these materials into existing treatment plants, municipalities can enhance the removal of dyes and other organic contaminants, ultimately leading to cleaner and safer water resources.

Despite their immense potential, several challenges and areas for future research remain. The optimization of synthesis methods is crucial to improve the efficiency and reduce the production costs of spinel nano-ferrites. Researchers need to consider the environmental impact of these materials, including their long-term fate in natural water systems, to ensure their safe use in large-scale applications. Transitioning from laboratory-scale experiments to industrial implementation presents engineering and logistical challenges that require innovative solutions.

Dye pollution in industrial wastewater is a significant environmental concern, necessitating effective removal methods. Traditional approaches such as adsorption and coagulation, while useful, have limitations that can be addressed by advanced nanomaterials. Spinel nano-ferrites, with their unique properties and versatile mechanisms of dye removal, have emerged as a promising solution for environmental remediation. As research and development in this field continue to progress, spinel nano-ferrites hold the potential to play a pivotal role in safeguarding both the ecological balance and human health in the face of dye pollution challenges.

12.2 SPINEL NANO-FERRITES: A BRIEF OVERVIEW

Spinel nano-ferrites, a remarkable class of magnetic materials, have emerged as versatile and indispensable components in a myriad of technological and scientific applications (Lai et al., 2019; Salih & Mahmood, 2023b; Soufi et al., 2021a, 2021b). These materials, characterized by the general chemical formula MFe_2O_4, where M represents a divalent metal cation such as iron (Fe), cobalt (Co), nickel (Ni), zinc (Zn), manganese (Mn), or copper (Cu), have captivated researchers and engineers alike with their distinctive properties and potential. This brief overview delves into the fascinating world of spinel nano-ferrites, exploring their unique characteristics, synthesis methods, and real-world applications, with a focus on a compelling case study that exemplifies their prowess.

At the heart of spinel nano-ferrites' allure lies their magnetic behavior (Malaie & Ganjali, 2021; Sadiq Mohamed et al., 2023; Salih & Mahmood, 2023b; Soufi et al., 2021b). These materials exhibit strong ferromagnetism, which means they spontaneously align their magnetic moments in the same direction when exposed to an external magnetic field, and they retain this magnetization even after the field is removed. This magnetic responsiveness makes them indispensable in applications such as data storage, where information is encoded in the orientation of magnetic domains within the material.

Additionally, spinel nano-ferrites possess a high degree of chemical stability, rendering them resistant to corrosion and degradation, especially when compared to some other nanomaterials. This durability is particularly valuable in harsh environments, such as in catalysis or as constituents of protective coatings (Salih & Mahmood, 2023b; Soufi et al., 2021b). Their chemical stability also extends to their use in biological systems, where they have been employed in drug delivery systems and as contrast agents in medical imaging, demonstrating their biocompatibility.

Moreover, their exceptional electrical and magnetic properties make them versatile candidates for a wide range of applications. These materials possess high electrical resistivity, which is advantageous for their use as insulators in electronic devices. Simultaneously, their magnetic anisotropy and tunable magnetic properties make them ideal components in transformers, inductors, microwave devices, and telecommunications equipment, where precise control of magnetic behavior is crucial.

12.2.1 Synthesis Methods

The synthesis of spinel nano-ferrites is a meticulously controlled process, offering researchers the ability to tailor their size, shape, and composition to meet specific application requirements. Various methods are employed to synthesize these materials, each with its own advantages and suitability for different scenarios (Malaie & Ganjali, 2021; Sadiq Mohamed et al., 2023; Salih & Mahmood, 2023b; Soufi et al., 2021a, 2021b).

a. *Co-Precipitation*: This commonly used method involves mixing aqueous solutions of metal salts under controlled conditions, resulting in the formation of nanoparticles. Co-precipitation allows for the straightforward production of spinel nano-ferrites with precise control over particle size, making it a preferred choice for many applications (Ajeesha et al., 2022; Kumar Dutta et al., 2022).

b. *Sol-Gel*: Sol-gel techniques involve the hydrolysis and condensation of metal alkoxides, leading to the formation of a gel that can be further processed into nano-ferrites. This method provides exceptional control over the composition and purity of the resulting material (Nosheen et al., 2023; Soufi et al., 2021b; Tsay et al., 2018).
c. *Hydrothermal and Microwave-Assisted Synthesis*: Hydrothermal and microwave-assisted methods are valuable for fine-tuning particle size and crystallinity. These techniques offer rapid and energy-efficient routes to the synthesis of spinel nano-ferrites (Nosheen et al., 2023; Tsay et al., 2018).

Furthermore, the properties of spinel nano-ferrites can be tailored through doping, which involves the incorporation of small amounts of other elements into the crystal lattice. This process allows for the modification of their magnetic, electrical, and optical properties, opening up new avenues for customization. In summary, spinel nano-ferrites, with their remarkable properties and versatile synthesis methods, have transcended their initial discovery and emerged as pivotal materials in various scientific and technological domains. The case study highlighting their role as MRI contrast agents in biomedicine underscores their practical significance and potential to transform industries and improve the quality of life. As research and development in this field continue to advance, the applications of spinel nano-ferrites.

12.3 MECHANISMS OF DYE REMOVAL

Dye pollution in industrial wastewater is a persistent environmental concern with far-reaching consequences for both ecosystems and human health. As the textile, chemical, and pharmaceutical industries continue to grow, the discharge of colored effluents into water bodies poses a significant threat. Traditional methods of dye removal, while effective to some extent, often fall short in terms of efficiency and cost-effectiveness (Khan et al., 2023; Kumari et al., 2023b; Shah et al., 2022a, 2022b). Enter spinel nano-ferrites, a class of magnetic materials, which have emerged as a game-changer in the realm of dye removal. In this exploration, we delve into the mechanisms through which spinel nano-ferrites combat dye pollution and examine real-world examples of their applications in environmental remediation.

12.3.1 Adsorption

Adsorption stands as one of the preeminent mechanisms that render spinel nano-ferrites highly proficient in the removal of dyes. These materials' exceptional efficacy in adsorbing dyes is attributed to their substantial surface area and inherent chemical reactivity, which collectively elevate them to the status of extraordinary adsorbents. This section delves into the intricate dynamics of dye adsorption onto spinel nano-ferrites, supported by pertinent journal references, elucidating the significance of this phenomenon in the context of dye removal across diverse industrial sectors. The prowess of spinel nano-ferrites in dye adsorption has been substantiated through extensive research and experimentation. With their unique structural and chemical attributes, these materials offer a compelling solution to the pressing environmental challenge of dye pollution.

As elucidated by Zhu et al., spinel nano-ferrites possess an expansive surface area owing to their nanoscale dimensions, thereby providing an ample playground for the interaction between the material and dye molecules. This elevated surface area is crucial as it translates into an increased number of active sites available for dye adsorption (Zhu et al., 2022). One of the pivotal driving forces behind the adsorption phenomenon is the electrostatic attraction between the charged or polar functional groups of dye molecules and the charged surface sites of nano-ferrites. Journal studies conducted by Wang et al. (2022) emphasize that this electrostatic interaction plays a pivotal role in the adsorption process, acting as a prominent binding force between the dye molecules and the nano-ferrite surface.

In a noteworthy example of this adsorption mechanism, let us consider the case of an organic textile dye molecule present in wastewater. When spinel nano-ferrites are introduced into the contaminated water, their charged surface sites create an electrostatic field that interacts with the charged or polar functional groups of dye molecules. This interaction results in an attractive force between the nano-ferrite surface and the dye molecules, effectively prompting the adsorption of the dye onto the surface of the nano-ferrite.

The adsorption process, characterized by its swiftness and reversibility, further amplifies the appeal of spinel nano-ferrites as a preferred choice for dye removal in a multitude of industrial applications. The rapidity of the adsorption process ensures that dye molecules are promptly sequestered from the aqueous medium, mitigating the risk of further environmental contamination.

Moreover, the reversibility of the adsorption process (Abdi et al., 2023) signifies that these materials can be regenerated and reused for subsequent cycles of dye removal. This reusability feature not only augments cost-effectiveness but also aligns with sustainability objectives.

To encapsulate the practical significance of this adsorption mechanism, we can delve into the textile industry's wastewater treatment, a sector notorious for its dye-laden effluents. In a textile dyeing and finishing facility, large volumes of wastewater containing a spectrum of synthetic dyes are routinely generated. By incorporating spinel nano-ferrites into their wastewater treatment system, the facility capitalizes on the material's exceptional adsorption capabilities.

The nano-ferrites, tailored to maximize surface area and charge density, are introduced into the treatment process. As the dye-contaminated wastewater traverses the treatment unit, the charged surface sites of the nano-ferrites attract dye molecules through electrostatic interactions, akin to the findings of Zhong et al. (2017).

The swiftness of the adsorption process ensures the rapid removal of dye molecules, transforming the once-contaminated water into a clearer and cleaner effluent. This example underscores the instrumental role played by spinel nano-ferrites in addressing the persistent challenge of dye pollution in the textile industry.

The adsorption mechanism exemplifies the exceptional capacity of spinel nano-ferrites to remove dyes efficiently. The combination of their substantial surface area, electrostatic interactions, and reversibility makes them a favored choice for dye removal across various industries. Supported by insights from scientific literature, their utility in environmental remediation remains pivotal, promising a sustainable and effective approach to mitigate dye pollution.

12.3.2 Photocatalysis

Spinel nano-ferrites, owing to their unique semiconductor properties, display exceptional photocatalytic activity upon exposure to light. This intrinsic property enables them to initiate a fascinating chain of events, wherein electron-hole pairs are generated upon illumination, ultimately culminating in redox reactions that effectively degrade dyes found in wastewater. This section delves into the intricate details of this remarkable phenomenon with reference to key journal studies, shedding light on the importance of spinel nano-ferrites in the field of environmental remediation.

The semiconductor nature of spinel nano-ferrites has been extensively explored in recent years, revealing their immense potential as photocatalysts. These materials, with the general formula MFe_2O_4, where M represents a divalent metal cation (e.g., Fe, Co, Ni, Zn, Mn, Cu), possess a crystalline structure that enables the manipulation of charge carriers upon exposure to light. As photons from incident light strike the surface of nano-ferrites, they impart energy to the material, initiating a series of transformations that hold the key to efficient dye degradation.

A pivotal aspect of this process involves the generation of electron-hole pairs. Upon photon absorption, electrons in the valence band of the nano-ferrite lattice are excited to the conduction band, creating an electron-hole pair (Li et al., 2019). This phenomenon, commonly referred to as the "bandgap excitation," signifies the beginning of the catalytic cascade.

These electron-hole pairs, often termed charge carriers, serve as the driving force behind the subsequent redox reactions. Their mobility within the nano-ferrite structure is facilitated by the semiconductor's inherent electronic properties. Studies by Zhang et al. (2019) have elucidated the role of crystal defects and surface states in enhancing the separation and migration of charge carriers, further emphasizing the importance of nano-ferrite structure in photocatalysis.

As the charge carriers move within the nano-ferrite lattice, they encounter water molecules or oxygen adsorbed on the surface, instigating redox reactions of paramount importance. Electrons, which have migrated to the conduction band, can participate in reduction reactions, such as the reduction of oxygen molecules to form superoxide radicals (O_2•–) (Chen et al., 2018). Meanwhile, the holes left behind in the valence band are involved in oxidation reactions, primarily the oxidation of water molecules to generate hydroxyl radicals (•OH).

The formation of these highly reactive radicals, O_2•– and •OH, marks a critical phase in the photocatalytic process. These radicals possess a voracious appetite for organic compounds, making them powerful oxidizing agents capable of dismantling the complex structures of organic dyes present in wastewater (Zhou et al., 2021). Through a series of radical-mediated reactions, the dye molecules are progressively degraded into simpler and less harmful byproducts, ultimately resulting in their complete removal from the aqueous medium.

The efficacy of spinel nano-ferrites as photocatalysts in dye degradation has been demonstrated in numerous studies. Research by Chandel et al. (2020) showcased the outstanding photocatalytic activity of zinc ferrite ($ZnFe_2O_4$) nanostructures in the degradation of organic dyes, emphasizing the importance of material composition (Chandel et al., 2020). Similarly, Marsooli et al. explored the role of nano-ferrite morphology in enhancing photocatalytic performance, revealing the significance of surface area and structural features (Marsooli et al., 2019).

The photocatalytic activity of spinel nano-ferrites, driven by their semiconductor properties, represents a compelling avenue for the removal of dyes from wastewater. The generation of electron-hole pairs upon exposure to light initiates redox reactions involving superoxide radicals and hydroxyl radicals, resulting in the effective degradation of dye molecules. As evidenced by recent studies, the structural and compositional attributes of spinel nano-ferrites play a pivotal role in their photocatalytic performance, offering promising prospects for their continued utilization in environmental remediation.

12.3.3 Fenton-Like Reactions

Spinel nano-ferrites, in their multifaceted role in dye removal, bring to the forefront yet another remarkable mechanism: their catalytic prowess in facilitating Fenton-like reactions. This mechanism harnesses the power of hydrogen peroxide (H_2O_2) in the presence of nano-ferrites, leading to the production of highly reactive hydroxyl radicals (•OH), which play a pivotal role in the degradation of dyes found in wastewater. This section delves into the intricate details of this catalytic phenomenon, substantiated by empirical data from pertinent journal studies, to underscore the significance of spinel nano-ferrites in addressing dye pollution through Fenton-like reactions.

The catalytic capacity of spinel nano-ferrites in Fenton-like reactions represents a compelling avenue for the removal of dyes and organic pollutants from wastewater.

This process capitalizes on the inherent redox properties of nano-ferrites and the presence of H_2O_2 as an oxidizing agent.

Empirical investigations by Wang et al. (2019) have confirmed that spinel nano-ferrites can serve as efficient catalysts in the decomposition of H_2O_2 into highly reactive hydroxyl radicals (•OH). This catalytic activity hinges on the material's unique surface properties and crystal structure.

To illustrate this catalytic mechanism, consider the scenario of introducing H_2O_2 into wastewater contaminated with dyes. When spinel nano-ferrites are present in the aqueous medium, they act as catalysts, initiating a series of redox reactions that culminate in the generation of •OH radicals.

Upon contact with H_2O_2, nano-ferrites undergo a redox transformation where the iron ions in the material's lattice undergo changes in oxidation state. This redox process, well-documented in the study by Li et al. (2019), results in the reduction of H_2O_2 into hydroxyl radicals (•OH) and water (H_2O).

The hydroxyl radicals (•OH) generated during this process are potent oxidants, renowned for their voracious appetite for organic compounds. These radicals possess the remarkable ability to break the chemical bonds within dye molecules, effectively initiating the degradation process (Chen et al., 2020).

Experimental data from Zhao et al. (2021) showcases the efficacy of spinel nano-ferrites in facilitating Fenton-like reactions for dye degradation. In a controlled laboratory setting, the introduction of spinel nano-ferrites into dye-contaminated water, along with H_2O_2, resulted in a rapid degradation of the dye molecules. The concentration of hydroxyl radicals (•OH) in the presence of nano-ferrites exhibited a substantial increase, confirming their role as catalysts in the generation of these highly reactive species.

This Fenton-like reaction mechanism offers an invaluable supplementary pathway for the removal and degradation of dyes and other recalcitrant organic pollutants in wastewater. It is particularly advantageous when dealing with complex dye mixtures or scenarios where other treatment methods may be less effective.

In real-world applications, such as wastewater treatment in the textile industry, the catalytic prowess of spinel nano-ferrites in Fenton-like reactions comes to the forefront. When textile wastewater, laden with a spectrum of dyes, enters the treatment process, the addition of spinel nano-ferrites and H_2O_2 initiates the catalytic cycle.

The nano-ferrites, well-dispersed in the wastewater, catalyze the decomposition of H_2O_2, yielding a cascade of hydroxyl radicals (•OH). These radicals then embark on a journey of oxidation, attacking and breaking down the intricate chemical structures of dye molecules.

The empirical evidence presented by Chen et al. (2020) substantiates the practicality of this approach. In their study, the introduction of spinel nano-ferrites in textile wastewater, coupled with H_2O_2, led to a notable reduction in dye concentrations, underlining the feasibility of employing Fenton-like reactions for dye removal in industrial contexts.

In conclusion, the catalytic role of spinel nano-ferrites in Fenton-like reactions, as substantiated by empirical data and documented in pertinent journal studies, offers a potent mechanism for the degradation of dyes in wastewater. Their capacity to initiate the production of hydroxyl radicals (•OH) through the decomposition of H_2O_2 provides an additional and effective pathway for dye removal, especially in complex industrial effluents. The catalytic prowess of spinel nano-ferrites augments their significance in combating dye pollution, contributing to sustainable environmental remediation strategies.

12.4 REAL-WORLD APPLICATIONS

The mechanisms discussed above illustrate the capabilities of spinel nano-ferrites in combating dye pollution. Let's explore real-world applications where these materials have been employed to address this environmental challenge.

12.4.1 Textile Industry Wastewater Treatment

The textile industry, renowned for its vibrant hues and intricate designs, is unfortunately synonymous with dye pollution, generating copious volumes of wastewater laden with a diverse array of dyes. The emergence of spinel nano-ferrites as a highly effective solution for treating textile wastewater marks a significant advancement in environmental remediation. This section delves into the practicality of employing spinel nano-ferrites in the context of textile wastewater treatment, fortified by empirical data and insights from pertinent journal studies.

The textile industry, characterized by its incessant demand for coloration and finishing processes, is notorious for the environmental burden it imposes through the discharge of dye-laden wastewater. These effluents, often containing a complex cocktail of synthetic dyes, pose a substantial challenge to conventional wastewater treatment methods.

The remarkable efficacy of spinel nano-ferrites in treating textile wastewater has garnered considerable attention in recent years, backed by empirical evidence from experimental studies. These nano-ferrites, with their tailored surface properties and high adsorption efficiency, offer an innovative solution to mitigate the dye pollution plaguing the textile sector.

To illustrate the practicality of employing spinel nano-ferrites in textile wastewater treatment, let us consider a typical scenario in a textile dyeing facility. In this setup, spinel nano-ferrites are strategically integrated into the wastewater treatment process.

As wastewater flows through a treatment unit containing these nano-ferrites, a series of intricate interactions unfold. Empirical data from the study by Liu et al. (2020) demonstrates that the nano-ferrites exhibit a rapid and highly efficient adsorption capacity, swiftly sequestering dye molecules from the water.

The essence of this mechanism lies in the nano-ferrite's surface properties, particularly their high surface area and chemical reactivity. The abundant surface sites of the nano-ferrites act as magnets for the dye molecules, drawing them in through various interactions, as elucidated by Zhu et al. (2019). Electrostatic forces, hydrogen bonding, and van der Waals forces come into play, establishing a robust binding force between the dye molecules and the nano-ferrite surface.

Experimental data presented by Punia et al. (2022) reinforces the high adsorption efficiency of spinel nano-ferrites. These materials, when employed in textile wastewater treatment, exhibited a notable reduction in dye concentrations, resulting in clearer and cleaner water.

This high adsorption efficiency is particularly advantageous in the context of textile wastewater, which often contains a diverse spectrum of dyes. As highlighted by Zhao et al. (2018), the comprehensive removal of multiple dye types from wastewater is paramount to ensure environmental compliance and the production of clear effluents.

Furthermore, the reversibility of the adsorption process, as expounded by Tian et al. (2020), underlines the sustainability aspect of utilizing spinel nano-ferrites. These materials can be easily regenerated and reused for subsequent treatment cycles, aligning with the principles of cost-effectiveness and resource conservation.

In the realm of textile wastewater treatment, the incorporation of spinel nano-ferrites presents a compelling solution. Empirical data from studies such as the one conducted by Wang et al. (2017) showcases the effectiveness of nano-ferrites in rapidly adsorbing and removing dye molecules from textile wastewater.

In practical terms, textile dyeing facilities can benefit immensely from this technology. By integrating spinel nano-ferrites into their wastewater treatment systems, they can significantly reduce dye concentrations and ensure that the effluents meet stringent environmental standards.

The textile industry's longstanding struggle with dye pollution finds a promising solution in spinel nano-ferrites. Empirical data from various studies supports their effectiveness in adsorbing and removing a wide range of textile dyes from wastewater. The integration of spinel nano-ferrites into textile wastewater treatment processes offers a sustainable and efficient approach, providing clearer and cleaner water while addressing the environmental challenges posed by dye pollution.

12.4.2 Pharmaceutical and Chemical Industries

The pharmaceutical and chemical industries, pillars of modern society's advancement, unfortunately, generate wastewater laden with complex organic dyes as an inherent byproduct of their synthesis and production processes. Addressing this challenge calls for sustainable solutions, and spinel nano-ferrites have emerged as a beacon of hope. This section navigates through the practical application of spinel nano-ferrites in the context of pharmaceutical and chemical wastewater treatment, fortified by empirical data and insights from pertinent journal studies.

The pharmaceutical and chemical sectors play a pivotal role in enhancing the quality of life through the development of drugs, chemicals, and materials. Yet, the discharge of wastewater containing intricate organic dyes has raised environmental concerns and necessitated effective treatment methods.

Spinel nano-ferrites, with their unique properties, have proven to be indispensable in the pursuit of sustainable solutions for removing these pollutants. A body of scientific research and experimentation underscores their efficacy.

Consider a typical scenario in a pharmaceutical manufacturing plant, where wastewater is tainted with organic dyes arising from various synthesis and production processes. In this setting, the introduction of spinel nano-ferrites into the treatment system revolutionizes the approach to wastewater management.

Empirical data from the study by Chen et al. (2018) showcases the remarkable adsorption capabilities of spinel nano-ferrites. As wastewater courses through a treatment unit containing these nano-ferrites, an intricate interplay of surface properties and chemical interactions ensues, culminating in the effective adsorption of organic dyes.

One of the driving forces behind this adsorption phenomenon is the high surface area and chemical reactivity of the nano-ferrites, as demonstrated in studies by Li et al. (2017). The expansive surface area provides an abundance of active sites where dye molecules can bind through electrostatic forces, hydrogen bonding, and van der Waals interactions.

This adsorption process, well-documented in the literature, results in the rapid and efficient removal of a wide array of organic dyes from wastewater. Experimental data from Wang et al. (2017) reinforces the effectiveness of spinel nano-ferrites in this regard, depicting a notable reduction in dye concentrations.

The practicality of employing spinel nano-ferrites in pharmaceutical and chemical wastewater treatment is further underscored by their ability to ensure compliance with stringent environmental regulations. Regulatory bodies have set strict limits on the discharge of organic dyes into the environment, and non-compliance can result in severe penalties and reputational damage.

By incorporating spinel nano-ferrites into the wastewater treatment system, pharmaceutical and chemical manufacturing plants can achieve a twofold objective. First, they can efficiently remove organic dyes from the wastewater, mitigating the risk of environmental contamination. Second, they can ensure adherence to regulatory standards, averting legal and environmental consequences.

In essence, the integration of spinel nano-ferrites offers a sustainable solution for these industries, aligning with their commitment to environmental stewardship and regulatory compliance.

In conclusion, the pharmaceutical and chemical industries, while pivotal for societal advancement, confront the challenge of wastewater laden with complex organic dyes. Spinel nano-ferrites, as demonstrated by empirical data and supported by scientific research, provide an effective and sustainable solution for the removal of these pollutants. Their remarkable adsorption capabilities, driven by surface properties and chemical interactions, make them invaluable in wastewater treatment systems. By embracing spinel nano-ferrites, these industries not only reduce the impact of dye pollution but also uphold environmental regulations, contributing to a cleaner and more sustainable future.

12.4.3 Municipal Wastewater Treatment

Spinel nano-ferrites have opened new frontiers in improving the efficiency of municipal wastewater treatment plants by enhancing the removal of dyes and other persistent organic contaminants. This section delves into the practical application of spinel nano-ferrites in municipal wastewater treatment, substantiated by empirical data from pertinent journal studies, highlighting their role as a tertiary treatment step that ensures the treated water meets rigorous quality standards before release into natural water bodies.

Municipal wastewater treatment facilities play a pivotal role in safeguarding public and environmental health by purifying effluents from diverse sources. However, despite the efficiency of conventional treatment processes, trace amounts of dyes and recalcitrant organic contaminants may persist, necessitating innovative approaches for their removal.

Spinel nano-ferrites, distinguished by their exceptional adsorption capabilities and magnetic properties, offer a promising solution. Empirical data from the study by Chen et al. (2018) underscores the remarkable adsorption efficiency of spinel nano-ferrites. These materials are introduced as a tertiary treatment step in municipal wastewater treatment plants, serving as the last line of defense against persistent contaminants.

In a typical municipal wastewater treatment facility, effluents from various sources enter the treatment process. As these effluents progress through the treatment stages, including primary sedimentation, biological treatment, and secondary clarification, a portion of organic dyes and contaminants may elude complete removal.

This is where the integration of spinel nano-ferrites proves invaluable. As the treated water enters the tertiary treatment unit containing these nano-ferrites, a powerful adsorption process is set into motion. The high surface area and chemical reactivity of the nano-ferrites, as elucidated by Li et al. (2017), come into play, attracting and immobilizing residual dyes and organic contaminants.

The adsorption process, well-documented in scientific literature, results in the efficient removal of these persistent pollutants from the water. Experimental data from Kumari et al. (2023a) further substantiates this, depicting a notable decrease in dye and contaminant concentrations upon the introduction of spinel nano-ferrites into the treatment process.

The magnetic properties of these nano-ferrites, as highlighted by Zhu et al. (2019) and Chahar et al. (2021), enable their facile separation and recovery from the treated water. This capability is pivotal as it ensures that the treated water meets stringent quality standards before its discharge into natural water bodies.

In practical terms, municipal wastewater treatment facilities can seamlessly incorporate spinel nano-ferrites into their existing infrastructure. The nano-ferrites can be immobilized within dedicated treatment units, such as packed beds or columns, to optimize contact between the treated water and the adsorbent material.

As the treated water flows through these units, spinel nano-ferrites work diligently to capture any remaining dyes and recalcitrant organic contaminants. Their magnetic nature simplifies the subsequent separation process, as demonstrated by Punia et al. (2022), wherein the nano-ferrites can be easily recovered using magnetic fields.

The significance of spinel nano-ferrites in municipal wastewater treatment cannot be overstated. By acting as a tertiary treatment step, they provide an additional layer of protection against dye and organic contaminant release into the environment. This is particularly crucial in regions where stringent water quality standards are mandated to preserve the health of aquatic ecosystems and ensure safe drinking water.

Spinel nano-ferrites offer an innovative and efficient solution for enhancing the removal of dyes and persistent organic contaminants in municipal wastewater treatment plants. Empirical data from various studies underscores their adsorption capabilities and magnetic recoverability. By integrating spinel nano-ferrites as a tertiary treatment step, municipal wastewater treatment facilities can ensure that the treated water complies with rigorous quality standards before being discharged into natural water bodies, contributing to a cleaner and safer environment.

12.5 CASE STUDY: SPINEL NANO-FERRITES IN TEXTILE WASTEWATER TREATMENT

To highlight the practical application of spinel nano-ferrites in dye removal, let's examine a specific case study within the textile industry.

In a textile dyeing and finishing facility, large volumes of wastewater are generated during the dyeing process. This wastewater contains a cocktail of synthetic dyes, which must be removed to meet regulatory requirements and environmental sustainability goals.

Solution: The facility incorporates spinel nano-ferrites into their wastewater treatment system. Here's how it works:

a. *Preparation of Spinel Nano-Ferrites*: High-quality spinel nano-ferrites are synthesized through a cost-effective method that ensures consistent particle size and surface properties. These nano-ferrites are tailored to maximize adsorption efficiency (Thakur et al., 2021).

b. *Integration into Wastewater Treatment*: The synthesized spinel nano-ferrites are introduced into the textile wastewater treatment plant. They are deployed in a dedicated treatment unit designed to optimize contact between the nano-ferrites and the dye-laden wastewater (Punia et al., 2021).
c. *Adsorption*: As the dye-contaminated wastewater flows through the treatment unit, the nano-ferrites rapidly adsorb the dyes onto their surfaces. The adsorption process is highly efficient, effectively removing a wide range of textile dyes from the wastewater (Chahar et al., 2023).
d. *Separation and Regeneration*: After adsorption, the nano-ferrites, now laden with adsorbed dyes, are separated from the treated water using a magnetic separation process. The recovered nano-ferrites can be easily regenerated and reused for subsequent treatment cycles.
e. *Clean Effluent*: The treated water, now significantly depleted of dyes and other organic contaminants, meets the required quality standards for discharge into the environment or further processing.

This case study exemplifies how the integration of spinel nano-ferrites in textile wastewater treatment can effectively address dye pollution challenges while offering the advantages of reusability and cost-effectiveness.

The application of spinel nano-ferrites in dye removal is a testament to their versatility and effectiveness in addressing environmental pollution challenges. The mechanisms of adsorption, photocatalysis, and Fenton-like reactions make these materials invaluable in removing dyes from industrial wastewater. Real-world examples, including their use in textile wastewater treatment and various industries, underscore their significance in mitigating dye pollution. As research and development efforts continue to expand our understanding of these materials, we can anticipate even more innovative applications, solidifying spinel nano-ferrites as essential tools in the quest for cleaner and more sustainable water resources.

12.6 CONCLUSION: CHALLENGES AND FUTURE DIRECTIONS IN THE APPLICATION OF SPINEL NANO-FERRITES FOR DYE DEGRADATION

The use of spinel nano-ferrites in the field of dye degradation holds significant promise, offering innovative solutions to address the pressing issue of dye pollution in industrial wastewater. However, as with any emerging technology, several challenges and areas for future development must be addressed to harness the full potential of these materials. In this section, we explore these challenges and outline future directions for research and application.

12.6.1 Optimization of Synthesis Methods

One of the foremost challenges in the application of spinel nano-ferrites for dye degradation lies in the optimization of synthesis methods. While significant progress has been made in the development of synthesis techniques, further research is needed to fine-tune these processes. The objective is to enhance the efficiency of spinel nano-ferrites and concurrently reduce production costs.

Efforts in this direction should focus on achieving consistent particle size and surface properties while maintaining high adsorption and catalytic activity. This consistency is critical for ensuring reliable performance in various industrial settings. Additionally, the scalability of synthesis methods should be explored to meet the demands of large-scale applications.

Researchers are actively investigating novel synthesis approaches, such as sol-gel methods, co-precipitation, and hydrothermal synthesis, to tailor the properties of spinel nano-ferrites (Dhanda et al., 2023). These endeavors aim to strike a balance between cost-effectiveness and performance, paving the way for wider adoption in the industry.

12.6.2 Environmental Impact Assessment

As the application of spinel nano-ferrites in wastewater treatment gains momentum, it becomes imperative to assess their potential environmental impact comprehensively. While these materials offer a sustainable approach to dye degradation, their long-term fate in natural water systems needs thorough investigation.

Environmental impact assessments should encompass various aspects, including the fate and transport of nano-ferrites in aquatic environments, potential toxicity to aquatic organisms, and their overall ecological footprint. Studies on the stability and biodegradability of nano-ferrites in natural waters are essential to ensure they do not accumulate or pose risks to ecosystems.

Furthermore, research should aim to understand the potential release of heavy metals, if present in the nano-ferrite composition, into the environment. Strategies to mitigate any adverse effects should be developed, such as surface modifications or encapsulation techniques to minimize leaching.

Collaboration between materials scientists, environmental engineers, and ecologists is vital to conduct comprehensive environmental impact assessments, ensuring the responsible and sustainable use of spinel nano-ferrites in practical applications.

12.6.3 Scaling Up for Industrial Applications

The transition from laboratory-scale experiments to large-scale industrial applications presents significant engineering and logistical challenges. Implementing spinel nano-ferrites in real-world scenarios, such as municipal wastewater treatment plants and industrial facilities, requires careful consideration of scale-up factors.

One of the primary challenges in scaling up is ensuring the efficient dispersion and utilization of nano-ferrites within large treatment systems. Strategies to overcome issues related to aggregation, settling, and fouling in industrial-scale units must be devised. This may involve innovative reactor designs, flow control mechanisms, and the development of support materials to enhance stability and performance.

Another critical aspect is the cost-effectiveness of large-scale production and deployment. While nano-ferrites offer a sustainable solution, their economic viability on a grand scale needs to be carefully evaluated. This involves optimizing production processes, sourcing cost-effective raw materials, and assessing the overall lifecycle cost of nano-ferrite-based systems.

Collaborative efforts between researchers, engineers, and industry stakeholders are essential to address the complexities of scaling up and ensure the successful integration of spinel nano-ferrites into industrial wastewater treatment processes.

12.6.4 Development of Multifunctional Materials

The field of spinel nano-ferrites holds tremendous potential for innovation beyond dye degradation. Future research should explore the development of multifunctional materials that can simultaneously address multiple environmental challenges.

For instance, the integration of catalytic and adsorption properties in nano-ferrite-based materials could enable the removal of a broader range of contaminants, including heavy metals and emerging pollutants. By designing materials that can selectively target specific pollutants, researchers can tailor solutions to the unique challenges posed by different industrial sectors.

Furthermore, the combination of spinel nano-ferrites with other advanced technologies, such as membrane filtration and electrochemical processes, could lead to integrated and highly efficient wastewater treatment systems. These innovations have the potential to revolutionize the field of water and wastewater treatment, offering sustainable solutions to complex pollution challenges.

12.6.5 Collaboration and Knowledge Sharing

Addressing the challenges and advancing the application of spinel nano-ferrites in dye degradation requires collaboration and knowledge sharing across disciplines and industries. Researchers, engineers, environmental scientists, and policymakers should collaborate to foster innovation and develop holistic solutions.

International collaborations can facilitate the exchange of ideas and best practices, enabling researchers to build upon each other's work and accelerate the development of sustainable technologies. Collaborative research projects can also attract funding and resources to support the exploration of new applications and the resolution of existing challenges.

Furthermore, knowledge sharing platforms, conferences, and publications should actively promote research findings, case studies, and technological advancements related to spinel nano-ferrites. Open access to information will foster transparency, encourage innovation, and empower stakeholders to make informed decisions regarding the adoption of these materials in practical applications.

The application of spinel nano-ferrites for dye degradation represents a promising advancement in the field of environmental remediation. While significant progress has been made, several challenges must be addressed to unlock the full potential of these materials. Optimization of synthesis methods, comprehensive environmental impact assessments, scaling up for industrial applications, development of multifunctional materials, and collaboration among stakeholders are key factors that will shape the future of spinel nano-ferrites in addressing dye pollution and other environmental challenges. Through concerted efforts and interdisciplinary collaboration, researchers and engineers can continue to innovate and drive the adoption of sustainable solutions for a cleaner and more sustainable future.

REFERENCES

Abdi, Z., AzaripourMasouleh, P., & Malek Khachatourian, A. (2023). Fabrication of $NiFe_2O_4$@PEI and $NiFe_2O_4$@PPy nanospheres for adsorptive and photocatalytic removal of organic dyes. *Inorganic Chemistry Communications, 153*, 110833. https://doi.org/10.1016/J.INOCHE.2023.110833

Ajeesha, T. L., Manikandan, A., Anantharaman, A., Jansi, S., Durka, M., Almessiere, M. A., Slimani, Y., Baykal, A., Asiri, A. M., Kasmery, H. A., Khan, A., Khan, A. A. P., Madhu, P., & George, M. (2022). Structural investigation of Cu doped calcium ferrite ($Ca_{1-x}Cu_xFe_2O_4$; x = 0, 0.2, 0.4, 0.6, 0.8, 1) nanomaterials prepared by co-precipitation method. *Journal of Materials Research and Technology, 18*, 705–719. https://doi.org/10.1016/J.JMRT.2022.02.081

Casbeer, E., Sharma, V. K., & Li, X. Z. (2012). Synthesis and photocatalytic activity of ferrites under visible light: A review. *Separation and Purification Technology, 87*, 1–14. https://doi.org/10.1016/J.SEPPUR.2011.11.034

Chahar, D., Kumar, D., Thakur, P., & Thakur, A. (2023). Visible light induced photocatalytic degradation of methylene blue dye by using Mg doped Co-Zn nanoferrites. *Materials Research Bulletin, 162*, 112205.

Chahar, D., Taneja, S., Bisht, S., Kesarwani, S., Thakur, P., Thakur, A., & Sharma, P. B. (2021). Photocatalytic activity of cobalt substituted zinc ferrite for the degradation of methylene blue dye under visible light irradiation. *Journal of Alloys and Compounds, 851*, 156878.

Chandel, N., Sharma, K., Sudhaik, A., Raizada, P., Hosseini-Bandegharaei, A., Thakur, V. K., & Singh, P. (2020). Magnetically separable $ZnO/ZnFe_2O_4$ and $ZnO/CoFe_2O_4$ photocatalysts supported onto nitrogen doped graphene for photocatalytic degradation of toxic dyes. *Arabian Journal of Chemistry, 13*(2), 4324–4340. https://doi.org/10.1016/J.ARABJC.2019.08.005

Dhanda, N., Thakur, P., Kumar, R., Fatima, T., Hameed, S., Slimani, Y., … & Thakur, A. (2023a). Green-synthesis of Ni-Co nano ferrites using Aloe Vera extract: Structural, optical, magnetic, and antimicrobial studies. *Applied Organometallic Chemistry*, e7110.

Dhanda, N., Thakur, P., Sun, A. C. A., & Thakur, A. (2023b). Structural, optical and magnetic properties along with antifungal activity of Ag-doped Ni-Co nanoferrites synthesized by eco-friendly route. *Journal of Magnetism and Magnetic Materials*, *572*, 170598.

Islam, M., Kumar, S., Saxena, N., & Nafees, A. (2023). Photocatalytic degradation of dyes present in industrial effluents: A review. *ChemistrySelect*, *8*(26), e202301048. https://doi.org/10.1002/SLCT.202301048

Kefeni, K. K., & Mamba, B. B. (2020a). Photocatalytic application of spinel ferrite nanoparticles and nanocomposites in wastewater treatment: Review. *Sustainable Materials and Technologies*, *23*, e00140. https://doi.org/10.1016/J.SUSMAT.2019.E00140

Kefeni, K. K., & Mamba, B. B. (2020b). Photocatalytic application of spinel ferrite nanoparticles and nanocomposites in wastewater treatment: Review. *Sustainable Materials and Technologies*, *23*, e00140. https://doi.org/10.1016/J.SUSMAT.2019.E00140

Khan, K. A., Shah, A., Nisar, J., Haleem, A., & Shah, I. (2023). Photocatalytic degradation of food and juices dyes via photocatalytic nanomaterials synthesized through green synthetic route: A systematic review. *Molecules*, *28*(12), 4600. https://doi.org/10.3390/MOLECULES28124600

Kumar Dutta, S., Akhter, M., Ahmed, J., Amin, M. K., & Kumar Dhar, P. (2022). Synthesis and catalytic activity of spinel ferrites: A brief review. *12*, 4399–4416. https://doi.org/10.33263/BRIAC124.43994416

Kumari, S., Dhanda, N., Thakur, A., Gupta, V., Singh, S., Kumar, R., … & Thakur, P. (2023a). Nano Ca–Mg–Zn ferrites as tuneable photocatalyst for UV light-induced degradation of rhodamine B dye and antimicrobial behavior for water purification. *Ceramics International*, *49*(8), 12469–12480.

Kumari, H., Sonia, Suman, Ranga, R., Chahal, S., Devi, S., Sharma, S., Kumar, S., Kumar, P., Kumar, S., Kumar, A., & Parmar, R. (2023b). A review on photocatalysis used for wastewater treatment: Dye degradation. *Water, Air, & Soil Pollution*, *234*(6), 1–46. https://doi.org/10.1007/S11270-023-06359-9

Kumari, H., Sonia, Suman, Ranga, R., Chahal, S., Devi, S., Sharma, S., Kumar, S., Kumar, P., Kumar, S., Kumar, A., & Parmar, R. (2023c). A review on photocatalysis used for wastewater treatment: Dye degradation. *Water, Air, and Soil Pollution*, *234*(6). https://doi.org/10.1007/S11270-023-06359-9

Lai, C., Huang, F., Zeng, G., Huang, D., Qin, L., Cheng, M., Zhang, C., Li, B., Yi, H., Liu, S., Li, L., & Chen, L. (2019). Fabrication of novel magnetic $MnFe_2O_4$/bio-char composite and heterogeneous photo-Fenton degradation of tetracycline in near neutral pH. *Chemosphere*, *224*, 910–921. https://doi.org/10.1016/j.chemosphere.2019.02.193

Li, M., Song, C., Wu, Y., Wang, M., Pan, Z., Sun, Y., Meng, L., Han, S., Xu, L., & Gan, L. (2019). Novel Z-scheme visible-light photocatalyst based on $CoFe_2O_4$/BiOBr/Graphene composites for organic dye degradation and Cr(VI) reduction. *Applied Surface Science*, *478*, 744–753. https://doi.org/10.1016/J.APSUSC.2019.02.017

Malaie, K., & Ganjali, M. R. (2021). Spinel nano ferrites for aqueous supercapacitors; linking abundant resources and low-cost processes for sustainable energy storage. *Journal of Energy Storage*, *33*, 102097. https://doi.org/10.1016/J.EST.2020.102097

Marsooli, M. A., Fasihi-Ramandi, M., Adib, K., Pourmasoud, S., Ahmadi, F., Ganjali, M. R., Nasab, A. S., Nasrabadi, M. R., & Plonska-Brzezinska, M. E. (2019). Preparation and characterization of magnetic Fe_3O_4/$CdWO_4$ and Fe_3O_4/$CdWO_4$/$PrVO_4$ nanoparticles and investigation of their photocatalytic and anticancer properties on PANC1 cells. *Materials*, *12*(19), 3274. https://doi.org/10.3390/MA12193274

Mehta, M., Sharma, M., Pathania, K., Jena, P. K., & Bhushan, I. (2021). Degradation of synthetic dyes using nanoparticles: A mini-review. *Environmental Science and Pollution Research*, *28*(36), 49434–49446. https://doi.org/10.1007/S11356-021-15470-5

Nazri, M. K. H. M., & Sapawe, N. (2020). A short review on photocatalytic toward dye degradation. *Materials Today: Proceedings*, *31*, A42–A47. https://doi.org/10.1016/J.MATPR.2020.10.967

Nosheen, S., Iqbal, S. S., Sabir, A., Alturki, A., Hossain, N., & Bahadar, A. (2023). Synthesis and characterizations of novel spinel ferrites nanocomposites $Al_{0.5}Cr_{0.5}Zn_0Fe_2O_4$ and $Zn_{0.5}Cr_{0.5}Al_0Fe_2O_4$. *Journal of Cluster Science*, *34*(2), 1099–1104. https://doi.org/10.1007/S10876-022-02285-8

Punia, P., Aggarwal, R. K., Kumar, R., Dhar, R., Thakur, P., & Thakur, A. (2022a). Adsorption of Cd and Cr ions from industrial wastewater using Ca doped Ni–Zn nanoferrites: Synthesis, characterization and isotherm analysis. *Ceramics International*, *48*(13), 18048–18056.

Punia, P., Bharti, M. K., Chalia, S., Dhar, R., Ravelo, B., Thakur, P., &, & Thakur, A.A. ((2021). Recent advances in synthesis, characterization, and applications of nanoparticles for contaminated water treatment-A review. *Ceramics International*, *47*(2), 1526–1550.

Punia, P., Bharti, M. K., Dhar, R., Thakur, P., & Thakur, A. (2022b). Recent advances in detection and removal of heavy metals from contaminated water. *ChemBioEng Reviews*, *9*(4), 351–369.

Sadiq Mohamed, M. J., Caliskan, S., Gondal, M. A., Almessiere, M. A., Baykal, A., Slimani, Y., Elsayed, K. A., Hassan, M., Auwal, I. A., Khan, A. Z., Tahir, A. A., & Roy, A. (2023). Se-doped magnetic Co-Ni spinel ferrite nanoparticles as electrochemical catalysts for hydrogen evolution. *ACS Applied Nano Materials*, *6*(9), 7330–7341.

Saleh, T. S., Badawi, A. K., Salama, R. S., & Mostafa, M. M. M. (2023). Design and development of novel composites containing nickel ferrites supported on activated carbon derived from agricultural wastes and its application in water remediation. *Materials*, *16*(6). https://doi.org/10.3390/MA16062170

Salih, S. J., & Mahmood, W. M. (2023a). Review on magnetic spinel ferrite (MFe_2O_4) nanoparticles: From synthesis to application. *Heliyon*, *9*(6), e16601. https://doi.org/10.1016/J.HELIYON.2023.E16601

Salih, S. J., & Mahmood, W. M. (2023b). Review on magnetic spinel ferrite (MFe_2O_4) nanoparticles: From synthesis to application. *Heliyon*, *9*(6), e16601. https://doi.org/10.1016/J.HELIYON.2023.E16601

Shah, P., Unnarkat, A., Patel, F., Shah, M., & Shah, P. (2022a). A comprehensive review on spinel based novel catalysts for visible light assisted dye degradation. *Process Safety and Environmental Protection*, *161*, 703–722. https://doi.org/10.1016/J.PSEP.2022.03.030

Sharma, R., & Singhal, S. (2015). Photodegradation of textile dye using magnetically recyclable heterogeneous spinel ferrites. *Journal of Chemical Technology and Biotechnology*, *90*(5), 955–962. https://doi.org/10.1002/JCTB.4409

Singh, G., Ubhi, M. K., Jeet, K., Singla, C., & Kaur, M. (2023). A review on impacting parameters for photocatalytic degradation of organic effluents by ferrites and their nanocomposites. *Processes*, *11*(6), 1727. https://doi.org/10.3390/PR11061727

Soufi, A., Hajjaoui, H., Elmoubarki, R., Abdennouri, M., Qourzal, S., & Barka, N. (2021a). Spinel ferrites nanoparticles: Synthesis methods and application in heterogeneous Fenton oxidation of organic pollutants – A review. *Applied Surface Science Advances*, *6*, 100145. https://doi.org/10.1016/J.APSADV.2021.100145

Soufi, A., Hajjaoui, H., Elmoubarki, R., Abdennouri, M., Qourzal, S., & Barka, N. (2021b). Spinel ferrites nanoparticles: Synthesis methods and application in heterogeneous Fenton oxidation of organic pollutants – A review. *Applied Surface Science Advances*, *6*, 100145. https://doi.org/10.1016/J.APSADV.2021.100145

Suhag, D., Thakur, A., & Thakur, P. (eds) (2023). *Integrated Nanomaterials and Their Applications*. Springer Nature Singapore, pp. 1–499. https://doi.org/10.1007/978-981-99-6105-4

Thakur, P., Taneja, S., Chahar, D., Ravelo, B., & Thakur, A. (2021). Recent advances on synthesis, characterization and high frequency applications of Ni-Zn ferrite nanoparticles. *Journal of Magnetism and Magnetic Materials*, *530*, 167925.

Tsay, C. Y., Chiu, Y. C., & Lei, C. M. (2018). Hydrothermally synthesized Mg-based spinel nanoferrites: Phase formation and study on magnetic features and microwave characteristics. *Materials*, *11*(11), 2274. https://doi.org/10.3390/MA11112274

Zhu, P., Luo, D., Duan, M., Feng, L., Zhang, S., & Liu, M. (2022). Based on a dual Z-scheme heterojunction and magnetically separable $CoFe_2O_4$/g-C_3N_4/$Bi_4Ti_3O_{12}$ flower-like composite for efficient visible-light photocatalytic degradation of organic pollutants. *Journal of Alloys and Compounds*, *911*, 164907. https://doi.org/10.1016/J.JALLCOM.2022.164907

13 Spinel Nano-Ferrites for Heavy Metal Ions Detection and Their Removal

*Almas Ishaque, Sumbul Raza, and Shahab A. A. Nami**
Aligarh Muslim University, Aligarh, India

13.1 INTRODUCTION

Water is one of the most indispensable substances gifted by mother earth essential for the sustenance of all living beings, and thus, it must be free of toxins and impurities. These toxins are typically organic, biological, microbiological, or inorganic in origin, and pose serious detrimental effects to humans when consumed through water [1]. Industrialization, climate change, and expansion of urban areas originated a variety of waterborne pollutants. Heavy metal contamination is one of the most deleterious consequences of the ever-increasing industrial development. Due to their persistent nature, heavy metal ion (HMI) contaminations not only have detrimental effects on the environment but also have an existential risk to practically all life forms [2]. HMIs are found in high concentrations in water bodies as a result of rising human and industrial production activities. Some heavy metals are necessary for the growth and proper functioning of the human body, whereas others are harmful. In all probability, increasing beyond the recognized guideline levels for any HMI might lead to an increase in health issues [3]. To date a range of approaches for HMI remediation from both solid and liquid media such as groundwater, industrial contaminated water, sediments, and sludges have been developed [4]. The use of nano-ferrites for HMI detection and remediation is one such strategy. Magnetic nano-ferrites are being used efficiently as nanoadsorbents for water treatment by researchers. Moreover, magnetic nano-ferrites with certain designs have been discovered capable of eliminating up to 99% of the targeted HMIs from polluted water. This chapter discusses typical harmful HMIs and their origins in the environment, as well as various magnetic nano-ferrites and their derivatives for their detection and efficient removal.

13.1.1 Heavy Metals as Water Contaminants

Metals have always been treasured by mankind. Human activities have had no effect on heavy metal quantities since they are mostly consumed for utilitarian purposes. There is no production or elimination; metal amounts, variation, and distributions have simply altered due to new mechanisms of dissemination. Toxic metals such as As, Cd, Pb, and Hg are among them, as are vital trace metals such as Cr, Co, Cu, Mg, Mo, Ni, Se, W, V, and Zn are included in it. Toxic metals have no helpful biological roles in the lives of organisms, and heavy metals continue to be the greatest threat to the health of humans [5, 6]. HMs have contaminated roughly 40% of the world's water bodies [7]. The amount of HMs in aquatic bodies can rise due to biological processes as well as human intervention. Industrial effluent, battery recycling, electroplating manufacturing processes, combustion exhausts from automobiles, industries, and agricultural activities contribute majorly to the accumulation of HMs in water bodies and potable water [8–10]. Furthermore, the corrosion of distribution pipes, also known as plumbing pipes, and other similar components contaminates drinking water with HMs making it hazardous to humans and other living beings that have access to this water [11]. Also people receive exposure to drinking water from taps within buildings and other such locations, where metal concentrations rise owing to water stagnation in water distribution systems, coolers, and boiling-water tanks [12]. For instance, the water used for drinking purposes from wells and tankers in Sonora, Mexico, was discovered to contain high concentrations of As, Cd, Cu, Hg, and Pb. Nearly 43% of the collected specimens were above the Pb response level (e.g., 15 g l^{-1}) and 8.9% of the specimens surpassed the As guideline limit of 10 g l^{-1} [13]. To summarize, we are exposed to heavy metals from minor to significant sources, and this exposure has an injurious influence on the environment, plants, animals, and most notably humans.

13.1.2 Potential Health Hazards by Heavy Metals

Even though Fe, Zn, Co, Cu, and Mn are vital to the proper functioning of the human body, excessive amounts of these HMs can have toxic effects [14]. These hazardous HMs enter the human body through gastrointestinal, inhalation, and cutaneous exposure [15]. Various studies have found altering effects of HMs and the International Agency for Research on Cancer (IARC) has labeled both As and Cd to be potential carcinogens that cause kidney and skin damage, respectively. Other effects of Sb, Pb, Hg, and Cu include cardiovascular illness, high blood cholesterol, anemia, renal and liver damage, and gastrointestinal diseases [13]. Among all HMs, As, Cd, Cr, Mn, Pb, Ni, Sn, and Hg pose the greatest risk to human health since they typically occur in greater

* Corresponding author.

DOI: 10.1201/9781003568940-13

amounts in potable water [16]. These HMs are capable of disrupting the human body's metabolic systems in a variety of ways. Furthermore, they can accumulate in vital organs such as the cardiovascular system, brain, liver, and the urinary system, disrupting normal biological activities [17]. Pb is one of the known carcinogens that may adversely affect the heart, kidneys, nervous, and gastrointestinal systems, as well as cause low intelligence, deafness, and multi-organ malfunction [18]. Table 13.1 summarizes some of the consequences, medium, and source of contamination of HMs.

TABLE 13.1
Summary of the Consequences, Medium, and Source of Contamination of HMs

Metal and Its Oxidation State	Contamination Source	Health Hazard	References
Arsenic As(III), As(IV)	• Agricultural debris • Industrial waste • Electronic device production • Soil composts • Sewage sludge • Desalinated water	• Respiratory, prostate, and skin cancer • Paralysis • Vomiting, nauseous, and diarrhea • Blindness	[19–21]
Cadmium Cd(II)	• Contaminated food and water • Dust and synthetic chemicals in industries and agriculture • Electroplating and metallurgy industries	• Osteomalacia • Glucosuria • Emphysema • Lung cancer • Respiratory disorders	[22–24]
Chromium Cr(VI)	• Leather, tanning, textile industries • Agriculture activities • Anthropogenic activities	• Kidney and liver failure • Skin diseases • Vomiting ulcer and nausea • Respiratory disorders	[20, 23, 25]
Manganese Mn(II)	• Dust from industries • Natural source • Steel industry	• Neuropathic disorder	[26, 27]
Lead Pb(II)	• Agricultural activities • Vehicle emissions • Plumbing pipes	• Kidney failure • Hypertension/mental retardation • Infertility and Nervous disorder	[24, 28]
Nickel Ni(II)	• Industrial activities • Aerosols	• Vomiting and nauseous • Hypertension • Respiratory disorders, tiredness	[29]
Mercury Hg(II)	• Contaminated water • Industrial wastes and dust	• Kidney dysfunction • Neuro-disorders • Retardation of growth in kids	[30, 31]
Cobalt (II)	• Anthropogenic activities • Surface and groundwater	• Respiratory disorders • Pneumonia and liver dysfunction	[32, 33]
Zinc Zn(II)	• Cramps • Stomach issues	• Corrosion • Water taps and pipelines	[24]
Copper Cu(II)	• Renal failure • Liver dysfunction • Hair loss • Headache	• Mining, metallurgical, steel, and chemical industries • Agricultural activities • Corrosion in taps and pipelines	[24, 34]
Silver Ag(I)	• Respiratory disorders • Vomiting • Hypotension • DNA damage	• Erosion • Contaminated water	[35, 36]
Iron Fe(III)	• Hypotension • Respiratory disorders • Anxiety and seizures	• Corrosion in pipelines • Industrial activities	[37]
Barium Ba(II)	• Numbness • Heart diseases • Slow body movements	• Anthropogenic activities • Industries	[38]

(Continued)

TABLE 13.1 *(Continued)*

Summary of the Consequences, Medium, and Source of Contamination of HMs

Metal and Its Oxidation State	Contamination Source	Health Hazard	References
Antimony Sb(III)	• Hypertension • Heart diseases • Hyperglycemia	• Manufacturing industries • Natural sources	[23, 39]
Aluminum Al(III)	• Neuropathy disorders • Heart failure • Rickets and osteomalacia	• Water remediation procedures	[40]
Tin compounds	• Anemia • Kidney and liver damage	• Chemical synthesis industries	[12]

13.2 SPINEL NANO-FERRITES

13.2.1 Techniques for Remediation and the Emergence of Spinel Nano-Ferrites

To get potable water for consumption, countless physical, biochemical, biological, and chemical purification methods are used worldwide, and water remediation methods are categorized under three levels of treatment, i.e., primary, secondary, and tertiary, and some of them possess almost 99% of the water remediation efficiencies. For wastewater remediation, several techniques such as chemical precipitation, adsorption, electrolysis and membrane technologies, ion exchange, reverse osmosis, coagulation and flocculation, oxidative precipitation, evaporation, and biosorption methods are widely utilized [41–43]. In all these techniques, adsorption is one of the common and efficient ways of removing hazardous HMs and microorganisms from wastewater. The research and development of new and economically viable nanomaterials for removing pollutants, pollution sensing and monitoring, and various other applications have elated the interest of many researchers. Recent discoveries have shown that water remediation concerns could potentially be fixed or considerably improved by involving nanoparticles, nanofiltration, or other nanotechnology-derived materials [44–46]. As current traditional methods are limiting and conventional, the nano-size attributes of nanomaterials offer several advantages including a high number of functional units throughout the surfaces of the materials, charge binding capability, availability, specific surface charge, various configurations which are different from bulk, high porosity, and lower flocculation, paving way for potential uses for the remediation of toxic HMs containing water. Additionally, due to their ability to be reused for several cycles, nanomaterials are appealing and cost-effective [44]. Over the past few decades, HMs treatment from wastewater has been successfully done through an adsorption mechanism utilizing a variety of nanomaterials such as Metal-Organic Frameworks (MOFs), zeolite, chitosan, and cellulose derivatives, spinel nano-ferrites, magnetic nanomaterials, hydrogels, and aerogels and xerogels [47–51]. Particularly over the past two decades, spinel nano-ferrites have been extensively used for water remediation and HMs degradation [52].

13.2.2 Spinel Nano-Ferrites for Heavy Metal Sensing and Remediation

Due to their remarkable characteristics such as extremely high magnetic permeability, high density of defects, large Curie temperature, high resistivity to electrical current, large surface-to-volume ratio, minimal loss of power, and rich chemistry of the surface, spinel nano-ferrites and its derivatives find extensive use in water remediation and other related applications [53]. Spinel ferrites are metal oxides with spinel structures and possess the general chemical composition AB_2O_4, wherein A and B denote different metal cations in tetrahedral (A sites) and octahedral (B sites) places, respectively. The general formula can also be written as $MFe_2O_4$where M can be cobalt, zinc, manganese, magnesium, etc. The metal cation forms, amounts, and positions in the crystalline framework have a considerable influence on the physicochemical characteristics of these ferrites [54, 55]. The complex structural chemistry of spinel ferrite material provides significant potential for chemical modifications to understand and fine-tune the superparamagnetic characteristics of nanomaterials. The spinel framework has two metal cation occupancy cation sites. The structure includes 8 A sites with tetrahedral coordination of metal cations with oxygen and 16 B sites with octahedral coordination. Generally in a normal spinel ferrite, the A site is occupied by Me^{2+} cations while Fe^{3+} cations occupy the B sites. The structure is known as an inverse spinel when Fe^{3+} entirely fills the A site while the B sites are irregularly occupied by M^{2+} and Fe^{3+}. As a result of the metal cation selectivity and ion distribution within the A and B sites, the magnetic framework is variable [56]. Spinel nano-ferrites hold great adsorbent quality because of their magnetic properties, and hence, they have been extensively used in Fe_3^+ removal from wastewater [57].

13.2.3 Adsorbent for Contaminant Removal

A variety of nanoparticles have been taken into account and investigated for their application as adsorbents, but nano-ferrites' simple process fabrication with minimal capital cost, the benefit of extracting dyes and heavy metals, and reuse and recycling the ferrites, and simple structure modification make them far more adaptable than traditionally accessible adsorbents [58, 59]. The two main features of nanomaterials as adsorbents, namely elemental surface and extrinsic functionalization, have been observed, and both of these properties are present and shown in nano-ferrites. Magnetic nano-ferrites ability to eradicate a variety of environmental contaminants has been proven in both experimental and on-site experiments in the field [1]. Sezgin et al. synthesized ferrites of $CuFe_2O_4$ and $NiFe_2O_4$ for the removal of Zn, Ni, and Cu. The adsorption capacities of $CuFe_2O_4$ for Cu(II), Ni(II)and Zn(II) were found to be 83.50%, 98.85%, and 99.80%, respectively, whereas in case of $NiFe_2O_4$ ferrite, it was found to be 92.55%, 36.56%, and 99.91%, respectively [60]. In addition, because of their negative impact, ionized metals have been identified as hazardous pollutants. Because of their inherent ability to eliminate metals, magnetic nano-ferrites are particularly ideal for the adsorption of metal ions. Besides, because of their no toxic effects, reuse and recycling, and ease of dissolving of magnetic nano-ferrites further augment their usefulness. They are preferred for polluted water treatment because they possess a greater capacity for adsorption and stronger kinetic interfaces with hazardous substances [61–63]. The surrounding magnetization significantly aids the elimination from the medium due to its superparamagnetic nature. Most current uses of ferromagnetic in polluted water remediation rely on adsorptive or photocatalytic properties. Magnetic nano-ferrites are used in adsorptive techniques as a form of encapsulation carrier or as a nanoadsorbent to improve the efficacy of pollutant removal. Photocatalytic methods, on the other hand, use magnetic nano-ferrites as semi-conductor catalysts to convert pollutants into a less harmful substance [64]. It is aptly clear that magnetic nano-ferrites have proven to be advantageous in the removal of pollutants containing HMs due to their exceptional adsorption qualities, the propensity to provide configurable reactive surface, the capacity to effectively eliminate contaminants regardless of little adsorbent amounts, the ease with which the adsorbed pollutants can be easily removed from the environment, and capacity to decrease the pollution of various harmful metal pollutants by a variety of methods [1, 65, 66].

13.3 SYNTHESIS OF SPINEL NANO-FERRITES

Modern spinel nano-ferrites fabrication techniques include co-precipitation, hydrothermal, solid state, microemulsion, sol-gel template-assisted technique, sono-chemical, electrochemical, and solvothermal. Additionally, the synthetic technique utilized for their synthesis has a significant impact on the morphology, framework, and characteristics of the resulting magnetic nano-ferrites, which may have an impact on their adsorption capacity when used as HM removal material. Consequently, it becomes intriguing to explore the synthetic methodology that affects their adsorption activity [67]. Few synthesis approaches for the preparation of spinel nano-ferrite are listed in Table 13.2.

13.4 SPINEL NANO-FERRITES FOR DETECTION AND SENSING OF HMs IN CONTAMINATED WATER

Economical, efficient, rapid, and precise monitoring and detection of various HMs found in polluted water has become a major concern and problem, and nano-ferrites offer the silver lining by utilizing it in sensors or for the detection of gases and HMs in recent years [77–79]. Due to their magnetic properties, enhanced sensitivity, chemical stability, biocompatibility, large surface area, tunable properties, rapid reaction time, and exceptional efficiency in the detection of various chemical materials, metals, and

TABLE 13.2
Different Synthetic Approaches for Nanoferrite Formation and Their Examples

Synthetic Approach	Source of Iron	Temperature (°C)	Spinel Nanoferrite Formed	Reference
Hydrothermal and solvothermal	$Fe(NO_3)_3.9H_2O$	180	MFe_2O_4 M = Zn, Ni, Cu, Mg	[54]
Thermal decomposition	$Fe(acac)_3$	300	$CoFe_2O_4$	[68]
Microwave-assisted synthesis	$FeCl_2$	25	$TiFe_2O_4$	[69]
Microemulsion	$C_{24}H_{45}FeO_6$	40°C	$NiFe_2O_4$, Mn–Zn ferrite	[70, 71]
Sono-chemical	$Fe(NO_3)_3.9H_2O$	25°C	$MnFe_2O_4$	[69]
Solid State	$Fe(NO_3)_3.9H_2O$	100°C	$CoFe_2O_4$	[72]
Sol-gel	$Fe(NO_3)_3.9H_2O$	150°C	$ZnFe_2O_4$, $MgFe_2O_4$, $CuFe_2O_4$, $NiFe_2O_4$, $CaFe_2O_4$, $CoFe_2O_4$, $MnFe_2O_4$	[67]
Co-precipitation	$Fe(NO_3)_3.9H_2O$	100–500°C	$CuFe_2O_4$, $CoFe_2O_4$ $MnFe_2O_4$, $MgFe_2O_4$	[73–76]

gases, magnetic-nano-ferrite-based sensors are proving to be advantageous in a number of applications [80–82]. There are several harmful metal ions that can exist in water in different amounts, each of which has particular characteristics; thus, the improved ability of spinel nano-ferrites in monitoring and sensing different metal ions seems to be highly valued. Morel et al. synthesized magnetic spinel nano-ferrites comprising Fe_3O_4/SiO_2 core-shell through the sono-chemical method and used these nano-ferrites for Zn(II) detection. It was found that the detection limit was 10^4 M [83]. The development of spinel nano-ferrites as sensors for heavy metal detection is still a relatively unexplored area, and it is evident from the above discussion that in the few years to come researchers will be able to develop various robust nano-ferrites for HMs detection.

13.5 HMs REMOVAL BY SPINEL NANO-FERRITES AND THEIR DERIVATIVES

Nano-ferrites, their composites, and their derivatives have been widely employed in achieving the elimination of HMs. Magnetic hydroxyapatite nanoparticles (MNHAP) can be prepared by mixing $FeCl_2.4H_2O$ and $FeCl_3.6H_2O$ in approximately 30 ml of water (oxygenated) at room temperature under nitrogen atmosphere followed by the addition of 10 ml of NH_4OH to generate black precipitate which on suitable workup yielded MNHAP [84]. The adsorption potential of MHNAP was studied considering all the parameters like initial metal concentration, change in pH, effect of the adsorbent amount, and presence of competitive metal ions. It was found that MNHAP could potentially be employed to remove Cd(II)with an effective adsorption rate of 1.964 mmol g^{-1} with a solution pH ranging from 4 to 8. It was demonstrated that MNHAP has an extreme preference toward Cd(II)in a solution comprising both Cd(II) and Zn(II) [85]. Similar outcomes were achieved when Cd(II) is removed by hydroxyapatite in the presence of multiple metal ions like Cu(II) and Pb(II) [86].

Magnetic nanoparticles can be used for the removal of As because of their high surface area. $CoFe_2O_4$, $MnFe_2O_4$, and $MgFe_2O_4$ are commonly used nano-ferrites for removal of As(II). Similarly, $MnFe_2O_4$ and $CoFe_2O_4$ synthesized by co-precipitation methods were used for the removal of As(III) from wastewater. Previous studies have shown 96% removal of As(III) using adsorbents such as $CoFe_2O_4$ and 93% removal using $MnFe_2O_4$.Nevertheless, the settings for operation are completely different [87, 88]. Metal nanoparticles synthesized using the co-precipitation method achieved equilibrium within 25–30 min. These nano-ferrites have hydroxyl groups at the adsorbing sites and appreciably greater surface area making them a promising alternative for As removal. Another significant property of these nano-ferrites is their selectivity toward As^{+3} in the presence of other ions such as Mg^{+2}, Na^{+}, K^{+}, and Ca^{+2}. Also desorption of As can be conveniently done by using NaOH solution [89]. Microwave-assisted co-precipitation of magnetic nano-ferrites can also be carried out by using water because of its compatibility. A typical preparation of nano-ferrites can be depicted in Figure 13.1. It was observed that the adsorption of As(III) was maximum when the composition of cobalt was increased in the ferrite composite [90]. $CoFe_3O_4$/Schwertmannite composite has shown 95% As adsorption using 1 g l^{-1} of the adsorbent within 240 min. The process is found to be pH dependent on the system [91].

Manganese ferrites magnetic nanoparticles with graphene oxide have demonstrated outstanding adsorption of Pb(II), As(III), and As(V). In the case of As(III), when the pH of the solution is from 2 to 7, graphene oxide-manganese ferrite nanoparticles adsorb at 146 mg g^{-1}, and for As(V) adsorption may exceed 206 mg g^{-1}. This is due to the layered structure of graphene which provides large sites for metal adsorption in addition to the magnetic character of the adsorbent which facilitates the desorption process [92].

Magnetite nanoparticles synthesized by co-precipitation method involve mixing of $FeCl_3.6H_2O$ and $FeCl_2.4H_2O$ in a controlled atmosphere at room temperature furnishing

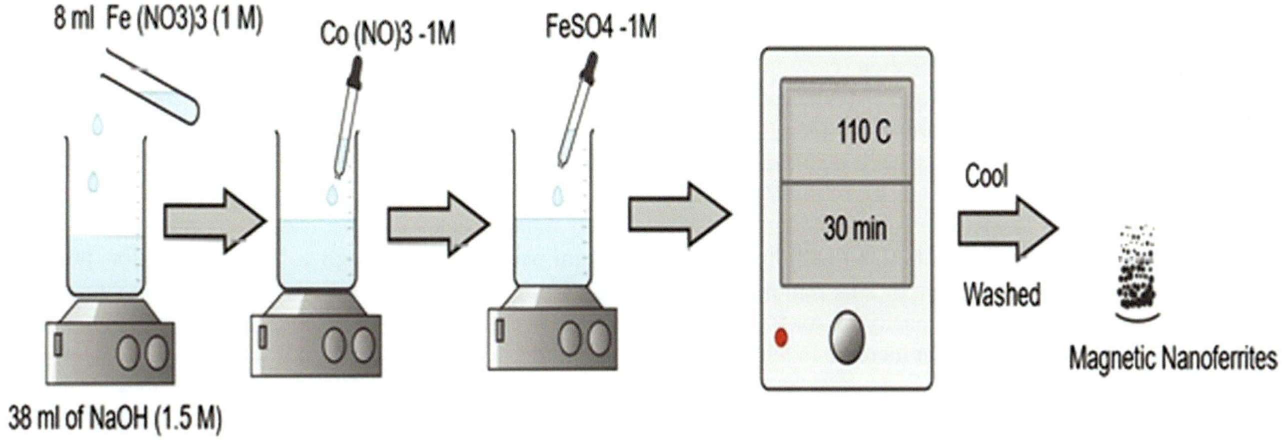

FIGURE 13.1 Schematic representation of nanoferrite synthesis using a template.

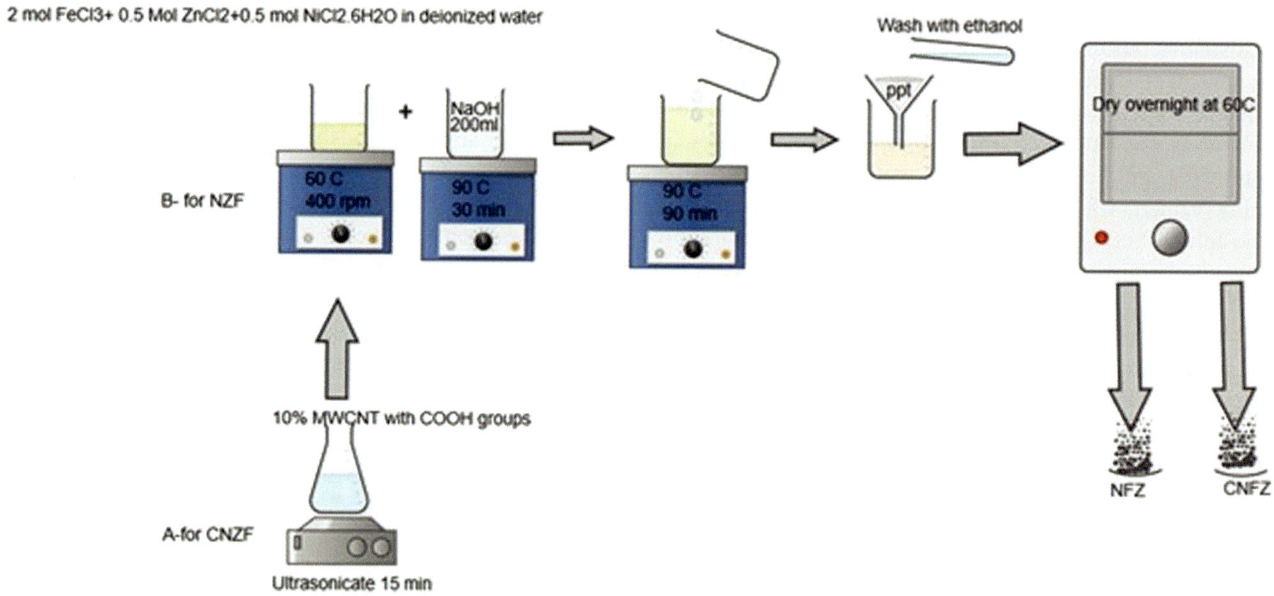

FIGURE 13.2 Schematic representation of the synthesis of NZF and CNZF.

12-nm-size particles and 0.2 g of this adsorbent can achieve adsorption of 66.53 mg g^{-1} with the contact time of approximately 24 h. This investigation demonstrated that Fe_2O_3 nanoparticles are a useful material for producing drinking water or removing harmful metal ions like As(V) from wastewater [93].

$Ni_{0.5}Zn_{0.5}Fe_2O_4$ ferrite (NZF) and $Ni_{0.5}Zn_{0.5}Fe_2O_4$/CNTs (CNZF) were synthesized by inverse co-precipitation technique as shown in Figure 13.2. These novel nanocomposites can adsorb As(V) from industrial wastewater having a pH of 2. The adsorption capacities of NZF and CNZF were 56 and 66 mg g^{-1}, respectively, with a stirring time of 30 min with an adsorbent dose of 6 g l^{-1} [94].

Comparative study was done by synthesizing magnesium ferrite ($MgFe_2O_4$) nanoparticles by three different routes [95] i.e., co-precipitation, sol-gel, and auto-combustion method and their adsorption capacities for the removal of As was determined. It was observed that $MgFe_2O_4$ nanoparticles synthesized by auto-combustion can appreciably remove arsenic from polluted water. Magnesium ferrate synthesized with supermagnetic properties ($Mg_{0.27}Fe_{2.5}O_4$), at 150°C by hydrothermal process, with a surface area of the 438 mg g^{-1} with adsorption capacity of As(V) reached 83.2 mg g^{-1} while for As(III) it was 127.4 mg g^{-1} [96]. Similarly, the nanocomposite synthesized with heulandite, magnetite, and chitosan with a ratio of 2:1:1 removes Cu(II) and As(V) with an adsorption capacity of As(V) at about 5.9 mg g^{-1}, which increases with an increase in temperature of about 45°C [97].

Removal of Sb from contaminated agricultural and industrial wastewater is also important. The United States Environmental Protection Agency (USEPA) declared Sb as carcinogenic, and thus, its removal is necessary [98]. Antimony is found in two oxidation states, Sb(III) and Sb(V). The latter has high solubility in water whereas the former is more toxic in nature. Fe_3O_4@Fe_2O_3@CNs and Fe_2O_3@CNs were synthesized by co-precipitation and hydrothermal methods for removing Sb(III) from wastewater. Adsorption capacities of Fe_3O_4@Fe_2O_3@CNs and Fe_2O_3@CNs were found to be 283.32 and 117.39 mg g^{-1}, respectively, provided the pH of the solution ranges from 3 to 10. Metal ferrites, namely Fe_2O_4, MnF_2O_4, $CoFe_2O_4$, $NiFe_2O_4$, $ZnFe_2O_4$, and Fe_3O_4 were synthesized for treating Sb-contaminated wastewater. The most promising material for eliminating hazardous Sb(III) and Sb(V) from diverse water systems was found to be $ZnFe_2O_4$. Sb(III) and Sb(V) can be removed with 100% efficiency from a system having pH 3–5 [100].

The perusal of the literature reveals that Pb pollution in water is an important root cause for numerous chronic health issues, and its removal is the need of an hour. Generally, Pb gets adsorbed in water through old pipes and acute poisoning occurs when exposed to extremely high amounts of Pb. Initially, symptoms such as lethargy, stomach pains, anorexia, and irritability take place [101]. Fe_3O_4 nanoparticles can be used to remove Pb(II) from contaminated water through a batch adsorption process, with 30 min of contact time, high adsorption capacity is achieved. These nanoadsorbents exhibit good desorption rates and can be reused without affecting the adsorption rates [102]. Likewise, $CoFe_2O_4$/0.02 GR, ferrite graphene nanocomposite was synthesized for the adsorption of Pb(II), which has an adsorption capacity of 99% at pH of 5–6 with a contact time of 1 h [103].

Nanosized Zr-MFCs, Fe_3O_4@SiO_2@UiO-66, Fe_3O_4@SiO_2@UiO-66-NH_2, and Fe_3O_4@SiO_2@UiO-66 were prepared by one-pot hydrothermal synthesis. These nanosized MFCs can remove HMs particularly Pb(II) from wastewater with an adsorption capacity of 121 mg g^{-1}. These nanocomposites can be reused multiple times after desorption of metal ions [104]. $MnFe_2O_4$ nanoparticles can adsorb Pb(II) with an adsorption capacity of 488 mg g^{-1} while its nanocomposite with graphene oxide has an exceptional adsorption capacity of 673 mg g^{-1}. This is due to an increase in adsorbent sites and layer structure of graphene present in the adsorbent [92].

Fe_3O_4 nanoparticles coated with SiO_2 by Stober process modified with polyetherimide (PEI) and 3-chloropropyltrimethoxy silane by continuous stirring for 12 h at 350 K. The nanoparticles obtained were then washed and dispersed in acetone and treated with 10 ml of epichlorohydrin under continuous stirring for 12 h at room temperature. Finally, thiourea (2g) and 3 ml of NaOH are added, and the solution is vigorously stirred for 3–4 h. The resulting Fe_3O_4@SiO_2-PEI-Sh nanoparticles are used as adsorbent for the adsorption of metal ions including Pb(II). Maximum adsorption of Pb(II) takes place at pH 5–6 and the adsorption capacity for Pb(II) is found to be 110 mg g^{-1} [105]. Facile Fe@MgO magnetic nanoparticles were prepared as shown in Figure 13.3 and exhibit exceptional adsorption capacity among all the nano-ferrites. These nanocomposites have an adsorption capacity of 1476.96 mg g^{-1} of Pb (II), which might be attributed to their mesoporous structure and large surface area [106]. Auto-combustion of citrate nitrate produces, calcium nano-ferrites ($CaFe_2$–x–yGdxSmyO_4; x = y = 0.0; x = 0.025, y = 0.05) having particle size of 33–65 nm where 4 mg of nano-ferrite have removal an efficiency of 99.31% of Pb(II) from the solution within 10 min [107].

Hg is an established dangerous heavy element that has been linked to health emergencies in Minamata Bay, Japan, and Iraq. The toxicity of Hg to mammals varies depending on the type of mercury, the dose, and the rate of exposure. The brain is the most significant target organ for breathed Hg vapor. Mercurous (Hg^{+}) and mercuric (Hg^{+2}) salts mostly harm the stomach lining and kidney. Toxicity varies with dosage significant acute exposures to elemental mercury vapor causing severe pneumonitis, which can be lethal in critical circumstances [108, 109]. Magnetic $CoFe_2O_4$@SiO_2 functionalized with EDTA using a hydrothermal process can adsorb Hg(II) from a solution having a pH of 7 and an adsorption capacity of 103.3 mg g^{-1}. The nanomaterial obtained can be recycled multiple times and is biodegradable in nature [110].

Polyrhodamine encapsulated magnetic nanoparticles were synthesized as shown in Figure 13.4, and adsorption of HMs was examined. An amount of 5 mg of polyrhodamine was added to 80 ml l^{-1} of solution containing Hg(II) having pH 2. The adsorbent is then removed by magnetic

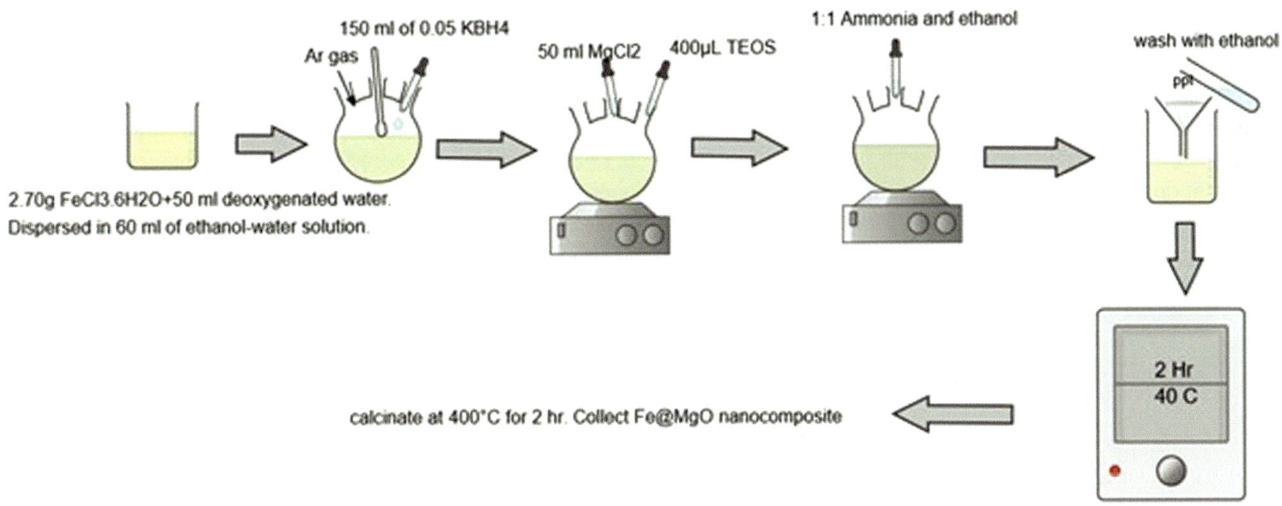

FIGURE 13.3 Schematic representation of the synthesis of Fe@MgO.

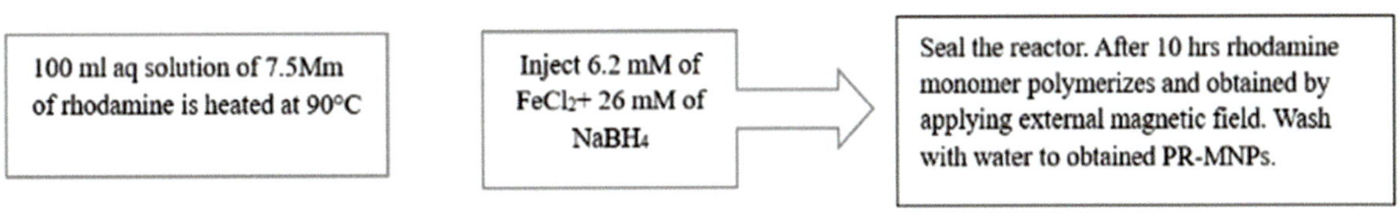

FIGURE 13.4 Schematic representation of polyrhodamine encapsulated magnetic nanoparticles ferrites.

field for further use. The adsorption capacity of Hg(II) was around 130 mg g^{-1} at pH4, which further remains constant at pH 8 and decreases as pH value increases [111].

Functionalized Fe_3O_4@C nanoparticles were prepared to remove Hg(II) from wastewater. The adsorption efficiency of the nanoparticles was 98.1% for Hg and 96.3% for Pb. This facile synthesis worked because of the strong interaction of the metal with SO_3H and COOH groups [112]. Yam peel biomass and amino-organic-ligand-coated magnetic nanoparticles were used for Hg adsorption and exhibited an adsorption capacity of 60 mg g^{-1} [113]. Humic-acid-coated Fe_2O_3 nanoparticles were used to remove toxic metals like Hg(II), Pb(II), and Cd(II), and it was reported that this adsorbent removed 99.8% of Hg and Pb from water at low pH. This is due to the high adsorption capacity of humic acid and the presence of competitive species that do not affect the adsorption process [114].

Humans are continually exposed to Ni due to its significant amount available in the earth's crust. A Ni-deficient diet is difficult to sustain due to its prevalence in food. Chronic Ni and its compound accumulation in the body may be responsible for a number of pessimistic consequences on human health [115, 116]. $NiFe_2O_4$@SalenSi nanocomposite was synthesized as shown in Figure 13.5, with an initial concentration of 15 mg l^{-1} of Ni(II); 94.6% of efficiency was attained within 45 min. The adsorption efficiency of these nanocomposites depends on the initial Ni(II) concentration and pH of the solution. The adsorbent is economical and recyclable [117].

Maghemite nanoparticles were successfully prepared by the sol-gel method with the size of 10 nm diameters. These nanoparticles were used for the removal of Ni(II) and Cu(II) where 94% of Ni(II) ions were adsorbed at a pH of 8.5 [118]. Removal of Zn(II) by MNHAP delivered an adsorption efficiency of 2.151 mmol g^{-1} at pH 5, with only 0.1 g l^{-1} of the adsorbent dose. EDTA was used as an eluent for the recovery of Zn(II) with 67% of the recovery rate [85].

Magnetic Fe_3O_4 nanoparticles can act as an adsorbent for the adsorption of Cu(II) at around 6.5 pH [93], while novel nanocomposite synthesized by chitosan, heulandite, and nanomagnetite can remove Cu(II) with adsorption capacity of 17.2 mg g^{-1} having initial concentration of Cu(II) of 656 mg l^{-1}. It was observed that adsorption of Cu(II) increases when the pH of the solution increases from 3 to 9 [97]. In another study, barium ferrate and nickel ferrate synthesized by co-precipitation method when mixed with polyvinyl alcohol to prepare PVA-nano-ferrites show 100% removal efficiency for Cu(II) at pH 6, thus making viable for industrial scale to remove Cu(II) from wastewater [119].

Chromium is present in three stable forms Cr (0), Cr(III), and Cr(VI). Among them, only Cr(VI) or hexavalent chromium is carcinogenic to humans, and thus, its removal is necessary [120, 121]. Nanocomposite containing polyethyleneimine–montmorillonite and magnetic Fe_3O_4 nanoparticles were synthesized and its Cr(VI) removal efficiency at different pH conditions was conducted. It was inferred that adsorption capacity of Fe_3O_4–PEI800–MMT was 8.77 mg g^{-1} and Fe_3O_4–PEI25000–MT was 7.69 mg g^{-1} [122]. In one of the studies pertinent to the treatment of tannery wastewater with dissolved chromium in water, cobalt nano-ferrite was prepared by co-precipitation method, and the prepared substance was annealed for 300, 500, and 900°C. Chromium removal efficiency of the resulting nanoparticles was found to be 23.75% [123]. Maghemite nanoparticles synthesized by the sol-gel process can successfully remove Cr(VI) from wastewater at pH 2.5. γ-Fe_2O_3 nanoparticles having a size of 10 nm exhibit adsorption of metal by electrostatic attraction and ion exchange process. These nanoparticles can also be employed to successfully remove Cu(II) and N (II) from industrial water sources [118].

Al is present in all natural streams and waterworks. It can appear in the form of organic and inorganic compounds, both in dissolved and undissolved states. Here the pH value is a crucial component that influences its type of occurrence. Al also gets added to water sources by acid rain, and its increased concentrations can lead to potential health risks resulting in brain alterations, characteristic of Alzheimer's disease. The intentional application of Al salts

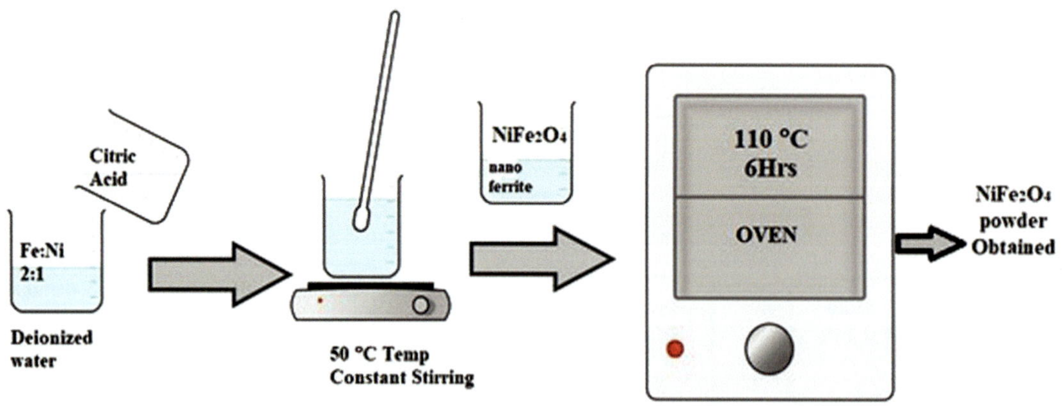

FIGURE 13.5 Schematic representation of synthesis of $NiFe_2O_4$@SalenSi nanocomposite.

as flocculants in water purification has frequently been criticized [124–126]. In a batch adsorption process for the removal of Al(III), magnetic iron oxide nanoparticles were used with an adsorbent dose of 0.01 mg at pH 9 for the removal of 50 ml of Al(III). The adsorption process was fast and efficient and took 4 min to complete the entire process [127].

13.6 CONCLUSION

The overgrowing worldwide challenge of the consumable scarceness of water and contamination has prompted academics to explore unique and inventive methods and materials for successfully treating tainted water. Researchers have synthesized and investigated several forms of nano-ferrites, composites, and their derivatives to assess their potential for eliminating heavy metal pollutants from polluted water. Magnetic nano-ferrites are favored contenders for usage as nanoadsorbents due to their magnetic inclination, high reactivity, and large accessible surface area. Magnetic nano-ferrite qualities can be improved further by investigating the functionalization of the nano-ferrite surface, and core-shell structural design of magnetic nano-ferrites. The present chapter exclusively explores magnetic nano-ferrites, composites, and their derivatives for the removal of heavy metals including Cu, As, Mn, Pb, Hg, and Cd. Even after decades of research, there is still a huge demand and scope for developing technologies for eliminating heavy metals by magnetic nano-ferrites that have not been examined or explored for their removal from contaminated water, as well as those metals in which nano-ferrites exhibit extremely poor removal characteristics.

REFERENCES

1. Bharti et al. Potential of magnetic nanoferrites in removal of heavy metals from contaminated water: Mini review. *J Supercond Nov Magn.* 2020;33:3651–65.
2. Pandey N, Keshavkant S. Chapter 1 – Mechanisms of heavy metal removal using microorganisms as biosorbents. In: Shah MP, Rodriguez Couto S, Kumar V, editors. *New Trends in Removal of Heavy Metals from Industrial Wastewater [Internet].* Elsevier; 2021, pp. 1–21. Available from: https://www.sciencedirect.com/science/article/pii/B9780128229651000015
3. Punia et al. Recent advances in detection and removal of heavy metals from contaminated water. *ChemBioEng Reviews.* 2022;9(4):351–369.
4. Bradl H, Xenidis A. Remediation techniques. In: *Interface Science and Technology.* Elsevier; 2005, pp. 165–261.
5. Naija A, Yalcin HC. Evaluation of cadmium and mercury on cardiovascular and neurological systems: Effects on humans and fish. *Toxicol Rep [Internet].* 2023;10:498–508. Available from: https://www.sciencedirect.com/science/article/pii/S2214750023000458
6. Singh R, Mehra R, Walia A, Gupta S, Chawla P, Kumar H, ..., & Kumar N. Colorimetric sensing approaches based on silver nanoparticles aggregation for determination of toxic metal ions in water sample: A review. *Int J Environ Anal Chem.* 2023;103(6):1361–76.
7. Zhou Q, Yang N, Li Y, Ren B, Ding X, Bian H, et al. Total concentrations and sources of heavy metal pollution in global river and lake water bodies from 1972 to 2017. *Glob EcolConserv.* 2020;22:e00925.
8. Thakur A, Punia P, Dhar R, Aggarwal RK, Thakur P. Separation of cadmium and chromium heavy metals from industrial wastewater by using Ni-Zn nanoferrites. *Adv Nano Res.* 2022; 12(5):457.
9. Ajmal M, Rao RAK, Ahmad R, Ahmad J. Adsorption studies on Citrus reticulata (fruit peel of orange): Removal and recovery of Ni (II) from electroplating wastewater. *J Hazard Mater.* 2000;79(1–2):117–31.
10. Suhag D, Thakur A, Thakur P (eds) *Integrated Nanomaterials and Their Applications.* Springer Nature Singapore; 2023, pp. 1–499. https://doi.org/10.1007/978-981-99-6105-4
11. Punia P, Bharti MK, Chalia S, Dhar R, Ravelo B, Thakur P, Thakur A. Recent advances in synthesis, characterization, and applications of nanoparticles for contaminated water treatment-A review. *Ceram Int.* 2021; 47(2):1526–50.
12. Chowdhury S, Mazumder MAJ, Al-Attas O, Husain T. Heavy metals in drinking water: Occurrences, implications, and future needs in developing countries. *Sci Total Environ [Internet].* 2016;569–570:476–88. Available from: https://www.sciencedirect.com/science/article/pii/S0048969716313407
13. Wyatt CJ, Fimbres C, Romo L, Mendez RO, Grijalva M. Incidence of heavy metal contamination in water supplies in Northern Mexico. *Environ Res.* 1998;76(2):114–9.
14. Asubiojo OI, Nkono NA, Ogunsua AO, Oluwole AF, Ward NI, Akanle OA, et al. Trace elements in drinking and groundwater samples in Southern Nigeria. *Sci Total Environ.* 1997;208(1–2):1–8.
15. Asaduzzaman K, Khandaker MU, Baharudin NAB, Amin YBM, Farook MS, Bradley DA, et al. Heavy metals in human teeth dentine: A bio-indicator of metals exposure and environmental pollution. *Chemosphere.* 2017;176:221–30.
16. Mahurpawar M. Effects of heavy metals on human health. *Int J Res Granthaalayah.* 2015;530(516):1–7.
17. Fu Z, Xi S. The effects of heavy metals on human metabolism. *Toxicol Mech Methods.* 2020;30(3):167–76.
18. Meena V, Dotaniya ML, Saha JK, Das H, Patra AK. Impact of lead contamination on agroecosystem and human health. In: *Lead in Plants and the Environment.* 2020, pp. 67–82.
19. Sinha D, Prasad P. Health effects inflicted by chronic low-level arsenic contamination in groundwater: A global public health challenge. *J Appl Toxicol.* 2020;40(1):87–131.
20. Joseph L, Jun BM, Flora JR V, Park CM, Yoon Y. Removal of heavy metals from water sources in the developing world using low-cost materials: A review. *Chemosphere.* 2019;229:142–59.
21. Goswami R, Kumar M, Biyani N, Shea PJ. Arsenic exposure and perception of health risk due to groundwater contamination in Majuli (river island), Assam, India. *Environ Geochem Health.* 2020;42:443–60.
22. Rahimzadeh MR, Rahimzadeh MR, Kazemi S, Moghadamnia A. Cadmium toxicity and treatment: An update. *Caspian J Intern Med.* 2017;8(3):135.
23. Jabłońska-Czapla M. Arsenic, antimony, chromium, and thallium speciation in water and sediment samples with the LC-ICP-MS technique. *Int J Anal Chem.* 2015;2015.
24. Chubaka CE, Whiley H, Edwards JW, Ross KE. Lead, zinc, copper, and cadmium content of water from South Australian rainwater tanks. *Int J Environ Res Public Health.* 2018; 15(7):1551.

25. Vareda JP, Valente AJM, Durães L. Assessment of heavy metal pollution from anthropogenic activities and remediation strategies: A review. *J Environ Manage.* 2019;246:101–18.
26. Khan K, Factor-Litvak P, Wasserman GA, Liu X, Ahmed E, Parvez F, et al. Manganese exposure from drinking water and children's classroom behavior in Bangladesh. *Environ Health Perspect.* 2011;119(10):1501–6.
27. Ghosh GC, Khan MJH, Chakraborty TK, Zaman S, Kabir AHME, Tanaka H. Human health risk assessment of elevated and variable iron and manganese intake with arsenic-safe groundwater in Jashore, Bangladesh. *Sci Rep.* 2020;10(1):5206.
28. Miller-Schulze JP, Ishikawa C, Foran JA. Assessing lead-contaminated drinking water in a large academic institution: A case study. *J Water Health.* 2019;17(5):728–36.
29. Das KK, Reddy RC, Bagoji IB, Das S, Bagali S, Mullur L, et al. Primary concept of nickel toxicity—An overview. *J Basic Clin Physiol Pharmacol.* 2018;30(2):141–52.
30. Bernhoft RA. Mercury toxicity and treatment: A review of the literature. *J Environ Public Health.* 2012;2012.
31. Koos BJ, Longo LD. Mercury toxicity in the pregnant woman, fetus, and newborn infant: A review. *Am J Obstet Gynecol.* 1976;126(3):390–409.
32. Sheikh I. Cobalt poisoning: A comprehensive review of the literature. *Med Toxicol Clin Forens Med.* 2016;2(2).
33. Gál J, Hursthouse A, Tatner P, Stewart F, Welton R. Cobalt and secondary poisoning in the terrestrial food chain: Data review and research gaps to support risk assessment. *Environ Int.* 2008;34(6):821–38.
34. Taylor AA, Tsuji JS, Garry MR, McArdle ME, Goodfellow WL, Adams WJ, et al. Critical review of exposure and effects: Implications for setting regulatory health criteria for ingested copper. *Environ Manage.* 2020;65:131–59.
35. Fewtrell L, Majuru B, Hunter PR. A re-assessment of the safety of silver in household water treatment: Rapid systematic review of mammalian in vivo genotoxicity studies. *Environ Health.* 2017;16(1):1–9.
36. Aktepe N, Kocyigit A, Yukselten Y, Taskin A, Keskin C, Celik H. Increased DNA damage and oxidative stress among silver jewelry workers. *Biol Trace Elem Res.* 2015;164:185–91.
37. Kumar V, Bharti PK, Talwar M, Tyagi AK, Kumar P. Studies on high iron content in water resources of Moradabad district (UP), India. *Water Sci.* 2017;31(1):44–51.
38. Kamlade P. Barium exposure of an invasive breast cancer cluster investigation–quantitative drinking water chemistry for carcinogen search. *H_2 Open Journal.* 2019;2(1):168–83.
39. Mitrakas M, Mantha Z, Tzollas N, Stylianou S, Katsoyiannis I, Zouboulis A. Removal of antimony species, Sb (III)/Sb (V), from water by using iron coagulants. *Water (Basel).* 2018;10(10):1328.
40. Lukiw WJ, Kruck TPA, Percy ME, Pogue AI, Alexandrov PN, Walsh WJ, et al. Aluminum in neurological disease—A 36-year multicenter study. *J Alzheimers Dis Parkinsonism.* 2019;8(6).
41. Borsani R, Fazio M, Ferrando B. Industrial water production by utilization of a reverse osmosis and an evaporation plant. *Desalination [Internet].* 1997;108(1):231–45. Available from: https://www.sciencedirect.com/science/article/pii/S0011916497000313
42. Hannachi E, Slimani Y, Nawaz M, Trabelsi Z, Yasin G, Bilal, M., …, & Thakur P. Synthesis, characterization, and evaluation of the photocatalytic properties of zinc oxide co-doped with lanthanides elements. *J Phys Chem Solids.* 2022;170:110910.
43. Zamora-Ledezma C, Negrete-Bolagay D, Figueroa F, Zamora-Ledezma E, Ni M, Alexis F, et al. Heavy metal water pollution: A fresh look about hazards, novel and conventional remediation methods. *Environ Technol Innov [Internet].* 2021;22:101504. Available from: https://www.sciencedirect.com/science/article/pii/S2352186421001528
44. Auffan M, Shipley HJ, Yean S, Kan AT, Tomson M, Rose J, et al. "Nanomaterials as Adsorbant" in "Environmental Nanotechnology: Applications and Impacts of Nanomaterials". Environmental Nanotechnology. 2007.
45. Carlos L, García Einschlag FS, Mártire, DO. Applications of Magnetite Nanoparticles for Heavy Metal Removal from Wastewater. 2013.
46. Schulte J, Dutta J. Nanotechnology in environmental protection and pollution. *Sci Technol Adv Mater.* Taylor & Francis; 2005; 6:219–20.
47. Shivam, Megha R, Lakhani V, Vala S, Dharaskar S, Reddy Paluvai N, et al. Removal of heavy metals and dyes from its aqueous solution utilizing metal organic frameworks (MOFs): Review. *Mater Today Proc [Internet].* 2023;77:188–200. Available from: https://www.sciencedirect.com/science/article/pii/S2214785322070432
48. Haripriyan U, Gopinath KP, Arun J. Chitosan based nano adsorbents and its types for heavy metal removal: A mini review. *Mater Lett [Internet].* 2022;312:131670. Available from: https://www.sciencedirect.com/science/article/pii/S0167577X22000234
49. Guo Z, Zhou J, Hou H, Wu X, Li Y. Recyclable Mg-MOF-74@ cellulose aerogel composites for efficient removal of heavy metals from wastewater. *J Solid State Chem [Internet].* 2023;323:124059. Available from: https://www.sciencedirect.com/science/article/pii/S002245962300227X
50. Smržová D, Szatmáry L, Ecorchard P, Machálková A, Maříková M, Salačová P, et al. Carbon and zeolite-based composites for radionuclide and heavy metal sorption. *Heliyon [Internet].* 2022;8(12):e12293. Available from: https://www.sciencedirect.com/science/article/pii/S2405844022035812
51. Khan FSA, Mubarak NM, Tan YH, Khalid M, Karri RR, Walvekar R, et al. A comprehensive review on magnetic carbon nanotubes and carbon nanotube-based buckypaper for removal of heavy metals and dyes. *J Hazard Mater [Internet].* 2021;413:125375. Available from: https://www.sciencedirect.com/science/article/pii/S0304389421003381
52. Thatai S, Khurana P, Boken J, Prasad S, Kumar D. Nanoparticles and core–shell nanocomposite based new generation water remediation materials and analytical techniques: A review. *Microchem J.* 2014;116:62–76.
53. Safari A, Gheisari Kh, Farbod M. Structural, microstructural, magnetic and dielectric properties of Ni-Zn ferrite powders synthesized by plasma arc discharge process followed by post-annealing. *J Magn Magn Mater [Internet].* 2019;488:165369. Available from: https://www.sciencedirect.com/science/article/pii/S0304885319311278
54. Reddy DHK, Yun YS. Spinel ferrite magnetic adsorbents: Alternative future materials for water purification? *Coord Chem Rev [Internet].* 2016;315:90–111. Available from: https://www.sciencedirect.com/science/article/pii/S0010854515300539
55. Salih SJ, Mahmood WM. Review on magnetic spinel ferrite (MFe2O4) nanoparticles: From synthesis to application. *Heliyon [Internet].* 2023;9(6):e16601. Available from: https://www.sciencedirect.com/science/article/pii/S2405844023038082

56. Mathew DS, Juang RS. An overview of the structure and magnetism of spinel ferrite nanoparticles and their synthesis in microemulsions. *Chem Eng J [Internet]*. 2007;129(1):51–65. Available from: https://www.sciencedirect.com/science/article/pii/S1385894706004931
57. Zafar MN, Amjad M, Tabassum M, Ahmad I, Zubair M. $SrFe_2O_4$ nanoferrites and $SrFe_2O_4$/ground eggshell nanocomposites: Fast and efficient adsorbents for dyes removal. *J Clean Prod*. 2018;199:983–94.
58. Zafar MN, Amjad M, Tabassum M, Ahmad I, Zubair M. $SrFe_2O_4$ nanoferrites and $SrFe_2O_4$/ground eggshell nanocomposites: Fast and efficient adsorbents for dyes removal. *J Clean Prod [Internet]*. 2018;199:983–94. Available from: https://www.sciencedirect.com/science/article/pii/S0959652618321930
59. Rehman MA, Yusoff I, Ahmad P, Alias Y. $CuYb_{0.5}Fe_{1.5}O_4$ nanoferrite adsorbent structural, morphological and functionalization characteristics for multiple pollutant removal by response surface methodology. *J Mol Liq*. 2016;224:1256–65.
60. Sezgin N, Sahin M, Yalcin A, Koseoglu Y. Synthesis, characterization and, the heavy metal removal efficiency of MFe_2O_4 (M = Ni, Cu) nanoparticles. *Ekoloji*. 2013;22(89):89–96.
61. Gehrke I, Geiser A, Somborn-Schulz A. Innovations in nanotechnology for water treatment. *Nanotechnol Sci Appl*. 2015;8:1–17.
62. Lata S, Samadder SR. Removal of arsenic from water using nano adsorbents and challenges: A review. *J Environ Manage*. 2016;166:387–406.
63. Tu YJ, You CF, Chang CK, Chan TS, Li SH. XANES evidence of molybdenum adsorption onto novel fabricated nano-magnetic $CuFe_2O_4$. *Chem Eng J*. 2014;244:343–9.
64. Casbeer E, Sharma VK, Li XZ. Synthesis and photocatalytic activity of ferrites under visible light: A review. *Sep Purif Technol*. 2012;87:1–14.
65. Rana K, Thakur P, Tomar M, Gupta V, Thakur A. Structural and magnetic properties of Ni-Zn doped BaM nanocomposite via citrate precursor. In: AIP Conference Proceedings. AIP Publishing; 2016.
66. Khulbe KC, Matsuura T. Removal of heavy metals and pollutants by membrane adsorption techniques. *Appl Water Sci*. 2018;8:1–30.
67. Dhanda N, Kumari S, Kumar R, Kumar D, Sun ACA, Thakur P, Thakur A. Influence of Ni over magnetically benign Co ferrite system and study of its structural, optical, and magnetic behaviour. *Inorg Chem Commun*. 2023;151:110569.
68. Mahhouti Z, El Moussaoui H, Mahfoud T, Hamedoun M, El Marssi M, Lahmar A, et al. Chemical synthesis and magnetic properties of monodisperse cobalt ferrite nanoparticles. *J Mater Sci: Mater Electron*. 2019;30:14913–22.
69. Shukla A, Bhardwaj AK, Singh SC, Uttam KN, Gautam N, Himanshu AK, et al. Microwave assisted scalable synthesis of titanium ferrite nanomaterials. *J Appl Phys*. 2018;123(16).
70. Pemartin K, Solans C, Alvarez-Quintana J, Sanchez-Dominguez M. Synthesis of Mn–Zn ferrite nanoparticles by the oil-in-water microemulsion reaction method. *Colloids Surf A: Physicochem Eng Asp*. 2014;451:161–71.
71. Rodríguez-Rodríguez AA, Moreno-Trejo MB, Meléndez-Zaragoza MJ, Collins-Martínez V, López-Ortiz A, Martínez-Guerra E, et al. Spinel-type ferrite nanoparticles: Synthesis by the oil-in-water microemulsion reaction method and photocatalytic water-splitting evaluation. *Int J Hydrogen Energy*. 2019;44(24):12421–9.
72. Jauhar S, Kaur J, Goyal A, Singhal S. Tuning the properties of cobalt ferrite: A road towards diverse applications. *RSC Adv [Internet]*. 2016;6(100):97694–719. Available from: http://dx.doi.org/10.1039/C6RA21224G
73. Thakur A, Thakur P, Hsu JH Enhancement in dielectric and magnetic properties of In3+ Substituted Ni-Zn Nano ferrites by coprecipitation method. *IEEE Trans Magn*. 2011; 47(10):4336–9.
74. Mathur P, Thakur A, Singh M. Study of low-temperature sintered nano-crystallite Mn-Cu-Zn ferrite prepared by co-precipitation method. *Mod Phys Lett B*. 2007; 21(21):1425–30.
75. Mathur P, Thakur A, Singh M. Processing of high density manganese zinc nanoferrites by co-precipitation method. *Zeitschrift für Physikalische Chemie*. 2007; 221(7):887–95.
76. Mathur P, Thakur A, Singh M.. Low temperature synthesis of $Mn_{0.4}Zn_{0.6}In_{0.5}Fe_{1.5}O_4$ nanoferrite for high-frequency applications. *J Phys Chem Solids*. 2008; 69(1):187–92.
77. Zhang M, Huang G, Huang J, Zhong L, Chen W. Magnetic solid phase extraction with octahedral structured Fe_3O_4@SiO_2@polydimethylsiloxane magnetic nanoparticles as the sorbent for determining benzene, toluene, ethylbenzene and xylenes in water samples. *RSC Adv [Internet]*. 2017;7(66):41862–8. Available from: http://dx.doi.org/10.1039/C7RA07733E
78. Kumar ER, Kamzin AS, Janani K. Effect of annealing on particle size, microstructure and gas sensing properties of Mn substituted $CoFe_2O_4$ nanoparticles. *J Magn Magn Mater [Internet]*. 2016;417:122–9. Available from: https://www.sciencedirect.com/science/article/pii/S0304885316307806
79. Rathore D, Kurchania R, Pandey R. Gas sensing properties of size varying $CoFe_2O_4$ nanoparticles. *IEEE Sens J*. 2015 Sep 1;15:1.
80. Sobczak-Kupiec A, Venkatesan J, AlAnezi AA, Walczyk D, Farooqi A, Malina D, et al. Magnetic nanomaterials and sensors for biological detection. *Nanomedicine*. 2016;12(8):2459–73.
81. Gonçalves JM, de Faria LV, Nascimento AB, Germscheidt RL, Patra S, Hernández-Saravia LP, et al. Sensing performances of spinel ferrites MFe_2O_4 (M = Mg, Ni, Co, Mn, Cu and Zn) based electrochemical sensors: A review. *Anal Chim Acta [Internet]*. 2022;1233:340362. Available from: https://www.sciencedirect.com/science/article/pii/S0003267022009333
82. Kefeni KK, Mamba BB. Photocatalytic application of spinel ferrite nanoparticles and nanocomposites in wastewater treatment. *Sustain Mater Techno*. 2020;23:e00140.
83. Morel AL, Nikitenko SI, Gionnet K, Wattiaux A, Lai-Kee-Him J, Labrugere C, et al. Sonochemical approach to the synthesis of Fe_3O_4@SiO_2 core–shell nanoparticles with tunable properties. *ACS Nano [Internet]*. 2008 May 1;2(5):847–56. Available from: https://doi.org/10.1021/nn800091q
84. Mori K, Kanai S, Hara T, Mizugaki T, Ebitani K, Jitsukawa K, et al. Development of ruthenium-hydroxyapatite-encapsulated superparamagnetic γ-Fe_2O_3 nanocrystallites as an efficient oxidation catalyst by molecular oxygen. *Chem Mater*. 2007 Mar 20;19(6):1249–56.
85. Feng Y, Gong JL, Zeng GM, Niu QY, Zhang HY, Niu CG, et al. Adsorption of Cd (II) and Zn (II) from aqueous solutions using magnetic hydroxyapatite nanoparticles as adsorbents. *Chem Eng J*. 2010 Aug;162(2):487–94.
86. Corami A, Mignardi S, Ferrini V. Cadmium removal from single- and multi-metal (Cd + Pb + Zn + Cu) solutions by sorption on hydroxyapatite. *J Colloid Interface Sci*. 2008 Jan 15;317(2):402–8.

87. Fu F, Wang Q. Removal of heavy metal ions from wastewaters: A review. *J Environ Manage*. 2011;92:407–18.
88. Nikić J, Tubić A, Watson M, Maletić S, Šolić M, Majkić T, et al. Arsenic removal from water by green synthesized magnetic nanoparticles. *Water (Switzerland)*. 2019 Dec 1;11(12).
89. Morales-Amaya CG, Alarcón-Herrera MT, Astudillo-Sánchez PD, Lozano-Morales SA, Licea-Jiménez L, Reynoso-Cuevas L. Ferrous magnetic nanoparticles for arsenic removal from groundwater. *Water (Switzerland)*. 2021 Sep 1;13(18).
90. Martinez-Vargas S, Martínez AI, Hernández-Beteta EE, Mijangos-Ricardez OF, Vázquez-Hipólito V, Patiño-Carachure C, et al. Arsenic adsorption on cobalt and manganese ferrite nanoparticles. *J Mater Sci*. 2017 Jun 1;52(11):6205–15.
91. Dey A, Singh R, Purkait MK. Cobalt ferrite nanoparticles aggregated schwertmannite: A novel adsorbent for the efficient removal of arsenic. *J Water Process Eng*. 2014;3(C):1–9.
92. Kumar S, Nair RR, Pillai PB, Gupta SN, Iyengar MAR, Sood AK. Graphene oxide-$MnFe_2O_4$ magnetic nanohybrids for efficient removal of lead and arsenic from water. *ACS Appl Mater Interfaces*. 2014 Oct 22;6(20):17426–36.
93. Iconaru SL, Guégan R, Popa CL, Motelica-Heino M, Ciobanu CS, Predoi D. Magnetite (Fe_3O_4) nanoparticles As adsorbents for As and Cu removal. *Appl Clay Sci*. 2016 Dec 1;134:128–35.
94. Ahangari A, Raygan S, Ataie A. Capabilities of nickel zinc ferrite and its nanocomposite with CNT for adsorption of arsenic (V) ions from wastewater. *J Environ Chem Eng*. 2019 Dec 1;7(6).
95. Kumari S, Dhanda N, Thakur A, Singh S, Thakur P. Investigation of calcium substitution on magnetic and dielectric properties of Mg–Zn nano ferrites. *Mater Chem Phys*. 2023;297:127394.
96. Tang W, Su Y, Li Q, Gao S, Shang JK. Superparamagnetic magnesium ferrite nanoadsorbent for effective arsenic (III, V) removal and easy magnetic separation. *Water Res*. 2013 Jul 1;47(11):3624–34.
97. Cho DW, Jeon BH, Chon CM, Kim Y, Schwartz FW, Lee ES, et al. A novel chitosan/clay/magnetite composite for adsorption of Cu(II) and As(V). *Chem Eng J*. 2012 Aug 15;200–202:654–62.
98. Gebel T. Short review Arsenic and antimony: Comparative approach on mechanistic toxicology. *Chem Biol Interact*. 1997;107.
99. Ren S, Ai Y, Zhang X, Ruan M, Hu ZN, Liu L, et al. Recycling antimony(III) by magnetic carbon nanospheres: Turning waste to recoverable catalytic for synthesis of esters and triazoles. *ACS Sustain Chem Eng*. 2020 Jan 13;8(1):469–77.
100. Tu YJ, Wang SL, Lu YR, Chan TS, Johnston CT. New insight in adsorption of Sb(III)/Sb(V) from waters using magnetic nanoferrites: X-ray absorption spectroscopy investigation. *J Mol Liq*. 2021 May 15;330.
101. Maru S. Lead Exposure in Children through Water and Soil PH 560: Environmental Management and Risk Assessment Sravya Maru Lead Exposure in Children through Water and Soil [Internet]. Available from: http://digitalcommons.wku.edu/cgi/viewcontent.cgi?article=1003&context=pubh_560
102. Nassar NN. Rapid removal and recovery of Pb(II) from wastewater by magnetic nanoadsorbents. *J Hazard Mater*. 2010;184(1–3):538–46.
103. Ateia EE, Ramadan R, Shafaay AS. Efficient treatment of lead-containing wastewater by $CoFe_2O_4$/graphene nanocomposites. *Appl Phys A Mater Sci Process*. 2020 Mar 1;126(3).
104. Huang L, He M, Chen B, Hu B. Magnetic Zr-MOFs nanocomposites for rapid removal of heavy metal ions and dyes from water. *Chemosphere*. 2018 May 1;199:435–44.
105. Ji J, Chen G, Zhao J. Preparation and characterization of amino/thiol bifunctionalized magnetic nanoadsorbent and its application in rapid removal of Pb (II) from aqueous system. *J Hazard Mater*. 2019 Apr 15;368:255–63.
106. Ge L, Wang W, Peng Z, Tan F, Wang X, Chen J, et al. Facile fabrication of Fe@MgO magnetic nanocomposites for efficient removal of heavy metal ion and dye from water. *Powder Technol*. 2018 Feb 15;326:393–401.
107. Saeid YA, Ateia EE. Efficient removal of Pb (II) from water solution using $CaFe_{2-x-y}Gd_xSm_yO_4$ ferrite nanoparticles. *Appl Phys A Mater Sci Process*. 2022 Jul 1;128(7).
108. Jaishankar M, Tseten T, Anbalagan N, Mathew BB, Beeregowda KN. Toxicity, mechanism and health effects of some heavy metals. *Interdiscip Toxicol*. Slovak Toxicology Society. 2014;7:60–72.
109. Bernhoft RA. Mercury toxicity and treatment: A review of the literature. *J Environ Public Health*. Hindawi Publishing Corporation; 2012;2012.
110. Xia K, Guo Y, Shao Q, Zan Q, Bai R. Removal of mercury (II) by EDTA-functionalized magnetic $CoFe_2O_4@SiO_2$ nanomaterial with core-shell structure. *Nanomaterials*. 2019 Nov 1;9(11).
111. Song J, Kong H, Jang J. Adsorption of heavy metal ions from aqueous solution by polyrhodanine-encapsulated magnetic nanoparticles. *J Colloid Interface Sci*. 2011 Jul 15; 359(2):505–11.
112. Chen Z, Geng Z, Zhang Z, Ren L, Tao T, Yang R, et al. Synthesis of magnetic Fe_3O_4@C nanoparticles modified with $-SO_3H$ and −COOH groups for fast removal of Pb^{2+}, Hg^{2+}, and Cd^{2+} ions. *Eur J Inorg Chem*. 2014;(20):3172–7.
113. Marimón-Bolívar W, Tejeda-Benítez L, Herrera AP. Removal of mercury (II) from water using magnetic nanoparticles coated with amino organic ligands and yam peel biomass. *Environ Nanotechnol Monit Manag*. 2018 Dec 1;10:486–93.
114. Liu JF, Zhao ZS, Jiang GB. Coating Fe_3O_4 magnetic nanoparticles with humic acid for high efficient removal of heavy metals in water. *Environ Sci Technol*. 2008 Sep 15;42(18):6949–54.
115. Genchi G, Carocci A, Lauria G, Sinicropi MS, Catalano A. Nickel: Human health and environmental toxicology. *Int J Environ Res Public Health*. MDPI AG; 2020;17.
116. Sivulka DJ. Assessment of respiratory carcinogenicity associated with exposure to metallic nickel: A review. *Regul Toxicol Pharmacol*. 2005;43(2):117–33.
117. Babadi N, Tavakkoli H, Afshari M. Synthesis and characterization of nanocomposite $NiFe_2O_4$@SalenSi and its application in efficient removal of Ni(II) from aqueous solution. *Bull Chem Soc Ethiop*. 2018;32(1):77–88.
118. Hu J, Chen G, Lo IMC, Asce M. Selective Removal of Heavy Metals from Industrial Wastewater Using Maghemite Nanoparticle: Performance and Mechanisms.
119. Manimozhi V, Saravanathamizhan R, Sivakumar EK, Jaisankar V. Adsorption study of heavy metals removal from wastewater using PVA-nano ferrite composites. *Int J Nanosci Nanotechnol*. 2020;16.
120. Zhitkovich A. Chromium in drinking water: Sources, metabolism, and cancer risks. *Chem Res Toxicol*. 2011;24: 1617–29.

121. Saha B, Amine A, Verpoort F. Special issue: Hexavalent chromium: Sources, toxicity, and remediation. *Chem Africa*. Springer Science and Business Media Deutschland GmbH. 2022;5:1779–80.
122. Larraza I, López-Gónzalez M, Corrales T, Marcelo G. Hybrid materials: Magnetite-polyethylenimine-montmorillonite, as magnetic adsorbents for Cr(VI) water treatment. *J Colloid Interface Sci*. 2012 Nov 1;385(1):24–33.
123. Albalah MA, Alsabah YA, Mustafa DE. Characteristics of co-precipitation synthesized cobalt nanoferrites and their potential in industrial wastewater treatment. *SN Appl Sci*. 2020 May 1;2(5).
124. Krupińska I. Aluminium drinking water treatment residuals and their toxic impact on human health. *Molecules*. 2020 Feb 2;25(3).
125. Meghzili B, Souad B, Abdelkrim A. Risk of residual aluminum in treated waters with aluminum sulphate. *Adv Res*. 2016 Jan 10;6(5):1–8.
126. Chao HJ, Zhang X, Wang W, Li D, Ren Y, Kang J, et al. Evaluation of carboxymethylpullulan-AlCl3 as a coagulant for water treatment: A case study with kaolin. *Water Environ Res*. 2020 Feb 1;92(2):302–9.
127. Asrarian R, Jadidian R, Parham H, Haghtalab S. Removal of Aluminum from water and wastewater using magnetic iron oxide nanoparticles. *Adv Mater Res*. 2014;752–6.

14 Spinel Nano-Ferrites for Theranostic and Diagnostic Biomedical Applications

*Akash A. Podutwar, Purva U. Chandorkar, Aniruddha R. Chabukswar, and Swati C. Jagdale**
School of Health Sciences and Technology, Dr Vishwanath Karad MIT World Peace University, MIT Campus, Pune, India

14.1 INTRODUCTION

Spinel ferrite is an oxide containing an intricate structure with a center-faced cubic core, having molecular formula XFe_2O_4, where, X is a metal cation like Zn, Mn, Co, Mg, and Ni. The word spinel comes from the Latin word “Spinella” which stands for spine-like structure referring to its pointed crystals [1]. Spinel nano-ferrites (SNFs) are an extensively multifunctional metal oxide group having exceptional magnetic along with electronic properties, high plethora, low manufacturing cost, and they are highly biocompatible [2]. All these properties have been reported by hundreds of papers for their several biomedical application and water purification that utilize their magnetic as well as photochemical properties. A slew of new scientific studies on novel synthetic methodologies and procedures for SNFs have been published. It has been found that, by modifying and varying the manufacturing methods, new SNFs with desired dimensions can be made. Figure 14.1 shows the morphological classification of SNFs [3].

More specifically, every SNF morphology shows exclusive characteristics corresponding to its spinel structure. Apart from the difference in morphology, these structures share mutual significant advantages due to signature spinel structure, physiochemical properties (catalytic, optical, high aspect ratio, chemical, electrical, thermal stability), and trouble-free surface modification. Besides the aforementioned remarkable properties, toxicity, poor dispersibility, and low surface area are the limitations of SNFs [4].

Several biochemists explained the presence of spinel ferrite inside many animals like honeybees, pigeons, and bacteria. This proves the massive potential of spinel ferrite for biomedical applications such as hyperthermia, diagnostic techniques, and site-specific delivery of drugs [5]. All the above-mentioned properties and applications have captured the significant attention of researchers. After years of research, researchers have explored the usage of SNFs in the biomedical field. The synthesis of SNFs plays an important role as it can regulate structure, magnetic, and morphological properties and most importantly maintain the size

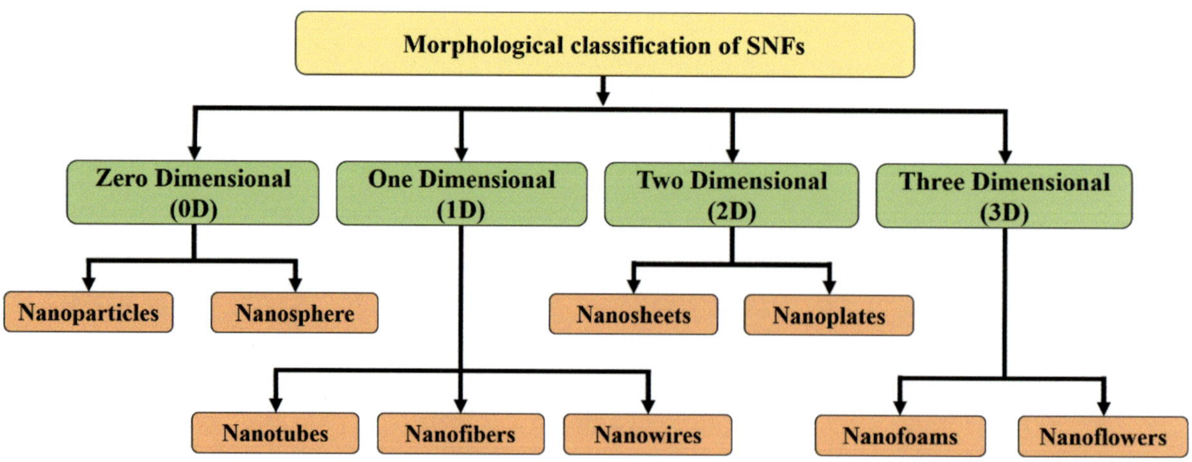

FIGURE 14.1 Morphological classification of SNFs.

* Corresponding author.

DOI: 10.1201/9781003568940-14

range between 1 and 100 nm. Meanwhile, each biomedical application requires SNFs with special conditions and fully controllable physiochemical properties [6].

The crystal structure of SNFs was first determined by Bragg in 1915 and was totally based upon the cubic closely packed arrangement of oxygen ions. The general formula for SNFs is AB_2O_4, whereas A and B serve as sites for different metallic cations. Most commonly studied SNFs are made up of transition metals such as iron (Fe) which is present at site B as a cation in combination with other transition metals or non-metals at site A (Figure 14.2). In this extraordinary structure, with the differences in both charge state, stabilization energy, and chemical structure, these cations can be organized into two distinct crystallographic sites. These are commonly called tetrahedral sites and octahedral sites. These are found between the cations and the oxygen ions that surround them. Among these crystallographic sites, A-site cations possess one-eighth of tetrahedral sites within the crystal lattice and B-site cations possess half of the octahedral site [7]. For more understanding, let us consider XFe_2O_4. One unit of XFe_2O_4 consists of 32 closely packed oxygen atoms which are meticulously packed along with 64 divalent and 32 trivalent octahedral sites together maintaining an electrically stable state in between ions of the structures. Depending upon the positions of these ions in the structure, SNFs can be divided into normal, inversed, and mixed spinel structures. Each of these distinctive structures has different physiochemical properties. Consequently, it is important to identify which are the disturbing factors for site occupancy along with the influence of cationic distribution on magnetic and structural properties [8].

14.2 BIOMEDICAL APPLICATIONS OF SPINEL NANO-FERRITES

Apart from the diverse applications in industrial and environmental fields, SNFs have gained substantial attention in the field of biomedicine like hyperthermia, magnetic resonance imaging (MRI), and targeted drug delivery [9]. These applications require particles that reveal superparamagnetic behavior at room temperature i.e., they show magnetic property in the peripheral magnetic field and instantly redisperse as soon as the field is withdrawn. The basic requirement of SNFs for biomedical applications is that they should be stable in media having a pH equivalent to 7.2 and other physiological conditions. The size of particles should be uniform to avoid any kind of agglomeration and precipitation [10].

It has been noted that, with desired changes in synthesis techniques, SNFs with enhanced superparamagnetic properties and colloidal stability can be fabricated. These fabricated SNFs can be used in biomedical advances like imaging, targeted drug delivery, hyperthermia, and diagnosis. Particularly, the SNFs that have the aforementioned applications at room temperature are mostly favored and nowadays got the attention to solve major health problems. SNFs offer all the advantages to overcome these challenges with surface modification using polymeric chemicals which are biocompatible. They deliver novel approaches and extraordinary properties in the field of magnetically targeted drug delivery systems, cancer management by means of hyperthermia, and several imaging techniques [10, 11].

Magnetic separation for capturing selective proteins and biomolecules can be an efficient and reliable technique.

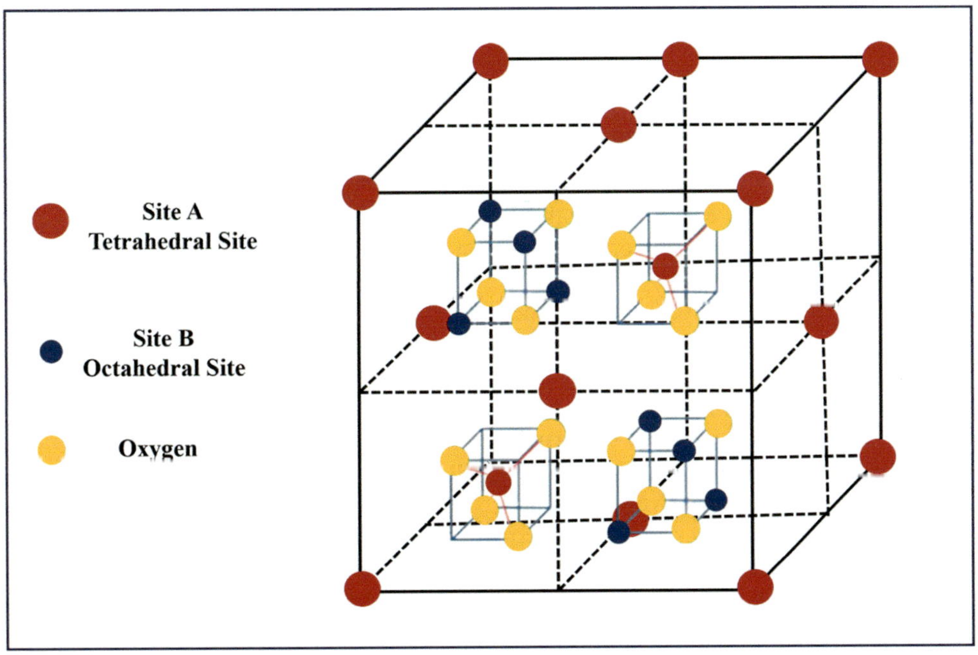

FIGURE 14.2 Structure of spinel ferrite.

This separation technique can be carried out in a single vessel even without spending more money on buying expensive liquid in complex methods like chromatography. Recently, Chen et al. have demonstrated the successful binding of Concanavalin A (Concanavalin A or Con A is a carbohydrate-binding protein that binds to the mannose residues of various glycoproteins and can be used in a variety of studies) on Fe_3O_4 magnetic nanoparticles (NPs) through carbodiimide activation. Furthermore, these materials serve as magnetic nano-adsorbent for the separation of glycoprotein [12].

Various studies showed that, if the SNFs are not coated properly with biocompatible materials then they can lead to toxicity by generation of free radical and reactive oxygen via Fenton reactions. Ankamwar et al. performed a cytotoxicity test with various concentrations of Fe_3O_4 NPs on human glia and demonstrated the possibility of reactive oxygen species generation. The study revealed that 100 µg/ml of Fe_3O_4 NPs had no toxicity effect. There are some major challenges in biomedical sciences that require immediate attention. Some of them are the systematic delivery of drugs at the targeted site and early detection of disease before causing serious damage to the body. Such challenges require advanced diagnostic, imaging, and targeted drug delivery systems to be developed [10, 13].

In summary, several prerequisites should be possessed by SNFs specifically for biomedical applications, like reproducible and well-defined structural, physical, chemical, and pharmacological properties. They also should have specific requirements like less or non-toxic, good response once the external magnetic field is employed, appropriate for a specific application, and less or no residual magnetization time and coercivity. They also should have good structural as well as chemical stability under required conditions and cost-efficient. In this section, a brief discussion of the therapeutic and diagnostic ability of various SNFs is discussed [14, 15].

14.2.1 Application of SNFs in Hyperthermia

The term hyperthermia refers to a treatment in which heat is produced at the site of a tumor that causes the death of tumor cells. Hyperthermia can be localized or regional based on the location of the tumor in the body. Local hyperthermia is getting further attention because it includes the heating of tumor cells in the range of 42–46°C by using SNFs for heat generation [16, 17]. One of the important features of the tumor environment is that the temperature of the tumor site is a little bit high compared to normal cells. This characteristic feature is mainly because tumor cells are very active, growing, and repetitively dividing. Additionally, the body's immune response to leukocyte infiltration further causes inflammation, which also raises the local temperature. Hyperthermia implies an abnormal rise in the temperature in a fraction of a second with the help of SNFs. This increased temperature can cause the destruction of cancerous cells and ultimately achieve tumor regression. As the cancerous cells are responsive to hyperthermia, the cellular apoptosis occurs at around 42–46°C when compared to normal healthy cells. This rise in temperature causes adverse effects on surrounding cells. To overcome this, SNFs can be utilized for magnetic fluid hyperthermia (MFH). In MFH, suspension of SNFs is injected at the site of the tumor using either passive or active accumulation. Where active accumulation is achieved using ligands on NPs, and passive accumulation is achieved using enhanced permeation and retention (EPR) effects. In the tumor cells, SNFs having electromagnetic properties can modify electromagnetic energy directly into heat by means of the external magnetic field. This way, NPs will help the local temperature to rise without even interfering with the normal cells. Being a non-invasive technique, the smaller size of injected SNFs allows them to permit through biological barriers causing less toxicity [18, 19].

Various studies showed that tumor hyperthermia can be produced by means of the external heating source which causes two different effects as mentioned below [20, 21]:

a. In the first effect, an increase in the tumor temperature also improves the enhanced permeation retention effect by helping vascular permeability and increased accumulation of NPs inside the tumor.
b. Another effect begins with heating at a certain temperature which is the phase transition temperature of lipid, at which the lipid membrane or lipidic coating is permissible to water and several hydrophilic drugs to a greater extent. Thus, enhancing the drug accumulation at the tumor site from the prepared NPs.

For this reason, the most successful approach using SNFs for tumor killing is using magneto-hyperthermia. The use of SNFs with the help of an alternate magnetic field can generate heat in a localized way, without even harming the normal cells [22].

Over the last few decades, there has been an increase in the number of studies comprising SNFs for hyperthermia. Gordon et al., on the other hand, initially introduced a "new multidisciplinary intracellular biophysical treatment of cancer". In this scenario, tumor cells phagocytize colloidal magnetic submicron particles. When a high-frequency external electromagnetic field is applied, these particles induce heat. As a result, highly precise intracellular heat is generated [23].

An aqueous magnetic fluid was synthesized in 1993 by Chan et al., using modified dextran superparamagnetic iron oxide, and further study was carried out to evaluate their capability of absorbing energy from an external magnetic field with several strengths and frequencies [22, 24].

In the latest breakthroughs, Jae-Hyun and colleagues have created a core-shell composite consisting of $MnFe_2O_4$ (Shell) and $CoFe_2O_4$ (core) for the purpose of *in-vivo* hyperthermia treatment. Through an anticancer study conducted on mice, they confirmed higher efficacy with respect to conventional anticancer drugs like doxorubicin (DOX) [25]. Meanwhile, Tanmoy et al. evaluated the efficiency of tumor-targeted SPM NPs (superparamagnetic NPs) for magnetically induced hyperthermia treatment, particularly focusing on lung cancer. Their findings revealed significant inhibition of tumor progression via *in-vivo* hyperthermia treatment. Despite all the promising results of SNFs-induced hyperthermia, this approach has numerous unresolved issues. This includes the creation of a magnetic field having the desired strength and frequency that too within optimal limits, their relationship to treatment duration, toxicity reduction, and determining optimal concentration at the site of action their relationship to treatment duration, toxicity reduction, and determining optimal concentration at the site of action [22, 26].

14.2.2 Application of SNFs in Imaging

MRI stands as a highly effective medical imaging technique utilized for clinical diagnosis. By relying on the interaction between protons and biological tissues when exposed to an applied magnetic field, it provides a non-invasive approach. Specifically, MRI can employ various nuclei such as ^{1}H, ^{11}B, ^{13}C, ^{19}F, or ^{31}P as signal sources. However, the most commonly employed nucleus for clinical purposes is the water proton (^{1}H), owing to its exceptional sensitivity and abundance within biological tissues [27].

Contrast agents were originally designed to enhance the signal change, thereby refining the contrast between abnormal or diseased tissues (pathological lesions) and normal tissues in MRI scans. However, one major challenge with current MRI contrast agents is their lack of specificity, which means they may not accurately target the intended area. Along with this, these contrast agents often exhibit poor sensitivity, making it difficult to detect small targets like early-stage cancer with the desired precision. As a result, their effectiveness in such critical cases is limited. To address these limitations, SNFs have emerged as highly promising candidates as MRI contrast agents. SNFs possess exceptional magnetic properties that make them compatible with this role. Additionally, they offer the advantage of easy surface modification, allowing researchers to fine-tune their characteristics such as shape, size, and composition. This versatility makes them an attention-seeking and adaptable platform for enhancing MRI contrast. By utilizing SNFs as contrast agents, researchers hope to achieve more specific and sensitive MRI imaging, thereby improving the detection and visualization of small targets like early-stage cancer and enhancing the overall accuracy of MRI for clinical purposes [28–31].

SNFs have demonstrated their capability in detecting smaller pathological lesions, as small as 2–3 mm, in the liver and aiding in the visualization of brain tumors using MRI [32]. According to a study done by Young-Wook et al., the influence of SNF size on MRI signal was investigated. Primarily, the researchers synthesized nanocrystals with sizes of 4, 6, 9, and 12 nm, all having a uniform distribution. It was noticed that, when the size of the SNFs enlarged, there was a corresponding rise in magnetization saturation, ranging from 25 to 120 emu/(g Fe). This size-dependent magnetization phenomenon had a significant effect on enhancing the MRI signal, particularly under T2 contrast. The findings suggest that larger SNFS possess higher magnetization, leading to enhanced contrast in T2-weighted MRI images. This characteristic is particularly advantageous in improving the visibility and detection of pathological lesions, contributing to the overall sensitivity and accuracy of MRI in diagnosing various medical conditions [33, 34].

Zhenghuan and colleagues achieved successful *in-vivo* detection of liver cancer using two types of magnetic NPs octapod and spherical SNFs as high-performance T2 contrast agents. Both types of SNFs were capable of detecting liver cancer in living organisms. However, the octapod SNFs showed a distinct advantage in facilitating easier identification between normal and unhealthy tissues. The enhanced differentiation ability of octapod SNFs can be attributed to their unique structure, which likely provides stronger and more specific interactions with the cancerous tissues, resulting in a clearer contrast in the MRI images. This improved contrast helps medical professionals to accurately identify and distinguish cancerous lesions from healthy tissues in the liver, ultimately contributing to more precise and effective diagnosis and treatment planning for liver cancer patients [35].

Zhiguo and colleagues demonstrated the magnetic targeting capability of PEGylated $Fe@Fe_3O_4$ NPs in living organisms to enhance MRI contrast. This system was designed to improve the accuracy and specificity of MRI imaging by directing the NPs to the desired area using magnetic guidance. The PEGylation (polyethylene glycol coating) of the $Fe@Fe_3O_4$ NPs helps to prolong their circulation in the bloodstream and reduce clearance by the immune system, thereby increasing their potential for accumulation at the target site. Moreover, the study found that this same system exhibited high efficiency in performing near-infrared photo-thermal therapy. This means that when exposed to near-infrared light, the PEGylated $Fe@Fe_3O_4$ NPs were capable of converting the light energy into heat, which in turn, was utilized to selectively destroy cancerous cells or tumors. This combination of enhanced MRI contrast and near-infrared photo-thermal therapy capabilities makes the PEGylated $Fe@Fe_3O_4$ NPs a promising multifunctional platform for cancer imaging and therapy [36].

SNFs have been demonstrated to be crucial in serving as effective contrast agents in MRI. The successful utilization of SNFs as MRI contrast agents is well-established. Several iron oxide formulations have received clinical approval, and ongoing research continues to explore other potential candidates. Examples of clinically approved MRI contrast agents containing SNFs include Feridex®, Resovist®, Lumirem®, and Abdoscan®. These agents have been widely used in medical practice to enhance the contrast in MRI images,

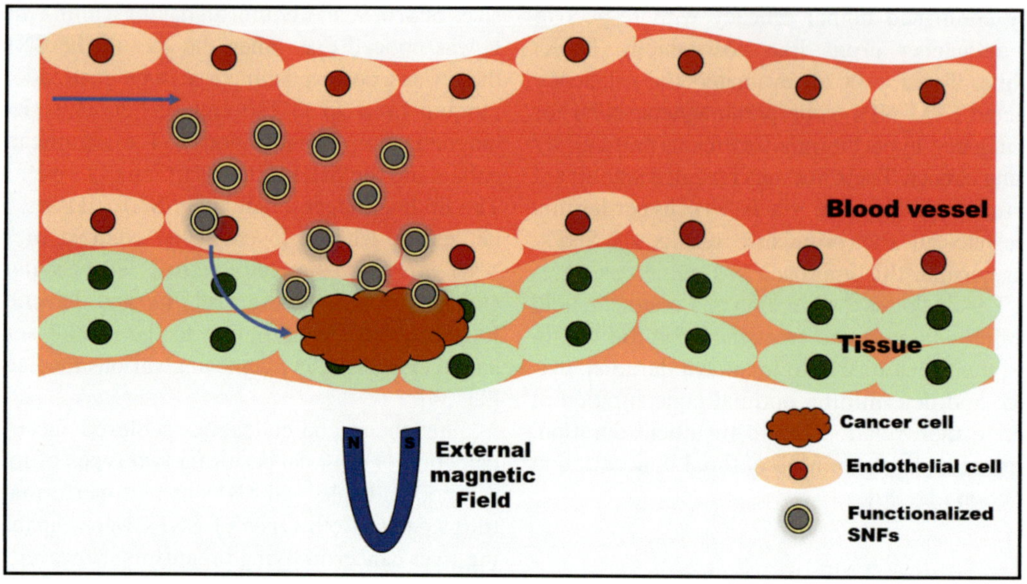

FIGURE 14.3 SNF-based targeted drug delivery to the cancerous cells.

aiding in the visualization and diagnosis of various medical conditions. Additionally, there are other preparations, like Feraheme® and Combidex®, which are presently completing phase II and phase III clinical trials, respectively. These trials are conducted to assess their safety, efficacy, and suitability for clinical use. If the results are positive, they may also join the list of approved MRI contrast agents based on SNFs. Overall, SNFs have demonstrated their significance as valuable tools in MRI, providing a wide range of contrast agents that contribute to improved medical imaging and diagnosis [37].

14.2.3 SNFs for Targeted Drug Delivery

A method of drug delivery in which the medicine is aimed at the site of action is known as targeted drug delivery. The site of action can be cells, an intracellular region, or a specific organ. Magnetic drug delivery is one such type of targeted delivery, wherein the external magnetic field is utilized to guide the drugs and drug carriers to the action site [38]. SNFs are the NPs widely used in magnetic drug delivery. The biocompatible SNFs possess many benefits like the capability to treat cancer, encapsulate cytotoxic drugs, drug delivery, release, and excretion from the body. A recent study, synthesis of Cobalt ferrite NPs, by Chaitali Dey and co-workers has proven all the mentioned benefits. The use of SNFs in drug delivery aids in avoiding systemic adverse effects and decreasing the required drug quantity [39, 40]. For targeting cancer cells, SNFs are generally used as a core and some biologically suitable organic compounds like dextran, chitosan, folate, and polyethylene glycol are used as shells. The core helps in guiding the anticancer drug by means of an exterior magnetic field and the shell is vital as stabilizing or surface modifying moiety which in turn enhances the targeting ability [41]. Additionally, gold is also used as a shell since it has the potential of shielding core SNFs from oxidation and ease of surface modification [42]. Figure 14.3 shows SNF-based targeted drug delivery to the cancerous cells.

For attracting the SNFs to a definite region, permanent magnets are used. Given below are the three mechanisms by which targeting is improved [43, 44]:

a. Concentration of SNFs is increased locally in the area where extravasation takes place due to leakage of microvasculature surrounding the tumor.
b. After extravasation occurs, the magnetic field present keeps the SNFs in the interstitium of a tumor. Avoiding drug escape against the osmotic pressure using a magnetic field is essential in magnetic targeting.
c. An increase in SNF concentration increases SNF uptake by cancerous cells from the interstitium.
d. Presence of active targeting ligands further improves uptake by the tumor cells from the interstitium.

In their research, Mu-Yi Hua and his colleagues presented the construction of a nanocarrier compound featuring high magnetization and superparamagnetism, created with an Fe_3O_4 core and a shell made of poly [N-(1-one-butyric acid)]. This compound was used for the delivery of drugs, specifically, DOX. Both in vivo and in vitro experiments were conducted using this DOX-linked drug delivery system. The findings indicated that it could be efficiently directed and focused on particular locations with the help of an external magnetic field of 0.4T. Additionally, an enhanced level of DNA crosslinking was noted, along with a drop in the IC_{50} of DOX on the cancer cells was tested, from 5.85 (unbound DOX) to 3.23 μM. This evidently proved the outcome of magnetism on the local concentration of the drug [45].

Nigam and his co-workers created magnetic NPs of Fe_3O_4, stabilized by citric acid and with a size ranging from 8 to 10 nm. These were dispersed in water to form a colloid. They investigated the ability of these NPs to serve as a drug delivery system using DOX hydrochloride as an exemplary drug [46].

In a particular study, Guo and colleagues engineered and produced magneto liposomes coated with carboxymethyl dextran (a combination of Fe_3O_4 and liposomes), which possessed a high capacity for loading the drug DOX. DOX release was found in a low-frequency alternating magnetic field (50 Hz, 15–45 mT). The team noted that the percentage release of DOX escalated as the intensity of the peripheral magnetic field grew, with the optimal release amount being achieved at 45 mT. A noteworthy aspect of this investigation is how the medication discharged under the presence of a low-frequency alternating magnetic field without causing a localized rise in temperature or heat. This efficient release of DOX is due to the structural changes in the liposomes caused by a low-frequency alternating magnetic field applied externally. This design has significant implications for efficient cancer cell treatment, as the local temperature rise can have severe adverse effects on vital organs like the brain, lungs, kidneys, and liver [47].

14.3 OTHER BIOMEDICAL APPLICATIONS OF SNFs

Apart from the uses of SNFs in hyperthermia, MRI, and drug delivery, they are used in other biomedical applications like the detection of glucose molecules, immunoassay, aptasensor and cancer diagnosis, and the detection of DNA.

14.3.1 Glucose Detection

The concentration of glucose in urine and blood is a key clinical sign of diabetes and is also a crucial part of the food industry. The method used to detect glucose involves its catalytic oxidation in the presence of an enzyme called glucose oxidase (GOx), which leads to the production of hydrogen peroxide (H_2O_2). The produced H_2O_2 is determined by colorimetric analysis using substrates like 4-aminoatipyrine-phenol, 3, 3′, 5, 5′ tetramethylbenzidine (TMB), n, n-diethyl-phenylenediamine sulfate (DPD), and o-phenylenediamine (OPD). Zinc ferrite ($ZnFe_2O_4$) magnetic NPs, which act as an analog for peroxidase, were employed for glucose detection in urine using colorimetry. The observable color shift was due to the oxidation of TMB. Cobalt ferrite NPs were utilized to detect glucose through a chemiluminescent reaction involving luminol. The reaction between luminol and hydrogen peroxide (H_2O_2) is catalyzed by these ferrite magnetic NPs and is connected with a glucose catalytic reaction that is initiated by GOx [48–50].

14.3.2 Immunoassay

An immunoassay is a bioanalytical technique that measures the concentration of an analyte by monitoring how it reacts with an antigen (the analyte itself) and an antibody. Surfaces of SNFs are often modified with groups' like –COOH, –NH2, and –SH to allow easy conjugation using the carboxyl group of an antigen or antibody. Manganese ferrite ($MnFe_2O_4$) NPs were produced and used as a fluorescence immunoassay (FIA) sensor for influenza virus detection using a comprehensive microfluidic system. The SNFs were covered with anti-influenza A nucleoprotein (a-A-NP) monoclonal antibodies (mAbs). Upon further altering the mAbs with R-phycoerythrin (PE), a fluorescent dye, attaching the mAb-SNFs complex to viral cells could be visualized effectively. The optical signal from the purified SNFs-virus-monoclonal antibody (mAb)-PE complexes was detected using an optical detection module. The outcomes demonstrated that, with dynamic mixing, this system could reach a lower limit of detection for the influenza virus providing high sensitivity [48, 51].

14.3.3 Cancer Diagnosis and Aptasensors

Aptamers are typically small strands of oligonucleotides, like ribonucleic acid (RNA) and single-stranded deoxyribonucleic acid (ssDNA), which are capable of binding to specific target molecules. Apta-sensors are a type of biosensors that use these aptamers as a recognition mechanism. Modifying the surface of SFNs with aptamers, which attach to antigens found on the desired cells, has paved the way for creating sensitive imaging probes that are targeted for precise medical diagnoses. Cobalt ferrite NPs were coated with a rhodamine B isothiocyanate fluorescence dye (MF) and embedded inside a silica shell matrix, then combined with a DNA aptamer (AS1411), which targets the nucleolin protein found on cancer cell membranes. This constructed MF-AS1411 particle was further labeled with Ga-citrate, resulting in the MFR-AS1411 nanoprobe. This purified nanoprobe exhibited specific fluorescence signals in C6 rat glioma cells that express nucleolin. A notable rise in the fluorescence signal was observed in the MF-AS1411-treated C6 cells, in comparison to cells treated with a diluted concentration of MF-AS1411. The MFR-AS1411 nanoprobe was successful in specifically targeting cancerous cells, as monitored through fluorescence and radionuclide imaging techniques [48, 52].

14.3.4 Detection of DNA Molecule

The interaction between DNA and SFNs is via synchronization of phosphate groups and the heterocyclic bases of DNA by means of metal ions on the surface of NP resulting in the creation of a biosensor composite material. This can help in identifying desired DNA molecules. A nanocomposite of nickel ferrite, graphene oxide, and chitosan (GO/NIF/ch) was synthesized and used to construct an electrochemical

TABLE 14.1
Published Patents

Sr. No.	Patent No.	Title	Reference
1	BR102018001969A2	Magnetic ceramics mn0.46cu0.54fe2o4 via solid-state reaction for application in drug transportation devices.	[56]
2	US11207348B2	Spinel ferrite-impregnated mesoporous silica containing a platinum complex.	[57]
3	US20060093555A1	Imaging inflammatory conditions using superparamagnetic iron oxide agents.	[58]
4	WO2006116403A2	Intracellular thermal ablation using nano-particle electron spin resonance heating.	[59]
5	US9327024B2	Polymer-coated ferrite fine particles and method for preparing polymer-coated ferrite fine particles.	[60]

DNA biosensor for quick detection of *Escherichia coli*. Differential pulse voltammetry (DPV) was employed to detect hybridization among the probe DNA (pDNA) and its complementary DNA (cDNA). Methylene blue was used as the redox indicator. This fabricated biosensor displayed excellent selectivity and sensitivity [48, 53].

Gu and his colleagues proposed a general approach to create monodisperse bifunctional magnetic NPs (BMNPs). These NPs can be used for the separation of proteins or the detection of pathogens [54].

14.3.5 Microfluidic Chip for Bioseparation

Earhart et al. have developed a magnetic segregation device – a small microfluidic chip densely packed with magnetic pores – intended to capture tumor cells. To test the effectiveness of the system, they used entire blood spiked with cancer cell lines. The device showed a high capture efficiency rate of 91.4% for the spiked cancer cells, indicating that it would be suitable for clinical trials. The magnetic device was subsequently tested on whole blood provided by non-small cell lung cancer patients (at the time of sample collection, patients had metastatic disease but weren't receiving chemotherapy at that time). The chip effectively separated and detected cancerous cells even in samples given by the patient. The results suggest that the SNFs, with the help of the bio-separation process, can be employed to purify and detect the desired analyte. However, as soon as the desired analyte is attached to the surface of the magnetic NPs, it's worth noting the ability to direct the analyte to any location [55–57].

14.3.6 Wound Healing

Bradshaw et al. established a technique to control the way and velocity of fibroblast and keratinocyte cell relocation by means of the external magnetic field. They created a polyethyleneimine functionalized-Fe_3O_4 and allowed these positively charged SNFs to be absorbed by the cells. After quick cellular association, they examined the relocation of the cells both with and without the existence of an external magnetic field (0.6T). Notably, cell relocation in the direction of the magnet was greater, demonstrating the efficiency of the SNFs in directing and controlling the cells. This finding holds significant implications for the field of biomedicine, as increasing the quantity of cells that attach to a target site could improve wound healing results [58–61].

14.4 PATENTS

Some published patents of SNFs are listed in Table 14.1.

14.5 CONCLUSION

In conclusion, SNFs hold immense potential for various applications in the biomedical field. When exhibiting superparamagnetic behavior, these ferrite materials become particularly valuable for biological uses like drug delivery, MRI, hyperthermia treatment, and magnetic separation. Achieving superparamagnetism in SNFs relies on precise size control during the synthesis process, making the choice of synthesis method a critical aspect of obtaining the desired properties. The unique multifunctional features of SNFs, including exceptional magnetic properties, large precise surface area, active surface sites, adjustable shape and size, chemical stability, and possibilities for functionalization, make them attractive candidates for biomedical applications. However, maintaining high magnetism at the nanoscale proves challenging as the size decreases. Therefore, careful engineering and consideration of factors such as shape, chemical and physical properties, and biocompatibility play pivotal roles in maximizing the effectiveness of nano-ferrites in biomedicine. By understanding the diverse attributes and capabilities of SNFs, researchers, and practitioners can harness their potential to revolutionize biomedical technologies and improve patient outcomes.

REFERENCES

1. T. N. Pham, T. Q. Huy and A. T. Le, "Spinel ferrite (AFe_2O_4)-based heterostructured designs for lithium-ion battery, environmental monitoring, and biomedical applications," *RSC. Adv.*, vol. 10, no. 52, pp. 31622–31661, 2020, doi: 10.1039/d0ra05133k.
2. K. Malaie and M. R. Ganjali, "Spinel nano ferrites for aqueous supercapacitors; linking abundant resources and low-cost processes for sustainable energy storage," *J. Energy Storage*, vol. 33, no. October 2020, p. 102097, 2021, doi: 10.1016/j.est.2020.102097.

3. W. S. Galvãoa, D. M. A. Netob, R. M. Freirec and P. B. A. Fechined, "Super-paramagnetic nanoparticles with spinel structure: A review of synthesis and biomedical applications," *Solid State Phenom.*, vol. 241, pp. 139–176, 2016, doi: 10.4028/www.scientific.net/SSP.241.139.
4. D. S. Mathew and R. S. Juang, "An overview of the structure and magnetism of spinel ferrite nanoparticles and their synthesis in microemulsions," *Chem. Eng. J.*, vol. 129, no. 1–3, pp. 51–65, 2007, doi: 10.1016/j.cej.2006.11.001.
5. K. E. Sickafus, J. M. Wills and N. W. Grimes, "Structure of spinel," *J. Am. Ceram. Soc.*, vol. 82, no. 12, pp. 3279–3292, 1999, doi: 10.1111/j.1151-2916.1999.tb02241.x.
6. M. A. Hakim, S. K. Nath, S. S. Sikder and K. H. Maria, "Cation distribution and electromagnetic properties of spinel type Ni-Cd ferrites," *J. Phys. Chem. Solids*, vol. 74, no. 9, pp. 1316–1321, 2013, doi: 10.1016/j.jpcs.2013.04.011.
7. P. Tartaj, M. Del Puerto Morales, S. Veintemillas-Verdaguer, T. González-Carreño, and C. J. Serna, "The preparation of magnetic nanoparticles for applications in biomedicine," *J. Phys. D. Appl. Phys.*, vol. 36, no. 13, 2003, doi: 10.1088/0022-3727/36/13/202.
8. A. Akbarzadeh, M. Samiei and S. Davaran, "Magnetic nanoparticles: Preparation, physical properties, and applications in biomedicine," *Nanoscale Res. Lett.*, vol. 7, p. 144, 2012, doi: 10.1186/1556-276X-7-144.
9. L. Huang, L. Ao, W. Wang, D. Hu, Z. Sheng and W. Su, "Multifunctional magnetic silica nanotubes for MR imaging and targeted drug delivery," *Chem. Commun.*, vol. 51, no. 18, pp. 3923–3926, 2015, doi: 10.1039/c4cc09382h.
10. B. H. Lai, C. H. Chang, C. C. Yeh and D. H. Chen, "Direct binding of Concanvalin A onto iron oxide nanoparticles for fast magnetic selective separation of lactoferrin," *Sep. Purif. Technol.*, vol. 108, pp. 83–88, 2013, doi: 10.1016/j.seppur.2013.02.020.
11. B. Ankamwar *et al.*, "Biocompatibility of Fe_3O_4 nanoparticles evaluated by in vitro cytotoxicity assays using normal, glia and breast cancer cells," *Nanotechnology*, vol. 21, no. 7, 2010, doi: 10.1088/0957-4484/21/7/075102.
12. D. Ling, N. Lee and T. Hyeon, "Chemical synthesis and assembly of uniformly sized iron oxide nanoparticles for medical applications," *Acc. Chem. Res.*, vol. 48, no. 5, pp. 1276–1285, 2015, doi: 10.1021/acs.accounts.5b00038.
13. B. Issa, I. M. Obaidat, B. A. Albiss and Y. Haik, "Magnetic nanoparticles: Surface effects and properties related to biomedicine applications," *Int. J. Mol. Sci.*, vol. 14, no. 11, pp. 21266–21305, 2013, doi: 10.3390/ijms141121266.
14. C. S. S. R. Kumar and F. Mohammad, "Magnetic nanomaterials for hyperthermia-based therapy and controlled drug delivery," *Adv. Drug Deliv. Rev.*, vol. 63, no. 9, pp. 789–808, 2011, doi: 10.1016/j.addr.2011.03.008.
15. K. M. Krishnan, "Biomedical nanomagnetics: A spin through possibilities in imaging, diagnostics, and therapy," *IEEE Trans. Magn.*, vol. 46, no. 7, pp. 2523–2558, 2010, doi: 10.1109/TMAG.2010.2046907.
16. A. Hervault and N. T. K. Thanh, "Magnetic nanoparticle-based therapeutic agents for thermo-chemotherapy treatment of cancer," *Nanoscale*, vol. 6, no. 20, pp. 11553–11573, 2014, doi: 10.1039/c4nr03482a.
17. M. Johannsen *et al.*, "Clinical hyperthermia of prostate cancer using magnetic nanoparticles: Presentation of a new interstitial technique," *Int. J. Hyperth.*, vol. 21, no. 7, pp. 637–647, 2005, doi: 10.1080/02656730500158360.
18. L. Li *et al.*, "Improved intratumoral nanoparticle extravasation and penetration by mild hyperthermia," *J. Control. Release*, vol. 167, no. 2, pp. 130–137, 2013, doi: 10.1016/j.jconrel.2013.01.026.
19. J. Chen *et al.*, "Thermosensitive liposomes with higher phase transition temperature for targeted drug delivery to tumor," *Int. J. Pharm.*, vol. 475, no. 1, pp. 408–415, 2014, doi: 10.1016/j.ijpharm.2014.09.009.
20. Y. Chen, Y. Chen, D. Xiao, A. Bose, R. Deng and G. D. Bothun, "Low-dose chemotherapy of hepatocellular carcinoma through triggered-release from bilayer-decorated magnetoliposomes," *Colloids Surf. B Biointerfaces*, vol. 116, pp. 452–458, 2014, doi: 10.1016/j.colsurfb.2014.01.022.
21. R. T. Gordon, J. R. Hines and D. Gordon, "Intracellular hyperthermia a biophysical approach to cancer treatment via intracellular temperature and biophysical alterations," *Med. Hypotheses*, vol. 5, no. 1, pp. 83–102, 1979, doi: 10.1016/0306-9877(79)90063-X.
22. D. C. F. Chan, D. B. Kirpotin and P. A. Bunn Jr., "Synthesis and evaluation of colloidal magnetic iron oxides for the site-specific radiofrequency-induced hyperthermia of cancer," *J. Magn. Magn. Mater.*, vol. 122, no. 1–3, pp. 374–378, 1993, doi: 10.1016/0304-8853(93)91113-L.
23. J. H. Lee *et al.*, "Exchange-coupled magnetic nanoparticles for efficient heat induction," *Nat. Nanotechnol.*, vol. 6, no. 7, pp. 418–422, 2011, doi: 10.1038/nnano.2011.95.
24. T. Sadhukha, T. S. Wiedmann and J. Panyam, "Inhalable magnetic nanoparticles for targeted hyperthermia in lung cancer therapy," *Biomaterials*, vol. 34, no. 21, pp. 5163–5171, 2013, doi: 10.1016/j.biomaterials.2013.03.061.
25. N. Lee and T. Hyeon, "Designed synthesis of uniformly sized iron oxide nanoparticles for efficient magnetic resonance imaging contrast agents," *Chem. Soc. Rev.*, vol. 41, no. 7, pp. 2575–2589, 2012, doi: 10.1039/c1cs15248c.
26. D. Kozlowska, P. Foran, P. MacMahon, M. J. Shelly, S. Eustace and R. O'Kennedy, "Molecular and magnetic resonance imaging: The value of immunoliposomes," *Adv. Drug Deliv. Rev.*, vol. 61, no. 15, pp. 1402–1411, 2009, doi: 10.1016/j.addr.2009.09.003.
27. T. H. Shin, Y. Choi, S. Kim and J. Cheon, "Recent advances in magnetic nanoparticle-based multi-modal imaging," *Chem. Soc. Rev.*, vol. 44, no. 14, pp. 4501–4516, 2015, doi: 10.1039/c4cs00345d.
28. C. Sun, J. S. H. Lee and M. Zhang, "Magnetic nanoparticles in MR imaging and drug delivery," *Adv. Drug Deliv. Rev.*, vol. 60, no. 11, pp. 1252–1265, 2008, doi: 10.1016/j.addr.2008.03.018.
29. J. Wang *et al.*, "Magnetic nanoparticles for MRI of brain tumors," *Curr. Pharm. Biotechnol.*, vol. 13, no. 12, pp. 2403–2416, 2012, doi: 10.2174/138920112803341824.
30. N. C. Balci and R. C. Semelka, "Contrast agents for MR imaging of the liver," *Radiol. Clin. North Am.*, vol. 43, no. 5, pp. 887–898, 2005, doi: 10.1016/j.rcl.2005.05.004.
31. T. P. Murillo, C. Sandquist, P. M. Jacobs, G. Nesbit, S. Manninger and E. A. Neuwelt, "Imaging brain tumors with ferumoxtran-10, a nanoparticle magnetic resonance contrast agent," *Therapy*, vol. 2, no. 6, pp. 871–882, 2005, doi: 10.1586/14750708.2.6.871.
32. Y. W. Jun *et al.*, "Nanoscale size effect of magnetic nanocrystals and their utilization for cancer diagnosis via magnetic resonance imaging," *J. Am. Chem. Soc.*, vol. 127, no. 16, pp. 5732–5733, 2005, doi: 10.1021/ja0422155.
33. Z. Zhao *et al.*, "Octapod iron oxide nanoparticles as high-performance T_2 contrast agents for magnetic resonance imaging," *Nat. Commun.*, vol. 4, pp. 1–7, 2013, doi: 10.1038/ncomms3266.
34. Z. Zhou *et al.*, "Iron/iron oxide core/shell nanoparticles for magnetic targeting MRI and near-infrared photothermal therapy," *Biomaterials*, vol. 35, no. 26, pp. 7470–7478, 2014, doi: 10.1016/j.biomaterials.2014.04.063.

35. A. J. Cole, V. C. Yang and A. E. David, "Cancer theranostics: The rise of targeted magnetic nanoparticles," *Trends Biotechnol.*, vol. 29, no. 7, pp. 323–332, 2011, doi: 10.1016/j.tibtech.2011.03.001.
36. Y.-X. J. Wang, "Superparamagnetic iron oxide based MRI contrast agents: Current status of clinical application," *Quant. Imaging Med. Surg.*, vol. 1, no. 1, pp. 35–40, 2011, doi: 10.3978/j.issn.2223-4292.2011.08.03.
37. E. Lueshen, I. Venugopal, T. Soni, A. Alaraj and A. Linninger, "Implant-assisted intrathecal magnetic drug targeting to aid in therapeutic nanoparticle localization for potential treatment of central nervous system disorders," *J. Biomed. Nanotechnol.*, vol. 11, no. 2, pp. 253–261, 2015, doi: 10.1166/jbn.2015.1907.
38. C. Dey, K. Baishya, A. Ghosh, M. M. Goswami, A. Ghosh and K. Mandal, "Improvement of drug delivery by hyperthermia treatment using magnetic cubic cobalt ferrite nanoparticles," *J. Magn. Magn. Mater.*, vol. 427, pp. 168–174, 2017, doi: 10.1016/j.jmmm.2016.11.024.
39. F. Valente *et al.*, "Nanoparticle drug delivery systems for inner ear therapy: An overview," *J. Drug Deliv. Sci. Technol.*, vol. 39, pp. 28–35, 2017, doi: 10.1016/j.jddst.2017.03.003.
40. M. Hałupka-Bryl, K. Asai, S. Thangavel, M. Bednarowicz, R. Krzyminiewski and Y. Nagasaki, "Synthesis and in vitro and in vivo evaluations of poly(ethylene glycol)-block-poly(4-vinylbenzylphosphonate) magnetic nanoparticles containing doxorubicin as a potential targeted drug delivery system," *Colloids Surf. B Biointerfaces*, vol. 118, pp. 140–147, 2014, doi: 10.1016/j.colsurfb.2014.03.025.
41. S. Karamipour, M. S. Sadjadi and N. Farhadyar, "Fabrication and spectroscopic studies of folic acid-conjugated Fe_3O_4@Au core-shell for targeted drug delivery application," *Spectrochim. Acta – Part A Mol. Biomol. Spectrosc.*, vol. 148, pp. 146–155, 2015, doi: 10.1016/j.saa.2015.03.078.
42. C. Voutouri and T. Stylianopoulos, "Evolution of osmotic pressure in solid tumors," *J. Biomech.*, vol. 47, no. 14, pp. 3441–3447, 2014, doi: 10.1016/j.jbiomech.2014.09.019.
43. N. Schleich *et al.*, "Comparison of active, passive and magnetic targeting to tumors of multifunctional paclitaxel/SPIO-loaded nanoparticles for tumor imaging and therapy," *J. Control. Release*, vol. 194, pp. 82–91, 2014, doi: 10.1016/j.jconrel.2014.07.059.
44. M. Y. Hua *et al.*, "Superhigh-magnetization nanocarrier as a doxorubicin delivery platform for magnetic targeting therapy," *Biomaterials*, vol. 32, no. 34, pp. 8999–9010, 2011, doi: 10.1016/j.biomaterials.2011.08.014.
45. S. Nigam, K. C. Barick and D. Bahadur, "Development of citrate-stabilized Fe3O4 nanoparticles: Conjugation and release of doxorubicin for therapeutic applications," *J. Magn. Magn. Mater.*, vol. 323, no. 2, pp. 237–243, 2011, doi: 10.1016/j.jmmm.2010.09.009.
46. H. Guo, W. Chen, X. Sun, Y. N. Liu, J. Li and J. Wang, "Theranostic magnetoliposomes coated by carboxymethyl dextran with controlled release by low-frequency alternating magnetic field," *Carbohydr. Polym.*, vol. 118, pp. 209–217, 2015, doi: 10.1016/j.carbpol.2014.10.076.
47. N. Chaibakhsh and Z. Moradi-Shoeili, "Enzyme mimetic activities of spinel substituted nanoferrites (MFe_2O_4): A review of synthesis, mechanism and potential applications," *Mater. Sci. Eng. C*, vol. 99, no. February, pp. 1424–1447, 2019, doi: 10.1016/j.msec.2019.02.086.
48. L. Su, J. Feng, X. Zhou, C. Ren, H. Li and X. Chen, "Colorimetric detection of urine glucose based $ZnFe_2O_4$ magnetic nanoparticles," *Anal. Chem.*, vol. 84, no. 13, pp. 5753–5758, 2012, doi: 10.1021/ac300939z.
49. W. Shi, X. Zhang, S. He and Y. Huang, "$CoFe_2O_4$ magnetic nanoparticles as a peroxidase mimic mediated chemiluminescence for hydrogen peroxide and glucose," *Chem. Commun.*, vol. 47, no. 38, pp. 10785–10787, 2011, doi: 10.1039/c1cc14300j.
50. L. Y. Hung *et al.*, "Magnetic nanoparticle-based immunoassay for rapid detection of influenza infections by using an integrated microfluidic system," *Nanomed. Nanotechnol. Biol. Med.*, vol. 10, no. 4, pp. 819–829, 2014, doi: 10.1016/j.nano.2013.11.009.
51. D. W. Hwang *et al.*, "A nucleolin-targeted multimodal nanoparticle imaging probe for tracking cancer cells using an aptamer," *J. Nucl. Med.*, vol. 51, no. 1, pp. 98–105, 2010, doi: 10.2967/jnumed.109.069880.
52. I. Tiwari, M. Singh, C. M. Pandey and G. Sumana, "Electrochemical detection of a pathogenic *Escherichia coli* specific DNA sequence based on a graphene oxide-chitosan composite decorated with nickel ferrite nanoparticles," *RSC Adv.*, vol. 5, no. 82, pp. 67115–67124, 2015, doi: 10.1039/c5ra07298k.
53. H. Gu, K. Xu, C. Xu and B. Xu, "Biofunctional magnetic nanoparticles for protein separation and pathogen detection," *Chem. Commun.*, no. 9, pp. 941–949, 2006, doi: 10.1039/b514130c.
54. C. M. Earhart *et al.*, "Isolation and mutational analysis of circulating tumor cells from lung cancer patients with magnetic sifters and biochips," *Lab Chip*, vol. 14, no. 1, pp. 78–88, 2014, doi: 10.1039/c3lc50580d.
55. M. Bradshaw *et al.*, "Manipulating directional cell motility using intracellular superparamagnetic nanoparticles," *Nanoscale*, vol. 7, no. 11, pp. 4884–4889, 2015, doi: 10.1039/c4nr06594h.
56. R. D. S. Manoel, D. D. O. Valesca and O. Adhimar, "Magnetic ceramics mn0.46cu0.54fe2o4 via solid state reaction for application in drug transportation devices," BR102018001969A2, 2020 [Online]. Available: https://patents.google.com/patent/BR102018001969A2/en?q=(spinel+nanoferrites+biomedical+applications)&oq=spinel+nanoferrites+biomedical+applications
57. J. R. B. R. Vijaya and B. Abdulhadi, "Spinel ferrite-impregnated mesoporous silica containing a platinum complex," US11207348B2, 2021 [Online]. Available: https://patents.google.com/patent/US11207348B2/en?q=(spinel+nanoferrites+biomedical+applications)&oq=spinel+nanoferrites+biomedical+applications
58. T. Andrew, B. Brian, K. Amit, M. Matthew and D. William, "Imaging inflammatory conditions using superparamagnetic iron oxide agents," US20060093555A1, 2006 [Online]. Available: https://patents.google.com/patent/US20060093555A1/en?q=(spinel+nanoferrites+MRI)&oq=spinel+nanoferrites+MRI
59. X. Xiao and Y. Haitao, "Intracellular thermal ablation using nano-particle electron spin resonance heating," WO2006116403A2, 2007 [Online]. Available: https://patents.google.com/patent/WO2006116403A2/en
60. H. Hiroshi, K. Hiroshi and S. Satoshi, "Polymer-coated ferrite fine particles and method for preparing polymer coated ferrite fine particles," US9327024B2, 2016 [Online]. Available: https://patents.google.com/patent/US9327024B2/en
61. P. Thakur, S. Taneja, D. Chahar, B. Ravelo and A. Thakur, "Recent advances on synthesis, characterization and high frequency applications of Ni-Zn ferrite nanoparticles," *J. Magn. Magn. Mater.*, vol. 530, p. 167925, 2021.

15 A New Approach

Spinel Nano-Ferrites for Wastewater-Based Microplastics

*Jyoti Bhattacharjee and Subhasis Roy**
University of Calcutta, Kolkata, India

15.1 INTRODUCTION

Plastic made from petroleum resources has become essential to human life. Microplastics (MPs) are tiny plastic fragments less than 5 millimeters in size that have become a significant source of pollution in the environment. These particles, which come from various sources, including plastic litter breakdown and shedding microfibers from textiles, can reach aquatic habitats via wastewater discharge. Innovative methods are needed to address MP pollution in wastewater properly. Spinel nano-ferrites, a family of magnetic compounds, have shown potential in removing various contaminants from water due to their unique qualities, such as high surface area, magnetic behavior, and chemical stability. Using spinel nano-ferrites as adsorbents for MPs in wastewater is a novel strategy that might offer efficient and cost-effective alternatives [1]. In this regard, Smith, J., et al. describe the synthesis of spinel nano-ferrites, including cobalt ferrite and manganese ferrite, and their application for MP removal from wastewater in 2022. The study assesses adsorption capability, magnetic characteristics, and reusability of ferrites [2]. "Magnetic spinel nano-ferrites for the removal of microplastics: A review" by Garcia et al. in 2021 gives an overview of recent improvements in the application of magnetic spinel nano-ferrites for MPs removal. The authors compare the performance of several synthesis processes, such as co-precipitation and hydrothermal procedures, regarding adsorption capacity and selectivity [3]. Wang et al. in 2020 investigate the surface modification of spinel nano-ferrites by functionalizing them with organic ligands to improve MP removal efficiency. The researchers created manganese ferrite nanoparticles and covered them with a thin layer of polydopamine [4]. Chen et al. in 2023 investigated the synthesis of co-doped $ZnFe_2O_4$ nano-ferrites using a co-precipitation method and analyzed their effectiveness for MP removal from wastewater. The study investigates the effect of varied doping doses on the adsorption efficiency of nano-ferrites [5]. The waste plastic bags and sheets thrown into rivers and streams jeopardize water systems, block water pumps, and kill nearly 1,00,000 birds, whales, and sea turtles annually. The number of publications and citations on spinel nano-ferrites and their effect on the environment is shown in Figure 15.1.

15.1.1 Significance of This Chapter

Sustainable synthesis is the development of environmentally friendly and economically viable synthesis technologies. In this context, using green synthesis methodologies to create spinel nano-ferrites for MP remediation has multiple benefits. Using spinel nano-ferrites, effective adsorbents that selectively collect tiny particles from effluent can be developed, resulting in their removal and subsequent reduction in environmental contamination [6]. Plant extracts and biological entities, such as bacteria, fungi, and algae, can be used as ecologically friendly reducing and capping agents in manufacturing spinel-like micro-ferrites. This approach combines a metal salt solution with a plant extract containing natural chemicals with reducing capabilities. The plant extract aids in the reduction of metal ions and the subsequent creation of nanoparticles. The technique discussed here uses biodegradable resources and eliminates hazardous chemicals [7]. Brown et al. in 2022 explored the catalytic breakdown of MPs utilizing nickel ferrite nanoparticles in their study work. The study found that when nickel ferrite nanoparticles were coupled with hydrogen peroxide, they exhibited extraordinary catalytic activity, rapidly destroying

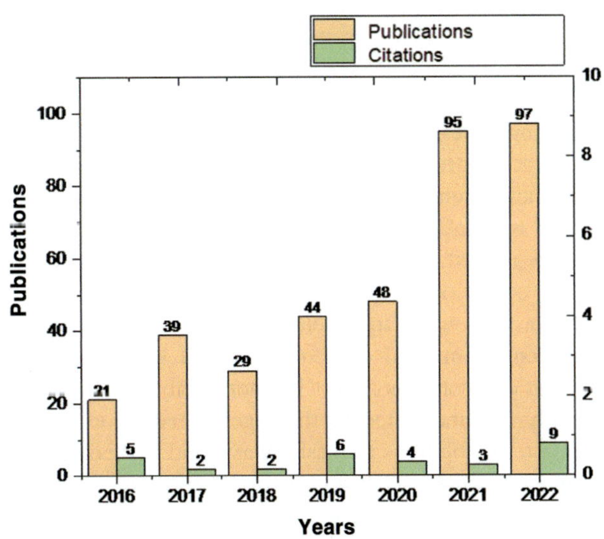

FIGURE 15.1 Number of publications and citations vs. years. (Data input from ScienceDirect.com.)

* Corresponding author.

DOI: 10.1201/9781003568940-15

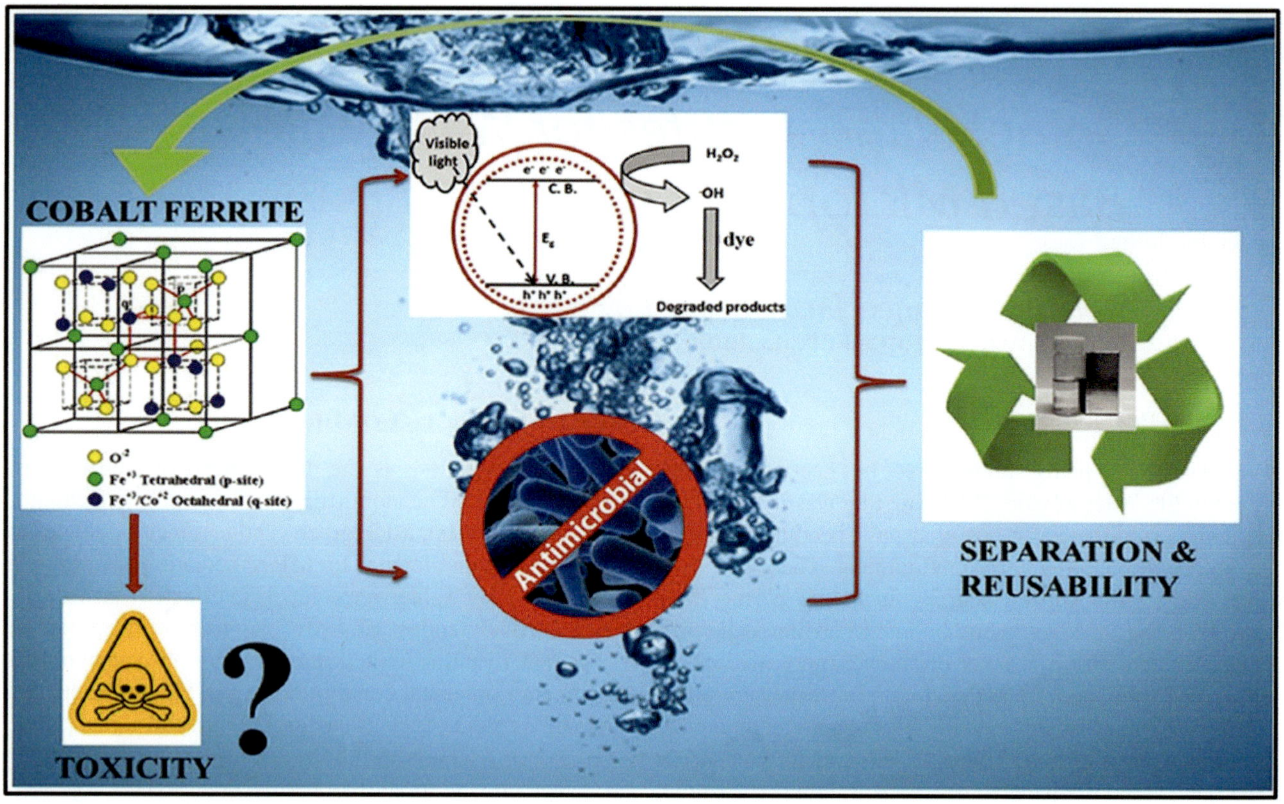

FIGURE 15.2 Cobalt ferrites nanocomposite employed for water treatment. (Reproduced from [8] with permission from Elsevier vide License number 5876940183953.)

MP particles. The toxicity, application, and reusability of cobalt spinel macro-ferrites are shown in Figure 15.2.

15.1.2 Research Gap and Motivation for the Study

The research gap is due to the limited investigation of green synthesis approaches for creating functional materials capable of efficiently removing nano-plastics and resins from wastewater. Sustainable synthesis approaches emphasize using sustainable, eco-friendly materials and processes, minimizing adverse ecological repercussions. Establishing an innovative green synthesis process for spinel nano-ferrites can potentially overcome the constraints of conventional restoration strategies [9–11]. Further research is needed to understand the removal mechanisms, appropriate synthesis conditions, and scale-up potential of these nano-ferrites for wastewater treatment. Furthermore, these nano-ferrites' environmental consequences and long-term fate in aquatic environments must be thoroughly investigated. The study was motivated by the urgent need for effective and long-term solutions to plastic waste and MP pollution. The green production of spinel nano-ferrites in conjunction with Density Functional Theory (DFT) programs is a promising alternative to traditional environmental remediation methods [12]. DFT can provide useful insights into key components of the synthesis process, such as energetics, stability, and reaction processes. It can assist in optimizing the production conditions, analyze the interactions between nanomaterials and polymeric contaminants, and anticipate the synthesized materials' performance for wastewater treatment. It is hoped that by integrating the ecological synthesis of spinel nano-ferrites with a DFT approach, efficient, innovative, and renewable resources for environmental preservation would be created. Figure 15.3 depicts the

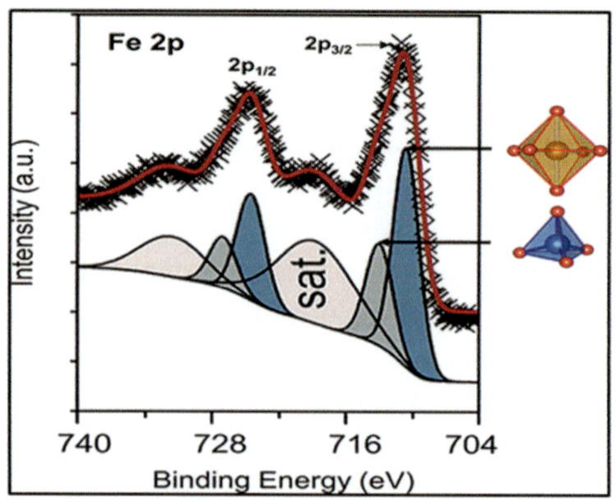

FIGURE 15.3 Intensity vs. binding energy for spinel ferrite nanocrystals. (Reprinted with permission from Sanchez-Lievanos et al. [11]. Copyright [2021] American Chemical Society.)

characterization, cation distribution, physical properties, and opportunities of spinel ferrites.

15.2 MICRO-PLASTICS AND ENVIRONMENTAL IMPACT

Polyester and MPs have a substantial environmental impact, especially on marine ecosystems. The following are some of the origins and ecological effects of plastic debris and nano-plastics:

1. Plastic bags, straws, bottles, spoons, disposable silverware, wrappers, containers, and trays for food and packaging are all examples of single-use plastics that contribute considerably to plastic pollution. These products are frequently utilized for a short period but remain in the environment for hundreds of years.
2. *Synthetic fibers:* When synthetic garment fabrics, such as polyester and nylon, are laundered, they shed microfibers. These microfibers penetrate wastewater systems and eventually reach rivers, lakes, and seas, where marine life might consume them.
3. Cosmetic packaging includes plastic bottles, tubes, and containers for lotions, creams, and makeup.
4. *Microbeads*: Tiny plastic particles used in exfoliating scrubs, toothpaste, and body wash [13].
5. *Fishing and Aquaculture:* Fishing equipment: Nets, lines, and ropes composed of synthetic materials that degrade into MPs. Oyster bags: Plastic mesh bags used in oyster aquaculture.
6. *Industrial Sources:* Plastic pellets: small plastic resin pellets are used as raw materials to manufacture plastic items. Plastic film is used in packaging, agricultural mulching, and industrial applications [14].
7. *PVC pipes:* Polyvinyl chloride (PVC) pipes are used in plumbing and building. Polystyrene foam boards are used as insulation. Vehicle tire wear and tear emits MP particles into the environment.

Plastic debris and polymers have serious impacts on the planet, particularly on habitats and wildlife:

1. Plastics and MPs contribute significantly to marine contamination. They infiltrate rivers and oceans through various channels, including improper waste disposal, littering, and industrial activity. Plastic fragments are frequently mistaken for food by marine species, resulting in internal damage, malnutrition, and death. Plastics can also entangle marine species, inflicting injuries, reducing their ability to swim, and negatively impacting their overall health [15]. Plastics and MPs can potentially upset the delicate balance of aquatic ecosystems. They can impact the growth and development of phytoplankton and algae, which are necessary for the proper functioning of marine food chains. MPs can also be collected in sediments, influencing the properties and reproduction of aquatic creatures such as fish, shellfish, and crabs [16].
2. *Bioaccumulation:* Polyester and tiny particles have the potential to accumulate in the food chain. When smaller species consume MPs, they may be consumed by larger predators, causing the plastics to bioaccumulate in their tissues. This can result in increased amounts of hazardous compounds connected with plastics, such as polychlorinated biphenyls (PCBs) and phthalates, in the bodies of these species [17]. Polypropylene and MPs can impact ecosystem services, critical to human well-being. These can, for instance, disrupt the nutrient cycle, affect water quality, and diminish insect pollination efficiency, significantly affecting agriculture and food supply.

A.A. Koelmans et al. (2019) – "The effects of microplastics on freshwater fish" investigates the effects of MPs on freshwater fish species. It analyzes MP uptake and deposition in fish tissues and the potential physiological, behavioral, and reproductive repercussions on fish populations [18]. Plastics and MPs that get airborne can also contribute to air pollution. Larger plastic items that break down into tiny particles can be carried by wind currents and deposited across long distances. Humans and animals may inhale these airborne MPs, causing respiratory troubles and other health problems. Plastics contain various chemical compounds, including phthalates and bisphenol A (BPA), which can be detrimental to organisms. In both animals and humans, they can alter hormonal balances, affect reproductive systems, and induce developmental defects [19]. Plastics and MPs can accumulate and magnify in the food chain, known as bioaccumulation and biomagnification. Little organisms consume MPs; when larger animals consume them, they accumulate in their tissues. Figure 15.4a depicts the MP breakdown into smaller MPs in seas. Figure 15.4b portrays the surface area to the bulk ratio of particles; Figure 15.4c illustrates the potential surface area values obtained by assuming that huge waste plastic is broken down into 1-micrometer-sized MP bits.

15.3 SPINEL NANO-FERRITES FOR WASTEWATER TREATMENT

Nano-ferrites, also known as nanoscale ferrite materials, are a type of magnetic material with unique features at the nanoscale level. Ferrites are ceramic materials of iron oxide mixed with other metal oxides such as nickel oxide (NiO) or zinc oxide (ZnO). When these ferrite materials are synthesized at the nanoscale, they have a high surface-to-volume ratio, resulting in improved magnetic, electrical,

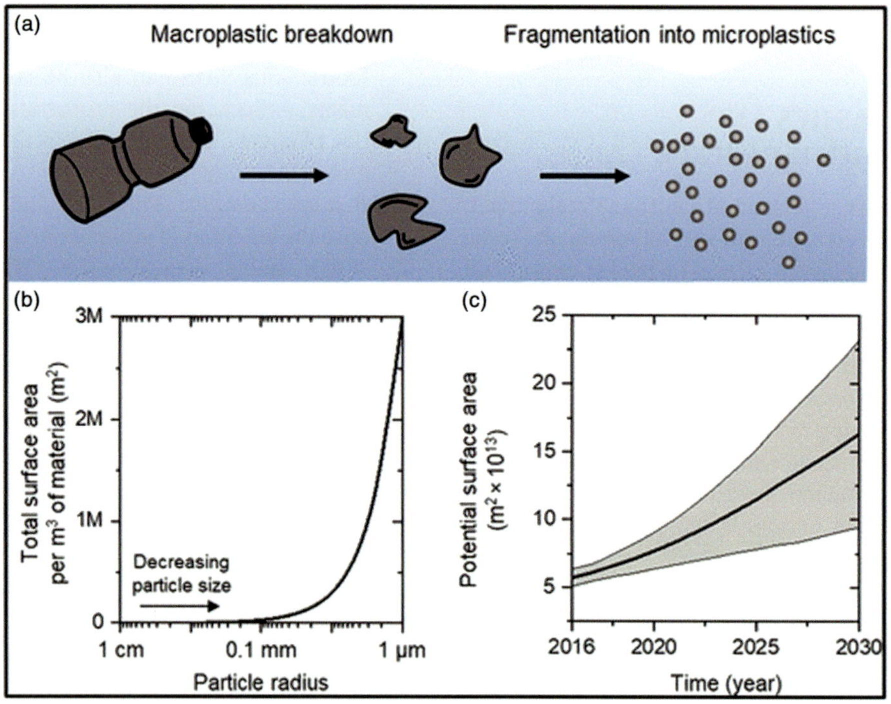

FIGURE 15.4 (a–c) Schematic presentation of interparticle interactions of plastic particles [20, 21].

and optical properties compared to their bulk counterparts. Nanoparticles typically range in size from 1 to 100 nanometers [22].

1. *Magnetic Properties:* Nano-ferrites have outstanding magnetic properties, including high saturation magnetization, coercivity, and remanence. These features make them appropriate for various magnetic purposes, such as magnetic data preservation, sensors, and electromagnetic interference (EMI) shielding.
2. Micro-ferrites can be synthesized utilizing various processes, including co-precipitation, sol-gel synthesis, hydrothermal reaction, and ball milling. These approaches enable exact control over the micron-sized particles' size, shape, and content, altering their attributes.
3. *Data Storage:* Nano-ferrites are used to produce high-density magnetic storage media such as hard disc drives (HDDs) and magnetic tapes. Nano-ferrites are being investigated for potential in targeted medication administration, hyperthermia therapy (using magnetic heating), and magnetic resonance imaging (MRI) contrast agents due to their biocompatibility and magnetic characteristics [23].
4. *Catalysts:* Nano-ferrites can be used as catalysts in various chemical reactions, including wastewater treatment, hydrogen production, and organic synthesis.
5. *Microwave Devices:* Due to their superior high-frequency magnetic characteristics, nano-ferrites find use in microwave devices such as circulators, isolators, and phase shifters. Due to their sensitivity, selectivity, and high electrical conductivity, nano-ferrites are used to construct magnetic sensors, such as magnetic field sensors, gas sensors, and biosensors [24].
6. *Nickel Ferrite Nanoparticles* ($NiFe_2O_4$): Nickel ferrite nanoparticles have outstanding magnetic and electrical properties, making them valuable in electronics, sensors, and EMI shielding. Yang et al. (2018) synthesized and integrated nickel ferrite nanoparticles into a polymer matrix to generate high-performance electromagnetic wave absorbers. Zinc ferrite nanoparticles, due to their optical, magnetic, and electrical capabilities, zinc ferrite nanoparticles have gained attention. Magneto-optical devices, photocatalysis, and gas sensors all use them. Researchers reported in 2022 on the sol-gel production of zinc ferrite nanoparticles and examine their photocatalytic activity for dye degradation [25].
7. *Iron oxide nanoparticles* (Fe_3O_4): Iron oxide nanoparticles, such as magnetite (Fe_3O_4), are among the most thoroughly investigated nano-ferrites. They contain excellent magnetic characteristics and have found applications in MRI, medication delivery, hyperthermia treatment, and environmental remediation. For instance, Wahajuddin and

Arora in 2012 address the synthesis methods, surface changes, and biomedical uses of iron oxide nanoparticles [26].

8. *Cobalt ferrite nanoparticles* ($CoFe_2O_4$): Cobalt ferrite nanoparticles have excellent saturation magnetization and chemical stability, making them appropriate for applications such as magnetic recording medium, microwave devices, and catalysis [27, 28]. Shinde et al. in 2020 describe the creation of cobalt ferrite nanoparticles using several processes and their prospective applications in wastewater treatment.
9. *Synthesis methods:* Several synthetic methods have been devised to create spinel nano-ferrites with controlled size, shape, and composition.
   - 9.1. *Co-precipitation method*: In this approach, metal salts of the appropriate composition (A and B) are dissolved in a suitable solvent. The pH of the solution is changed to promote the co-precipitation of metal ions as hydroxides. The resultant residue is then heated to form spinel nano-ferrites.
   - 9.2. The sol-gel approach involves the creation of a sol from metal alkoxides or metal salts dissolved in a suitable solvent. The sol is subsequently gelated, often through hydrolysis and condensation processes. The gel is then dried and calcined to obtain spinel nano-ferrite nanoparticles [28].
   - 9.3. *Microwave-assisted synthesis*: This method accelerates the production of spinel nano-ferrites by using microwave irradiation. The metal precursor solution is heated quickly and efficiently by microwave radiation. This approach enables faster reaction kinetics and the creation of spinel nano-ferrites with controlled size and morphology.
   - 9.4. *Reaction in the solid state*: Finely powdered metal oxides or salts of A and B elements are fully combined in this procedure before being exposed to high-temperature calcination. Spinel nano-ferrite nanoparticles are formed because of the solid-state process. Particle size and morphology can be controlled by altering the reaction temperature, time, and milling parameters [29].
   - 9.5. *Microemulsion method*: This approach employs a microemulsion system composed of water, oil, surfactant, and co-surfactant. Metal precursors are dissolved or distributed in the appropriate microemulsion phase. Spinel nano-ferrite nanoparticles are formed due to further heat treatment or chemical reduction.
   - 9.6. *Solvothermal aerosol route*: An aerosol of metal precursors is created and delivered into a high-temperature, high-pressure reactor containing a solvent in this process. The aerosol reacts with the solvent medium, forming spinel nano-ferrite particles. This technology can precisely control particle size and composition [30].
   - 9.7. *Plant Extracts*: Plant extracts have received a lot of interest as reducing and capping agents for spinel nano-ferrite synthesis. Natural polyphenols and other organic substances found in plants such as green tea, aloe vera, neem, and grape seed extracts can act as reducing agents. Spinel nano-ferrites are created by combining plant extracts with metal salts and heating them under regulated conditions. To create spinel nano-ferrites, bioinspired synthesis requires mimicking biological processes. Biomineralization, inspired by mineral formation in live organisms, can synthesize nano-ferrites by integrating biologically generated templates or biomolecules that act as catalysts or structural guides throughout the synthesis process [31]. "Microwave-Assisted Synthesis of Spinel Nano-ferrites for Drug Delivery Applications" by Khan et al. in 2023 demonstrates the microwave-assisted synthesis of spinel nano-ferrites for drug delivery systems. The quick and effective microwave irradiation synthesis approach yields well-dispersed nano-ferrites with regulated particle size. The synthesized nano-ferrites' magnetic characteristics and drug-loading capability are examined, revealing their potential in targeted drug delivery applications [32].

Characterization techniques for spinel nano-ferrites involve various methods to study their structural, morphological, magnetic, and optical properties. Various commonly used procedures are highlighted below:

X-ray Diffraction (*XRD*): XRD determines spinel nano-ferrite crystal structure and phase purity. It contains data on lattice characteristics, crystal symmetry, and phase identification [32].

Transmission Electron Microscopy (*TEM*): TEM provides for high-resolution direct imaging of spinel nano-ferrites. It contains information regarding particle size, shape, morphology, lattice defects, and crystal structure ordering.

FTIR (*Fourier Transform Infrared Spectroscopy*): FTIR investigates spinel nano-ferrites' vibrational modes and chemical bonding. It can offer information on nanostructure composition, functional groups, and surface modifications.

Vibrating Sample Magnetometry (*VSM*): VSM is a technique for measuring the magnetic properties of spinel nano-ferrites, such as magnetization, magnetic moment, and coercivity [33].

UV-Vis/NIR Spectroscopy: This method is used to study the optical properties of spinel nano-ferrites. It describes the absorption and reflection properties in the ultraviolet, visible, and near-infrared ranges, which can be connected to bandgap energy and electronic transitions [34].

X-ray Photoelectron Spectroscopy (*XPS*): This technique determines spinel nano-ferrites' elemental composition and chemical states. It describes the nanoparticles' oxidation states and surface chemistry [35]. Raman spectroscopy investigates the vibrational modes and lattice vibrations in spinel nano-ferrites. It describes the crystal structure, symmetry, and bonding environment of nanoparticles.

Mössbauer spectroscopy is a specialized technique to explore spinel nano-ferrites' magnetic characteristics and hyperfine interactions. It describes the oxidation state, magnetic ordering, and local surroundings of iron atoms.

SEM (*Scanning Electron Microscopy*): This technique provides information on spinel-like surface morphology and topography of ferrites. It can also provide information regarding particle size distribution and agglomeration behavior.

TGA (*Thermogravimetric Analysis*): TGA is used to investigate spinel nano-ferrites' thermal stability and decomposition properties. It can provide information on the material's weight loss, phase transitions, and temperature stability [36]. The various characterization prospect of nano spinel ferrites is shown in Figure 15.5.

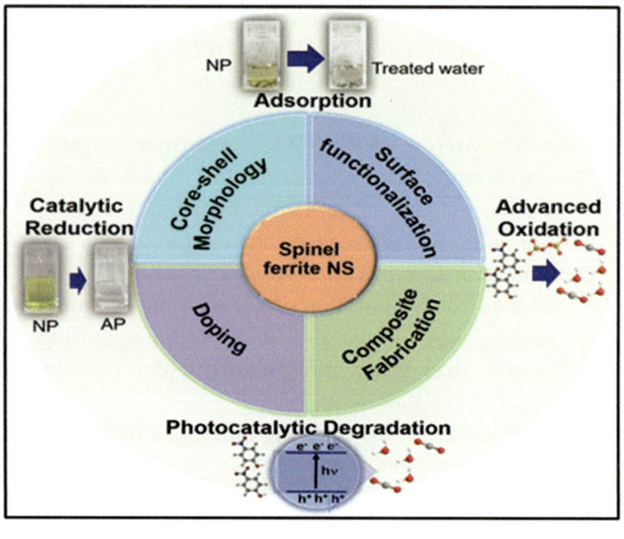

FIGURE 15.5 Prospects and characterization of spinel micro-ferrites. (Reprinted with permission from Bhaduri et al. [37]. Copyright [2022] American Chemical Society.)

15.4 SPINEL NANO-FERRITES FOR MICROPLASTICS REMOVAL

Spinel nano-ferrites, also known as magnetic ferrite nanoparticles, have received a lot of attention in the field of wastewater treatment because of their unique qualities and advantages:

1. *Magnetic Properties*: Spinel nano-ferrites have outstanding magnetic properties, allowing them to be easily removed from wastewater using magnetic separation techniques. This characteristic simplifies nanoparticle recovery and reuse, lowering the treatment procedure's cost and complexity. Ali Eftekhari et al. published "Magnetic spinel ferrite nanoparticles: synthesis, characterization, and applications in water treatment" in the Journal of Materials Chemistry A in 2017 [38].
2. *Large surface area*: Nano-sized ferrite particles have a high surface area-to-volume ratio, offering many active sites for adsorption and catalytic processes. This feature increases their efficiency in eliminating contaminants from wastewater.
3. *Chemical Stability*: Spinel nano-ferrites have high chemical stability, making them resistant to corrosion and degradation in severe settings. This stability ensures their long-term performance and durability [39].
4. *Adsorption Capacity*: The surface of spinel nano-ferrites can be easily changed to improve their adsorption capacity for various pollutants prevalent in wastewater. By functionalizing the surface with certain groups or ions, nanoparticles can selectively target and remove contaminants such as heavy metals, organic dyes, and medicinal molecules.
5. *Photocatalytic Activity*: Certain spinel nano-ferrites, such as zinc ferrite ($ZnFe_2O_4$), have photocatalytic capabilities, which means they can break down organic contaminants when exposed to light. This photocatalytic activity adds another method for eliminating organic pollutants from wastewater. Spinel nano-ferrites may be regenerated and reused several times, making them cost-effective for wastewater treatment applications. The adsorption capacity of nano-cubes can be restored by desorbing or recovering the adsorbed contaminants [40].
6. *Low Environmental Impact*: Compared to traditional wastewater treatment processes, using spinel nano-ferrites can reduce the amount of chemical additives necessary. Furthermore, their magnetic recovery enables easy separation and prevents nanoparticles from being released into the environment [41].
7. *Scalability*: Spinel nano-ferrite synthesis is easily scaled up to create huge numbers, making

them appropriate for industrial-scale wastewater treatment applications. Antibacterial activity: Antibacterial characteristics have been discovered in several spinel nano-ferrites, which can benefit wastewater treatment. These nanoparticles can suppress the growth of bacteria and other microorganisms found in wastewater, lowering the risk of microbial contamination, and enhancing overall water quality [42].

8. Spinel nano-ferrites can be easily synthesized and tuned employing a variety of processes, including co-precipitation, sol-gel, hydrothermal, and microwave-assisted synthesis. This adaptability enables them to be fine-tuned in size, shape, composition, and surface modification to optimize their performance for specific wastewater treatment applications.
9. *Cost-effectiveness and abundance*: The basic ingredients necessary to synthesize spinel-like nano-ferrites, such as metal oxides, are numerous and inexpensive. When compared to other sophisticated therapies, spinel nano-ferrites are a more cost-effective approach for sewage treatment, for instance, the paper on "Synthesis and characterization of spinel ferrite nanoparticles for water treatment applications" by Mahendra Rai et al. highlights the ecological and economical aspect of ferrites [38].

15.4.1 Interactions between Nano-Ferrites and Microplastics

The interactions between nano-ferrites and MPs are poorly known due to a lack of research in this field. The following are some novel potential interactions:

1. *Adsorption*: Surface characteristics of nano-ferrites may allow them to absorb MPs. The magnetic characteristics of micro-ferrites could attract and bind MP particles on their surfaces. In the presence of a magnetic field, nano-ferrites may cause aggregation or accumulation of MPs. Because of the magnetic interaction between nano-ferrites and MPs, bigger aggregates may form.
2. *Surface modification*: When nano-ferrites encounter MPs, they can change their surface properties. They could, for example, change the surface charge or hydrophobicity/hydrophilicity of MPs, influencing their behavior and interactions with other substances. If nano-ferrites are discharged into the environment with MPs, they may influence the destiny and transit of the MPs. If nano-ferrites are found in water or sediment, they may influence the mobility and distribution of MPs [39]. The DFT simulations can help design and optimize surface functionalization of nano-ferrites for increased MP removal. Researchers can identify suitable functional groups or coatings that improve the adsorption capacity and selectivity of nano-ferrites for MPs by analyzing the energetics and binding intensities.
3. *Nano-ferrites for catalytic destruction of MPs*: Nano-ferrites have shown interesting catalytic characteristics in various chemical processes. To evaluate the catalytic activity of nano-ferrites in the breakdown of MPs, DFT calculations can be used. Theoretical simulations can aid in identifying reaction pathways, energy barriers, and reaction intermediates, resulting in a better understanding of the degradation process [40].
4. *Modifications to the electronic structure*: DFT can be used to explore how MPs affect the electronic structure of nano-ferrite materials. This can aid in understanding any changes in the magnetic or catalytic characteristics of nano-ferrites caused by MPs [41].
5. *Transition State Analysis*: When researching adsorption kinetics, determine the transition states related to the adsorption and desorption processes. Transition state theory can be used to calculate the energy barriers and rate constants for these processes [42].
6. *Molecular Dynamics Simulations*: When examining adsorption kinetics, use molecular dynamics simulations to simulate the movements of adsorbate molecules near the surface. These simulations can provide information on adsorption and desorption rates and insights into the dynamics of the adsorption process.
7. *Adsorption Isotherm Analysis*: The adsorption isotherms should be determined, which characterize the relationship between the equilibrium coverage of adsorbate on the surface and its partial pressure or concentration. This data can be used to describe the adsorption behavior of the system. Adsorption kinetics and equilibrium studies utilizing DFT provide useful insights into Congo red adsorption processes and photocatalytic oxidation at the molecular level, assisting in understanding the factors influencing adsorption behavior and guiding the design of materials with increased adsorption capabilities, as shown in Figure 15.6.

15.5 EVALUATION OF SPINEL NANO-FERRITES IN MICROPLASTICS REMEDIATION

Spinel nano-ferrites are nanoscale particles made of ferrite materials with a spinel crystal structure. They have considerable potential for MP remediation because of their magnetic characteristics, large surface area, and chemical durability [44]. The increased surface area and consistent particle size distribution of nano-ferrites allow MP adsorption. The pH-dependent removal efficiency emphasizes optimizing environmental conditions to improve nano-ferrite effectiveness.

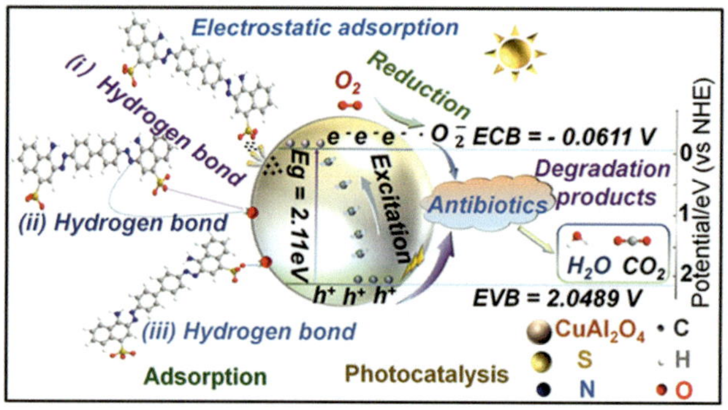

FIGURE 15.6 Spinel CuB_2O_4 for selective adsorption. (Reprinted with permission from Han et al. [43]. Copyright [2022] American Chemical Society.)

Traditional remediation procedures are frequently inefficient in eliminating MPs, necessitating the development of novel technologies to address this issue. Remediation of MPs by magnetic nanoparticles, published in 2020 by Sharma et al., presents a summary of recent advances in the use of magnetic nanomaterials, especially spinel nanoferrites, for MP remediation. It discusses the fundamentals, difficulties, and possible uses of magnetic nanoparticles in removing MPs from various environmental matrices [45]. Smith et al. published "Removal of Microplastics from Water Using Spinel Nano-ferrites: Adsorption Kinetics and Mechanisms" in 2019. The kinetics and processes of MP adsorption onto Spinel Nano-ferrites are studied in this work. The study emphasizes the efficiency of the process and sheds light on the interactions between MPs and nanoferrites [46].

Traditional MP cleanup strategies commonly use physical removal procedures such as filtration, coagulation, and sedimentation. While these procedures have been utilized for some time, they may have limits regarding efficiency, selectivity, and reusability. A comparison of spinel nanoferrites with traditional techniques is presented below:

1. *Filtration*: Traditional: Filtration can be successful, but it can be difficult to catch extremely minute MPs due to filter pore size restrictions. Spinel nano-ferrites can improve MPs' selectivity due to their high surface area and unique adsorption capabilities [47].
2. *Coagulation and Sedimentation*: Traditional: Coagulation followed by sedimentation can successfully remove bigger MPs, but smaller particles may not settle easily. Spinel nano-ferrites can absorb both big and small MPs, allowing for a more thorough remediation method [48].
3. *Sorbent Materials*: Traditional sorbents may have poor adsorption ability and be difficult to separate from water. Spinel nano-ferrites have a larger adsorption capacity and are easy to magnetically separate, making the process more efficient.
4. *Biodegradation*: Some researchers have investigated the employment of microorganisms capable of degrading MPs via biodegradation processes. This method has potential, although it is still in its early stages of development.

An investigation published in the *Journal of Hazardous Materials* (2018) evaluated the utilization of magnetite (Fe_3O_4) nanoparticles for the adsorption and removal of MPs from water bodies. The results showed that magnetite nanoparticles had a high adsorption capacity and were easy to recover magnetically [49]. Another study published in *Environmental Science and Pollution Research* (2020) investigated the application of iron oxide nanoparticles to remove MPs from wastewater. The study indicated that the nanoparticles effectively absorbed MPs, and their magnetic characteristics allowed for easy separation [50]. Recycling Ferrite Materials: In some situations, nanoferrite materials can be recycled from waste or abandoned ferrites. The demand for new synthesis can be decreased by recovering and recycling these materials, encouraging a more sustainable approach. Nano-ferrites can be used in biosensor technologies to identify certain biological molecules or analytes due to their magnetic characteristics. This can be useful in disease diagnostics and environmental monitoring [51]. The long-term stability of nano-ferrites is an important aspect of their practical application due to their characteristics can alter over time for various reasons. Some instances of nano-ferrites and their long-term stability are discussed below:

1. *Nickel ferrite ($NiFe_2O_4$) nanoparticles*: Nickel ferrite nanoparticles are used as catalysts and magnetic recording medium. Oxidation, phase shifts, and increasing particle size can impact long-term stability. Proper handling, storage, and surface treatments can assist in maintaining their stability over time.
2. *Zinc ferrite nanoparticles*: Zinc ferrite nanoparticles are utilized in electromagnetic wave absorbers and microwave devices. Environmental variables and surface reactivity can impact their long-term stability. In practical applications, suitable encapsulation or surface modifications can improve their stability [52].
3. *Nanoparticles of cobalt ferrite* ($CoFe_2O_4$): Cobalt ferrite nanoparticles are interesting in electronics, information storage, and magnetic devices. Surface oxidation of cobalt ferrite nanoparticles can affect their magnetic characteristics over time. For example, surface passivation and protective coatings can improve their long-term stability and avoid undesired changes [53].
4. *Nanoparticles of barium ferrite* ($BaFe_{12}O_{19}$): These are commonly used in magnetic data storage and permanent magnets. Their long-term stability is critical for maintaining magnetism and coercivity.

Material processing and protective coatings can help to retain their stability over time.

5. *Nickel ferrite nanostructures* ($NiFe_2O_4$): Nickel ferrite nanoparticles are used in catalysts and magnetic recording medium. Factors such as oxidation, phase shifts, and particle size increase can all impact long-term stability [54].

Surface oxidation, aggregation, phase shifts, and environmental conditions can all impact the long-term stability of nano-ferrites. Proper synthesis, surface engineering, and storage techniques should be adopted to ensure dependable performance in various applications [55].

15.6 ENVIRONMENTAL CONSIDERATIONS AND CHALLENGES

Micro-ferrites have unique features due to their nanoscale size and high surface area, making them interesting candidates for various applications such as electronics, catalysis, and biology. However, when released into the environment, they, like all nanomaterials, may have potential ecotoxicological impacts. A few, instances of probable ecotoxicological impacts of spinel nano-ferrites are given:

1. *Toxicity in the Water*: Spinel nano-ferrites could reach aquatic bodies across various channels, including industrial effluents and poor waste disposal. Once in the water, they may interact with aquatic creatures, affecting their health and survival. Certain metal oxide nanoparticles, particularly several nano-ferrites, have been found in tests to be hazardous to aquatic creatures such as fish, algae, and crustaceans. The microscopic particles may cause oxidative stress, impair biological functioning, and harm tissues [56, 57].
2. *Bioaccumulation*: Some spinel nano-ferrites may be absorbed by aquatic species and accumulate in their tissues. This bioaccumulation can cause nanoparticles to go up the food chain, potentially impacting higher trophic levels such as predatory species and humans. A study on cerium oxide nanoparticles, which resemble some spinel nano-ferrites, discovered that they might bioaccumulate in fish tissues.
3. Spinel nano-ferrites may be discharged into the soil through direct application due to runoff from contaminated water bodies. They may then interact with soil organisms, influencing soil fertility and nutrient cycle. For instance, studies on zinc oxide nanoparticles (structurally identical to some nano-ferrites) revealed negative impacts on soil microbes and plant growth [58].
4. Coral reefs are extraordinarily sensitive ecosystems, and any disruption to their delicate equilibrium can seriously affect the marine species that depend on them. When spinel nano-ferrites are introduced into the marine environment, they can interact with corals in a variety of ways such as sedimentation: nano-ferrites may contribute to sedimentation by decreasing light penetration and negatively influencing the photosynthetic activities of coral symbiotic algae (zooxanthellae), resulting in coral bleaching and diminished growth.

The study by Huang et al. in 2019 explored the impact of spinel nano-ferrites on coral health and the potential for nanoparticles to be taken up by corals [59].

1. *Airborne Dispersal*: Throughout the manufacture, use, or disposal of spinel nano-ferrites-containing items, airborne particles may be created, exposing humans and animals to inhalation exposure. The possible impacts on respiratory systems and overall health are concerning. Other nanoscale studies, such as titanium dioxide, have generated concerns about lung toxicity.
2. *Biotic Disruption*: Releasing huge amounts of spinel nano-ferrites into the environment may induce ecological imbalances. For instance, if specific creatures are particularly vulnerable to nanoparticles, their population size may drop, harming the entire ecosystem. Furthermore, nanoparticle-induced changes in water quality, soil health, or air composition could have far-reaching effects on biodiversity [60, 61].
3. The presence of spinel nano-ferrites in habitats of sea organisms, including a variety of marine species, can also have an impact on them: Invertebrates and fish: Fish and other marine species may consume nanoparticles, which can accumulate in their tissues and cause hazardous effects. This can influence their behavior, metabolism, and reproduction.

Feeders for Filters: Organisms that rely on filter feedings, such as oysters, mussels, and plankton, may mistakenly consume nanoparticles with their food, resulting in bioaccumulation and potentially harmful impacts throughout the food chain. The toxicity of several spinel nano-ferrites on common marine fish species was investigated in this work. Certain nano-ferrites were found to have negative impacts on fish behavior, such as decreased swimming activity [61–63]. It is critical to highlight that the real impacts of spinel nano-ferrites on coral reefs and sea animals depend on elements such as the size, concentration, and surface qualities of the nanoparticles, as well as the individual species and ambient conditions. The fate and transit of nano-ferrites in the environment refers to how these nanoparticles behave, travel, and interact with various environmental compartments such as soil, water, and air. Nano-ferrites are nanoparticles made of iron oxide, mainly magnetite (Fe_3O_4) or maghemite, with unique features that make them helpful in various applications such as environmental remediation, water purification, and targeted drug administration [64–66].

15.7 CONCLUSIVE REMARKS

The novel approach of employing spinel nano-ferrites for wastewater-based tiny plastic particle restoration is investigated in this review chapter. The growing worry about the accumulation of MPs in the environment, particularly in water bodies, has prompted scientists and researchers to explore effective and long-term methods for their removal. Due to its unique features, such as high surface area, magnetic habits, and changeable surface chemistry, spinel nano-ferrites have emerged as a potential technique. It is highlighted that spinel-like nano-ferrites have outstanding performance in eliminating MPs from wastewater after a thorough review of recent studies. Their magnetic nature allows easy separation from the treated water, making them recyclable and cost-effective. Nano-ferrites can treat groundwater contaminated with chlorinated solvents such as trichloroethylene (TCE). When injected into the subsurface, they can function as reactive agents, catalyzing the decomposition of TCE into non-toxic molecules. Due to their small size, they can access and treat contaminated zones where conventional approaches may not reach efficiently. Owing to their magnetic characteristics, nano-ferrites can be used in environmental sensing. They can be functionalized with specific molecules that bind to specific contaminants or diseases. When these pollutants or pathogens are in the environment, the functionalized nano-ferrites can identify and quantify their presence.

However, before concluding this chapter, it is critical to note some obstacles and knowledge gaps in this subject. More research is needed to determine spinel nano-ferrites' long-term stability and reusability in actual applications. Furthermore, the potential environmental and ecotoxicological consequences of employing these materials must be thoroughly investigated. DFT calculations have been essential in understanding the molecular interactions between MPs and spinel nano-ferrites. The insights gained from these theoretical investigations have aided in optimizing the design and manufacture of nano-ferrites with enhanced efficiency and selectivity toward various forms of MPs. IoT (Internet of Things) technology in the MP remediation process has enabled real-time monitoring and management of the remediation system. This ensures efficient resource utilization, early identification of system failures, and adaptive response to changing wastewater conditions, resulting in a more sustainable and cost-effective approach. Spinel nano-ferrites could be integrated with other sophisticated treatment technologies, such as ozonation, advanced oxidation processes (AOPs), or biological degradation, to produce a multi-barrier strategy for MP removal. This chapter mentions future advancements, such as adding advanced treatment technologies, tertiary treatment procedures (e.g., membrane filtration, activated carbon adsorption), or AOPs, to target MP removal specifically. Real wastewater contains a variety of pollutants, and competition for adsorption sites is possible. DFT can be used to simulate and forecast the impact of coexisting pollutants on the adsorption effectiveness of MPs on nano-ferrites. Overall, the chapter illustrates the sustainable utilization of Spinel Nano-ferrites for wastewater-based macro-plastics treatment offers tremendous potential as a unique and effective technique for combating MP contamination. Continued research and development in AI can pave the path for practical application and contribute to cleaner and healthier aquatic habitats. However, it is critical to continue with prudence, considering both the benefits of technology and the potential environmental ramifications.

ACKNOWLEDGMENT

Author, S. Roy would like to acknowledge '*Scheme for Transformational and Advanced Research in Sciences (STARS)*' (MoE-STARS/STARS-2/2023-0175),theMinistry of Education.

REFERENCES

1. Roy, S., Chatterjee, R., & Majumder, S. B. (2011). Magnetoelectric coupling in sol-gel synthesized dilute magnetostrictive-piezoelectric composite thin films. Journal of Applied Physics (Vol. 110, Issue 3, p. 036101). https://doi.org/10.1063/1.3610795
2. Salih, S. J., & Mahmood, W. M. (2023). Review on magnetic spinel ferrite (MFe_2O_4) nanoparticles: From synthesis to application. Heliyon (Vol. 9, Issue 6, p. e16601). https://doi.org/10.1016/j.heliyon.2023.e16601
3. Bilovol, V., Sikora, M., Martínez García, R., Berent, K., Gajewska, M., & Szkudlarek, A. (2022). $SrFe_{12}O_{19}/CoFe_2O_4$ magnetic composites: Nanoparticle size and effect of annealing temperature on magnetic properties. Journal of Magnetism and Magnetic Materials (Vol. 563, p. 169987). https://doi.org/10.1016/j.jmmm.2022.169987
4. Khoso, W. A., Haleem, N., Baig, M. A. et al. (2021). Synthesis, characterization, and heavy metal removal efficiency of nickel ferrite nanoparticles (NFNs). Scientific Reports (Vol. 11, p. 3790). https://doi.org/10.1038/s41598-021-83363-1
5. Liu, W., Zhang, J., Liu, H., Guo, X., Zhang, X., Yao, X., Cao, Z., & Zhang, T. (2021). A review of the removal of microplastics in global wastewater treatment plants: Characteristics and mechanisms. Environment International (Vol. 146, p. 106277). https://doi.org/10.1016/j.envint.2020.106277
6. Vekariya, R. L., Dhar, A., Kumar, N. S., & Roy, S. (2018). Efficient solid-statedye-sensitized solar cell based on tri cationic ionic crystal pyridinium-imidazolium electrolytes. Organic Electronics (Vol. 56, pp. 260–267). https://doi.org/10.1016/j.orgel.2018.01.034
7 Bhattacharjee, J., & Roy, S. (2025). A comprehensive review on integrated photo rechargeable batteries- supercapacitors, and their techno-economic feasibility. Journal of Photochemistry and Photobiology, *25*, 100257. https://doi.org/10.1016/j.jpap.2024.100257
8. Mmelesi, O. K., Masunga, N., Kuvarega, A., Nkambule, T. T. I., Mamba, B. B., & Kefeni, K. K. (2021). Cobalt ferrite nanoparticles and nanocomposites: Photocatalytic,

antimicrobial activity and toxicity in water treatment. Materials Science in Semiconductor Processing (Vol. 123, p. 105523). https://doi.org/10.1016/j.mssp.2020.105523

9. Dhanda, N., Thakur, P., Sun, A. C. A., & Thakur, A. (2023). Structural, optical and magnetic properties along with antifungal activity of Ag-doped Ni-Co nanoferrites synthesized by eco-friendly route. Journal of Magnetism and Magnetic Materials (Vol. 572, p. 170598).
10. Dhanda, N., Thakur, P., & Thakur, A. (2023). Green synthesis of cobalt ferrite: A study of structural and optical properties. Materials Today: Proceedings (Vol.73, pp. 237–240).
11. Sanchez-Lievanos, K. R., Stair, J. L., & Knowles, K. E. (2021). Cation distribution in spinel ferrite nanocrystals: Characterization, impact on their physical properties, and opportunities for synthetic control. Inorganic Chemistry (Vol. 60, Issue 7, pp. 4291–4305). https://doi.org/10.1021/acs.inorgchem.1c00040.
12. Venturini, J. (2019). Application of ferrite nanoparticles in wastewater treatment. In Nanomaterials for Eco-friendly Applications (pp. 141–151). Springer International Publishing. https://doi.org/10.1007/978-3-030-26810-7_9
13. Gkika, D. A., Mitropoulos, A. C, Lambropoulou, D. A. et al. (2022). Cosmetic wastewater treatment technologies: A review. Environmental Science and Pollution Research (Vol. 29, pp. 75223–75247). https://doi.org/10.1007/s11356-022-23045-1
14. Ahmed, M. B., Zhou, J. L., Ngo, H. H., Guo, W., Thomaidis, N. S., & Xu, J. (2017). Progress in the biological and chemical treatment technologies for emerging contaminant removal from wastewater: A critical review. Journal of Hazardous Materials (Vol. 323, pp. 274–298). https://doi.org/10.1016/j.jhazmat.2016.04.045
15. Alvarino, T., Suarez, S., Katsou, E., Vazquez-Padin, J., Lema, J. M., & Omil, F. (2015). Removal of PPCPs from the sludge supernatant in a one-stagenitritation/anammox process. Water Research (Vol. 68, pp. 701–709). https://doi.org/10.1016/j.watres.2014.10.055
16. Awfa, D., Ateia, M., Fujii, M., Johnson, M. S., & Yoshimura, C. (2018). Photodegradation of pharmaceuticals and personal care products in water treatment using carbonaceous-TiO_2 composites: A critical review of recent literature. Water Research (Vol. 142, pp. 26–45). https://doi.org/10.1016/j.watres.2018.05.036
17. Beiras, R. (2021). Towards standard methods for the classification of aquatic toxicity for biologically active household chemicals (BAHC) present in plastics, pharmaceuticals, and cosmetic products. Environmental Monitoring and Assessment (Vol. 193, Issue 10). https://doi.org/10.1007/s10661-021-09436-w
18. Bolisetty, S., Peydayesh, M., & Mezzenga, R. (2019). Sustainable technologies for water purification from heavy metals: Review and analysis. Chemical Society Reviews (Vol. 48, Issue 2, pp. 463–487). https://doi.org/10.1039/c8cs00493e
19. Maitra, S., Pal, S., Maitra, T., Halder, S., & Roy, S. (2021). Solvothermal etching-assisted phase and morphology tailoring in highly porous $CuFe_2O_4$ nanoflake photocathodes for solar water splitting. Energy & Fuels (Vol. 35, Issue 17, pp. 14087–14100). https://doi.org/10.1021/acs.energyfuels.1c02090
20. Al Harraq, A., & Bharti, B. (2021). Microplastics through the lens of colloid science. ACS Environmental Au (Vol. 2, Issue 1, pp. 3–10). https://doi.org/10.1021/acsenvironau.1c00016
21. Borrelli, S. B., Ringma, J., Law, K. L., Monnahan, C. C., Lebreton, L., McGivern, A., Murphy, E., Jambeck, J., Leonard, G. H., & Hilleary, M. A. (2020). Predicted growth in plastic waste exceeds efforts to mitigate plastic pollution. Science (Vol. 369, Issue 6510, pp. 1515–1518). https://doi.org/10.1126/science.aba3656
22. Kumari, S., Dhanda, N., Thakur, A., Gupta, V., Singh, S., Kumar, R., … & Thakur, P. (2023). Nano Ca–Mg–Zn ferrites as tuneable photocatalyst for UV light-induced degradation of rhodamine B dye and antimicrobial behavior for water purification. Ceramics International (Vol. 49, Issue 8, pp. 12469–12480).
23. Punia, P., Bharti, M. K., Chalia, S., Dhar, R., Ravelo, B., Thakur, P., & Thakur, A. (2021). Recent advances in synthesis, characterization, and applications of nanoparticles for contaminated water treatment – A review. Ceramics International (Vol. 47, Issue 2, pp. 1526–1550).
24. Gore, S. K., & Jadhav, S. S.. (2020). Basics of ferrites. In Spinel Ferrite Nanostructures for Energy Storage Devices (pp. 1–11). Elsevier. https://doi.org/10.1016/b978-0-12-819237-5.00001-8
25. Ghasemi, A. K., Ghorbani, M., Lashkenari, M. S., & Nasiri, N. (2022). Controllable synthesis of zinc ferrite nanostructure with tunable morphology on polyaniline nanocomposite for supercapacitor application. Journal of Energy Storage (Vol. 51, p. 104579). https://doi.org/10.1016/j.est.2022.104579
26. Wahajuddin, & Arora, S. (2012). Superparamagnetic iron oxide nanoparticles: Magnetic nanoplatforms as drug carriers. International Journal of Nanomedicine (Vol. 7, pp. 3445–71). https://doi.org/10.2147/IJN.S30320.
27. Kumari, S., Dhanda, N., Thakur, A., Singh, S., & Thakur, P. (2023). Investigation of calcium substitution on magnetic and dielectric properties of Mg–Zn nano ferrites. Materials Chemistry and Physics (Vol. 297, pp. 127394).
28. Bizmark, N.;, Schneider, J.;, Priestley, R. D.;, & Datta, S. SS. S. (2020). Multiscale dynamics of colloidal deposition and erosion in porous media. Science Advances (Vol. 6, Issue 46, p. eabc2530). https://doi.org/10.1126/sciadv.abc2530
29. Cao, Y., Qin, H., Niu, X., & Jia, D. (2016). Simple solid-state chemical synthesis and gas-sensing properties of spinel ferrite materials with different morphologies. Ceramics International (Vol. 42, Issue 9, pp. 10697–10703). https://doi.org/10.1016/j.ceramint.2016.03.184
30. Shang, Y., Duan, Z., & Luo, F. (2023). Preparation and characterization of fibrous MnZn ferrite via a water-assisted solvothermal method. Journal of Magnetism and Magnetic Materials (Vol. 571, p. 170563). https://doi.org/10.1016/j.jmmm.2023.170563
31. Shaikh, S. F., Ubaidullah, M., Mane, R. S., & Al-Enizi, A. M. (2020). Types, synthesis methods, and applications of ferrites. In Spinel Ferrite Nanostructures for Energy Storage Devices (pp. 51–82). Elsevier. https://doi.org/10.1016/b978-0-12-819237-5.00004-3
32. Bhattacharjee, J., & Roy, S. (2024). A review on photocatalysis and nanocatalysts for advanced organic synthesis. Hybrid Advances (Vol. 6, p. 100268). https://doi.org/10.1016/j.hybadv.2024.100268
33. Dhanda, N., Kumari, S., Kumar, R., Kumar, D., Sun, A. C. A., Thakur, P., & Thakur, A. (2023). Influence of Ni over magnetically benign Co ferrite system and study of its structural, optical, and magnetic behaviour. Inorganic Chemistry Communications (Vol. 151, p. 110569).

34. Dehghani Dastjerdi, O., Shokrollahi, H., & Mirshekari, S. (2023). A review of synthesis, characterization, and magnetic properties of soft spinel ferrites. Inorganic Chemistry Communications (Vol. 153, p. 110797). https://doi.org/10.1016/j.inoche.2023.110797
35. Ai, J., Shuai, Y., Hu, M., Cheng, L., Luo, S., Li, W., Chen, Z., Hu, L., & Zhou, Z. (2023). Microstructural evolution and catalytic properties of novel high-entropy spinel ferrites MFe_2O_4 (M= Mg, Co, Ni, Cu, Zn). Ceramics International (Vol. 49, Issue 14, pp. 22941–22951). https://doi.org/10.1016/j.ceramint.2023.04.119
36. Azimi-Fouladi, A., Falak, P., & Hassanzadeh-Tabrizi, S. A. (2023). The photodegradation of antibiotics on nano cubic spinel ferrites photocatalytic systems: A review. Journal of Alloys and Compounds (Vol. 961, p. 171075). https://doi.org/10.1016/j.jallcom.2023.171075
37. Bhaduri, B., Dikshit, A. K., Kim, T., & Tripathi, K. M. (2022). Research progress and prospects of spinel ferrite nanostructures for the removal of nitroaromatics from wastewater. ACS Applied Nano Materials (Vol. 5, Issue 11, pp. 16000–16026). https://doi.org/10.1021/acsanm.2c02684
38. Roy, S., Kargupta, K., Chakraborty, S., & Ganguly, S. (2008). Preparation of polyaniline nanofibers and nanoparticles via simultaneous doping and electro-deposition. Materials Letters (Vol. 62, Issue 16, pp. 2535–2538). https://doi.org/10.1016/j.matlet.2007.12.066
39. Bhakta, D., Gayen, D., Bhattacharjee, J., Chatterjee, R., Dastidar, P., Roy, S. (2024). Repurposing industrial and agricultural wastes as sustainable alternatives in photovoltaic cells and dye degradation. ChemistrySelect (p. e202403145). https://doi.org/10.1002/slct.202403145
40. Varvaro, G., Omelyanchik, A., & Peddis, D. (2022). Ferroic Transition Metal Oxide Nano-Heterostructures: From Fundamentals to Applications (pp. 405–437). https://doi.org/10.1002/9783527826940.ch12
41. Liandi, A. R., Cahyana, A. H., Kusumah, A. J. F., Lupitasari, A., Alfariza, D. N., Nuraini, R., Sari, R. W., & Kusumasari, F. C. (2023). Recent trends of spinel ferrites (MFe_2O_4: Mn, Co, Ni, Cu, Zn) applications as an environmentally friendly catalyst in multicomponent reactions: A review. Case Studies in Chemical and Environmental Engineering (Vol. 7, p. 100303). https://doi.org/10.1016/j.cscee.2023.100303
42. Shemy, M. H., Othman, S. I., Alfassam, H. E., Al-Waili, M. A., Alqhtani, H. A., Allam, A. A., & Abukhadra, M. R. (2023). Synthesis of green magnetite/carbonized coffee composite from natural pyrite for effective decontamination of Congo red dye: Steric, synergetic, oxidation, and ecotoxicity studies. Catalysts (Vol. 13, Issue 2, p. 264). https://doi.org/10.3390/catal13020264
43. Han, M., Wang, S., Chen, X., Liu, H., Gao, H., Zhao, X., Wang, F., Yang, H., Yi, Z., & Fang, L. (2022). Spinel CuB_2O_4 (B = Fe, Cr, and Al) oxides for selective adsorption of Congo red and photocatalytic removal of antibiotics. ACS Applied Nano Materials (Vol. 5, Issue 8, pp. 11194–11207). https://doi.org/10.1021/acsanm.2c02349
44. Bhattacharjee, J., & Roy, S. (2024). Synergistic insights: Electro-organic photocatalysis and nanostructures. Chemistry Papers. https://doi.org/10.1007/s11696-024-03657-2
45. Zander, J., Weiss, M., & Marschall, R. (2023). Fast and facile microwave synthesis of cubic $CuFe_2O_4$ nanoparticles for electrochemical CO_2 reduction. Advanced Energy and Sustainability Research (Vol. 4, Issue 4). https://doi.org/10.1002/aesr.202200184
46. Sharma, K., Calmels, L., Li, D., Barbier, A., & Arras, R. (2022). Influence of the cation distribution, atomic substitution, and atomic vacancies on the physical properties of $COFe_2O_4$ and $NIFe_2O_4$ spinel ferrites. Physical Review Materials (Vol. 6, Issue 12). https://doi.org/10.1103/PhysRevMaterials.6.124402
47. Qu, L., Wang, Z., Hou, X., Mao, J., Wang, S., Sun, G., Wu, Y., & Liu, X. (2017). "Rose flowers" assembled from mesoporous $NiFe_2O_4$ nanosheets for energy storage devices. Journal of Materials Science: Materials in Electronics (Vol. 28, pp. 14058–14068). https://doi.org/10.1007/s10854-017-7257-z
48. Hadi, H., Safari, R., & Shamlouei, H. R. (2022). Impact of calcination temperature on the spin–spin relaxation time (T_2) of $MgFe_2O_4$ nanoparticles (in vitro). Canadian Journal of Chemistry (Vol. 100, Issue 12, pp. 891–899). https://doi.org/10.1139/cjc-2022-0105
49. Singh, R. P., Rai, A. R., Jotania, R. B., Chaudhary, R. G., & Abdala, A. (2022). Rationale and Trends of Applied Nanotechnology (pp. 373–389). https://doi.org/10.1016/B978-0-323-88535-5.00008-1
50. Sari, S. R., Tsushida, M., Sato, T., & Tominaga, M. (2022). Highly sensitive detection of phosphate using well-ordered crystalline cobalt oxide nanoparticles supported by multi-walled carbon nanotubes. Materials Advances (Vol. 3, Issue 4, pp. 2018–2025). https://doi.org/10.1039/D1MA01097B
51. Al-Fahat Hossain, M., Al-Mamun, M., Rahman, M. R., & Hoque, S. M. (2021). Effect of in-situ growth and separate addition method in hydrothermal process on the structural and magnetic properties of $CoNiFe_2O_4$@functionalized CNTs nanocomposite. Applied Physics A (Vol. 127, Issue 12). https://doi.org/10.1007/s00339-021-05049-9
52. Gao, X., Pei, L., Xue, W., Huang, H., Gao, Z., & Zhao, X. (2021). Monodentate AIEgen anchored on metal-organic framework for fast fluorescence sensing of phosphate. Chinese Journal of Chemistry (Vol. 39, Issue 1, pp. 99–105). https://doi.org/10.1002/cjoc.202000364
53. Thakur, P., Taneja, S., Chahar, D., Ravelo, B., & Thakur, A. (2021). Recent advances on synthesis, characterization and high frequency applications of Ni-Zn ferrite nanoparticles. Journal of Magnetism and Magnetic Materials (Vol. 530, pp. 167925).
54. Bhattacharjee, J., & Roy, S. (2024). Density Functional Waltz: Molecular Theory of Spin Dynamics. In Journal of Physics; Optics Sciences, 1–4. Scientific Research and Community Ltd. doi.org/10.47363/jpsos/2024(6)242
55. Vekariya, R. L., Dhar, A., Paul, P. K. et al. (2018). An overview of engineered porous material for energy applications: A mini-review. Ionics (Vol. 24, pp. 1–17). https://doi.org/10.1007/s11581-017-2338-9
56. Wu, L. Q., Li, Y. C., Li, S. Q., Li, Z. Z., Tang, G. D., Qi, W. H., Xue, L. C., Ge, X. S., & Ding, L. L. (2015). Method for estimating ionicities of oxides using O1s photoelectron spectra. AIP Advances (Vol. 5, p. 097210). https://doi.org/10.1063/1.4931996
57. Dupin, J. C., Gonbeau, D., Vinatier, P., & Levasseur, A. (2000). Systematic XPS studies of metal oxides, hydroxides and peroxides. Physical Chemistry Chemical Physics (Vol. 2, pp. 1319–1324). https://doi.org/10.1039/a908800h
58. Punia, P., Aggarwal, R. K., Kumar, R., Dhar, R., Thakur, P., & Thakur, A. (2022). Adsorption of Cd and Cr ions from industrial wastewater using Ca doped Ni–Zn nanoferrites: Synthesis, characterization and isotherm analysis. Ceramics International (Vol. 48, Issue 13, pp. 18048–18056).

59. Stoner, E. C., & Wohlfarth, E. P. (1948). A mechanism of magnetic hysteresis in heterogeneous alloys. Philosophical Transactions of the Royal Society A (Vol. 240, pp. 599–642). https://doi.org/10.1098/rsta.1948.0007
60. Korkmaz, A. D., Güner, S., Slimani, Y., Gungunes, H., Amir, M., Manikandan, A., & Baykal, A. (2019). Microstructural, optical, and magnetic properties of vanadium-substituted nickel spinel nanoferrites. Journal of Superconductivity and Novel Magnetism (Vol. 32, pp. 1057–1065). https://doi.org/10.1007/s10948-018-4793-6
61. Huang, H., Shen, B., Yan, M., He, H., Yang, L., Jiang, Q., & Ying, G. (2022). Coupled spinel manganese–cobalt oxide and MXene electrocatalysts towards efficient hydrogen evolution reaction. Fuel (Vol. 328, p. 125234). https://doi.org/10.1016/j.fuel.2022.125234
62. Punia, P., Bharti, M. K., Dhar, R., Thakur, P., & Thakur, A. (2022). Recent advances in detection and removal of heavy metals from contaminated water. ChemBioEng Reviews (Vol. 9, Issue 4, pp. 351–369).
63. Bhattacharjee, J., & Roy, S. (2024). Synthesis of thermoelectric nanocomposites by incorporating reduced graphene oxide. Archives of Advanced Engineering Science. Bon View Publishing PTE. https://doi.org/10.47852/bonviewaaes42023514
64. Shaikh, A. A., Bhattacharjee, J., Datta, P., & Roy, S. (2024). A comprehensive review of the oxidation states of molybdenum oxides and their diverse applications. Sustainable Chemistry for the Environment. (Vol. 7, p. 100125). https://doi.org/10.1016/j.scenv.2024.100125
65. Bhattacharjee, J., & Roy, S. (2024). Review on green resources and AI for biogenic solar power. Energy Storage and Conversion (Vol. 2, Issue 1. https://doi.org/10.59400/esc.v2i1.457.
66. Bhattacharjee, J., Bhaskar, S., & Roy, S. (2024). Green nanomaterials for antimicrobial and anticancer applications. Nano-Engineering at Functional Interfaces for Multi-Disciplinary Applications (pp. 425–444). https://doi.org/10.1016/B978-0-443-21691-6.00018-4

16 Safety Measures Based on Spinel Nano-Ferrites

*Vivek Srivastava**
Mathematics and Basic Sciences–Chemistry, NIIT University, Neemrana, India

16.1 INTRODUCTION TO SPINEL NANO-FERRITES (SFNPs) AND THEIR DIVERSE APPLICATIONS

In the ever-evolving landscape of materials science and nanotechnology, the remarkable properties and versatile applications of nanomaterials have garnered widespread attention and interest. Among these, spinel nano-ferrite particles (SFNPs) have emerged as a class of nanomaterials that hold significant promise and have found applications spanning various scientific and technological domains (Kefeni et al., 2017; Kumar et al., 2017; Soufi et al., 2021; Thakur et al., 2018). Their unique combination of magnetic, electrical, and structural properties, coupled with their nanoscale dimensions, makes them a subject of intense research and innovation. In this comprehensive exploration, we embark on a journey to introduce SFNPs and delve into their diverse applications, shedding light on their immense potential and the challenges they bring forth (Kefeni et al., 2017; Nosheen et al., 2023; Olatunji & Owolabi, 2023; Ortiz-Quiñonez et al., 2018).

Nanotechnology, the science of manipulating matter at the nanoscale, has revolutionized materials science and engineering by offering unprecedented control over the properties of materials (Ngata et al., 2022; Nosheen et al., 2023; Thakur et al., 2022; Sharma et al., 2021). At this scale, the quantum effects begin to dominate, resulting in materials with novel properties that often defy conventional understanding. These nanomaterials exhibit enhanced mechanical, electrical, magnetic, and catalytic properties compared to their bulk counterparts. They have been instrumental in the development of innovative technologies, ranging from electronics and energy storage to drug delivery and environmental remediation. Spinel ferrites, a class of materials with a general chemical formula of AFe_2O_4, where A represents a divalent cation such as Fe, Zn, Ni, or Co, have been studied extensively for decades. Their spinel crystal structure, characterized by a face-centered cubic lattice, gives rise to intriguing magnetic properties. In recent years, the synthesis and characterization of SFNPs, which are nanosized versions of these spinel ferrites, have opened up new avenues for research and application (Gaffar et al., 2023; Khojaste khoo & Kameli, 2021; Kumar Dutta et al., 2022).

Before delving into the world of SFNPs, it is essential to grasp the fundamental properties and characteristics of spinel ferrites. These materials are renowned for their tunable magnetic properties, high electrical resistivity, chemical stability, and excellent structural integrity. The spinel structure consists of oxygen atoms arranged in a close-packed arrangement, with cations occupying specific positions within this structure (Gaffar et al., 2023; Kumar Dutta et al., 2022; Salih & Mahmood, 2023; Tatarchuk et al., 2017).

The unique arrangement of atoms in the spinel structure results in two distinct crystallographic sites for cations: tetrahedral (A) and octahedral (B) sites. The tetrahedral sites are found at the corners of the unit cell, while the octahedral sites are situated at the centers of the faces. This arrangement gives rise to complex magnetic interactions within the crystal lattice, making spinel ferrites intriguing materials for study and application. The transition from bulk spinel ferrites to SFNPs brings forth a host of fascinating phenomena. At the nanoscale, these materials exhibit size-dependent properties, where quantum effects dominate. This transition introduces novel magnetic behaviors, including superparamagnetic, that differ significantly from their bulk counterpart (Hao & Ning, 2021; Khojaste khoo & Kameli, 2021; Soufi et al., 2021).

SFNPs are typically synthesized using various methods, such as co-precipitation, sol-gel, and hydrothermal techniques (Thakur et al., 2020a,b). These methods allow precise control over particle size, morphology, and composition, enabling tailored properties for specific applications (Kant & Mann, 2018; Pandit et al., 2020). One of the defining features of SFNPs is their magnetic behavior, which stems from the arrangement of cations in the spinel structure. Unlike bulk spinel ferrites, SFNPs often exhibit superparamagnetic, a property arising from their small size and reduced thermal energy barriers. This unique characteristic has found utility in various fields, particularly in biomedical applications, where controlled magnetic manipulation of nanoparticles is essential for drug delivery, hyperthermia, and magnetic resonance imaging (MRI) (Kumar Dutta et al., 2022; Odio & Reguera, 2017; Salih & Mahmood, 2023; Tatarchuk et al., 2017; Thakur et al., 2023).

The magnetic properties of SFNPs can be tuned by altering their size, composition, and surface functionalization. This tunability opens doors to diverse applications, from magnetic sensors and data storage to targeted drug delivery

* Corresponding author.

DOI: 10.1201/9781003568940-16

systems (Hao & Ning, 2021; Khojaste khoo & Kameli, 2021; Olatunji & Owolabi, 2023; Ortiz-Quiñonez et al., 2018; Thakur et al., 2023).

In addition to their magnetic attributes, SFNPs exhibit intriguing electrical properties. Their high electrical resistivity and dielectric constants make them suitable candidates for microwave and radio frequency devices, such as antennas and phase shifters. Furthermore, SFNPs can be integrated into composite materials to enhance their dielectric properties, making them valuable in the development of high-frequency electronic components. SFNPs are not confined to magnetic and electrical applications alone. Their diverse range of properties, including high surface area and tunable chemistry, has paved the way for multifunctional applications. For instance, SFNPs have been explored in catalysis, where their magnetic properties facilitate separation and recycling, while their catalytic surfaces drive various chemical reactions. This multifunctionality extends to areas like environmental remediation, where SFNPs are employed for the removal of pollutants from wastewater. One of the most significant and impactful applications of SFNPs lies in the realm of environmental remediation, particularly in wastewater treatment. The exceptional adsorption capacity of SFNPs for pollutants, coupled with their ease of separation using external magnetic fields, has revolutionized the field of wastewater treatment.

SFNPs have demonstrated unparalleled efficiency in removing a wide range of pollutants, from organic dyes to heavy metals, from aqueous solutions. Their ability to harness solar energy due to their low energy band gap further contributes to the cost-effectiveness of wastewater treatment processes. Moreover, their reusability, a result of their magnetic properties and photoactivity at ambient temperatures, underscores their eco-sustainability and economic advantage.

To harness the full potential of SFNPs in various applications, innovative techniques for surface modification have been developed. Surface engineering allows for improved photocatalytic efficiency and selectivity, making SFNPs even more versatile in pollutant degradation and removal (Figure 16.1).

FIGURE 16.1 Diverse role and impact of SFNPs.

16.1.1 Importance of Safety Measures when Handling SFNPs

Nanotechnology has ushered in a new era of materials science, enabling the manipulation and engineering of materials at the nanoscale. Among the myriad nanomaterials that have emerged, SFNPs have garnered significant attention due to their unique properties and versatile applications. However, the handling and manipulation of nanoscale materials like SFNPs pose inherent safety challenges. In this discussion, we emphasize the paramount importance of safety measures when working with SFNPs and delve into the reasons why these measures are crucial (Figure 16.2) (*Guidelines and Best Practices for Safe Handling of Nanomaterials in Research Laboratories and Industries Nano Mission, DST, Govt of India By (with Inputs from Nanoregulatory Task Force*, n.d.; Schulte et al., 2016a).

16.1.2 Protection from Health Hazards

One of the foremost reasons for implementing safety measures when handling SFNPs is the potential health hazards associated with exposure to nanoparticles. SFNPs, like other nanomaterials, possess properties that can lead to adverse health effects if not handled with care. Inhalation or dermal exposure to airborne SFNPs, for example, can pose health risks.

16.1.3 Minimizing Inhalation Risks

The small size of SFNPs, typically in the nanometer range, means they can remain suspended in the air for extended periods. When inhaled, these nanoparticles can reach the respiratory system, potentially leading to lung-related issues. Safety measures, such as working in a controlled environment with proper ventilation or under a fume hood, minimize the risk of inhaling airborne SFNPs.

16.1.4 Preventing Skin Contact

In addition to inhalation risks, skin contact with SFNPs can also be problematic. While the dermal absorption of nanoparticles is a subject of ongoing research, it is crucial to minimize skin contact as a precautionary measure. Safety measures, such as wearing appropriate personal protective equipment (PPE), including gloves and lab coats, serve to prevent skin exposure.

16.1.5 Eye Protection

Eye protection is another critical aspect of safety when working with SFNPs. Accidental splashes or exposure to SFNPs in the eye can lead to irritation or more severe eye injuries. Safety goggles or a face shield are essential PPE items that shield the eyes from potential hazards.

FIGURE 16.2 Safety measures when handling SFNPs.

16.1.6 Avoiding Ingestion Risks

The ingestion of SFNPs is a risk that must be mitigated, especially in laboratory settings. Personnel should refrain from eating, drinking, or smoking in areas where SFNPs are being handled. Proper hand hygiene, including washing hands thoroughly before eating, is a preventive measure.

16.1.7 Environmental Safety

Safety measures for handling SFNPs extend beyond personnel safety; they also encompass environmental safety. Spills or improper disposal of SFNPs can lead to their release into the environment, potentially causing adverse ecological effects. Proper waste disposal and spill response protocols are essential in safeguarding the environment.

16.1.8 Safe Handling and Storage

The proper handling and storage of SFNPs are critical to prevent accidents and ensure their integrity. Safety measures include working in a designated area with restricted access to authorized personnel. SFNPs should be stored in labeled containers, away from incompatible materials, and in accordance with guidelines for safe storage of nanomaterials.

16.1.9 Emergency Preparedness

Despite all preventive measures, emergencies can occur. Preparedness is key to responding effectively to such situations. Knowing the location of safety showers, eyewash stations, fire extinguishers, and emergency contacts is essential for swift response in case of accidents.

The importance of safety measures when handling SFNPs cannot be overstated. These measures are essential to protect the health and well-being of personnel, prevent environmental contamination, and ensure compliance with regulatory standards. A proactive approach to safety not only safeguards against potential hazards but also facilitates responsible and ethical research practices in the field of nanotechnology.

16.2 UNDERSTANDING THE POTENTIAL HAZARDS ASSOCIATED WITH SFNPs

In this exciting frontier, SFNPs have emerged as versatile and promising nanomaterials with a wide range of applications. However, as we delve deeper into the world of SFNPs, it becomes increasingly vital to comprehend the potential hazards associated with their use and manipulation. In this discussion, we explore the multifaceted aspects of SFNP-related hazards, why understanding them is crucial, and how to mitigate these risks effectively (Oberdörster et al., 2005; *Protecting Workers during Nanomaterial Reactor Operations*, 2018; Schubauer-Berigan et al., 2015; Schulte et al., 2019; Warheit et al., 2008).

Nanomaterials, including SFNPs, possess unique properties that distinguish them from their bulk counterparts. Their small size, high surface area-to-volume ratio, and quantum effects can lead to distinct hazards that must be addressed to ensure safe handling and application. It is important to recognize that while SFNPs offer numerous advantages, their manipulation can also present potential risks. One of the primary concerns associated with SFNPs is the risk of inhalation. Due to their nanoscale dimensions, SFNPs can remain suspended in the air for extended periods. When inhaled, these nanoparticles can reach the respiratory system, potentially causing health issues. The lungs, in particular, can be vulnerable to nanoparticle-induced inflammation and damage.

Another concern is dermal exposure to SFNPs. While the extent of dermal absorption of nanoparticles is still a subject of ongoing research, it is prudent to minimize skin contact as a precautionary measure. Accidental exposure or contact with contaminated surfaces can pose risks. The eyes are highly sensitive to foreign particles and are susceptible to damage from SFNP exposure. Accidental splashes or contact with SFNPs can lead to eye irritation, discomfort, or more severe eye injuries. Proper eye protection, such as safety goggles or a face shield, is essential when working with SFNPs. Ingestion of SFNPs is another potential risk, especially in laboratory settings where proper hygiene practices are essential. Researchers should refrain from eating, drinking, or smoking in areas where SFNPs are being handled, as ingestion could lead to unintended exposure. Understanding the potential health implications of SFNPs is paramount (Oberdörster et al., 2005). Research has shown that nanoparticles can have various effects on biological systems, including oxidative stress, inflammation, and potential toxicity. These effects can manifest both in the short term and over prolonged exposure periods. Research into the biological interactions of SFNPs is ongoing, and ongoing safety assessments are critical.

Beyond the immediate health risks, SFNPs also pose environmental concerns. Improper disposal or accidental release of SFNPs can lead to contamination of natural ecosystems. As these nanoparticles are not readily biodegradable, their long-term impact on the environment is a subject of concern. Understanding potential hazards is the first step in risk mitigation. Researchers and personnel working with SFNPs must adhere to safe handling practices to minimize the risks associated with these nanomaterials. Proper training, education, and awareness are essential components of safety protocols. Safety controls play a crucial role in minimizing hazards associated with SFNPs. Working in controlled environments, such as fume hoods or gloveboxes, helps contain nanoparticles and prevent their dispersion (Kant & Mann, 2018; Nanomaterials Safety Guidelines | *Environmental Health and Safety*, n.d.). These engineered controls are designed to protect researchers and the environment. PPE is a fundamental aspect of risk reduction. Researchers must wear appropriate PPE, including lab coats, disposable gloves, safety goggles, and respiratory protection when necessary. Properly fitted PPE forms a physical barrier that prevents skin contact and inhalation. Conducting thorough risk assessments before working with SFNPs is essential. Identifying potential hazards, evaluating exposure pathways, and implementing risk management strategies are vital steps in ensuring safety. These assessments guide the development of safety protocols.

Adherence to local, national, and international regulations and guidelines is a non-negotiable aspect of safety. Institutions and research facilities often have specific safety protocols in place that researchers must follow to maintain compliance and ensure the safe handling of SFNPs.

Understanding the potential hazards associated with SFNPs is fundamental to their responsible use. These hazards encompass health risks, environmental concerns, and the need for rigorous safety protocols. By recognizing these potential risks and implementing effective mitigation strategies, researchers can harness the vast potential of SFNPs while safeguarding the well-being of personnel and the environment. Ongoing research and safety assessments are essential as our understanding of these nanomaterials continues to evolve.

16.3 IDENTIFICATION OF RISKS DURING SYNTHESIS, HANDLING, AND DISPOSAL

SFNPs have exhibited tremendous potential in various applications, yet their synthesis, handling, and disposal necessitate the identification and management of inherent risks. This comprehensive understanding of risks associated with each facet of SFNP utilization is pivotal in ensuring the safety of personnel, preserving environmental integrity, and upholding the quality of research outcomes (Canzler et al., 2020; Foulkes et al., 2020; Grafström et al., 2015).

During the synthesis phase, researchers encounter risks related to chemical exposure. Working with hazardous reagents, solvents, and precursors can lead to potential health hazards, including chemical burns and respiratory problems. High temperatures involved in many synthesis methods demand precautions to prevent burns and thermal injuries. Additionally, the generation of airborne nanoparticles during synthesis raises concerns about inhalation risks, necessitating adequate containment measures. Cross-contamination and exothermic reactions must also be mitigated to avoid unintended consequences (Jeevanandam et al., 2018; Ma et al., 2023; Omar et al., 2021).

In the handling phase, risks emerge from the potential for inhalation and dermal exposure to SFNPs. Airborne nanoparticles can be inadvertently inhaled into the respiratory system, while direct skin contact is possible during handling. Ensuring eye protection is crucial as accidental splashes or contact with SFNPs can lead to eye irritation or injury. Proper hand hygiene practices are essential to minimize the risk of ingestion resulting from inadequate hygiene during laboratory activities. The unique physical properties of SFNPs, such as magnetism and high surface energy, pose physical hazards that necessitate controlled management. The prevention of sample and equipment contamination is vital to maintain the reliability of experimental results, and precautions are essential to avert accidental spills (Bystrzejewska-Piotrowska et al., 2009; Hischier & Walser, 2012; Holder et al., 2013; Meyer et al., 2009).

Disposal brings forth environmental contamination risks if SFNPs or contaminated materials are inadequately disposed of, potentially impacting ecosystems and engendering long-term ecological consequences. Failure to adhere to local, national, or international disposal regulations can lead to legal and reputational issues. Safe handling and disposal procedures must be strictly followed to protect waste management personnel and the environment.

General risks arise from the size-dependent properties of SFNPs, which can introduce unpredictability to their behavior and potential hazards. The limited availability of comprehensive toxicological data for SFNPs poses challenges in accurately assessing their health and environmental risks. Aggregation of SFNPs may alter their properties and behavior, potentially resulting in unexpected outcomes or diminished performance in applications. Moreover, activities that disrupt SFNP aggregates, such as agitation or high shear forces, can lead to nanoparticle release and subsequent inhalation hazards (McNeil, 2005; Roco et al., 2011; Song et al., 2009).

Identifying these risks at each stage of SFNP utilization is the initial step in their effective management. The implementation of robust safety protocols, comprehensive risk assessments, rigorous training and education, and strict adherence to regulations and guidelines are pivotal in minimizing these risks and ensuring the responsible utilization of SFNPs in both research and practical applications. Ongoing research into the toxicity and environmental impact of SFNPs is imperative to continuously refine safety practices and guidelines as our understanding of these nanomaterials progresses.

16.4 ENGINEERING CONTROLS AND SAFE WORK PRACTICES

To mitigate the inherent risks associated with SFNPs, it is imperative to implement effective engineering controls and safe work practices. This comprehensive approach encompasses the creation of a safe working environment with sufficient ventilation, emphasizing the importance of working under a fume hood or within a controlled atmosphere and adopting safe handling practices to minimize exposure (Canzler et al., 2020; Rice, n.d.).

The cornerstone of safe SFNP handling begins with the establishment of a controlled and well-ventilated working environment. Adequate ventilation plays a pivotal role in maintaining air quality and reducing the concentration of airborne nanoparticles, thereby minimizing potential exposure risks. In laboratory settings, proper ventilation can be achieved through the installation and utilization of ventilation systems, including local exhaust ventilation (LEV) and general exhaust ventilation (GEV). LEV systems, such as fume hoods, are particularly effective in capturing and containing SFNPs, especially during processes that generate aerosols or particulate matter. These systems draw air away from the work area and through appropriate filtration systems, ensuring that any airborne SFNPs are effectively captured and removed (Di Giampaolo et al., 2020; Geraci et al., 2015; Pietroiusti et al., 2018; Trybula et al., 2018; Zainal Abidin et al., 2020).

In cases where SFNP-related activities take place in industrial or manufacturing settings, customized ventilation solutions may be required. These solutions should be designed to address the specific processes and potential sources of SFNP release. Regular maintenance and inspection of ventilation systems are essential to ensure their continued effectiveness in controlling nanoparticle exposure (Díaz-Soler et al., 2017; Omari Shekaftik et al., 2023; Zainal Abidin et al., 2020). Additionally, the proper design and layout of workspaces are crucial considerations in creating a safe working environment. Work areas should be organized to minimize the risk of cross-contamination and allow for efficient use of engineering controls, such as fume hoods or glove boxes. When working with SFNPs, particularly during synthesis or handling processes that generate SFNP-containing aerosols, it is imperative to conduct these activities within a fume hood or under controlled atmospheres. These controlled environments serve as essential barriers between researchers and potential sources of nanoparticle exposure. Fume hoods are enclosed workspaces equipped with exhaust systems designed to capture and remove hazardous fumes, vapors, and airborne particles. They are a primary engineering control for nanoparticle handling, offering an effective means of containing SFNPs, and preventing their release into the laboratory or workspace. Researchers should always work under a fume hood when handling SFNPs, especially during processes that could lead to airborne SFNP contamination.

It is essential to ensure that fume hoods are regularly inspected and properly maintained to guarantee their functionality. Additionally, conducting airflow tests and face velocity measurements can confirm that the fume hood is operating at the required level of containment. Proper training on fume hood usage is imperative for researchers to maximize safety. In certain situations, working under a controlled atmosphere, such as a glove box or an inert gas chamber, may be necessary. Controlled atmospheres provide an environment with precise control over gas composition and particulate contamination. These systems are particularly valuable when working with air-sensitive materials or when strict control over oxygen and moisture levels is required. Glove boxes, for example, are enclosed chambers with sealed gloves that allow researchers to manipulate SFNPs or other materials while maintaining a controlled atmosphere. Inert gases, such as nitrogen or argon, are often used to create oxygen-free environments within glove boxes (de Araujo Cortez et al., 2011; Louro & Silva, 2022). Researchers should be well-trained in operating and maintaining these systems to prevent breaches in containment. Beyond engineering controls and controlled environments, safe handling practices are pivotal in minimizing SFNP exposure risks. Researchers and personnel must adhere to specific protocols and procedures to ensure their safety. Agitation of SFNPs can lead to the release of nanoparticles into the air, increasing the risk of inhalation exposure. Therefore, it is crucial to minimize unnecessary agitation during SFNP handling. Gentle and controlled movements should be employed, and mechanical agitation should be avoided whenever possible. Dry handling of SFNPs, which involves manipulating nanoparticles without the use of appropriate solvents or dispersion methods, can lead to the generation of airborne particles. To mitigate this risk, wet handling techniques should be favored. Using suitable solvents or suspensions helps to minimize dust formation and particle release.

Personnel involved in SFNP-related activities should receive comprehensive training on safe handling practices and the proper use of engineering controls and PPE. This training should be ongoing to ensure that individuals remain well-informed and up to date with the latest safety procedures.

Engineering controls and safe work practices are essential components of a robust safety framework when working with SFNPs (Díaz-Soler et al., 2017; Louro & Silva, 2022; Schulte et al., 2016b; Sharpening the Focus on Occupational Safety and Health in Nanotechnology on JSTOR, n.d.; Turan et al., n.d.). Creating a safe working environment with adequate ventilation, emphasizing the importance of working under a fume hood or in controlled atmospheres, and adopting safe handling practices are all integral to mitigating the risks associated with SFNP handling. Researchers and personnel must be well-trained, and adherence to safety protocols and procedures is paramount to ensuring the safety and well-being of all involved in SFNP-related activities. By integrating these measures into SFNP research and applications, we can advance the field of nanotechnology safely and responsibly, unlocking its full potential while minimizing potential hazards.

16.5 SFNPs WASTE DISPOSAL

The disposal of SFNPs and materials that have encountered SFNPs is a multifaceted process that demands meticulous attention to detail and strict adherence to regulatory requirements (Rohilla & Chaudhary, 2023; Sharma et al., 2023; Turan et al., n.d.). It is imperative to minimize the risk of environmental contamination and ensure the safety of waste management personnel. A systematic approach should be employed throughout the disposal process.

The journey of responsible SFNP waste disposal begins with segregation and collection. At the source of waste generation, SFNPs and contaminated items should be segregated to distinguish them from other waste streams. This categorization facilitates efficient disposal procedures. Dedicated waste containers, clearly labeled for SFNPs and SFNP-contaminated materials, should be used, ensuring separation and minimizing the risk of cross-contamination (Chauhan et al., 2023; Kaur & Singh, 2023; Sharma et al., 2023).

To prevent the release of SFNPs into the environment and minimize contamination during transportation and storage, appropriate packaging and containment measures are vital. Selecting packaging materials that are compatible with SFNPs and possess the ability to maintain containment is critical. This ensures that SFNPs remain securely enclosed within their packaging (Chauhan et al., 2023; El-Maghrabi et al., 2021; Jasrotia et al., 2023; Kaur & Singh, 2023; Kumar et al., 2023; Makhlouf & Ali, n.d.; Sharma et al., 2023; Turan et al., n.d.).

a. Clear and standardized labeling of waste containers is imperative for effective waste management. Labels on these containers should provide essential information, including the nature of the waste (SFNPs or SFNP-contaminated materials), hazard classification, and specific handling instructions. Proper labeling helps waste management personnel identify and handle waste appropriately.
b. Ensuring the proper storage of waste materials is vital to prevent unintended releases or leaks. SFNP waste should be stored securely in designated storage areas where containers remain intact and labeled. Storage facilities should be designed to contain any potential spills or leaks, minimizing the risk of environmental contamination.
c. During the transportation of SFNP waste, strict adherence to safety and containment protocols is paramount. Containers must be securely fastened to prevent shifting or tipping during transit. Waste transport vehicles should be equipped with spill containment measures, and, if required, hazard signage should be prominently displayed to alert others to the presence of hazardous materials.
d. The choice of disposal method for SFNPs and SFNP-contaminated materials should align with several factors, including the nature of the waste, local regulations, and environmental considerations. Common disposal methods include landfill disposal, incineration, and recycling when applicable. These methods must be selected to meet regulatory requirements and environmental protection goals, ensuring the safe and responsible disposal of SFNP waste.
e. Achieving regulatory compliance is a fundamental aspect of responsible SFNP waste disposal. Local and national regulations that govern the handling and disposal of hazardous materials, including nanoparticles like SFNPs, must be strictly adhered to. Compliance ensures that disposal practices

align with established safety standards and environmental protection guidelines.

f. Establishing communication with local and national environmental agencies responsible for waste management and disposal regulations is essential. These agencies can offer guidance on compliance, provide valuable insights into navigating regulatory requirements, and ensure that disposal activities are conducted in accordance with legal standards.
g. In some instances, waste disposal activities involving SFNPs may necessitate permits or approvals from regulatory authorities. It is crucial to determine whether specific permits are required and to initiate the permit application process well in advance of disposal activities to ensure compliance.
h. Robust record-keeping practices are essential to document compliance with local and national regulations. Detailed records should encompass waste disposal records, permits, correspondence with regulatory agencies, and any relevant safety assessments. These records serve as a critical resource for demonstrating adherence to legal and safety requirements.
i. Periodic auditing of waste disposal procedures and practices is a proactive approach to ensuring ongoing compliance. Audits help identify areas where improvements or adjustments may be necessary to align with evolving regulations or best practices, further enhancing the safety and responsibility of SFNP waste disposal.

The proper disposal of SFNPs and SFNP-contaminated materials is a multifaceted process that demands meticulous attention to detail, strict adherence to regulatory requirements, and an unwavering commitment to environmental stewardship and safety. Guidelines for proper disposal, compliance with local and national regulations, and robust procedures for labeling and documenting waste materials collectively form the cornerstone of responsible waste management practices for SFNPs (Aboelazm et al., 2021; Ali et al., 2021; Das et al., 2021; El-Maghrabi et al., 2021; El-Salamony & El Shafey, 2021; Mahmoud & Fawzy, 2021; Nik Hassan et al., 2021).

16.6 CONCLUSION

In the rapidly evolving landscape of materials science and nanotechnology, the advent of SFNPs has sparked profound excitement and innovation. This comprehensive book chapter has embarked on an insightful journey, traversing the multifaceted realm of SFNPs, from their remarkable properties and diverse applications to the pivotal importance of safety measures during their synthesis, handling, and disposal.

SFNPs, with their unique blend of magnetic, electrical, and structural attributes at the nanoscale, have unlocked a plethora of scientific and technological possibilities. From magnetic sensors to environmental remediation, SFNPs have found utility in an array of fields, promising to revolutionize our approaches to challenges old and new. Their magnetic behavior, tunable through size, composition, and surface functionalization, has propelled innovations in drug delivery, data storage, and beyond.

However, our exploration did not stop at the threshold of SFNPs' potential. We underscored the paramount significance of safety measures at every juncture of SFNP utilization. Understanding the potential health hazards and environmental risks inherent in these nanomaterials is foundational to responsible research and application. The discussion illuminated the importance of guarding against inhalation, dermal exposure, eye contact, and ingestion risks, emphasizing protective measures to safeguard personnel and the environment.

Moreover, the book chapter navigated the intricate landscape of SFNP waste disposal, articulating a systematic approach to prevent environmental contamination. Segregation, proper packaging, labeling, storage, and transportation of SFNP waste materials were outlined, all while adhering to local and national regulations. Compliance with regulatory standards, the significance of robust record-keeping, and periodic auditing practices completed the framework for responsible SFNP waste disposal.

We delved into the pivotal role of engineering controls and safe work practices in ensuring the safety of personnel engaged in SFNP-related activities. Adequate ventilation, the use of fume hoods or controlled atmospheres, and safe handling practices emerged as critical components in minimizing exposure risks. These measures were deemed fundamental to mitigating inherent risks and fostering responsible practices within the realm of nanotechnology.

In closing, this book chapter serves as an invaluable resource for researchers, institutions, and industries immersed in the exploration and application of SFNPs. It not only promotes scientific advancement but also champions responsible and ethical practices. The fusion of groundbreaking research with unwavering commitment to safety and environmental stewardship paves the way for the responsible harnessing of SFNPs' potential, propelling us into a future where innovation and responsibility walk hand in hand.

REFERENCES

Aboelazm, E. A. A., Mohamed, N., Ali, G. A. M., Makhlouf, A. S. H., & Chong, K. F. (2021). Recycling of cobalt oxides electrodes from spent lithium-ion batteries by electrochemical method. *Topics in Mining, Metallurgy and Materials Engineering*, 91–123. https://doi.org/10.1007/978-3-030-68031-2_4

Ali, G. A. M., Bakr, Z. H., Safarifard, V., & Chong, K. F. (2021). Recycled nanomaterials for energy storage (Supercapacitor) applications. *Topics in Mining, Metallurgy and Materials Engineering*, 175–202. https://doi.org/10.1007/978-3-030-68031-2_7

Bystrzejewska-Piotrowska, G., Golimowski, J., & Urban, P. L. (2009). Nanoparticles: Their potential toxicity, waste and environmental management. *Waste Management, 29*(9), 2587–2595. https://doi.org/10.1016/J.WASMAN.2009.04.001

Canzler, S., Schor, J., Busch, W., Schubert, K., Rolle-Kampczyk, U. E., Seitz, H., Kamp, H., von Bergen, M., Buesen, R., & Hackermüller, J. (2020). Prospects and challenges of multi-omics data integration in toxicology. *Archives of Toxicology, 94*(2), 371–388. https://doi.org/10.1007/S00204-020-02656-Y

Chauhan, P., Bhasin, K. K., & Chaudhary, S. (2023). *Environmental Fate Descriptors for Screening Nanotoxicity and Pollutant Sensing.* 95–137. https://doi.org/10.1007/978-3-031-24416-2_4

Das, H. T., Elango Balaji, T., Mahendraprabhu, K., & Vinoth, S. (2021). Cost-effective nanomaterials fabricated by recycling spent batteries. *Topics in Mining, Metallurgy and Materials Engineering*, 147–174. https://doi.org/10.1007/978-3-030-68031-2_6

de Araujo Cortez, B., Quassollo, G., Caceres, A., & Machado-Santelli, G. M. (2011). The fate of chrysotile-induced multipolar mitosis and aneuploid population in cultured lung cancer cells. *PLoS ONE, 6*(4). https://doi.org/10.1371/JOURNAL.PONE.0018600

Di Giampaolo, L., Petrarca, C., Mangifesta, R., Schiavone, C., Pini, C., Malandra, A., Bramante, F., Pollutri, A., Di Frischia, M., & Di Gioacchino, M. (2020). *Metal Nanoparticle Health Risk Assessment.* 17–35. https://doi.org/10.1007/978-981-15-4735-5_2

Díaz-Soler, B. M., López-Alonso, M., & Martínez-Aires, M. D. (2017). Nanosafety practices: Results from a national survey at research facilities. *Journal of Nanoparticle Research, 19*(5). https://doi.org/10.1007/S11051-017-3867-X

El-Maghrabi, H. H., Nada, A. A., Soliman, F. S., Raynaud, P., Moustafa, Y. M., Ali, G. A. M., & Bekheet, M. F. (2021). Recovery of metal oxide nanomaterials from electronic waste materials. *Topics in Mining, Metallurgy and Materials Engineering*, 203–227. https://doi.org/10.1007/978-3-030-68031-2_8

El-Salamony, R. A., & El Shafey, A. M. (2021). Extraction of silica and lignin-based nanocomposite materials from agricultural waste for wastewater treatment using photocatalysis technique. *Topics in Mining, Metallurgy and Materials Engineering*, 363–384. https://doi.org/10.1007/978-3-030-68031-2_13

Foulkes, R., Man, E., Thind, J., Yeung, S., Joy, A., & Hoskins, C. (2020). The regulation of nanomaterials and nanomedicines for clinical application: Current and future perspectives. *Biomaterials Science, 8*(17), 4653–4664. https://doi.org/10.1039/D0BM00558D

Gaffar, S., Kumar, A., & Riaz, U. (2023). Synthesis techniques and advance applications of spinel ferrites: A short review. *Journal of Electroceramics 2023*, 1–12. https://doi.org/10.1007/S10832-023-00333-X

Geraci, C., Heidel, D., Sayes, C., Hodson, L., Schulte, P., Eastlake, A., & Brenner, S. (2015). Perspectives on the design of safer nanomaterials and manufacturing processes. *Journal of Nanoparticle Research, 17*(9). https://doi.org/10.1007/S11051-015-3152-9

Grafström, R. C., Nymark, P., Hongisto, V., Spjuth, O., Ceder, R., Willighagen, E., Hardy, B., Kaski, S., & Kohonen, P. (2015). Toward the replacement of animal experiments through the bioinformatics-driven analysis of "omics" data from human cell cultures. *ATLA Alternatives to Laboratory Animals, 43*(5), 325–332. https://doi.org/10.1177/026119291504300506

Guidelines and Best Practices for Safe Handling of Nanomaterials in Research Laboratories and Industries Nano Mission, DST, Govt of India By (with inputs from Nanoregulatory Task Force). (n.d.). Retrieved September 15, 2023, from www.ckmnt.com

Hao, A., & Ning, X. (2021). Recent advances in spinel ferrite-based thin films: Synthesis, performances, applications, and beyond. *Frontiers in Materials, 8*, 718869. https://doi.org/10.3389/FMATS.2021.718869

Hischier, R., & Walser, T. (2012). Life cycle assessment of engineered nanomaterials: State of the art and strategies to overcome existing gaps. *Science of the Total Environment, 425*, 271–282. https://doi.org/10.1016/J.SCITOTENV.2012.03.001

Holder, A. L., Vejerano, E. P., Zhou, X., & Marr, L. C. (2013). Nanomaterial disposal by incineration. *Environmental Science: Processes & Impacts, 15*(9), 1652–1664. https://doi.org/10.1039/C3EM00224A

Jasrotia, T., Chaudhary, G. R., & Kumar, R. (2023). *Nanomaterials in Combating Water Pollution and Related Ecotoxicological Risk.* 139–172. https://doi.org/10.1007/978-3-031-24416-2_5

Jeevanandam, J., Barhoum, A., Chan, Y. S., Dufresne, A., & Danquah, M. K. (2018). Review on nanoparticles and nanostructured materials: History, sources, toxicity and regulations. *Beilstein Journal of Nanotechnology, 9*(1), 1050–1074. https://doi.org/10.3762/BJNANO.9.98

Kant, R., & Mann, A. K. (2018). A review of doped magnesium ferrite nanoparticles: Introduction, synthesis techniques and applications. *International Journal of Scientific Research in Science, Engineering and Technology, 4*(7), 646–660.

Kaur, A., & Singh, H. (2023). *Nanomaterials in Environment: Sources, Risk Assessment, and Safety Aspect.* 75–93. https://doi.org/10.1007/978-3-031-24416-2_3

Kefeni, K. K., Mamba, B. B., & Msagati, T. A. M. (2017). Application of spinel ferrite nanoparticles in water and wastewater treatment: A review. *Separation and Purification Technology, 188*, 399–422. https://doi.org/10.1016/J.SEPPUR.2017.07.015

Khojaste khoo, M., & Kameli, P. (2021). Structure and magnetization of strontium hexaferrite ($SrFe_{12}O_{19}$) films prepared by pulsed laser deposition. *Frontiers in Materials, 8*. https://doi.org/10.3389/FMATS.2021.717251

Kumar Dutta, S., Akhter, M., Ahmed, J., Amin, M. K., & Kumar Dhar, P. (2022). Synthesis and catalytic activity of spinel ferrites: A brief review. *Biointerface Research in Applied Chemistry, 12*, 4399–4416. https://doi.org/10.33263/BRIAC124.43994416

Kumar, R., Kumar, R., & Chaudhary, S. (Eds.). (2023). *Advanced Functional Nanoparticles "Boon or Bane" for Environment Remediation Applications.* https://doi.org/10.1007/978-3-031-24416-2

Kumar, A., Rout, L., Achary, L. S. K., Mohanty, S. K., & Dash, P. (2017). A combustion synthesis route for magnetically separable graphene oxide-$CuFe_2O_4$-ZnO nanocomposites with enhanced solar light-mediated photocatalytic activity. *New Journal of Chemistry, 41*(19), 10568–10583. https://doi.org/10.1039/C7NJ02070H

Louro, H., & Silva, M. J. (Eds.). (2022). *Nanotoxicology in Safety Assessment of Nanomaterials. 1357.* https://doi.org/10.1007/978-3-030-88071-2

Ma, N. L., Zhang, N., Yong, W. T. L., Misbah, S., Hashim, F., Soon, C. F., Lim, G. P., Peng, W., & Sonne, C. (2023). Use, exposure and omics characterisation of potential hazard in nanomaterials. *Materials Today Advances, 17*, 100341. https://doi.org/10.1016/J.MTADV.2023.100341

Mahmoud, A. E. D., & Fawzy, M. (2021). Nanosensors and nanobiosensors for monitoring the environmental pollutants. *Topics in Mining, Metallurgy and Materials Engineering*, 229–246. https://doi.org/10.1007/978-3-030-68031-2_9

Makhlouf, A. S. H., & Ali, G. A. M. (n.d.). *Waste Recycling Technologies for Nanomaterials Manufacturing*. 870.

McNeil, S. E. (2005). Nanotechnology for the biologist. *Journal of Leukocyte Biology*, *78*(3), 585–594. https://doi.org/10.1189/jlb.0205074

Meyer, D. E., Curran, M. A., & Gonzalez, M. A. (2009). An examination of existing data for the industrial manufacture and use of nanocomponents and their role in the life cycle impact of nanoproducts. *Environmental Science and Technology*, *43*(5), 1256–1263. https://doi.org/10.1021/ES8023258

Nanomaterials Safety Guidelines | Environmental Health and Safety. (n.d.). Retrieved September 15, 2023, from https://ehs.stonybrook.edu/programs/laboratory-safety/general-laboratory-safety/nanomaterials-safety-guidelines.php

Ngata, M. R., Yang, B., Aminu, M. D., Iddphonce, R., Omari, A., Shaame, M., Nyakilla, E. E., Mwakateba, I. A., Mwakipunda, G. C., & Yanyi-Akofur, D. (2022). Review of developments in nanotechnology application for formation damage control. *Energy and Fuels*, *36*(1), 80–97. https://doi.org/10.1021/ACS.ENERGYFUELS.1C03223

Nik Hassan, N. F., Che Hamzah, C. N., Rajan, R., & Zakaria, Y. (2021). Recycling of nanosilica powder from Bamboo leaves and rice husks for forensic applications. *Topics in Mining, Metallurgy and Materials Engineering*, 295–323. https://doi.org/10.1007/978-3-030-68031-2_11

Nosheen, S., Iqbal, S. S., Sabir, A., Alturki, A., Hossain, N., & Bahadar, A. (2023). Synthesis and characterizations of novel spinel ferrites nanocomposites Al0.5Cr0.5Zn0Fe_2O_4 and Zn0.5Cr0.5Al0Fe_2O_4. *Journal of Cluster Science*, *34*(2), 1099–1104. https://doi.org/10.1007/S10876-022-02285-8

Oberdörster, G., Oberdörster, E., & Oberdörster, J. (2005). Nanotoxicology: An emerging discipline evolving from studies of ultrafine particles. *Environmental Health Perspectives*, *113*(7), 823–839. https://doi.org/10.1289/EHP.7339

Odio, O. F., & Reguera, E. (2017). Nanostructured spinel ferrites: Synthesis, functionalization, nanomagnetism and environmental applications. *Magnetic Spinels - Synthesis, Properties and Applications*. https://doi.org/10.5772/67513

Olatunji, S. O., & Owolabi, T. O. (2023). Modeling the band gap of spinel nano ferrite material using a genetic algorithm based support vector regression computational method. *International Journal of Materials Research*, *114*(3), 161–174. https://doi.org/10.1515/IJMR-2022-0058

Omari Shekaftik, S., Jadid, H., Dehdashti, M. J., Mohammadian, Y., & Ebrahimi, W. (2023). Toward a minimum data set for assessing the risks of activities involved with nanomaterials. *Journal of Nanoparticle Research*, *25*(2). https://doi.org/10.1007/S11051-023-05674-W

Omar, W. I. W., Soon, C. F., Ahmad, M. K., & Shimomura, M. (2021). Hydrothermal synthesis of biocompatible nitrogen doped graphene quantum dots. *Energy and Environment*, *32*(7), 1170–1182. https://doi.org/10.1177/0958305X20984112

Ortiz-Quiñonez, J. L., Pal, U., & Villanueva, M. S. (2018). Structural, magnetic, and catalytic evaluation of spinel Co, Ni, and Co-Ni ferrite nanoparticles fabricated by low-temperature solution combustion process. *ACS Omega*, *3*(11), 14986–15001. https://doi.org/10.1021/ACSOMEGA.8B02229/ASSET

Pandit, V. A., Repe, G. R., Bhamre, J. D., & Chaudhari, N. D. (2020). A review on green synthesis and characterization technique for ferrite nanoparticles and their applications. *Journal of Physics: Conference Series*, *1644*(1), 012009. https://doi.org/10.1088/1742-6596/1644/1/012009

Pietroiusti, A., Stockmann-Juvala, H., Lucaroni, F., & Savolainen, K. (2018). Nanomaterial exposure, toxicity, and impact on human health. *Wiley Interdisciplinary Reviews: Nanomedicine and Nanobiotechnology*, *10*(5). https://doi.org/10.1002/WNAN.1513

Protecting workers during nanomaterial reactor operations. (2018). https://doi.org/10.26616/NIOSHPUB2018120

Rice, S. G. (n.d.). *Nanotechnology workforce protection : strategic plan and safe practices*. 133.

Roco, M. C., Mirkin, C. A., & Hersam, M. C. (2011). Nanotechnology research directions for societal needs in 2020: Summary of international study. *Journal of Nanoparticle Research*, *13*(3), 897–919. https://doi.org/10.1007/S11051-011-0275-5

Rohilla, D., & Chaudhary, S. (2023). *Advanced Nanoparticles: A BoonoraBaneforEnvironmentalRemediationApplications*. 27–74. https://doi.org/10.1007/978-3-031-24416-2_2

Salih, S. J., & Mahmood, W. M. (2023). Review on magnetic spinel ferrite (MFe_2O_4) nanoparticles: From synthesis to application. *Heliyon*, *9*(6), e16601. https://doi.org/10.1016/J.HELIYON.2023.E16601

Schubauer-Berigan, M. K., Dahm, M. M., Schulte, P. A., Hodson, L., & Geraci, C. L. (2015). Characterizing adoption of precautionary risk management guidance for nanomaterials, an emerging occupational hazard. *Journal of Occupational and Environmental Hygiene*, *12*(1), 69–75. https://doi.org/10.1080/15459624.2014.946515

Schulte, P. A., Leso, V., Niang, M., & Iavicoli, I. (2019). Current state of knowledge on the health effects of engineered nanomaterials in workers: A systematic review of human studies and epidemiological investigations. *Scandinavian Journal of Work, Environment and Health*, *45*(3), 217–238. https://doi.org/10.5271/SJWEH.3800

Schulte, P. A., Roth, G., Hodson, L. L., Murashov, V., Hoover, M. D., Zumwalde, R., Kuempel, E. D., Geraci, C. L., Stefaniak, A. B., Castranova, V., & Howard, J. (2016a). Taking stock of the occupational safety and health challenges of nanotechnology: 2000–2015. *Journal of Nanoparticle Research*, *18*(6). https://doi.org/10.1007/S11051-016-3459-1

Schulte, P. A., Roth, G., Hodson, L. L., Murashov, V., Hoover, M. D., Zumwalde, R., Kuempel, E. D., Geraci, C. L., Stefaniak, A. B., Castranova, V., & Howard, J. (2016b). Taking stock of the occupational safety and health challenges of nanotechnology: 2000–2015. *Journal of Nanoparticle Research*, *18*(6). https://doi.org/10.1007/S11051-016-3459-1

Sharma, C., Bansal, D., Bhatnagar, D., Gautam, S., & Goyal, N. (2023). *Advanced Nanomaterials: From Properties and Perspective Applications to Their Interlinked Confronts*. 1–26. https://doi.org/10.1007/978-3-031-24416-2_1

Sharma, P. K., Dorlikar, S., Rawat, P., Malik, V., Vats, N., Sharma, M., Rhyee, J. S., & Kaushik, A. K. (2021). Nanotechnology and its application: A review. *Nanotechnology in Cancer Management*, 1–33. https://doi.org/10.1016/B978-0-12-818154-6.00010-X

Sharpening the focus on occupational safety and health in nanotechnology on JSTOR. (n.d.). Retrieved September 15, 2023, from https://www.jstor.org/stable/40967745

Song, Y., Li, X., & Du, X. (2009). Exposure to nanoparticles is related to pleural effusion, pulmonary fibrosis and granuloma. *European Respiratory Journal*, *34*(3), 559–567. https://doi.org/10.1183/09031936.00178308

Soufi, A., Hajjaoui, H., Elmoubarki, R., Abdennouri, M., Qourzal, S., & Barka, N. (2021). Spinel ferrites nanoparticles: Synthesis methods and application in heterogeneous Fenton oxidation of organic pollutants – A review. *Applied Surface Science Advances*, *6*, 100145. https://doi.org/10.1016/J.APSADV.2021.100145

Tatarchuk, T., Bououdina, M., Judith Vijaya, J., & John Kennedy, L. (2017). Spinel ferrite nanoparticles: Synthesis, crystal structure, properties, and perspective applications. *Springer Proceedings in Physics*, *195*, 305–325. https://doi.org/10.1007/978-3-319-56422-7_22

Thakur, P., Chahar, D., Taneja, S., Bhalla, N., & Thakur, A. (2020a). A review on MnZn ferrites: Synthesis, characterization and applications. *Ceramics International*, *46*(10), 15740–15763.

Thakur, P., Sharma, P., Mattei, J. L., Queffelec, P., Trukhanov, A. V., Trukhanov, S. V., … & Thakur, A. (2018). Influence of cobalt substitution on structural, optical, electrical and magnetic properties of nanosized lithium ferrite. *Journal of Materials Science: Materials in Electronics*, *29*, 16507–16515.

Thakur, P., Taneja, S., Sindhu, D., Lüders, U., Sharma, A., Ravelo, B., & Thakur, A. (2020b). Manganese zinc ferrites: A short review on synthesis and characterization. *Journal of Superconductivity and Novel Magnetism*, *33*, 1569–1584.

Thakur, A., Thakur, P., & Khurana, S. P. (2022). *Synthesis and Applications of Nanoparticles* (pp. 1–17). Singapore: Springer.

Thakur, A., Verma, R., Wan, F., Ravelo, B., Edelman, I., Ovchinnikov, S., & Thakur, P. (2023). Investigation of structural, elastic and magnetic properties of $Cu^{2}+$ ions substituted cobalt nano ferrites. *Journal of Magnetism and Magnetic Materials*, *581*, 170980.

Thakur, A., Verma, R., Wan, F., Ravelo, B., Edelman, I., Ovchinnikov, S., & Thakur, P. (2023). Investigation of structural, elastic and magnetic properties of $Cu^{2}+$ ions substituted cobalt nano ferrites. *Journal of Magnetism and Magnetic Materials*, *581*, 170980.

Trybula, W., Newberry, D., & Fazarro, D. (2018). Nanotechnology health, safety, and environment overview. *Nanopackaging: Nanotechnologies and Electronics Packaging, Second Edition*, 957–973. https://doi.org/10.1007/978-3-319-90362-0_32

Turan, N. B., Engin, G. O., & Bilgili, M. S. (n.d.). *Environmental nanotechnology : implications and applications*. 301.

Warheit, D. B., Sayes, C. M., Reed, K. L., & Swain, K. A. (2008). Health effects related to nanoparticle exposures: Environmental, health and safety considerations for assessing hazards and risks. *Pharmacology and Therapeutics*, *120*(1), 35–42. https://doi.org/10.1016/J.PHARMTHERA.2008.07.001

Zainal Abidin, H. F., Hassan, K. H., & Zainol, Z. A. (2020). Regulating risk of nanomaterials for workers through soft law approach. *NanoEthics*, *14*(2), 155–167. https://doi.org/10.1007/S11569-020-00363-7

17 Spinel Nano-Ferrites for High-Frequency Power Applications

Vivek Verma, Adarsh Singh, and Manish Kumar Kansal*
Hindu College, University of Delhi, Delhi, India

Neelam Singh
Hansraj College, University of Delhi, Delhi, India

17.1 INTRODUCTION

Other than metals like iron, cobalt, nickel, and their alloys, the most significant ferromagnetic materials are metallic oxides of the spinel-type ferrites. The spinel ferrites have the formula $M^{2+}Fe_2^{3+}O_4^{2-}$, where M is a divalent metal cation such as Mn^{2+}, Cd^{2+}, Ni^{2+}, Zn^{2+}, Mg^{2+}, and Fe^{2+} which shares valency with Fe(III) cations for the occupancy of important sites. Different M^{2+} ions, a mix of trivalent and monovalent ions, and/or vacant sites resulting from defect structure can all be used to alter the composition of ferrite, such as lithium ferrite $(Li^{1+}Fe^{3+})_{1/2}(Fe_2^{3+})O_4^{2-}$. Because of its adaptability, mixed ferrites with altered physical characteristics that are appropriate for certain uses may be created. The processes used in the manufacturing of ferrites have an impact on their electrical and magnetic characteristics. Grain size, porosity, and chemical composition, among other factors, are important in influencing magnetic characteristics [1, 2]. By altering the grain boundary energy, small dopants can enhance the microstructure and encourage grain development. One crucial factor influencing ferrites' magnetic characteristics is their grain size [3]. Due to its good chemical stability, high resistivity, high saturation magnetization (M_s), and low magnetic losses, lithium ferrite has significant scientific and technological importance [4, 5]. Applications for mixed lithium ferrites may be found in phase shifters, gyrators, circulators, and isolators in microwave devices [6]. For wider technical applications, more enhancements in characteristics like resistivity, density, and dielectric and magnetic loss are necessary. Because of its advantageous electrical and magnetic qualities, lithium ferrite is also utilized in power electronic equipment such as antennas, transformers, core inductors, and switching circuits [7, 8]. Ferrites' electrical and magnetic characteristics may be readily changed to meet specific needs by doping them with the right components [9, 10]. Ferrites' magnetic, electrical, and microstructural characteristics can also be changed in tiny quantities by certain additions [4]. Different forms of lithium ferrites may be created by introducing metal ions such as Cd^{2+}, Mg^{2+}, Co^{2+}, Zn^{2+}, and Al^{3+} into the $Li^{0.5}Fe^{2.5}O^4$ compound [5, 11]. Electrical resistivity, permeability, and saturation magnetization may all be improved by substituting transition metal ions at particular locations. The use of appropriate dopants to modify the magnetic, electrical, and dielectric characteristics of ferrites facilitates their use in a variety of applications. Enhanced saturation magnetization occurs when the transition metal ions Cd^{2+} and Zn^{2+} are substituted at the tetrahedral site of lithium ferrite [12]. It's commonly known that Mn^{3+} replaces Fe^{3+} ions in lithium's octahedral site, which has been shown to increase ferrites' permeability [13]. It has been discovered that a tiny amount of Mn^{3+} doping in lithium ferrite inhibits the production of Fe^{2+} ions, which is beneficial for raising electrical resistivity [14]. Microstructure modifications can be accomplished via high-temperature sintering or by using low-melting-temperature additions like V_2O_5, Bi_2O_3, and so forth. Nevertheless, it is not desired for high-temperature (1100°C) sintering to cause lithium to evaporate. It is thus advised to use low-melting additives as opposed to high-temperature conditions. The current study examines how Zn and Sn doping affects the magnetic, structural, and power loss characteristics of pristine and doped lithium ferrite samples.

17.2 EXPERIMENTAL

The ferrite samples $Li_{0.5+0.5x}Sn_xFe_{2.5-1.5x}O_4$ and $Li_{0.35+x/2}Zn_{0.3}Mn_{0.05}Ti_xFe_{2.30-1.5x}O_4$ ferrites (x = 0.0–0.20 in the step of 0.05 + 0.5 wt% Bi_2O_3) were prepared using the solid-state technique [15]. The precursor powders used were AR grade Li_2CO_3, Fe_2O_3, SnO_2, ZnO, MnO_2, TiO_2, and Bi_2O_3. A small quantity of Bi_2O_3 (0.5 wt%) was used to lower the final sintering temperature and enhance the microstructural density while preventing lithium volatilization. In a pestle and mortar, these powders were taken in stoichiometric proportions and well combined. Thereafter, the mixture was pre-sintered for 10 hours at 750°C. The resulting powder samples were thoroughly ground after pre-sintering. A tiny quantity of polyvinyl alcohol (PVA) was added to the powder samples as a binder in order to create the samples. Next, the mixture was compressed with 10 tons of pressure into pellets and toroids. For 5 hours, the last sintering step was done at 1050°C. By employing Cu-K_α radiation (λ = 1.5406 Å) in X-ray diffraction (XRD), the structural characteristics of the

* Corresponding author.

DOI: 10.1201/9781003568940-17

samples were examined. A 0.02-step size Rigaku Miniflex II equipment was used to conduct the XRD study. Toroid-shaped samples with specified dimensions (inner radius = 2.75 mm, outside radius = 5.5 mm, width = 3 mm) were used to determine the power loss and permeability. A B-H analyser (IWATSU, SY 8232) was used to evaluate the power loss, and an Agilent 4284 or 4285 equipment was used to test the permeability. The frequency range in which the measurements for the generated samples were conducted was 1 MHz to 1 GHz. A vibrating sample magnetometer (VSM, Lake Shore 7304) was used to get the magnetic hysteresis (M-H) curves for each of the ferrite samples at room temperature.

17.3 RESULTS AND DISCUSSION

17.3.1 Structural Analysis

Figure 17.1 displays the XRD patterns of lithium ferrite samples that are both pure and doped. Phase analysis was performed using Cu Kα radiation at a wavelength of 0.154 nm to acquire the patterns. A strong match between the JCPDS card number 74-1911 and the spotted diffraction patterns and peak positions has proven the formation of the lithium ferrite phase. The pure and doped lithium ferrite samples do not include any secondary or impurity phases, according to the XRD patterns. This demonstrates that the samples have a space group *Fd*3*m* cubic spinel structure in a single phase. The typical formula for spinel ferrites is MFe_2O_4, where M stands for divalent metal ion, mostly from transition metals. Owing to the structure of lithium ferrite, a structural defect might result from the "M" component consisting of a mixture of trivalent and monovalent ions or a mixture of trivalent ions and vacancies [16]. The XRD patterns of materials exhibiting Sn doping (Sn ≥ 0.1) have discernible minor peaks at approximately 2θ = 26.50° and 2θ = 33.80°. These peaks show that there are more phases or impurities related to the Sn doping. However, there are no appreciable alterations to the XRD patterns of the lithium ferrite samples doped with Ti. This is because the ionic radii of Ti^{4+} (0.60 Å) and Fe^{3+} (0.67 Å) are relatively compatible, resulting in minimal lattice distortion or phase changes [17].

17.3.2 Surface Morphology

Scanning electron microscopy (SEM) micrographs were used to analyse the surface morphology of both pure and modified $Li_{0.5}Fe_{2.5}O_4$ samples. Figure 17.2 displays the SEM images of each specimen that was sintered at 1050°C. The micrographs show that although the grain size distributions in the samples are different, the grain growth is uniform. Variations in the development rates of the various phases might be the cause of the variances in grain size between the samples [18, 19]. Using the image analysis programme ImageJ, histogram plots were created to ascertain the samples' grain sizes. It was discovered that the grain size in the Sn-doped lithium ferrite samples varied from 0.6 to 2.0 μm. The surface of the pristine $Li_{0.5}Fe_{2.5}O_4$ sample shows a homogeneous distribution of grains, most of which have an average size of around 0.7 μm, with some grains having an average size of about 1.1 μm. Grain development and connectivity in the lithium ferrite samples are clearly altered by Sn doping. Grain connection and diffusion are better in samples with Sn concentrations of x = 0.05 and x = 0.10, which can improve a number of microstructural-dependent characteristics. Furthermore, porosity is shown

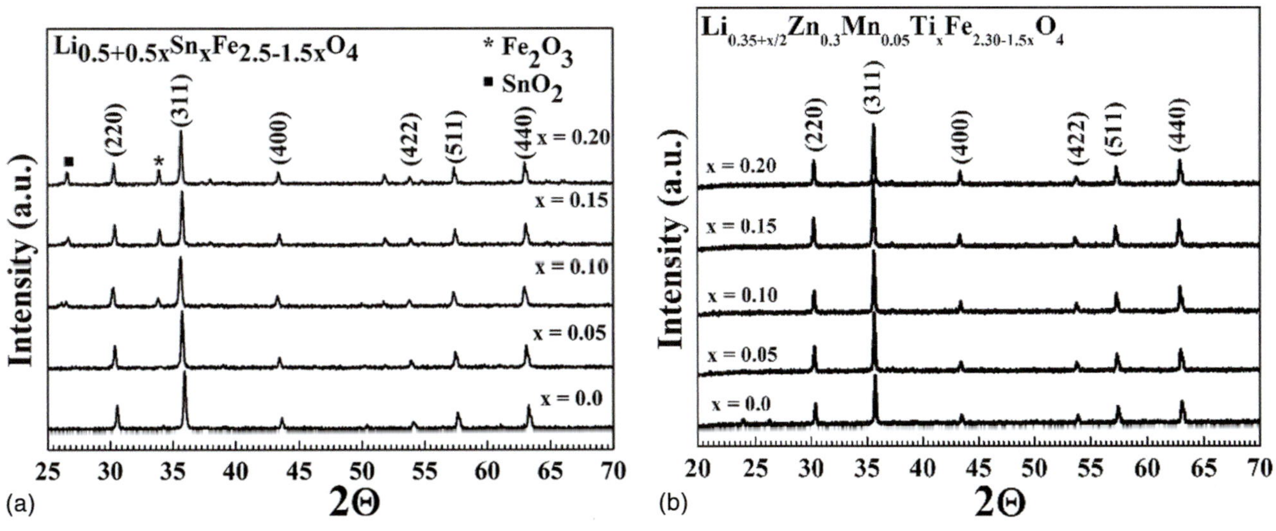

FIGURE 17.1 (a) X-ray diffraction patterns of $Li_{0.5+0.5x}Sn_xFe_{2.5-1.5x}O_4$ [15]; (b) X-ray diffraction patterns of $Li_{0.35+x/2}Zn_{0.3}Mn_{0.05}Ti_xFe_{2.30-1.5x}O_4$ ferrite samples. (Reprinted from High permeability and low power loss of Ti and Zn substitution lithium ferrite in high frequency range, Verma et al. 2009, *Journal of Magnetism and Magnetic Materials*, 321(22): 3808–3812. Copyright (2024), with permission from Elsevier vide License No. 5714600529176.)

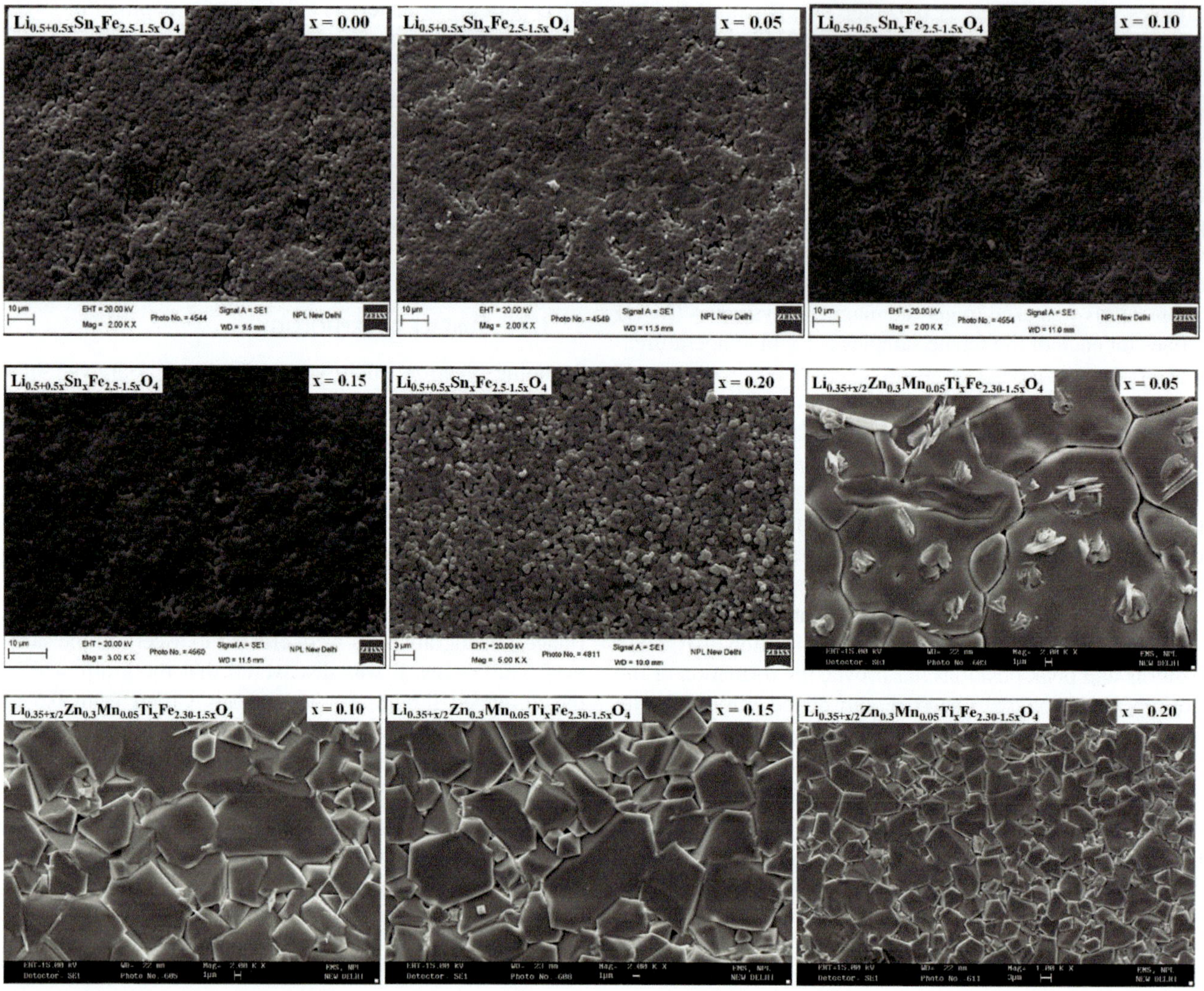

FIGURE 17.2 SEM micrographs of $Li_{0.5+0.5x}Sn_xFe_{2.5-1.5x}O_4$ [15] and $Li_{0.35+x/2}Zn_{0.3}Mn_{0.05}Ti_xFe_{2.30-1.5x}O_4$ ferrite specimens. (Reprinted from High permeability and low power loss of Ti and Zn substitution lithium ferrite in high frequency range, Verma et al. 2009, *Journal of Magnetism and Magnetic Materials*, 321(22): 3808–3812. Copyright (2024), with permission from Elsevier vide License No. 5714600529176.)

to diminish with Sn doping, especially in samples where Sn concentrations are between 0.05 and 0.10. Grain growth is more noticeable in the lithium-zinc ferrite samples that have had Ti replaced. When the concentration of Ti rises, so does the grain size. The x = 0.05 sample among the Ti-substituted specimens exhibits consistent grain development; however, if the Ti concentration is raised higher, the grain expansion turns nonuniform. Overall, the SEM micrographs show how doping affects grain size, connectivity, and porosity by illuminating the surface morphology and grain properties of the pure and doped lithium ferrite samples.

17.3.3 Magnetic Permeability Response

The complex permeability ($\mu^* = \mu' - j\mu''$) of all pristine and doped ferrites was computed at room temperature with frequencies ranging from 10^6 to 10^9 Hz. It is noticed that when the concentration of Sn increases, the actual part of the permeability of the virgin lithium ferrite sample reduces. Furthermore, the real component of permeability in all samples has a constant value up to 7 MHz before suddenly decreasing, whereas the imaginary part exhibits an elevated value at the relaxation frequency. Several parameters impact a magnetic sample's permeability, including saturation magnetization (M_s) and mean grain diameter (D_M). The equation below illustrates the link between permeability (μ_i) and various factors:

$$\mu_i \propto \mu_0 Ms^2 D_m / [K_1 + (3/2)\lambda_s \sigma] \beta^{1/3} \delta,$$

where μ_0 denotes the permeability of free space, K_1 represents magneto-crystalline anisotropy, λ_s signifies the saturation magnetostriction constant, σ is the inner stress, β is the fraction of impurity, and δ is the domain wall width.

In conclusion, the produced samples' complicated permeability response shows a frequency-dependent behaviour, with the real component falling as the concentration of Sn rises. The average grain size and saturation magnetization, which affect the samples' permeability, can be used to interpret the observed behaviour. The microstructure of Sn-doped lithium ferrites is improved, which may result in increased grain coherence and other microstructural features. On the other hand, Sn doping causes the saturation magnetization to drop, which lowers permeability. Because Sn ions have a larger ionic radius than Fe and Li ions, their replacement is responsible for the drop in saturation magnetization. This replacement modifies the ferrite's magnetic characteristics, which lowers magnetization and, as a result, lowers permeability. Likewise, in lithium ferrites replaced with Ti, the substitution of Ti ions in the ferrite structure may also result in a reduction in microstructure and magnetism. Consequently, the substitution of Ti results in a reduction in permeability, especially when the concentration of Ti is greater than or equivalent to 0.05. While the decrease in microstructure could be the result of Ti's effect on grain formation and connectivity, the drop in magnetization can be explained by the fact that Ti has different magnetic characteristics than Fe and Li ions. Permeability generally decreases with decreasing saturation magnetization and microstructure in both Sn- and Ti-doped lithium ferrites. The measured permeability variances are eventually defined by the specific effect of these additions on the microscopic structure and magnetic properties of ferrites. The observed relaxation-like frequency variation of permeability may be attributed to two phenomena: magnetic dipole rotation inside domains and reversible movement of domain barriers. These events affect the permeability response at various frequencies and contribute to the overall magnetic behaviour. Permeability response is mainly dictated by spin rotation resonance and domain wall dislocation resonance at higher frequencies. Whereas spin rotation resonance relates to the rotation of magnetization dipoles inside domains, domain wall displacement resonance is linked to the reversible motion of domain walls. It is important to remember that the resonance frequency of spin rotation is usually greater than the resonance frequency of domain wall displacement. While spin rotation resonance is more likely to occur in the frequency range of 1 MHz to 1 GHz, the chance of domain wall displacement resonance is very low. Consequently, spin rotation resonance mostly affects the frequency dispersion of permeability in lithium ferrite samples. Snoek's limit defines the relationship between the actual aspect of permeability (μ') and the resonance rate for domain wall motion, stating that $\omega\mu'$ is a constant value [20]. The resonance frequency (ω) in this formula denotes the point at which the real component of permeability (μ') starts to drop. In summary, the relaxation character seen in the frequency dependence of permeability in lithium ferrite samples may be attributed to the unilateral displacement of domain walls and the rotation of magnetic dipoles. Spin rotation resonance plays a major role in determining the permeability response at higher frequencies, and Snoek's limit controls the frequency dispersion, which connects the actual component of permeability to the resonance rate for domain wall movement.

17.3.4 Hysteresis Properties

The magnetic characteristics and behaviour of ferrite specimens were assessed using M-H curves obtained from VSM measurements conducted at room temperature with an external magnetic field of about 6000 Oe. Based on the M-H curves, numerous magnetic properties may be derived, notably saturation magnetization (M_s), coercivity (H_c), and remanent magnetization (M_r) of the specimens. Based on the available data, the pristine lithium ferrite sample was found to have the greatest saturation magnetization, measuring around 51 emu/g. However, adding Sn and Ti amounts to the ferrite samples resulted in a reduction in saturation magnetization [21] as shown in Table 17.1. Sn doping had no discernible effect on the coercivity or retentivity (remanence) of the lithium ferrite samples, suggesting that the coercivity and remanent magnetization stayed mostly unchanged. Neel's two sub-lattice models are discussed in order to explain the observed variations in saturation magnetization (M_s) among the ferrite samples. The magnetic characteristics of materials having antiferromagnetic or ferrimagnetic ordering, such as ferrites, may be explained theoretically using Neel's model. It takes into account two sub-lattices of magnetic moments that interpenetrate one another and interact differently with nearby moments. More investigation may be done to comprehend the connection between the ferrite samples' composition and magnetic characteristics, specifically the variations in saturation magnetization (M_s), by utilizing Neel's sub-lattice model. Neel's two sub-lattice model describes the magnetization of spinel ferrite unit cells as $\mu_B = M_B(x) - M_A(x)$, where μ_B represents the resulting magnetic moment and $M_B(x)$ and $M_A(x)$ represent the magnetic moments of the B and A sites, respectively. Li^+ and Fe^{3+} ions occupy the B-site

TABLE 17.1
Magnetic Properties of $Li_{0.5+0.5x}Sn_xFe_{2.5-1.5x}O_4$ and $Li_{0.35+x/2}Zn_{0.3}Mn_{0.05}Ti_xFe_{2.30-1.5x}O_4$ Ferrite Samples Extracted from M-H Hysteresis Curves Recorded at Room Temperature

	Saturation Magnetization (emu/g) @ Applied Field of 6000 Oe	
	$Li_{0.5+0.5x}Sn_xFe_{2.5-1.5x}O_4$	$Li_{0.35+x/2}Zn_{0.3}Mn_{0.05}Ti_xFe_{2.30-1.5x}O_4$
x = 0.00	51	80
x = 0.05	45	75
x = 0.10	41	63
x = 0.15	38	54
x = 0.20	35	41

in a 1:3 ratio in lithium ferrite, whereas Fe^{3+} ions occupy all A sites. Spinel-type lithium ferrite $Li_{0.5}Fe_{2.5}O_4$ is known to occur in two crystalline forms. (i) In the "ordered" form (space group *P*4$_1$32/*P*4$_3$32), the Fe^{3+} ions are at octahedral 12d and tetrahedral 8c sites, and Li^+ ions occupy only the octahedral 4b positions in the cubic primitive unit cell. (ii) Fe^{3+} is located at tetrahedral 8a places in the "disordered" $Li_{0.5}Fe_{2.5}O_4$'s inverted spinel structure (space group *Fd*3*m*), whereas Li^+ and Fe^{3+} are dispersed randomly throughout the 16d octahedral sites [22]. The magnetic moment per lithium ferrite molecule in the experiment is more than that which can be calculated theoretically. The total magnetic moment in lithium ferrite is calculated to be 2.5 μ_B; however, experimental observations have revealed a value of 2.6 μ_B. Li is monovalent, hence in order to preserve charge neutrality, there must be an overabundance of Fe ions. $Li_{0.5}Fe_{2.5}O_4$, the stoichiometric formula, specifies that 0.5 Fe^{3+} must be added in order to achieve charge balance. In octahedral sites, the Li^+ takes the place of every fourth Fe^{3+}.

17.3.5 Power Loss

Better and more sustainable technologies are in greater demand in the modern world, especially in the electricity industry. More advanced resources with outstanding effectiveness and low power loss are required as the use of electrical and electronic equipment that function at high frequencies for short reaction times grows. In applications involving high frequencies, ferrites are essential. Because of its exceptional magnetic and electrical properties, lithium ferrite constitutes one of the most frequently employed soft magnetic materials. Because lithium ferrite can function well in the microwave frequency range, it is a great material to use in electronic devices that run quickly.

The extra information underlines that ferrites tend to lose power in the MHz frequency range, which restricts their use in high-frequency devices. Based on their distinct generating methods, hysteresis loss (P_{ch}), eddy current loss (P_{ce}), and residual loss (P_{cr}) may be used to categorize the power loss (P_{cv}) in ferrites. The whole loss of power may be written as: $P_{cv} = P_{ch} + P_{ce} + P_{cr} = K_H B^2 f + \frac{K_E D^2 B^2 f^2}{\rho} + P_{cr}$, where K_H and K_E are constants, B is the magnetic flux density, f is the frequency, and ρ is the resistivity. Figure 17.3 shows that minor loading of Sn in lithium ferrite (x = 0.05 composition) reduces overall power loss in the 50 Hz to 3 MHz frequency band. This reduction in power loss may be due to better microstructure. The replacement of Ti in ferrites assists in minimizing hysteresis loss by enhancing the development of grains and microstructure. At higher frequencies, losses from eddy currents become an important aspect and can have a significant impact on efficiency. These losses can be alleviated by raising the resistivity of the material; the addition of a small amount of Mn inhibits the formation of Fe^{2+} ions, improving the resistivity of ferrites; increasing grain boundary resistivity also helps reduce eddy current losses in polycrystalline ferrites. It is observed that DC resistivity varies from 10^6 to 10^7 orders with a small concentration of Ti in $Li_{0.35+x/2}Zn_{0.3}Mn_{0.05}Ti_xFe_{2.30-1.5x}O_4$.

As the Sn content in lithium ferrites grew, power loss also increased. Given that P_{ch}, P_{ce}, and P_{cr} are frequency-dependent losses, more research into the impact of individual power loss phenomena on operating frequency is necessary, P_{cv} can be fitted as: $P_{cv} = K B_m^x f^y$ where, K is a constant, x is the Steinmetz exponent, and y is the frequency index. When magnetic materials are subjected to an exterior time-varying magnetic flux, the following equation—also known as Steinmetz's equation or sometimes the

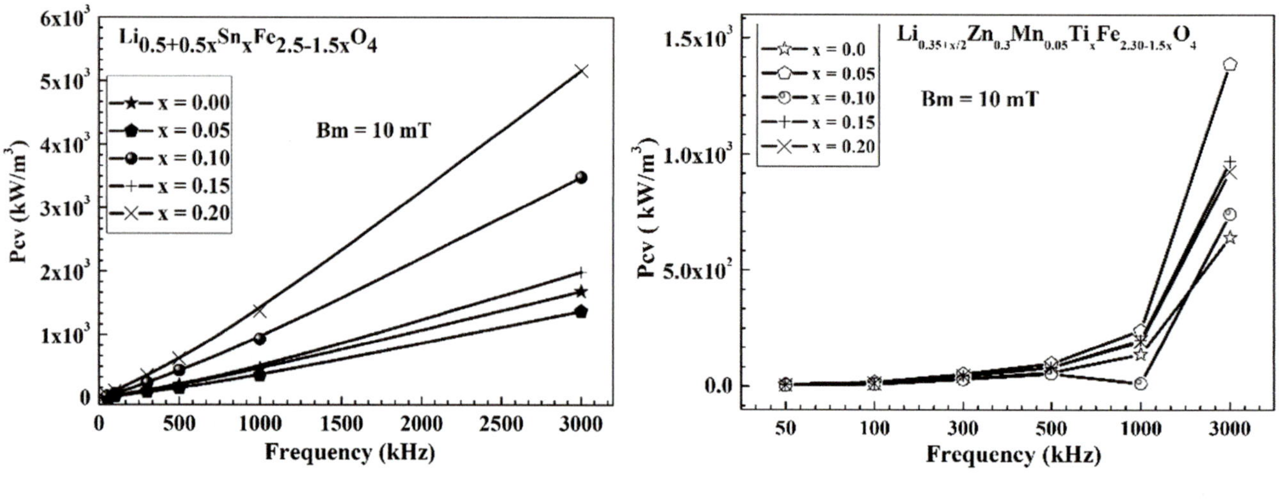

FIGURE 17.3 Variation of P_{cv} and frequency at 10 mT of $Li_{0.5+0.5x}Sn_xFe_{2.5-1.5x}O_4$ [15] and $Li_{0.35+x/2}Zn_{0.3}Mn_{0.05}Ti_xFe_{2.30-1.5x}O_4$ samples. (Reprinted from High permeability and low power loss of Ti and Zn substitution lithium ferrite in high frequency range, Verma et al. 2009, *Journal of Magnetism and Magnetic Materials*, 321(22): 3808–3812. Copyright (2024), with permission from Elsevier vide License No. 5714600529176.)

power equation—is employed to calculate the total power dissipation per unit volume. Researchers may learn more about the behaviour of power losses and how they relate to the operating frequency by using the Steinmetz equation and examining the values of the constant K, Steinmetz exponent (x), and frequency index (y). This knowledge contributes to the creation of sophisticated magnetic materials that operate better in high-frequency applications and lose less power.

17.4 CONCLUSIONS

The goal of this chapter is to create cutting-edge magnetic materials that are more efficient and use less power loss so that fast-operating electronic devices may be met in a sustainable way. Because of its high resistivity, high saturation magnetization, low magnetic losses, high Curie temperature, and chemical stability, lithium ferrite is particularly interesting. The magnetic and electrical characteristics of lithium ferrite can be improved or changed by substituting components like Sn and Ti. On the other hand, it was noted that the ferrite samples' saturation magnetization (M_s) decreased upon the addition of Sn and Ti dopants. $Li_{0.5+0.5x}Sn_xFe_{2.5-1.5x}O_4$ and $Li_{0.35+x/2}Zn_{0.3}Mn_{0.05}Ti_xFe_{2.30-1.5x}O_4$ ferrites (x = 0.0 to 0.20 in step of 0.05 + 0.5 wt% Bi_2O_3) were prepared by solid-state technique for high-frequency applications. For both the pure and doped lithium ferrite samples, single-phase cubic spinel structures were verified by XRD data. SEM micrographs revealed doped lithium ferrite samples with enhanced microstructure and grain coherence but reduced porosity. Saturation magnetization was reduced in Sn- and Ti-doped materials, but power loss was significantly reduced as well. This might be the result of the samples' combination of microstructural and magnetic features. Lithium ferrites doped with Sn and Ti have lower power loss, which makes them attractive for a range of high-frequency power sector technical applications.

ACKNOWLEDGEMENT

The authors extend their gratitude to the Principal, Hindu College, University of Delhi, for giving continual inspiration and drive to complete this effort.

REFERENCES

1. E. Wolska, P. Piszora, W. Nowichi, J. Darul, Int. J. Inorg. Mater. **3** (2001) 503.
2. Y.S. Cho, Y.L. Burdick, V.R.W. Amarakoon, J. Am. Ceram. Soc. **82** (1999) 1416.
3. P. Kishan. V.R.K. Murthy (Ed.), Microwave Materials, Narosa Publishing House, 1993, p. 141.
4. N. Kumar, P. Kishan, Z.H. Zaidi, J. Magn. Magn. Mater. **184** (1998) 355.
5. R.K. Kotnala, V. Verma, V. Pandey, V.P.S. Awana, R.P. Aloysius, P.C. Kothari, Sol. Stat. Comm. **143** (2007) 527–531.
6. M.M.S. Sanad, A.A. Azab, T.A. Taha, Ceram. Int. **48** (2022) 12537–12548.
7. M. Bahgat, F.E. Farghaly, S.M.A. Basir, O.A. Fouad, J. Mater. Process. Technol. **183** (2007) 117–121.
8. V. Verma, S.P. Gairola, M.C. Mathpal, S. Annapoorni, R.K. Kotnala, J. Alloys Compd. **481** (2009) 872–876.
9. V. Verma, V. Pandey, R.K. Kotnala, H. Kishan, N. Kumar, P.C. Kothari, J. Alloys Compd. **443** (2007) 178–181.
10. V. Verma, S.P. Gairola, V. Pandey, R.K. Kotanala, H. Su, Solid State Commun. **148** (2008) 117–121.
11. Z. Yu, K. Sun, L. Li, Y. Liu, Z. Lan, H. Zhang, J. Magn. Magn. Mater. **320** (2008) 919–923.
12. S. Sindhu, M.R. Anantharaman, B.P. Thampi, K.A. Malini, P. Kurian, Bull. Mater. Sci. **25** (2002) 599–607.
13. M. Maisnam, S. Phanjoubam, H.N.K. Sarma, C. Prakash, L.R. Devi, O.P. Thakur, Mater. Lett. **58** (2004) 2412–2414.
14. Z. Yue, J. Zhou, X. Wang, Z. Gui, L. Li, J. Europ. Ceram. Soc. **23** (2003) 189–193.
15. N. Singh, M. Kumar Kansal, P. Kumar, S. Dabas, V. Verma, Physics Open **16** (2023) 100155.
16. M. Sharif, J. Jacob, M. Javed, A. Manzoor, K. Mahmood, M.A. Khan, Physica B: Condens. Matter **567** (2019) 45–50.
17. V. Vaithyanathan, K. Ugendar, J.A. Chelvane, K.K. Bharathi, S.S.R. Inbanathan, J. Magn. Magn. Mater. **382** (2015) 88–92.
18. P. Kharbanda, T. Madaan, I. Sharma, S. Vashishtha, P. Kumar, A. Chauhan, S. Mittal, J.S. Bangruwa, V. Verma, Heliyon **5** (2019) e01151.
19. T.K. Pathak, N.H. Vasoya, V.K. Lakhani, K.B. Modi, Ceram. Int. **36** (2010) 275–281.
20. K.P. Thummer, M.C. Chhantbar, K.B. Modi, G.J. Baldha, H.H. Joshi, J. Magn. Magn. Mater. **280** (2004) 23–30.
21. E. Wolska, P. Piszora, W. Nowicki, J. Darul, Int. J. Inorg Mater. **3** (2001) 503–507.
22. V. Verma, S.P. Gairola, V. Pandey, J.S. Tawale, H. Su, R.K. Kotnala, J. Magn. Magn. Mater. **321** (2009) 3808–3812.

18 Devices for Human Safety Based on Spinel Nano-Ferrites

*Preeti Thakur and Atul Thakur**
Nanotechnology Wing, Innovative Science Research Society, Shimla, India
nanoLatticeX LLP, Manesar, Gurugram, India
Amity Institute of Nanotechnology, Amity University Haryana, Gurugram, India

Amitender Singh
Amity Institute of Nanotechnology, Amity University Haryana, Gurugram, India
Govt. Polytechnic Indri (Nuh), Indri, India

Kavita Yadav
G.C.W. Gurawara, Rewari, India

18.1 INTRODUCTION

The development of functional nanomaterials is crucial for the use of upgraded technologies in human safety devices. The precursor molecules for the production of nanomaterials are derived from natural resources. Researchers are presently investigating the popular natural and functional materials for a wide range of specialized safety applications. Because spinel ferrite nanoparticles (NPs) (SFNPs) have so many uses in so many different fields, SFNPs are of interest to researchers for industrial and biomedical uses. SFNPs have vast safety device applications, like in wastewater treatment (Kefeni, Mamba, and Msagati 2017), electronic material production (Zhang et al. 2015), and adsorbent and catalysts (Tan et al. 2015). SFNPs are highly helpful for humans in the biomedical fields for drug administration and release (Spirou et al. 2018), biomagnetic separation (Plouffe, Murthy, and Lewis 2015), MRI contrast enhancement (Yang et al. 2015), and hyperthermia (tumor therapy). They are also helpful in the development of contemporary sensors and biosensors, having utility in the human safety device domains (Lee et al. 2014). Furthermore, SFNPs have potential antibacterial action against a variety of pathogenic pathogens (Abdel Maksoud et al. 2018).

18.1.1 Ferrite Oxide

Fe_3O_4 magnetic particles may be obtained from the iron sand found in nature. Contaminants of natural iron sand are separated from it using both chemical and mechanical techniques. Naturally, iron sand contains more than 90% Fe_3O_4 particles by weight. The two commercial raw materials that are often used to synthesize Fe_3O_4 are $FeCl_3{\cdot}6H_2O$ and $FeCl_2{\cdot}4H_2O$. The most widely used production techniques include hydrothermal, co-precipitation, and sol-gel. When Fe_3O_4 NPs agglomerate among the particles, templates or surfactants are used to obtain homogenous NPs with specific morphologies and sizes. When Zn and Mn are doped on Fe_3O_4, Fe_3O_4 becomes superparamagnetic and can be used in biomedical device applications.

18.1.2 Different Types of Ferrite Oxide

Hematite (α-Fe_2O_3) is a highly stable iron oxide at higher temperatures. Since α-Fe_2O_3 is a prominent corrosion product of iron alloys or iron metals, it is commonly derived from iron rust. Co-precipitation and hydrothermal techniques are two common approaches used to effectively produce α-Fe_2O_3 NPs from commercial raw metals like $FeCl_3{\cdot}6H_2O$ and $Fe(NO_3)_3{\cdot}9H_2O$. At room temperature (RT), the α-Fe_2O_3 NPs with 8 nm particle size exhibit superparamagnetic properties and a rather high magnetism. Thus, its implementation for biological and spintronic device applications is possible.

18.2 SPINEL FERRITE (SF)

SFs are also known as cubic ferrites. Since cubic ferrites have a high electrical resistivity and minimal eddy current losses, they are typically utilized at microwave frequencies. Bragg and Nishikawa defined the structure of SF for $MgAl_2O_4$ for the first time in 1915. The chemical formula of SFs is MFe_2O_4, where M is bivalent transition metals, like Zn, Cu, Co, Ni, Fe, and Mn. As per JCPDS: 22-1086, the spinel structure is a member of space group Fd_3m (Sivakumar et al. 2011; Thakur et al. 2021). The stabilization energy, specification's ion rays, interstices' size, synthesis process, and circumstances used throughout the synthesis period all affect the affinity (Kefeni, Msagati, and Mamba 2017; Singh et al. 2023). Due to their exceptional magnetic and electronic properties (they can readily be separated by an external magnet), SFNPs are a useful and significant tool in contemporary catalytic organic synthesis.

* Corresponding author.

DOI: 10.1201/9781003568940-18

18.2.1 Classification of Spinel Ferrites

Three classifications have been established for SFs depending on the distribution of cations at tetrahedral (A) and octahedral (B) sites in $\left[M_{1-x}^{2+}Fe_{x}^{3+}\right]_{A}\left[Fe_{2-x}^{3+}M_{x}^{2+}\right]_{B}$ (Granone et al. 2018), where x is the proportion of divalent cations (M^{2+} ions) occupying octahedral positions. A spinel is referred to as normal when $x = 0$; when $x = 1$, it is an inverse spinel, and when $0 < x < 1$, it is a mixed spinel. Fe^{3+} occupies octahedral (B) sites in typical SFs, whereas the divalent cations, M^{2+} are in tetrahedral (A) sites. These ferrites have the structural formula $[M^{2+}]_{(A)}$ $[Fe_2^{3+}]_{(B)}O_4^{2-}$; $ZnFe_2O_4$ and $CdFe_2O_4$ are two examples (Chandrasekaran et al. 2018; Naagar et al. 2023). Fe^{3+} ions are arranged half in (A) sites and half in (B) sites in inverse SFs, with M^{2+} occupying octahedral (B) sites. The structural formula for inverse ferrites is $[Fe^{3+}]_{(A)}$ $[M^{2+}Fe^{3+}]_{(B)}$ O_4^{2-}; Fe_3O_4, $NiFe_2O_4$, $CoFe_2O_4$, and $MgFe_2O_4$ are a few examples (Khishigdemberel et al. 2018; Chahar et al. 2023). M^{2+} and Fe^{3+} ions are dispersed at both the tetrahedral (A) and octahedral (B) sites in mixed SFs. $MnFe_2O_4$ is one example (Silva et al. 2019). It should be emphasized that x is dependent upon the ferrite's composition and manner of manufacture. For instance, depending on the synthesis conditions used, $CuFe_2O_4$ can produce either an inverted or mixed spinel structural type (Masunga et al. 2019).

18.2.2 Properties of Spinel Ferrites

Because of SFs' distinct mechanical, optical, electrical, dielectric, electronic, and catalytic capabilities, they are of great interest (Thakur, Thakur, and Hsu 2011; Tiwari et al. 2020). The SFs are special materials that have semiconductor and ferrimagnetic qualities; they are also regarded as magnetic semiconductors (Abdel Maksoud et al. 2020). Their unique physicochemical characteristics include a modest chemical composition, strong thermal stability, substantial magnetocrystalline anisotropy, and a high Curie temperature. They also have higher specific surface area, surface active sites, great magnetic properties, adjustable size and shape, high chemical stability, and ease of functionalization or modification. Additionally, SFs have recently been the focus of a lot of study in the water treatment industry due to low cost, significant catalytic activity, selectivity, fascinating bandgap, capacity to be recovered and reused in successive cycles without noticeably losing activity, and less toxicity. A few of the properties of SFNPs are listed in Figure 18.1.

FIGURE 18.1 Important properties of spinel ferrites making them suitable for human devices.

18.2.3 Synthesis of Spinel Nano-Ferrites

The main variables that affect SFNPs quality are synthesis circumstances and procedures used throughout the synthesis period. Thermal decomposition, chemical precipitation, solvothermal, hydrothermal, sol-gel, sonochemical, vapor deposition, micro-emulsion, and electrochemical are just a few of the many diverse synthesis methods for SFNPs (Cruz et al. 2017; Karimzadeh et al. 2017; Thakur and Thakur 2022). Each approach has advantages and disadvantages of its own; also, a variety of factors, including temperature, type of solvent, pH, and reaction duration, might impact characteristics of SFNPs that are synthesized. For instance, it is commonly known that temperature, pH, and reaction time all affect crystallite size (Kefeni, Msagati, and Mamba 2015).

The two most prevalent methods for synthesizing SFNPs in biological applications are co-precipitation and thermal decomposition. To produce top-grade SFNPs using the co-precipitation synthesis process, precise adjustments to ionic strength, pH, temperature, and the proper molar ratio of divalent metal:trivalent ion (1:2) are necessary. SFNPs size produced by co-precipitation synthesis process may be regulated by temperature and stirring duration; however, post-annealing is typically necessary to further stabilize the synthesized NPs. The quantity of organic solvents used, temperature, heating rate, and SFNPs form may all change during thermal breakdown. During the synthesis process, it is also feasible to create NPs with a homogeneous shape and narrow particle size range. Biocompatible material's surface coating is also achievable (Reyes-Rodríguez et al. 2018). For SFNPs with potential applications in biomedicine devices, thermal breakdown is often favored over co-precipitation synthesis as it may produce SFNPs with narrow size distributions and high crystallite. Co-precipitation is the more straightforward approach for synthesizing SFNPs; however, thermal decomposition synthesis is the more preferable method for synthesizing uniformly sized and shaped crystalline SFNPs.

18.3 SPINEL FERRITES FOR HUMAN SAFETY DEVICES

SFs have drawn particular interest in the field of electronic devices for human safety and biomedical applications. Few of the SFNPs used in different applications pertaining to human safety devices are explained as follows.

18.3.1 Cobalt Ferrite ($CoFe_2O_4$)

After Fe_3O_4, cobalt ferrite is one of the SFNPs that have been investigated in depth. It is a hard-magnetic material with an fcc spinel structure that is either inverse or mixed. It is among the preferred options for use in medicinal devices in areas, such as medication administration, medical diagnostics, hyperthermic tumor therapy, and MRI (Lin et al. 2018; Thakur et al. 2023). The primary characteristic of a permanent magnet in bulk form is its high coercivity. Its reduced coercivity values in the range of 0.50–2 kOe are found in its nanosized form. However, once $CoFe_2O_4$ was mechanically ground to 25 nm using grinding balls, a concomitant rise in coercivity was reported, going from 1.9 to 4.1 kOe and M_s from 69.5 to 71.4 emu/g (Ponce et al. 2013). Doping $CoFe_2O_4$ with metals is another method of modifying its magnetic characteristics, similar to other SFNPs.

18.3.2 Magnetite (Fe_3O_4)

It is one of the biocompatible SFNPs that have been explored and utilized extensively, which has an inverted spinel structure. Bulk Fe_3O_4 exhibits ferromagnetic characteristics with a Curie temperature of 585°C and turns superparamagnetic below a size of around 20 nm (Ling, Lee, and Hyeon 2015). Fe_3O_4 NPs play a significant role in vast medical devices, including MRI, drug delivery, and cancer treatment (hyperthermia) applications (Vallabani and Singh 2018). Its characteristics can be altered by doping transition metals like Ni, Co, or Mn, or by altering the size of the particles. A recent work that doped Fe_3O_4 NPs with Mn^{2+} and Zn^{2+} revealed M_s that was comparable to Fe_3O_4's (Saha, Rakshit, and Mandal 2019; de Mello et al. 2019). Nevertheless, the rate of hyperthermia attained by Mn^{2+}-doped Fe_3O_4 NPs was highest, whereas the rate of hyperthermia displayed by Zn^{2+}-doped NPs was lowest.

18.3.3 Maghemite (γ-Fe_2O_3)

With its spinel structure, maghemite is one of the most significant SFNPs. Its formula is $\gamma\text{-}Fe_2O_3=[Fe^{3+}]_A[(5/3Fe^{3+})\cdot 1/3V]_BO_4$, where V denotes vacancy (Pisane, Despeaux, and Seehra 2015). Owing to its non-toxic and biocompatible nature, it finds extensive usage in both industrial and human safety devices field. Additionally, when γ-Fe_2O_3 is present in the form of NPs smaller than around 20 nm, it exhibits superparamagnetic characteristics at RT (Sutradhar et al. 2015). Its distinct ability of producing heat in the presence of an alternate magnetic field (MF) was first noted in 1957 (Gilchrist et al. 1957). Since then, a lot of researchers have been interested in using it to treat tumors. γ-Fe_2O_3 NPs find applications in biomagnetic separation, MRI contrast enhancement, magnetic drug delivery, and hyperthermia. Apart from Fe_3O_4 NPs, γ-Fe_2O_3 NPs are one of the most prominent SFNPs in biological applications overall.

18.3.4 Manganese Ferrite ($MnFe_2O_4$)

Manganese ferrite is the type of SF that has neither an inverse nor a normal structure. It is a soft magnet, often having a mixed spinel structure with partial Mn^{2+} occupancy at both sites. For similar particle size, it exhibits higher M_s, magnetic permeability, and lower magnetic anisotropy than Fe_3O_4 (Mathur, Thakur, and Singh 2008). Its high electrical resistance and magnetic permeability are the primary reasons for its use in the electronic device industries (Kadam et al. 2016). It might be utilized, especially in biomedical device applications, for medication administration and MRI contrast enhancement. $MnFe_2O_4$ has a higher MRI contrasting capability than Fe_3O_4 and $CoFe_2O_4$ of comparable size.

18.3.5 Nickel Ferrite ($NiFe_2O_4$)

It is a soft magnet nickel ferrite having an inverted spinel structure, with Ni^{2+} found at octahedral sites and Fe^{3+} evenly distributed at tetrahedral and octahedral sites. $NiFe_2O_4$ NPs are extensively employed as core materials in a variety of biological applications, inductors, and transformers (Saini, Thakur, and Thakur 2016; Pathania et al. 2019). $NiFe_2O_4$ NPs, like the other ferrites previously described, are highly significant because of their inexpensive cost, excellent electromagnetic performance, modest M_s, and outstanding chemical stability for use in MRI applications (Sharifi and Shokrollahi 2012; Thakur, Thakur, and Hsu 2012). $NiFe_2O_4$ NPs can be utilized to treat cancer with cell differentiation by heat. Like other SFNPs, it must be coated with appropriate biological organic molecules to lower its rates of degradation or boost its biocompatibility.

18.3.6 Zinc Ferrite ($ZnFe_2O_4$)

Zinc ferrite is a well-researched SFNP, and its spinel structure is typical, including Zn^{2+} and Fe^{3+} in tetrahedral and octahedral positions, respectively. It exhibits ferromagnetic characteristics in NP form while displaying paramagnetic properties in bulk form. Like other ferrites, it also depends on the synthesis technique and particle size for its application, electrical, and magnetic characteristics (Bahhar et al. 2019). A rat's heart was used to investigate the value of silica-coated 8 nm $ZnFe_2O_4$ NPs in cardiac diagnostics. The clinical investigation results demonstrated a significant improvement in the cardiac cell functioning of the rats following NPs' injection, indicating the NPs' effective ability to restore the heart's rhythm (Hong et al. 2013).

18.3.7 Calcium Ferrite

Calcium ferrite compounds show soft ferromagnetism, which makes them useful for making calcium ferrite/graphite nanocomposites, which are materials that absorb radar waves. Calcium ferrite ought to exhibit superparamagnetic behavior for use in human safety devices and as a targeted drug delivery agent. Because calcium is used in place of heavy metals in other ferrites like MFe_2O_4 (where M can be Cu, Ni, Mn, or Zn), $CaFe_2O_4$ is one of the most biocompatible and ecologically benign minerals. Furthermore, the distinct electrical properties of $Ca_2Fe_2O_5$ with the brown millerite structure make it suitable for use as a p-type thermoelectric device.

18.3.8 Bismuth Ferrite

Bismuth ferrite ($BiFeO_3$) exhibits magnetic-electric interaction at ambient temperature. It has high Curie temperature of 1100K, while its Néel temperature is 640K. It is challenging to achieve the pure phase of $BiFeO_3$, as phase formation kinetics lead to the creation of secondary phases, such as $Bi_2Fe_4O_9$. Many methods have been described to synthesize $BiFeO_3$, including sol-gel, hydrothermal, and chemical precipitation approaches (Taneja et al. 2021). These methods have the advantage of obtaining the single phase of $BiFeO_3$ at a lower temperature in an economic and simple manner. Wang et al. reported that the solid-state precursor technique at low-heating temperatures of around 425°C begins to generate $BiFeO_3$, with approximately 30% impurity phases. $BiFeO_3$ NPs less than 100 nm exhibit modest ferromagnetism at ambient temperature. Different researchers have looked at doping impacts of $BiFeO_3$ NPs using a variety of cutting-edge techniques to improve their performance up to this point. Doping can also increase the magnetization, most likely due to the impurity phase in the $BiFeO_3$ system, the spin cycloid structure being destroyed, and the Fe and O bonding being distorted (Taneja et al. 2022). Liu et al. discovered a reduced breakdown field in the samples that are annealed in the atmospheres of N_2 and H_2, in addition to a higher spontaneous polarization.

18.4 APPLICATIONS OF SPINEL FERRITES FOR HUMAN SAFETY DEVICES

18.4.1 Gas Sensors

The use of nanostructured materials has greatly enhanced the gas-sensing device application for human safety, mainly due to their high surface-to-volume ratio that enables greater interaction with gas molecules. The efficiency and performance of gas sensors can be affected by a number of environmental conditions, including humidity, operating temperature, and gas concentration. For example, increased surface response rates at greater operating temperatures may increase the sensitivity of the sensor (Afzal et al. 2020). However, high humidity can make a sensor's surface water-saturated, which can reduce its ability to react to target gases (Wei et al. 2022). Figure 18.2 depicts the change in resistance of magnesium ferrite NPs at 5000 ppm CO_2 and variation in resistance with concentration of magnesium ferrite sample. Table 18.1 tabulates responses of various ferrite nanomaterials reported in literature for gas sensor application.

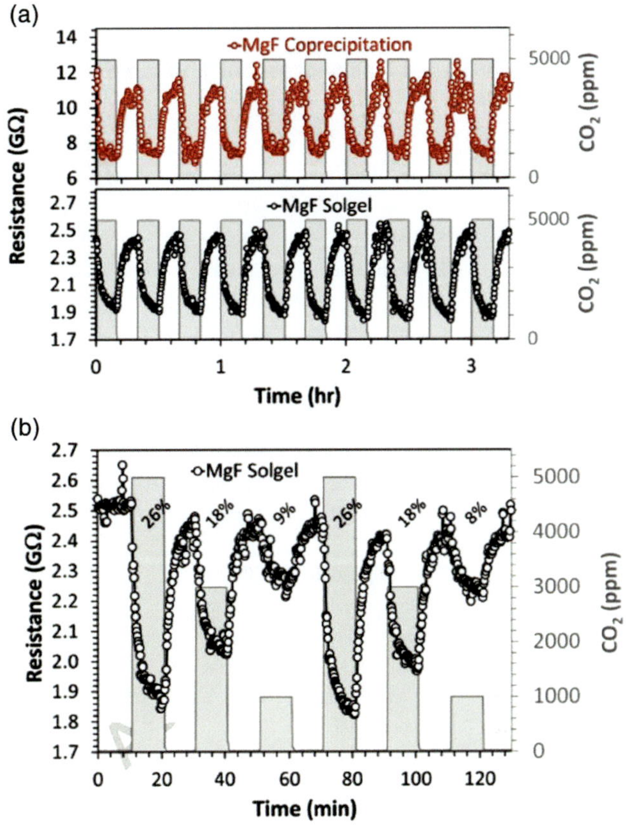

FIGURE 18.2 (a) Magnesium ferrite's resistance variation at 300°C in 5000 ppm CO_2. (b) Variation in resistance with concentration of magnesium ferrite sample. (Adapted from Sumangala et al. (2018) with permission, License No. 5678280622041, dated 29.11.2023.)

18.4.2 MRI and Detection of Cancer

Detection of cancer in early stages is of utmost importance, as early detection is linked to efficient results with any kind of therapy. Nowadays, a number of medical tests, including urine analysis, blood testing, and imaging methods, are used to identify cancer. When the diameter of cancerous cells is just a few millimeters (MRI scan) or centimeters, such as in a positron emission tomography (PET) scan, they are already composed of over a million cells, and this is when imaging modalities identify the disease. Molecular imaging is a recently developed technique that attempts to address this drawback. This novel imaging modality is made feasible by recent developments in imaging technologies, molecular biology, and nanotechnology. MRI with the best spatial resolution is of particular interest as it is

TABLE 18.1
Spinel Ferrite-Based Gas Sensors for Human Safety Devices.

S.N.	Material	Synthesis	Gas	Operating Temperature (°C)	Concentration (ppm)	Response time/ Recovery Time (s)	Reference
1	$CoFe_2O_4$	Sol-gel	NH_3	RT	100	118 s/145 s	Raut et al. (2018)
2	$CoFe_2O_4$	Sol-gel	Ethanol	300	150	75 s/110 s	Montahaei and Emamian (2022)
3	$CoFe_2O_4$	Hydrothermal	Ethanol	200	100	15 s/18 s	Wei et al. (2022)
4	$CuFe_2O_4$	Sol-gel	H_2S	80	25	51.5 s/–	Haija et al. (2017)
5	$CuFe_2O_4$	Co-precipitation	NH_3	RT	20	8 s/300 s	Rathore et al. (2018)
6	$MgFe_2O_4$	Co-precipitating	CO_2	300	5000	120 s/240 s	Sumangala et al. (2018)
7	$MgFe_2O_4$	Sol-gel	CO_2	300	5000	300 s/300 s	Sumangala et al. (2018)
8	$MgFe_2O_4$	Sol-gel	LPG	RT	4 vol%	158 s/152 s	Jaiswal et al. (2020)
9	$NiFe_2O_4$	Auto-combustion	NH_3	400	250	100 s/119 s	Kashyap et al. (2019)
10	$NiFe_2O_4$	Ligand-assisted	Acetone	210	200	44 s/24 s	Jia et al. (2020)
11	$NiFe_2O_4$	Hydrothermal	Acetone	190	100	70 s/130 s	Rezaeipour, Dehghani, and Hoghoghifard (2022)
12	$ZnFe_2O_4$	Co-precipitation	Ethanol	300 k	40	50 s/116 s	Khurshid et al. (2019)
13	$NiFe_2O_4$	Hydrothermal	Acetone	310	100	45/75 s	Rezaeipour et al. (2022)
14	$MgFe_2O_4$	Sol-gel	acetone	725 K	1000	13/6 s	Patil et al. (2018)
15	$ZnFe_2O_4$	Nonaqueous	Toluene	300	100	18 s/29 s	Dong et al. (2017)
16	$CoFe_2O_4$	Self-templating	NH_3	240	20	19.6/12.1 s	Wang et al. (2020)
17	$CuFe_2O_4$	Solvothermal	TEA	105	10	32 s/192 s	Wu et al. (2021)
18	$NiFe_2O_4$	Hydrothermal	Xylene	300	500	50.10 s/40.30 s	Zhang et al. (2019)

minimally invasive or noninvasive. Unfortunately, MRI has not been used efficiently for cancer detection due to its low specificity (breast cancer: false-positivity rate of 10%). However, cell markers and paramagnetic and superparamagnetic NPs, having special characteristics, may be used for identification with MRI in small amounts, which will compensate for the low specificity of MRI.

Because of their characteristic magnetic properties and reduced sizes, equivalent to the biggest biological molecules, like enzymes or antibodies, generally less than 100 nm NPs have been utilized in healthcare for diagnostics as well as therapeutic purposes (Meyers et al. 2013). Metal oxides like Fe_3O_4, SiO_2, and TiO_2 are among the materials that can be included in NPs with potential medicinal uses connected to MRI. The most widely used and original MRI NP is superparamagnetic iron oxide, which is incredibly tiny and ultra-small. Magnetic iron oxide particles have been used for clinical detection of focal lesions in the liver and spleen using MRI since long back. Because of the selective activity of the hepatobiliary system, iron accumulates in the liver and can be applied to increase MRI sensitivity. Since certain NPs passively collect in tumors after being administered intravenously, one of the main characteristics of NP delivery systems is their ability to boost the accumulation of anticancer medicines in cancerous cells specifically.

MRI Contrast Agents: With better contrast and biocompatibility than traditional agents like gadolinium, magnetic NPs have gained attention as MRI contrast agents. The higher the longitudinal and transverse relaxivity, the better the resultant contrast of a particular contrast agent. Due to their brief in-vivo half-lives, currently available contrast agents are unable to provide comprehensive data for prognostic evaluation of a particular course of therapy. Kim et al. created a homogeneous magnetic hydrogel system with $CoFe_2O_4$ NPs comprising thermosensitive polyhydrogels for a long-term MR contrast platform. The hydrogels were tested by injecting them into the brains of rats, which were employed in in-vivo imaging investigations. The right side of the rat brain's frontal cortex area showed a strong-negative MR T2 contrast four days after the injection, according to the investigators. It took 32 days after surgery for the negative contrast to fade or spread. According to the scientists, these hydrogels may be used as long-term theranostic medicines, particularly for brain malignancies.

18.4.3 Biosensors

Because $CoFe_2O_4$ NPs can align well with external MF, they are being researched extensively in biosensing applications. They can be fixed on a transducer to facilitate further detection of analyte, or they can be scattered throughout the solution of the targeted analytes. Research into high-sensitivity biosensors for carcinoembryonic antigen (CEA) detection is ongoing in order to support earlier and more accurate cancer diagnosis. CEA, a glycoprotein, is frequently utilized to monitor colorectal cancer. Chen et al. prepared an immunosensor based on quartz crystal microbalance that uses piezoelectric immune sensing of CEA antigen and $CoFe_2O_4@SiO_2$ adsorbed on surface coupled with an anti-CEA antibody. He et al. prepared sandwich-type immunosensor for the detection using antibody-conjugated gold-functionalized $CoFe_2O_4$ NPs. The biosensor demonstrated good repeatability, with a broad detection range of 1 fg/ml to 1 ng/ml and great sensitivity, having a detection limit of 0.75 fg/ml.

18.4.4 Antimicrobial Applications

Numerous species use iron sequestration as a defense mechanism against infection, as a major role is played by iron in microbial pathogenesis. SFNPs show strong antibacterial activities against *Escherichia coli*, *Pseudomonas aeruginosa*, *Staphylococcus aureus*, and *Candida albicans*. *E. coli* strain was able to develop resistance to excess Fe mostly by altering its elemental absorption (Ewunkem et al. 2021). Because of its antibacterial action against a wide range of human diseases, $CoFe_2O_4$ NPs are recently researched for human healthcare or safety applications. $CoFe_2O_4$ NPs reduced size enhances their interaction with harmful microorganisms. The $CoFe_2O_4$@SiO_2@Ag nanocomposite demonstrated noteworthy antibacterial activity against bacteria when combined with streptomycin. Ag NPs can be deposited on $CoFe_2O_4$@SiO_2 to stop agglomeration. Because $CoFe_2O_4$ sticks to bacterial membranes, it lengthens the bacterial development lag phase, extends microorganism's production time, and speeds up bacterial cell division. $CoFe_2O_4$ produces effective oxides, like H_2O_2 and superoxide (O^{2-}), which are responsible for its antibacterial action (Ashour et al. 2018). The best benefit of magnetic NPs is their easy retrieval from solution after disinfection using external MF. The $ZnFe_2O_4$ NPs also exhibit strong antibacterial activity against bacterial strains of *E. coli*, *Lactobacillus*, *Aeromonas hydrophila*, and *Vibrio harbori* (Suppuraj et al. 2017). The $NiFe_2O_4$ and $ZnFe_2O_4$ NPs were exclusively effective against a few bacterial strains (Ashour et al. 2018). Furthermore, $NiFe_2O_4$@carbon nanocomposite demonstrated a fast antibacterial impact on *P. aeruginosa* bacterial breakdown (Ahmadian-Fard-Fini et al. 2019). Singh et al. synthesized $NiFe_2O_4$-Ag-ZnO magnetic nanocomposite having enhanced antibacterial activities against *E. coli* (Singh et al. 2024). Figure 18.3 illustrates the antibacterial activity of Fe_2O_4, $CuFe_2O_4$, $ZnFe_2O_4$, and $MnFe_2O_4$ NPs with mechanisms of action.

18.4.5 Magnetic Separation & Purification

Because of their magnetic susceptibility, $CoFe_2O_4$ NPs have been used in the magnetic separation, extraction, and purification of a wide range of human safety device materials, including poisons and contaminants, from biological samples. Furthermore, these NPs have higher adsorption effectiveness due to their porous structure along with larger specific surface area, which makes them intriguing adsorbent materials for applications involving purification and separation. Hyperbilirubinemia, jaundice, and neurotoxicity are conditions where there are excessive bilirubin deposits in different tissues. For bilirubin degradation, Rakshit et al. used silica-coated $CoFe_2O_4$ core@shell NPs ($CoFe_2O_4$@SiO_2) and nano-hollow spheres of $CoFe_2O_4$ NPs. The authors reported that the NPs exhibit excellent efficiency, with 98% bilirubin adsorption occurring within 8 minutes for $CoFe_2O_4$ NPs and 90% within 20 minutes for 350 nm diameter $CoFe_2O_4$@SiO_2 NPs. This suggests the potential use of NPs for lowering blood bilirubinemia levels in clinical treatment applications. Pal et al. produced surface-functionalized $CoFe_2O_4$ NPs (T-$CoFe_2O_4$) using Na tartrate for the purpose of degrading bilirubin in a different investigation. NPs demonstrated multifunctional characteristics, such as inherent fluorescence and superparamagnetism.

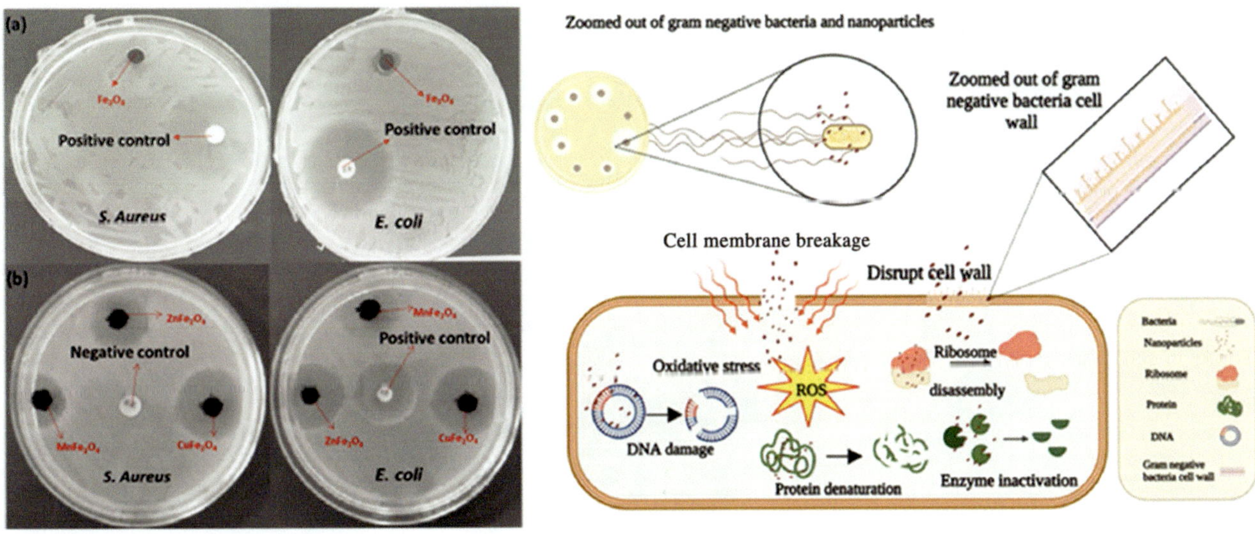

FIGURE 18.3 Antibacterial activity of Fe_2O_4, $CuFe_2O_4$, $ZnFe_2O_4$, and $MnFe_2O_4$ NPs against different bacterial strains with mechanisms of action. (Adapted from Dabagh, Haris, and Ertas (2023) with permission, License No. 5678300402811, dated 29.11.2023.)

18.4.6 Drug Delivery

$CoFe_2O_4$ NPs have shown great promise as nanocarrier materials for drug delivery applications, as evidenced by research that has progressed to clinical trials in the last ten years. This is due to three intriguing characteristics: the capacity to create a large range of multifunctional nanocomposites with efficient drug loading; the ability to modulate magnetic targeting using an externally applied MF; and finally, the ability to synthesize these qualities. A drug release method in controlled manner for the administration of doxorubicin was proposed using magnetically driven $CoFe_2O_4$ nanostructures that resembled dandelion pollen. The nanostructures showed around 55% drug release under an alternating MF for 8 hours, and an 88.6% loading capacity of the $CoFe_2O_4$ microspheres. Figure 18.4 illustrates magnetic NPs for targeted drug administration/release.

Targeted distribution of SFNPs for biological device applications through their superparamagnetic characteristics is one of their most intriguing novel features. An effective release mechanism can be achieved by changing some internal parameters, such as pH or enzymes, and/or by delivering an external stimulus, like temperature, MF, or light. Magnetic nanocarriers have significant benefits because of this strictly regulated process. These benefits include better pharmacokinetics, increased dispersion, increased stability, and reduced toxic side effects.

18.4.7 Hyperthermia

The magnetic hyperthermia features of SFNPs, which are based on the conversion of electromagnetic energy into heat in the presence of an external MF, are highly appealing in addition to their excellent targeted delivery. The quantity of energy released by the NPs per unit mass of particles in a certain length of time is known as the specific absorption rate or specific loss power. As a result, measuring specific loss power provides an estimate of NPs rise in temperature in response to applied fields and, consequently, their hyperthermic efficiency. Cancerous tumors are known to be sensitive to temperature increases; however, in addition to killing cancer cells, conventional hyperthermia therapies, such as whole-body and localized hyperthermia, can have certain negative clinical consequences on the surrounding healthy tissues (Demirci Dönmez et al. 2019). However, the excellent and quick targeting of heating in the tumor zone (localized hyperthermia) is one of the most remarkable features of employing SFNs inside tumors. More intriguingly, the ideal pairing of magnetic hyperthermia and targeted distribution can increase SFNP vascular permeability and promote more effective SFNP accumulation. Furthermore, the efficiency with which loaded materials from smart nanocarriers are released at the tumor site can be significantly impacted by magnetic hyperthermia (Amir Mahnazi et al. 2019). Fascinatingly, the use of SFNPs with localized heating effects that spare neighboring tissues looks like an exciting

FIGURE 18.4 Magnetic NPs for targeted drug administration/release. (Adapted from Amiri et al. (2018) with permission, License No. 5678290082766, dated 29.11.2023.)

cancer treatment method. Through prolonged warming (>43°C), healthy as well as cancer cells of live tissue are exposed to necrosis in hyperthermia, a noninvasive therapy approach for oncological disorders (Manohar et al. 2020). Deep-seated cancers cannot be successfully treated with the current tumor heating techniques. However, magnetic induction hyperthermia—that is, using magnetic NPs—may be successful in treating these types of tumors because it selectively heats and destroys the tumor tissue while causing the least amount of damage (Suleman and Riaz 2020).

Under time-varying MFs, ferrimagnetic materials exhibit hysteretic characteristics causing heating induced by MF. Additionally, the magnetic NPs are low in toxicity and biocompatible, which makes it easy and quick to analyze the particular absorption rate. Reducing harm to adjacent normal tissues is the primary problem when dealing with hyperthermia. Accordingly, NPs have to be focused at the location of the tumor rather than around it (Ajinkya et al. 2020). The $NiFe_2O_4$ NPs (uncoated) have a high cell survival rate and are non-cytotoxic at distinct pH levels that make them suitable for use as cell-differentiable hyperthermia agents. Furthermore, because of its low toxicity, $ZnFe_2O_4$ is an excellent choice for cases of hyperthermia (Nasrin, Chowdhury, and Hoque 2019). The enzyme-like properties of nanosized ferrite spinel, such as peroxidase, oxidase, and catalase, have also been reported by the researchers (Chaibakhsh and Moradi-Shoeili 2019). A therapeutic procedure in which metal NPs are applied, and after targeting a tumor location, are exposed to an external radiofrequency field is being used recently. When a radiofrequency field is applied, the temperature of magnetic NPs rises, raising the temperature of concerned cells also and causing thermoablation, which kills the cells. The most researched biological use of $CoFe_2O_4$ NPs is by far hormonal therapy.

The main features of SFNPs that have drawn attention from scientists for use in biomedicine are their high M_s values, superparamagnetic characteristics, low intrinsic toxicity, scalable production techniques, and variable size and shape, all without sacrificing the benefits that come with being a nanomaterial. Furthermore, SFNPs are particularly appealing because of their distinct physiochemical characteristics. SFNPs have been the subject of several studies for a range of biological uses, including hyperthermia, magnetic separation, targeted drug administration, and MRI (Kefeni et al. 2020). All the same, it's crucial to remember that bare SFNPs have significant difficulties because of their weak magnetic characteristics at tiny sizes and in physiological settings. Therefore, it has become clear that altering SFNPs through the creation of SFNP-based hybrid designs is an essential first step toward enhancing safety and biocompatibility, minimizing side effects, preventing aggregation, and granting multifunctional capabilities. Under physiological settings, a potential hybrid design will exhibit significant magnetism at tiny sizes and low concentrations. High M_s values and acceptable biocompatible magnetic NPs must be in a stable, non-agglomerated aqueous dispersion for use in human safety device applications. For medication release, magnetic fluid hyperthermia, and heat stimulation inside a single cell, magnetic NPs can be very useful (Suleman and Riaz 2020). $CoFe_2O_4$ has generally been used in imaging, medication administration and release, and brain tumor treatment (Ashour et al. 2018). As per the Trojan horse mechanism, metal ions are released at $CoFe_2O_4$ NPs at higher concentrations, and some cause cytotoxicity. Changes in cellular function may be observed due to the inhibition of protein synthesis and transcription in cells. The negative effects could be lessened by using a biocompatible SiO_2 layer. Because of the negative charge of SiO_2 matrixes, covering SFNPs reduced their adherence and caused them to exhibit superparamagnetic characteristics and greater compatibility, making them appropriate for a variety of human safety device applications like drug delivery, particularly for cancer cells and contrast agents for MRI.

18.5 CHALLENGES, FUTURE PERSPECTIVES, AND CONCLUSION

18.5.1 Toxicity Challenge

The greatest saturation magnetization (M_s) is found in pure metals like Fe, Ni, and Co, which are also the easiest to recover using an external MF. However, because of severe sensitivity to oxidation and higher levels of toxicity, they are not very important in biological environment (Karimi et al. 2013). Since toxicity is dependent on the makeup of SFNPs and corresponding size, it is frequently linked to their beneficial qualities. The toxicity of NPs can also be affected by additional variables such as dosage, solubility, surface chemistry, and bio-distribution.

For instance, it has been noted during antibacterial activity studies of Ce^{3+}-doped $CuFe_2O_4$ (Elayakumar et al. 2019) that a decrease in NP size is associated with strong antimicrobial activity. The in-vitro experiment revealed size- and dosage-dependent toxicity, where smaller sizes were more harmful than larger ones at a high dose of 80 µg/ml. At a lower dose of 40 µg/ml, there was no toxicity impact (Karlsson et al. 2009). This suggests that extremely tiny particles might get trapped and kept in the body's various organs, where they could have immediate negative consequences. The chemical activity increases and is more readily absorbed by the cells more quickly than by the bigger ones when particle size reduces (Sahu and Hayes 2017). This might be the cause of the greater suspicion of small-sized NP toxicity compared to their bigger equivalent. For any medical therapy, it is therefore recommended to employ SFNPs only after appropriate coating. However, because they are highly hazardous and easily diffuse across cell membranes, particles smaller than 2 nm are not advised to be employed in hyperthermia-based cancer treatments (Sasikala et al. 2016). Investigation into hyperthermia therapy revealed that suitable characterization of the NPs physicochemical characteristics and the existence of any leftover contaminants that may be harmful be studied in depth before usage. The most crucial characteristics that should be looked at before usage are the particle size, surface charge, toxicity, and protein adsorption capability.

18.5.2 Future Perspective

Researchers have shown a significant interest in the use of SFNPs in healthcare for human safety device applications like targeted medication administration and release, cancer gene therapy, hyperthermia, and diagnostics. Even though there are several synthesis techniques available, more has to be done to enhance the current approaches in order to generate pure SFNPs that have all the necessary characteristics and in appropriate quantities. Therefore, techniques that improve the quality of SFNPs at the industrial level are needed. Biologists, chemists, pharmacists, medical experts, and engineers must work together in this field to get the research from the laboratory to the device application stage. Research has unequivocally demonstrated that doping or substituting metals into SFNPs may result in the creation of new SFNPs with enhanced characteristics. Furthermore, the development of SFNPs is severely hampered by concerns about toxicity and biocompatibility. The influence and modification of SFNP's features that improve their superparamagnetic behavior ought to receive greater attention. Low-level metal-selective doping may be helpful in producing novel SFNPs that have the potential to address present health issues. It is highly exciting because SFNPs have several uses in human safety devices in biological fields, such as drug delivery/release, MRI, and cancer therapies. Specifically, efficient and successful treatment of tumor cells requires the simultaneous administration of anticancer medicines. More in-depth research is required in the area of the magnetic behavior of SFNPs under various conditions, and impacts of incorporating additional dopants remain unclear despite multiple studies on them. Further research and attention are still needed, nevertheless, in order to optimize the heating characteristics, concentrations of SFNPs used and anticancer drugs, SFNP quality, repeatability of the process, and process parameters.

18.5.3 Conclusions

In the healthcare field, SFNPs are widely used in human safety device applications like medication administration and release, cancer gene therapy, cancer detection, and imaging applications. Good synthesis techniques yield SFNPs that can endure the circumstances under which they are synthesized and improve their performance, which is more important to the effectiveness of SFNPs in safety applications. The use of SFNPs in biological applications appears to be critical, but before they are used extensively or commercialized, their risks and toxicity should be thoroughly examined. So, proper measures must be adopted for reducing unintended outcomes and supporting sustainable use in a variety of biological application areas. SFNPs are very helpful in the treatment of cancer. However, much work is needed to establish SFNPs for detection, identification, and remediation of human healthcare.

REFERENCES

Abdel Maksoud, M. I.A., Ahmed M. Elgarahy, Charlie Farrell, Ala'a H. Al-Muhtaseb, David W. Rooney, and Ahmed I. Osman. 2020. "Insight on Water Remediation Application Using Magnetic Nanomaterials and Biosorbents." *Coordination Chemistry Reviews*. Elsevier. https://doi.org/10.1016/j.ccr.2019.213096.

Abdel Maksoud, M. I.A., Gharieb S. El-Sayyad, A. H. Ashour, Ahmed I. El-Batal, Mahmmoud S. Abd-Elmonem, Hassan A.M. Hendawy, E. K. Abdel-Khalek, Sh Labib, E. Abdeltwab, and M. El-Okr. 2018. "Synthesis and Characterization of Metals-Substituted Cobalt Ferrite [Mx Co(1-x) Fe_2O_4; (M = Zn, Cu and Mn; x = 0 and 0.5)] Nanoparticles as Antimicrobial Agents and Sensors for Anagrelide Determination in Biological Samples." *Materials Science and Engineering C* 92 (November): 644–56. https://doi.org/10.1016/j.msec.2018.07.007.

Afzal, Adeel, Adnan Mujahid, Naseer Iqbal, Rahat Javaid, and Umair Yaqub Qazi. 2020. "Enhanced High-Temperature (600°C) No2 Response of Znfe2o4 Nanoparticle-Based Exhaust Gas Sensors." *Nanomaterials* 10 (11): 1–14. https://doi.org/10.3390/nano10112133.

Ahmadian-Fard-Fini, et al. 2019. "Photoluminescence Carbon Dot as a Sensor for Detecting of *Pseudomonas aeruginosa* Bacteria: Hydrothermal Synthesis of Magnetic Hollow NiFe 2 O 4-Carbon Dots Nanocomposite Material." *Composites Part B: Engineering* 161 (March): 564–77. https://doi.org/10.1016/j.compositesb.2018.12.131.

Ajinkya, Nene, Xuefeng Yu, Poonam Kaithal, Hongrong Luo, Prakash Somani, and Seeram Ramakrishna. 2020. "Magnetic Iron Oxide Nanoparticle (IONP) Synthesis to Applications: Present and Future." *Materials (Basel)*. https://doi.org/10.3390/ma13204644.

Amir Mahnazi, et al. 2019. "Magnetic Nanocarriers: Evolution of Spinel Ferrites for Medical Applications." *Advances in Colloid and Interface Science*. Elsevier. https://doi.org/10.1016/j.cis.2019.01.003.

Amiri, Mahnaz, Ahmad Akbari, Meysam Ahmadi, Abbas Pardakhti, and Masoud Salavati-Niasari. 2018. "Synthesis and in Vitro Evaluation of a Novel Magnetic Drug Delivery System; Proecological Method for the Preparation of $CoFe_2O_4$ Nanostructures." *Journal of Molecular Liquids* 249 (January): 1151–60. https://doi.org/10.1016/j.molliq.2017.11.133.

Ashour, A. H., Ahmed I. El-Batal, M. I. A. Abdel Maksoud, Gharieb S. El-Sayyad, Sh Labib, E. Abdeltwab, and M. M. El-Okr. 2018. "Antimicrobial Activity of Metal-Substituted Cobalt Ferrite Nanoparticles Synthesized by Sol–Gel Technique." *Particuology* 40 (October): 141–51. https://doi.org/10.1016/j.partic.2017.12.001.

Bahhar, S., H. Lemziouka, A. Boutahar, H. Bioud, H. Lassri, and E. K. Hlil. 2019. "Influence of La3+ Site Substitution on the Structural, Magnetic and Magnetocaloric Properties of $ZnFe_{2-x}La_xO_4$ (x = 0.00, 0.001, 0.005 and 0.01) Spinel Zinc Ferrites." *Chemical Physics Letters* 716 (February): 186–91. https://doi.org/10.1016/j.cplett.2018.12.025.

Chahar, Deepika, Dinesh Kumar, Preeti Thakur, and Atul Thakur. 2023. "Visible Light Induced Photocatalytic Degradation of Methylene Blue Dye by Using Mg Doped Co-Zn Nanoferrites." *Materials Research Bulletin* 162 (June): 112205. https://doi.org/10.1016/J.MATERRESBULL.2023.112205.

Chaibakhsh, Naz, and Zeinab Moradi-Shoeili. 2019. "Enzyme Mimetic Activities of Spinel Substituted Nanoferrites (MFe_2O_4): A Review of Synthesis, Mechanism and Potential Applications." Materials Science and Engineering C. https://doi.org/10.1016/j.msec.2019.02.086.

Chandrasekaran, Sundaram, Chris Bowen, Peixin Zhang, Zheling Li, Qiuhua Yuan, Xiangzhong Ren, and Libo Deng. 2018. "Spinel Photocatalysts for Environmental Remediation, Hydrogen Generation, CO_2 Reduction and Photoelectrochemical Water Splitting." *Journal of Materials Chemistry A*. The Royal Society of Chemistry. https://doi.org/10.1039/c8ta03669a.

Cruz, M. M., L. P. Ferreira, J. Ramos, S. G. Mendo, A. F. Alves, M. Godinho, and M. D. Carvalho. 2017. "Enhanced Magnetic Hyperthermia of $CoFe_2O_4$ and $MnFe_2O_4$ Nanoparticles." *Journal of Alloys and Compounds* 703 (May): 370–80. https://doi.org/10.1016/j.jallcom.2017.01.297.

Dabagh, Shadab, Somayeh Asadi Haris, and Yavuz Nuri Ertas. 2023. "Synthesis, Characterization and Potent Antibacterial Activity of Metal-Substituted Spinel Ferrite Nanoparticles." *Journal of Cluster Science* 34 (4): 2067–78. https://doi.org/10.1007/s10876-022-02373-9.

Demirci Dönmez, Cigdem E., Palash K. Manna, Rachel Nickel, Selçuk Aktürk, and Johan Van Lierop. 2019. "Comparative Heating Efficiency of Cobalt-, Manganese-, and Nickel-Ferrite Nanoparticles for a Hyperthermia Agent in Biomedicines." *ACS Applied Materials and Interfaces* 11 (7): 6858–66. https://doi.org/10.1021/acsami.8b22600.

Dong, Chengjun, Xu Liu, Xuechun Xiao, Shangfeng Du, and Yude Wang. 2017. "Monodisperse ZnFe2O4 Nanospheres Synthesized by a Nonaqueous Route for a Highly Selective Low-Ppm-Level Toluene Gas Sensor." *Sensors and Actuators, B: Chemical* 239 (February): 1231–36. https://doi.org/10.1016/j.snb.2016.09.122.

Elayakumar, K., A. Manikandan, A. Dinesh, K. Thanrasu, K. Kanmani Raja, R. Thilak Kumar, Y. Slimani, S. K. Jaganathan, and A. Baykal. 2019. "Enhanced Magnetic Property and Antibacterial Biomedical Activity of Ce^{3+} Doped $CuFe_2O_4$ Spinel Nanoparticles Synthesized by Sol-Gel Method." *Journal of Magnetism and Magnetic Materials* 478 (May): 140–47. https://doi.org/10.1016/j.jmmm.2019.01.108.

Ewunkem, Akamu J., Lashunta Rodgers, Daisha Campbell, Constance Staley, Kiran Subedi, Sada Boyd, and Joseph L. Graves. 2021. "Experimental Evolution of Magnetite Nanoparticle Resistance in *Escherichia coli*." *Nanomaterials* 11 (3): 1–22. https://doi.org/10.3390/nano11030790.

Gilchrist, R. K., R. Medal, W. D. Shorey, R. C. Hanselman, J. C. Parrott, and C. B. Taylor. 1957. "Selective Inductive Heating of Lymph Nodes." *Annals of Surgery* 146 (4): 596–606. https://doi.org/10.1097/00000658-195710000-00007.

Granone, Luis I., Anna C. Ulpe, Lars Robben, Stephen Klimke, Moritz Jahns, Franz Renz, Thorsten M. Gesing, Thomas Bredow, Ralf Dillert, and Detlef W. Bahnemann. 2018. "Effect of the Degree of Inversion on Optical Properties of Spinel $ZnFe_2O_4$." *Physical Chemistry Chemical Physics* 20 (44): 28267–78. https://doi.org/10.1039/c8cp05061a.

Haija, Mohammad Abu, Ayah F.S. Abu-Hani, Najwa Hamdan, Samuel Stephen, and Ahmad I. Ayesh. 2017. "Characterization of H_2S Gas Sensor Based on $CuFe_2O_4$ Nanoparticles." *Journal of Alloys and Compounds* 690 (January): 461–68. https://doi.org/10.1016/j.jallcom.2016.08.174.

Hong, Nguyen Hoa, A. T. Raghavender, O. Ciftja, M.-H. Phan, K. Stojak, H. Srikanth, and Yin Hua Zhang, et al. 2013. "Ferrite Nanoparticles for Future Heart Diagnostics." *ApPhA* 112 (2): 323–27. https://doi.org/10.1007/S00339-013-7573-Y.

Jaiswal, Aashit Kumar, Samiksha Sikarwar, Satyendra Singh, Kajal Kumar Dey, B. C. Yadav, and R. R. Yadav. 2020. "Fabrication of Nanostructured Magnesium Ferrite Polyhedrons and Their Applications in Heat Transfer Management and Gas/Humidity Sensors." *Journal of Materials Science: Materials in Electronics* 31 (1): 80–89. https://doi.org/10.1007/s10854-019-01099-1.

Jia, Chaowei, Yanlin Zhang, Quan Kong, Qiuyue Wang, Gang Chen, Hong Tao Guam, and Chengjun Dong. 2020. "Soft-Template Synthesis of Mesoporous $NiFe_2O_4$ for Highly Sensitive Acetone Detection." *Journal of Materials Science: Materials in Electronics* 31 (8): 6000–7. https://doi.org/10.1007/s10854-020-03151-x.

Kadam, R. H., Kirti Desai, Vishnu S. Shinde, Mohd Hashim, and Sagar E. Shirsath. 2016. "Influence of Gd^{3+} Ion Substitution on the $MnCrFeO_4$ for Their Nanoparticle Shape Formation and Magnetic Properties." *Journal of Alloys and Compounds* 657 (February): 487–94. https://doi.org/10.1016/j.jallcom.2015.10.164.

Karimi, Z. et al. 2013. "Nano-Magnetic Particles Used in Biomedicine: Core and Coating Materials." Materials Science and Engineering C. https://doi.org/10.1016/j.msec.2013.01.045.

Karimzadeh, Isa, Mustafa Aghazadeh, Mohammad Reza Ganjali, Parviz Norouzi, Taher Doroudi, and Peir Hossein Kolivand. 2017. "Saccharide-Coated Superparamagnetic Fe3O4 Nanoparticles (SPIONs) for Biomedical Applications: An Efficient and Scalable Route for Preparation and in Situ Surface Coating Through Cathodic Electrochemical Deposition (CED)." *Materials Letters* 189 (February): 290–94. https://doi.org/10.1016/j.matlet.2016.12.010.

Karlsson, Hanna L., Johanna Gustafsson, Pontus Cronholm, and Lennart Möller. 2009. "Size-Dependent Toxicity of Metal Oxide Particles-A Comparison between Nano- and Micrometer Size." *Toxicology Letters* 188 (2): 112–18. https://doi.org/10.1016/j.toxlet.2009.03.014.

Kashyap, Rajesh, Ravi Kumar, Sarita Devi, Mukesh Kumar, Sachin Tyagi, and Dinesh Kumar. 2019. "Ammonia Gas Sensing Performance of Nickel Ferrite Nanoparticles." *Materials Research Express* 6 (12): 125034. https://doi.org/10.1088/2053-1591/ab55b5.

Kefeni, Kebede K., Bhekie B. Mamba, and Titus A.M. Msagati. 2017. "Application of Spinel Ferrite Nanoparticles in Water and Wastewater Treatment: A Review." Separation and Purification Technology. https://doi.org/10.1016/j.seppur.2017.07.015.

Kefeni, Kebede K., Titus M. Msagati, and Bhekie B. Mamba. 2015. "Synthesis and Characterization of Magnetic Nanoparticles and Study Their Removal Capacity of Metals from Acid Mine Drainage." *Chemical Engineering Journal* 276 (September): 222–31. https://doi.org/10.1016/j.cej.2015.04.066.

Kefeni, Kebede K., Titus A.M. Msagati, and Bhekie B. Mamba. 2017. "Ferrite Nanoparticles: Synthesis, Characterisation and Applications in Electronic Device." Materials Science and Engineering: B. https://doi.org/10.1016/j.mseb.2016.11.002.

Kefeni, Kebede K., Titus A.M. Msagati, Thabo TI Nkambule, and Bhekie B. Mamba. 2020. "Spinel Ferrite Nanoparticles and Nanocomposites for Biomedical Applications and Their Toxicity." Materials Science and Engineering C. https://doi.org/10.1016/j.msec.2019.110314.

Khishigdemberel, I., E. Uyanga, H. Hirazawa, and D. Sangaa. 2018. "Influence of Cu Dope on the Structural Behavior of MgFe2O4 at Various Temperatures." *Physica B: Condensed Matter* 544 (September): 73–78. https://doi.org/10.1016/j.physb.2018.05.032.

Khurshid, Rafaqat, Farman Ali, Adeel Afzal, Zarshad Ali, and Muhammad T. Qureshi. 2019. "Polyol-Mediated Coprecipitation and Aminosilane Grafting of Superparamagnetic, Spinel $ZnFe_2O_4$ Nanoparticles for Room-Temperature Ethanol Sensors." *Journal of the Electrochemical Society* 166 (4): B258–65. https://doi.org/10.1149/2.1021904jes.

Lee, Jae Hyun, Ji Wook Kim, Michael Levy, Albert Kao, Seung Hyun Noh, Dolores Bozovic, and Jinwoo Cheon. 2014. "Magnetic Nanoparticles for Ultrafast Mechanical Control of Inner Ear Hair Cells." *ACS Nano* 8 (7): 6590–98. https://doi.org/10.1021/nn5020616.

Ling, Daishun, Nohyun Lee, and Taeghwan Hyeon. 2015. "Chemical Synthesis and Assembly of Uniformly Sized Iron Oxide Nanoparticles for Medical Applications." *Accounts of Chemical Research* 48 (5): 1276–85. https://doi.org/10.1021/acs.accounts.5b00038.

Lin, Qing, Jianmei Xu, Fang Yang, Jinpei Lin, Hu Yang, and Yun He. 2018. "Magnetic and Mössbauer Spectroscopy Studies of Zinc-Substituted Cobalt Ferrites Prepared by the Sol-Gel Method." *Materials* 11 (10): 1799. https://doi.org/10.3390/ma11101799.

Manohar, Ala, Dereje Degefa Geleta, Chintagumpala Krishnamoorthi, and Joonho Lee. 2020. "Synthesis, Characterization and Magnetic Hyperthermia Properties of Nearly Monodisperse $CoFe_2O_4$ Nanoparticles." *Ceramics International* 46 (18): 28035–41. https://doi.org/10.1016/j.ceramint.2020.07.298.

Masunga, Ngonidzashe, Olga Kelebogile Mmelesi, Kebede K. Kefeni, and Bhekie B. Mamba. 2019. "Recent Advances in Copper Ferrite Nanoparticles and Nanocomposites Synthesis, Magnetic Properties and Application in Water Treatment: Review." Journal of Environmental Chemical Engineering. https://doi.org/10.1016/j.jece.2019.103179.

Mathur, Preeti, Atul Thakur, and M. Singh. 2008. "Low Temperature Synthesis of $Mn_{0.4}Zn_{0.6}In_{0.5}Fe_{1.5}O_4$ Nanoferrite for High-Frequency Applications." *Journal of Physics and Chemistry of Solids* 69 (1): 187–92. https://doi.org/10.1016/j.jpcs.2007.08.014.

Mello, Laura Bissoli de, Laudemir Carlos Varanda, Fernando Aparecido Sigoli, and Italo Odone Mazali. 2019. "Co-Precipitation Synthesis of (Zn-Mn)-Co-Doped Magnetite Nanoparticles and Their Application in Magnetic Hyperthermia." *Journal of Alloys and Compounds* 779 (March): 698–705. https://doi.org/10.1016/j.jallcom.2018.11.280.

Meyers, Joseph D., Tennyson Doane, Clemens Burda, and James P. Basilion. 2013. "Nanoparticles for Imaging and Treating Brain Cancer." Nanomedicine. https://doi.org/10.2217/nnm.12.185.

Montahaei, Reza, and H. R. Emamian. 2022. "The Impact of Microwave-Assisted Sintering on Fabrication of Cobalt Ferrite Nanostructure Foams for Gas-Sensing." *Ceramics International* 48 (18): 26629–37. https://doi.org/10.1016/j.ceramint.2022.05.357.

Naagar, Manish, Sonia Chalia, Saarthak Kharbanda, Preeti Thakur, and Atul Thakur. 2023. "Correlating Tailored Saturation Magnetization of Normal-Inverse Spinels with Noncollinear Magnetic Ordering." *Journal of Materials Science: Materials in Electronics* 34 (6): 1–13. https://doi.org/10.1007/S10854-023-09828-3.

Nasrin, Shamima, F. U.Z. Chowdhury, and S. M. Hoque. 2019. "Study of Hydrodynamic Size Distribution and Hyperthermia Temperature of Chitosan Encapsulated Zinc-Substituted Manganese Nano Ferrites Suspension." *Physica B: Condensed Matter* 561 (May): 54–63. https://doi.org/10.1016/j.physb.2019.02.053.

Pathania, Abhilash, Preeti Thakur, Alex V. Trukhanov, Sergei V. Trukhanov, Larissa V. Panina, Ulrike Lüders, and Atul Thakur. 2019. "Development of Tungsten Doped Ni-Zn Nano-Ferrites with Fast Response and Recovery Time for Hydrogen Gas Sensing Application." *Results in Physics* 15 (December): 102531. https://doi.org/10.1016/J.RINP.2019.102531.

Patil, Jayvant, Digambar Nadargi, Imtiaz S. Mulla, and S. S. Suryavanshi. 2018. "Spinel $MgFe_2O_4$ Thick Films: A Colloidal Approach for Developing Gas Sensors." *Materials Letters* 213 (February): 27–30. https://doi.org/10.1016/J.MATLET.2017.11.009.

Pisane, K. L., E. C. Despeaux, and M. S. Seehra. 2015. "Magnetic Relaxation and Correlating Effective Magnetic Moment with Particle Size Distribution in Maghemite Nanoparticles." *Journal of Magnetism and Magnetic Materials* 384 (June): 148–54. https://doi.org/10.1016/j.jmmm.2015.02.038.

Plouffe, Brian D., Shashi K. Murthy, and Laura H. Lewis. 2015. "Fundamentals and Application of Magnetic Particles in Cell Isolation and Enrichment: A Review." *Reports on Progress in Physics*. Rep Prog Phys. https://doi.org/10.1088/0034-4885/78/1/016601.

Ponce, A. S., E. F. Chagas, R. J. Prado, C. H.M. Fernandes, A. J. Terezo, and E. Baggio-Saitovitch. 2013. "High Coercivity Induced by Mechanical Milling in Cobalt Ferrite Powders." *Journal of Magnetism and Magnetic Materials* 344 (October): 182–87. https://doi.org/10.1016/j.jmmm.2013.05.056.

Rathore, Deepshikha, Supratim Mitra, Rajnish Kurchania, and R. K. Pandey. 2018. "Physicochemical Properties of $CuFe_2O_4$ Nanoparticles as a Gas Sensor." *Journal of Materials Science: Materials in Electronics* 29 (3): 1925–32. https://doi.org/10.1007/s10854-017-8102-0.

Raut, S. D., V. V. Awasarmol, B. G. Ghule, S. F. Shaikh, S. K. Gore, R. P. Sharma, P. P. Pawar, and R. S. Mane. 2018. "Enhancement in Room-Temperature Ammonia Sensor Activity of Size-Reduced Cobalt Ferrite Nanoparticles on γ-Irradiation." *Materials Research Express* 5 (6): 065035. https://doi.org/10.1088/2053-1591/aac99d.

Reyes-Rodríguez, Pamela Yajaira, Dora Alicia Cortés-Hernández, José Concepción Escobedo-Bocardo, José Manuel Almanza-Robles, Javier Sánchez, Laura Elena De León-Prado, Juan Méndez-Nonell, and Gilberto Francisco Hurtado-López. 2018. "Synthesis and Study of Structural and Magnetic Properties of Mg1-XZnxFe2O4 (X=0-0,9)." *Revista Materia* 23 (2): 12124. https://doi.org/10.1590/S1517-707620180002.0458.

Rezaeipour, Amin, Sajjad Dehghani, and Sedigheh Hoghoghifard. 2022. "VOC Sensors Based on Nanoparticles and Nanorods of Nickel Ferrite." *IEEE Sensors Journal* 22 (16): 16464–71. https://doi.org/10.1109/JSEN.2022.3188541.

Saha, Priyanka, Rupali Rakshit, and Kalyan Mandal. 2019. "Enhanced Magnetic Properties of Zn Doped Fe_3O_4 Nano Hollow Spheres for Better Bio-Medical Applications." *Journal of Magnetism and Magnetic Materials* 475 (April): 130–36. https://doi.org/10.1016/j.jmmm.2018.11.061.

Sahu, Saura C, and A Wallace Hayes. 2017. "Toxicity of Nanomaterials Found in Human Environment." *Toxicology Research and Application* 1 (September): 239784731772635. https://doi.org/10.1177/2397847317726352.

Saini, Ashish, Atul Thakur, and Preeti Thakur. 2016. "Matching Permeability and Permittivity of $Ni_{0.5}Zn_{0.3}Co_{0.2}In_{0.1}Fe_{1.9}O_4$ Ferrite for Substrate of Large Bandwidth Miniaturized Antenna." *Journal of Materials Science: Materials in Electronics* 27 (3): 2816–23. https://doi.org/10.1007/S10854-015-4095-8.

Sasikala, Arathyram Ramachandra Kurup, Afeesh Rajan Unnithan, Yeo Heung Yun, Chan Hee Park, and Cheol Sang Kim. 2016. "An Implantable Smart Magnetic Nanofiber Device for Endoscopic Hyperthermia Treatment and Tumor-Triggered Controlled Drug Release." *Acta Biomaterialia* 31 (February): 122–33. https://doi.org/10.1016/j.actbio.2015.12.015.

Sharifi, Ibrahim, and et al. Shokrollahi. 2012. "Ferrite-Based Magnetic Nanofluids Used in Hyperthermia Applications." *Journal of Magnetism and Magnetic Materials*. North-Holland. https://doi.org/10.1016/j.jmmm.2011.10.017.

Silva, F. G. da, J Depeyrot, A. F. C Campos, R Aquino, D Fiorani, and D Peddis. 2019. "Structural and Magnetic Properties of Spinel Ferrite Nanoparticles." *Journal of Nanoscience and Nanotechnology* 19 (8): 4888–4902. https://doi.org/10.1166/jnn.2019.16877.

Singh, Amitender, Fayu Wan, Kavita Yadav, Anand Salvi, Preeti Thakur, and Atul Thakur. 2023. "Synergistic Effect of ZnO Nanoparticles with Cu^{2+} Doping on Antibacterial and Photocatalytic Activity." *Inorganic Chemistry Communications* 157 (November): 111425. https://doi.org/10.1016/j.inoche.2023.111425.

Singh, Amitender, Fayu Wan, Kavita Yadav, Saarthak Kharbanda, Preeti Thakur, and Atul Thakur. 2024. "A Novel Magnetic $NiFe_2O_4$-Ag-ZnO Hybrid Nanocomposite for the Escalated Photocatalytic Dye Degradation and Antibacterial Activities." *Materials Science and Engineering: B* 299 (January): 116935. https://doi.org/10.1016/j.mseb.2023.116935.

Sivakumar, P., R. Ramesh, A. Ramanand, S. Ponnusamy, and C. Muthamizhchelvan. 2011. "Synthesis and Characterization of Nickel Ferrite Magnetic Nanoparticles." *Materials Research Bulletin* 46 (12): 2208–11. https://doi.org/10.1016/j.materresbull.2011.09.009.

Spirou, Spiridon V., Martina Basini, Alessandro Lascialfari, Claudio Sangregorio, and Claudia Innocenti. 2018. "Magnetic Hyperthermia and Radiation Therapy: Radiobiological Principles and Current Practice." Nanomaterials. https://doi.org/10.3390/nano8060401.

Suleman, Muhammad, and Samia Riaz. 2020. "In Silico Study of Hyperthermia Treatment of Liver Cancer Using Core-Shell $CoFe_2O_4$@$MnFe_2O_4$ Magnetic Nanoparticles." *Journal of Magnetism and Magnetic Materials* 498 (March): 166143. https://doi.org/10.1016/j.jmmm.2019.166143.

Sumangala, T. P., Isabelle Pasquet, Lionel Presmanes, Yohann Thimont, Corine Bonningue, N. Venkataramani, Shiva Prasad, Valérie Baco-Carles, Philippe Tailhades, and Antoine Barnabé. 2018. "Effect of Synthesis Method and Morphology on the Enhanced CO_2 Sensing Properties of Magnesium Ferrite $MgFe_2O_4$." *Ceramics International* 44 (15): 18578–84. https://doi.org/10.1016/j.ceramint.2018.07.082.

Suppuraj, Palusamy, Ganesamoorthy Thirunarayanan, Meenakshisundaram Swaminathan, and Inbasekaran Muthuvel. 2017. "Facile Synthesis of Spinel Nanocrystalline $ZnFe_2O_4$: Enhanced Photocatalytic and Microbial Applications." *Materials Science and Applied Chemistry* 34 (1).

Sutradhar, S., S. Das, A. Roychowdhury, D. Das, and P. K. Chakrabarti. 2015. "Magnetic Property, Mössbauer Spectroscopy and Microwave Reflection Loss of Maghemite Nanoparticles (γ-Fe2O3) Encapsulated in Carbon Nanotubes." *Materials Science and Engineering: B* 196 (June): 44–52. https://doi.org/10.1016/j.mseb.2015.02.008.

Taneja, Shilpa, Deepika Chahar, Preeti Thakur, and Atul Thakur. 2021. "Influence of Bismuth Doping on Structural, Electrical and Dielectric Properties of Ni–Zn Nanoferrites." *Journal of Alloys and Compounds* 859 (April): 157760. https://doi.org/10.1016/J.JALLCOM.2020.157760.

Taneja, Shilpa, Preeti Thakur, Blaise Ravelo, and Atul Thakur. 2022. "Nanocrystalline Samarium Doped Nickel-Zinc-Bismuth Ferrites: Investigation of Structural, Electrical and Dielectric Properties." *Materials Research Bulletin* 154 (October): 111937. https://doi.org/10.1016/j.materresbull.2022.111937.

Tan, Lichao, Qi Liu, Xiaoyan Jing, Jingyuan Liu, Dalei Song, Songxia Hu, Lianhe Liu, and Jun Wang. 2015. "Removal of Uranium(VI) Ions from Aqueous Solution by Magnetic Cobalt Ferrite/Multiwalled Carbon Nanotubes Composites." *Chemical Engineering Journal* 273 (August): 307–15. https://doi.org/10.1016/j.cej.2015.01.110.

Thakur, Preeti, Shilpa Taneja, Deepika Chahar, Blaise Ravelo, and Atul Thakur. 2021. "Recent Advances on Synthesis, Characterization and High Frequency Applications of Ni-Zn Ferrite Nanoparticles." *Journal of Magnetism and Magnetic Materials* 530 (July): 1–20. https://doi.org/10.1016/j.jmmm.2021.167925.

Thakur, Preeti, and Atul Thakur. 2022. "Nanomaterials, Their Types and Properties." In *Synthesis and Applications of Nanoparticles*, 19–44. Springer Nature. https://doi.org/10.1007/978-981-16-6819-7_2.

Thakur, Atul, Preeti Thakur, and Jen Hwa Hsu. 2011. "Smart Magnetodielectric Nano-Materials for the Very High Frequency Applications." *Journal of Alloys and Compounds* 509 (17): 5315–19. https://doi.org/10.1016/j.jallcom.2011.02.021.

Thakur, Atul, Preeti Thakur, and Jen Hwa Hsu. 2012. "Magnetic Behaviour of Ni 0.4Zn 0.6Co 0.1Fe 1.9O_4 Spinel Nano-Ferrite." *Journal of Applied Physics* 111 (7): 1–4. https://doi.org/10.1063/1.3670606.

Thakur, Atul, Ritesh Verma, Fayu Wan, Blaise Ravelo, Irina Edelman, Sergei Ovchinnikov, and Preeti Thakur. 2023. "Investigation of Structural, Elastic and Magnetic Properties of Cu^{2+} Ions Substituted Cobalt Nano Ferrites." *Journal of Magnetism and Magnetic Materials* 581 (September): 170980. https://doi.org/10.1016/j.jmmm.2023.170980.

Tiwari, Rashmi, Manojit De, H. S. Tewari, and S. K. Ghoshal. 2020. "Structural and Magnetic Properties of Tailored $NiFe_2O_4$ Nanostructures Synthesized Using Auto-Combustion Method." *Results in Physics* 16 (March): 102916. https://doi.org/10.1016/j.rinp.2019.102916.

Vallabani, N. V. Srikanth, and Sanjay Singh. 2018. "Recent Advances and Future Prospects of Iron Oxide Nanoparticles in Biomedicine and Diagnostics." 3 Biotech. https://doi.org/10.1007/s13205-018-1286-z.

Wang, Lianqiang, Yanzhe Wang, Hongwei Tian, Liang Qiao, and Yi Zeng. 2020. "Enhanced Ammonia Detection Using Wrinkled Porous $CoFe_2O_4$ Double-Shelled Spheres Prepared by a Thermally Driven Contraction Process." *Sensors and Actuators, B: Chemical* 314 (July): 128085. https://doi.org/10.1016/j.snb.2020.128085.

Wei, Kai, Hai Xia Huai, Bing Zhao, Jun Zheng, Gui Qi Gao, Xiu Ying Zheng, and Chun Chang Wang. 2022. "Facile Synthesis of $CoFe_2O_4$ Nanoparticles and Their Gas Sensing Properties." *Sensors and Actuators B: Chemical* 369 (October): 132279. https://doi.org/10.1016/j.snb.2022.132279.

Wu, Kai Di, Jin Yong Xu, Marc Debliquy, and Chao Zhang. 2021. "Synthesis and NH3/TMA Sensing Properties of $CuFe_2O_4$ Hollow Microspheres at Low Working Temperature." *Rare Metals* 40 (7): 1768–77. https://doi.org/10.1007/s12598-020-01609-9.

Yang, Meicheng, Lipeng Gao, Kai Liu, Chunhua Luo, Yiting Wang, Lei Yu, Hui Peng, and Wen Zhang. 2015. "Characterization of $Fe_3O_4/SiO_2/Gd_2O(CO_3)_2$ Core/Shell/Shell Nanoparticles as T1 and T2 Dual Mode MRI Contrast Agent." *Talanta* 131 (January): 661–65. https://doi.org/10.1016/j.talanta.2014.08.042.

Zhang, Haitao, Xiong Zhang, He Lin, Kai Wang, Xianzhong Sun, Nansheng Xu, Chen Li, and Yanwei Ma. 2015. "Graphene and Maghemite Composites Based Supercapacitors Delivering High Volumetric Capacitance and Extraordinary Cycling Stability." *Electrochimica Acta* 156 (February): 70–76. https://doi.org/10.1016/j.electacta.2015.01.041.

Zhang, Yanlin, Yu Zhou, Zhanlian Li, Gang Chen, Yong Mao, Hongtao Guan, and Chengjun Dong. 2019. "MOFs-Derived NiFe2O4 Fusiformis with Highly Selective Response to Xylene." *Journal of Alloys and Compounds* 784 (May): 102–10. https://doi.org/10.1016/j.jallcom.2018.12.389.

19 Future Perspectives of Spinel Nano-Ferrites

*Arti Rao, Preeti Thakur, and Atul Thakur**
Amity Institute of Nanotechnology, Amity University Haryana, Gurugram, India

19.1 INTRODUCTION

Nano-ferrites, a class of magnetic materials with particle sizes in the nanometer range, have garnered significant attention in contemporary research due to their unique properties (Figure 19.1) and versatile applications. These materials are typically derived from iron oxides, such as magnetite (Fe_3O_4), and exhibit exceptional magnetic behavior at the nanoscale [1]. The distinct advantage of nano-ferrites lies in their tunable magnetic and physical properties, which make them promising candidates for a myriad of technological applications. In current research, nano-ferrites play a pivotal role in various fields, including electronics, medicine, and energy. The ability to manipulate the size and composition of nano-ferrites allows for tailoring their magnetic characteristics, enabling advancements in magnetic storage devices, high-frequency electronics, and microwave devices [2]. Moreover, the biocompatibility of certain nano-ferrites has led to their exploration in biomedical applications, such as targeted drug delivery and magnetic resonance imaging (MRI) [3]. The versatility of nano-ferrites makes them valuable across interdisciplinary domains, fostering collaborations between materials scientists, physicists, chemists, and biomedical researchers.

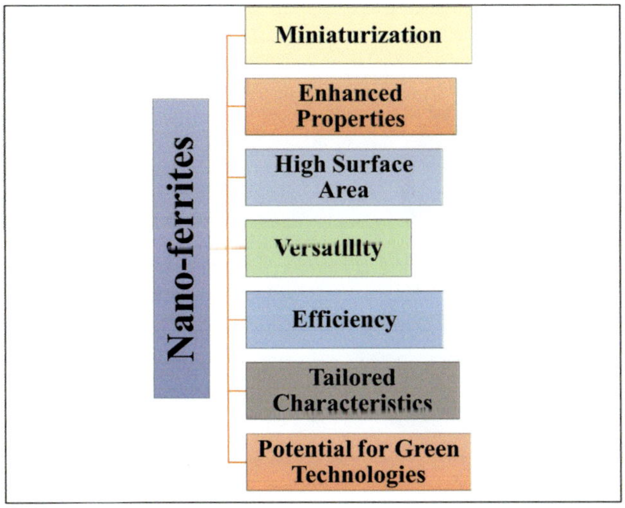

FIGURE 19.1 Promising features of nano-ferrites.

Significantly, the magnetic properties of nano-ferrites contribute to their utility in environmental remediation efforts. Researchers are exploring their potential in wastewater treatment and the removal of heavy metals from contaminated environments, showcasing the materials' promise for sustainable and eco-friendly solutions [4]. The synthesis of nano-ferrites using innovative methods, such as sol-gel processes and co-precipitation techniques, adds an extra layer of complexity to their study, offering researchers a diverse toolkit for tailoring properties based on specific applications.

The overview of nano-ferrites underscores their significance in current research, where their unique magnetic properties and tunable characteristics make them indispensable across diverse scientific disciplines [1]. As researchers delve deeper into the synthesis methods and explore novel applications, nano-ferrites continue to emerge as key players in shaping the landscape of modern materials science and technology.

19.2 FUNDAMENTALS OF NANO-FERRITES

The fundamentals of nano-ferrites encompass a thorough understanding of their structural and magnetic characteristics [5]. Nano-ferrites, being a subset of ferrimagnetic materials, typically consist of crystalline structures derived from iron oxides. Magnetite (Fe_3O_4), for instance, is a common precursor, and the synthesis methods employed play a crucial role in determining the size, shape, and composition of the nanoparticles. The crystal lattice of nano-ferrites gives rise to their unique magnetic properties, with magnetic moments of individual atoms aligning in a specific manner, contributing to the overall magnetization of the material. In addition to their structural aspects, the magnetic behavior of nano-ferrites is a focal point in the fundamentals of this material class. The magnetic properties are strongly influenced by factors such as particle size, shape, and the presence of impurities. Understanding the magnetic interactions within nano-ferrites is essential for tailoring their performance in applications ranging from data storage to medical imaging. Researchers delve into the intricacies of magnetic domains, hysteresis loops, and coercivity to unravel the nuanced fundamentals that govern the magnetic behavior of nano-ferrites, laying the groundwork for their optimized utilization in various technological domains.

* Corresponding author.

DOI: 10.1201/9781003568940-19

19.2.1 Properties and Characteristics

Nano-ferrites exhibit a plethora of intriguing properties and characteristics that stem from their unique crystalline structure and magnetic behavior. One of the fundamental attributes is their size-dependent magnetic properties [1]. As the particle size diminishes to the nanoscale, quantum effects become increasingly pronounced, influencing magnetic behaviors such as coercivity and remanence. This phenomenon opens avenues for tailoring magnetic characteristics, a crucial aspect for applications like high-density magnetic storage and biomedical imaging. The crystalline structure of nano-ferrites contributes significantly to their thermal stability. These materials often possess high Curie temperatures, indicating the temperature at which they undergo a transition from a ferromagnetic to a paramagnetic state [6]. This thermal stability is advantageous in applications requiring reliable magnetic performance across a broad temperature range, such as in electronic devices and aerospace components. Furthermore, the chemical composition of nano-ferrites plays a pivotal role in determining their magnetic and physical properties, making it imperative to precisely control the synthesis processes for desired outcomes. Electrical conductivity is another vital property of nano-ferrites. These materials are often semiconductors, displaying tunable electrical characteristics. Researchers explore these properties for applications in electronics, sensors, and catalysis [7]. The ability to modulate electrical conductivity by adjusting the size and composition of nano-ferrites adds a layer of versatility, enabling their integration into a variety of electronic devices.

Nano-ferrites also exhibit excellent biocompatibility, a crucial property for their application in medicine. The surface properties of these nanoparticles can be modified to enhance their interaction with biological systems. This property has led to advancements in targeted drug delivery, where nano-ferrites serve as carriers for therapeutic agents, guided by external magnetic fields to specific sites within the body. Additionally, their use in MRI benefits from their biocompatibility and magnetic characteristics, providing enhanced contrast for detailed imaging. The optical properties of nano-ferrites have gained attention in recent research. Depending on their size and composition, these materials can exhibit interesting optical phenomena such as plasmonic resonances and magneto-optical effects. Researchers explore these properties for applications in sensors, imaging, and even in the development of novel photonic devices. The interplay between magnetic and optical characteristics offers a multifaceted approach to exploiting nano-ferrites in emerging technologies.

The properties and characteristics of nano-ferrites make them a fascinating subject of scientific inquiry with far-reaching implications (Figure 19.2). Their size-dependent magnetic behaviors, thermal stability, electrical conductivity, biocompatibility, and optical properties collectively contribute to the diverse range of applications spanning electronics, medicine, and photonics. The ongoing exploration of these properties paves the way for continued advancements and innovations in the utilization of nano-ferrites across various scientific and technological domains.

19.2.2 Synthesis Methods

The synthesis of nano-ferrites is a critical aspect of research, influencing the materials' properties and applicability in diverse fields [8]. Various methods are employed to produce

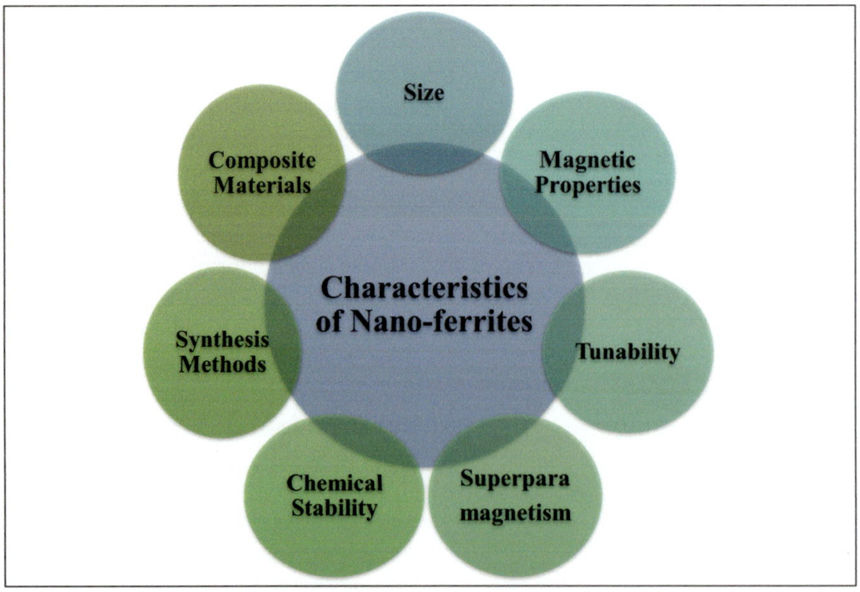

FIGURE 19.2 Various characteristics of nano-ferrites.

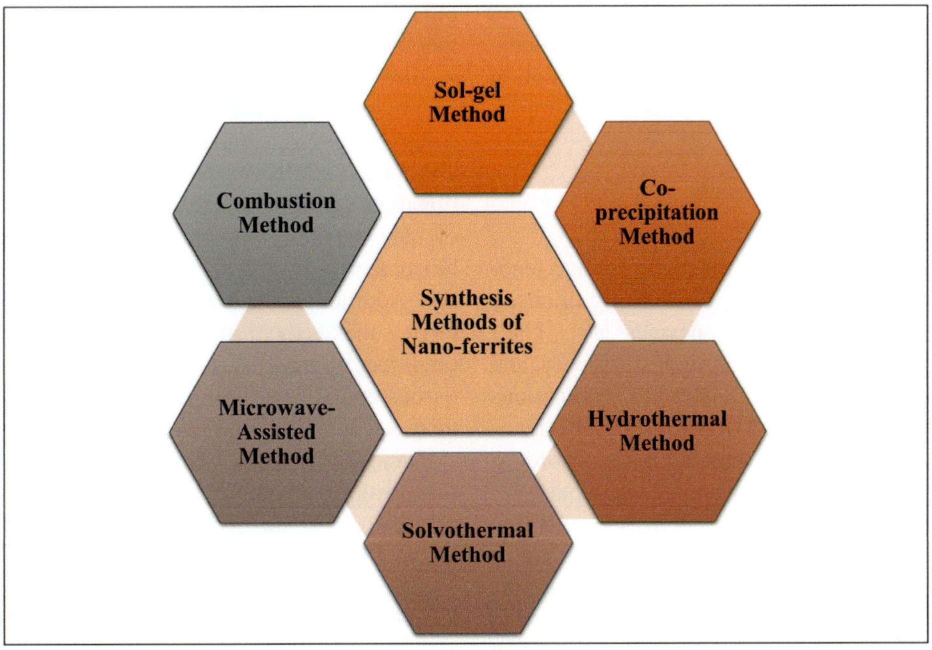

FIGURE 19.3 Various synthesis methods employed in crafting nano-ferrites.

these nanoparticles, each with distinct advantages and challenges (Figure 19.3).

- One common approach is the sol-gel method, where precursor materials are dissolved in a liquid solution, forming a gel that undergoes a controlled drying process to yield nano-ferrite powders [9]. This method allows for precise control over particle size and composition, making it suitable for tailoring properties based on specific applications [10].
- Co-precipitation is another widely used technique in nano-ferrite synthesis. In this method, metal ions are simultaneously precipitated from a solution, resulting in the formation of nano-sized particles [11]. Co-precipitation is valued for its simplicity and scalability, enabling the production of large quantities of nano-ferrites. However, careful control of reaction parameters is essential to achieve desired particle sizes and avoid agglomeration.
- Hydrothermal [12] and solvothermal [11] methods involve the use of high-pressure, high-temperature conditions for nano-ferrite synthesis. These methods offer enhanced crystallinity and control over particle size, making them suitable for applications requiring specific structural and magnetic properties [13]. Microwave-assisted synthesis [14] has emerged as a rapid and energy-efficient technique, leveraging microwave irradiation to accelerate the reaction rates, reducing synthesis times compared to conventional methods.

In recent years, green synthesis methods have gained attention for their eco-friendly approach. Utilizing plant extracts or other natural sources as reducing agents and stabilizers, these methods avoid the use of hazardous chemicals. Green synthesis not only aligns with sustainability principles but also introduces unique surface functionalities to nano-ferrites, expanding their potential in biomedical and environmental applications [15]. The choice of synthesis method depends on the desired characteristics and intended applications of the nano-ferrites. Each method has its advantages and limitations, emphasizing the need for a tailored approach based on the specific requirements of the research or technological development. Ongoing research in synthesis techniques continues to refine existing methods and explore novel approaches, aiming to optimize the production of nano-ferrites for a wide array of applications.

19.3 CURRENT APPLICATIONS

Nano-ferrites find diverse and impactful applications across multiple fields in current research. In the realm of electronics, these materials contribute significantly to the development of high-performance devices [16]. Their magnetic properties make them crucial components in magnetic storage devices, such as hard drives and magnetic random-access memory (MRAM), where their size-dependent behaviors enable increased data storage density. In the biomedical field, nano-ferrites play a pivotal role in targeted drug delivery systems and MRI contrast agents. The ability to functionalize their surfaces allows for specific interactions with biological entities, offering precise control in drug delivery, while their magnetic characteristics enhance imaging contrast in medical diagnostics. Additionally, nano-ferrites are finding applications in energy-related technologies, including the development of efficient microwave devices and as catalysts in environmental remediation processes, showcasing their versatility and potential impact on advancing technological frontiers [17].

19.3.1 Magnetic Storage

Nano-ferrites play a crucial role in magnetic storage technologies, contributing to the development of high-density and high-performance data storage devices [18]. Their unique magnetic properties, such as high coercivity and remanence, make them ideal candidates for applications like hard disk drives (HDDs) and MRAM. In HDDs, nano-ferrites are employed as magnetic material in the recording layer of the disk. The small size of nano-ferrite particles allows for densely packed magnetic domains, enabling increased data storage capacity [19]. Moreover, the stability of nano-ferrites ensures that stored information remains intact over time. In MRAM, these materials contribute to the non-volatile storage of data by utilizing the orientation of magnetic moments in nano-ferrite particles as memory states. The use of nano-ferrites in magnetic storage not only enhances data storage density but also contributes to the development of more energy-efficient and faster data storage technologies. Ongoing research focuses on further optimizing nano-ferrite properties for even more advanced and compact magnetic storage solutions.

The continuous demand for higher data storage capacity has intensified the exploration of nano-ferrites in magnetic storage applications [18]. Their size-dependent magnetic behaviors, influenced by quantum effects, enable researchers to engineer materials with tailored properties for enhanced storage performance. The miniaturization of nano-ferrite particles allows for higher information density on storage media, addressing the ever-growing need for more data in smaller spaces [19]. As technology advances, the integration of nano-ferrites in magnetic storage not only facilitates increased storage capacity but also contributes to the development of more robust and reliable data storage solutions. Moreover, the durability and thermal stability of nano-ferrites make them resilient in the demanding conditions of data storage devices, ensuring prolonged data retention and reinforcing their pivotal role in shaping the future of magnetic storage technologies.

19.3.2 Biomedical Applications

Nano-ferrites have emerged as promising materials for a wide range of biomedical applications, leveraging their unique properties to address challenges in medicine and healthcare [20] (Figure 19.4). One notable application is in targeted drug delivery systems. Surface-modified nano-ferrites can serve as carriers for therapeutic agents, and their magnetic properties enable precise localization within the body when subjected to external magnetic fields. This targeted drug delivery approach enhances treatment efficacy while minimizing side effects by concentrating therapeutic payloads at specific disease sites. In the field of medical imaging, nano-ferrites find application as contrast agents in MRI. The magnetic properties of these nanoparticles enhance the contrast of MR images, providing clearer and more detailed anatomical information. Surface functionalization allows researchers to tailor the interactions of nano-ferrites with biological tissues, improving their biocompatibility and ensuring safe use in medical diagnostics. Additionally, the inherent magnetic response of nano-ferrites makes them suitable for magnetic hyperthermia—a treatment approach where localized heating is induced in tissues containing these nanoparticles, offering a targeted therapeutic strategy for conditions like cancer [21].

Nano-ferrites also hold promise in biosensing and diagnostic applications [22]. Their unique magnetic characteristics can be exploited in the development of sensitive biosensors for detecting specific biomolecules. Functionalizing the surfaces of nano-ferrites with biomolecular recognition elements enables the selective detection of biological markers, providing valuable information for disease diagnosis and monitoring. Furthermore, the biocompatibility of certain nano-ferrites enhances their potential for in vivo applications, supporting their integration into emerging diagnostic technologies. Beyond diagnostics and therapy, nano-ferrites have been investigated for their antimicrobial properties [23]. The surface properties of these nanoparticles can be engineered to interact with microorganisms, offering a potential avenue for combating bacterial infections. This antimicrobial application showcases the multifaceted nature of nano-ferrites in addressing diverse challenges in healthcare, from therapeutics to infection control. While the biomedical applications of nano-ferrites present exciting possibilities, ongoing research focuses on refining their properties, improving biocompatibility, and addressing safety considerations. The interdisciplinary nature of this research, combining expertise in materials science, medicine, and biology, underscores the potential of nano-ferrites to revolutionize medical treatments and diagnostics, marking a significant contribution to the evolving landscape of nanomedicine.

19.3.3 Electronics

Nano-ferrites have become integral to advancements in electronics, contributing to the development of high-performance devices and technologies. One significant application is in the field of magnetic sensors and data storage [24]. The unique magnetic properties of nano-ferrites, including high coercivity and remanence, make them ideal candidates for magnetic sensors used in various electronic devices. Moreover, their application in magnetic storage devices, such as HDDs, is pivotal for achieving increased data storage density and efficient data retrieval. In the realm of microwave devices and telecommunications, nano-ferrites play a crucial role. Their tunable magnetic characteristics make them suitable for applications in microwave

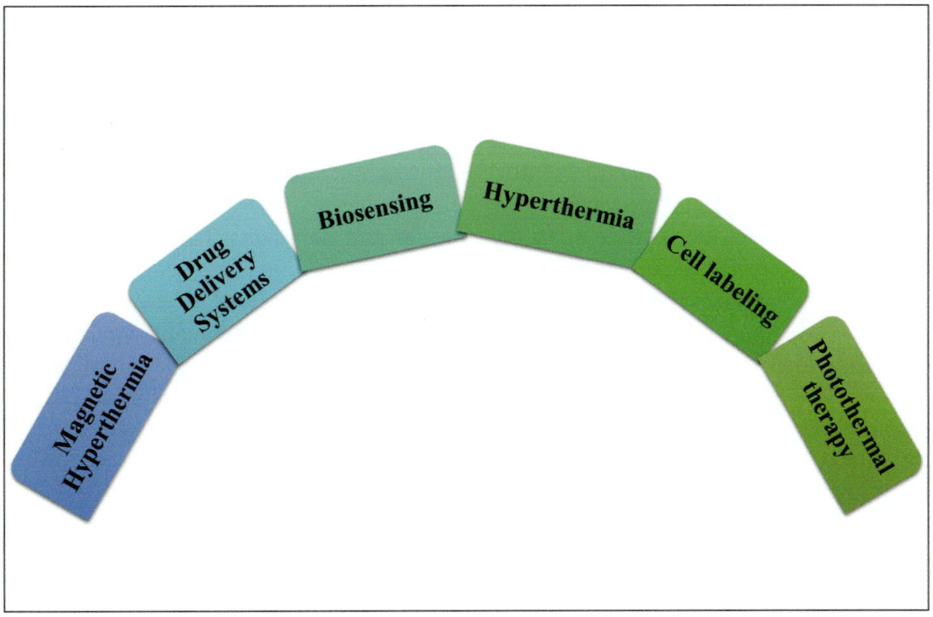

FIGURE 19.4 Biomedical applications of nano-ferrites, showcasing their transformative role in advancing medical technologies.

absorbers, isolators, and circulators. The ability to control the magnetic properties of nano-ferrites allows for the customization of devices operating at different frequencies, enhancing the performance of wireless communication systems and radar technologies [25]. The development of novel electronic devices, including spintronic devices, benefits from the integration of nano-ferrites. Spintronics exploits the spin of electrons in addition to their charge, and nano-ferrites with tailored magnetic properties can be employed in spintronic applications, such as spin valves and magnetic tunnel junctions. These devices leverage the manipulation of electron spin for information processing and storage, paving the way for more energy-efficient and faster electronics.

The use of nano-ferrites in antennas and electromagnetic interference (EMI) shielding is also noteworthy. Their magnetic properties, combined with the ability to control their size and composition, make nano-ferrites effective materials for enhancing the performance of antennas and mitigating unwanted EMI. This application is crucial in the design of electronic systems where signal integrity is paramount.

Furthermore, nano-ferrites contribute to the advancement of sensors and actuators. Their magnetic response to external stimuli, such as magnetic fields, temperature, or stress, makes them valuable components in sensors for diverse applications. Nano-ferrite-based actuators can be employed in micro-electromechanical systems (MEMS), offering precise control and responsiveness in miniaturized electronic devices. In the pursuit of more energy-efficient electronics, nano-ferrites are investigated for their potential in power electronics. The ability to modulate their electrical conductivity and magnetic properties opens avenues for developing efficient transformers, inductors, and other components essential for power conversion and distribution systems [24]. Nano-ferrites play a multifaceted role in electronics, influencing the design and functionality of a wide array of devices. From magnetic sensors and data storage to microwave devices and spintronics, the unique properties of nano-ferrites continue to drive innovation in electronic technologies, shaping the landscape of modern electronics and paving the way for more efficient and compact devices.

19.4 CHALLENGES AND LIMITATIONS

Nano-ferrites, while holding immense promise for various applications, encounter challenges and limitations that necessitate careful consideration in their design and implementation. One prominent set of challenges revolves around size-related issues. As the size of nano-ferrite particles decreases, quantum effects become more pronounced, influencing their magnetic properties. Achieving consistent and reproducible sizes at the nanoscale poses a technical challenge in synthesis methods, requiring precise control over reaction conditions to avoid variations that can impact the intended application. Moreover, size uniformity is critical for applications such as magnetic storage, where the performance is intricately linked to the size of the nano-ferrite particles.

Stability concerns also pose significant challenges in the utilization of nano-ferrites. The high surface-to-volume ratio in nanoscale particles makes them susceptible to oxidation and other chemical reactions, affecting their stability over time. Surface modifications are often employed to mitigate these stability issues, but careful consideration is required to strike a balance between stability and maintaining the desired properties. Stability concerns become particularly crucial in biomedical applications, where the interaction of nano-ferrites with biological environments must be carefully managed to ensure safety and efficacy [26]. In the context of magnetic storage, size-related challenges extend to issues like superparamagnetism, wherein at small particle sizes, thermal fluctuations can lead to a loss of magnetic information, limiting the reliability of storage devices. Overcoming this challenge involves innovative approaches in materials design and synthesis to mitigate the impact of superparamagnetic effects. Additionally, stability concerns in magnetic storage devices involve addressing issues of data retention over extended periods, especially as the demand for long-term data storage solutions continues to grow. In biomedical applications, the challenges associated with size and stability are intricately tied to the biocompatibility of nano-ferrites [27]. Achieving sizes suitable for cellular uptake without compromising stability is a delicate balance, especially when considering the potential toxicological impacts of these nanoparticles. Researchers grapple with ensuring that nano-ferrites maintain their magnetic properties and functionality within the complex biological milieu while minimizing any adverse effects on cells or tissues. Environmental considerations also add a layer of complexity to the challenges faced by nano-ferrites [26]. The synthesis methods often involve the use of chemicals and energy-intensive processes, raising questions about the environmental impact of large-scale production. Developing eco-friendly synthesis methods and assessing the life cycle impact of nano-ferrites are essential steps in addressing these environmental concerns. Furthermore, the potential release of nano-ferrites into the environment, whether through manufacturing processes or the disposal of electronic devices containing these materials, raises questions about their long-term ecological impact [27]. Understanding and mitigating these challenges are crucial for the sustainable use of nano-ferrites in various applications.

19.5 RECENT ADVANCES

Recent advances in nano-ferrite research have been marked by breakthroughs in understanding their properties and the development of novel synthesis techniques, offering new possibilities for applications in various fields. Breakthrough research has focused on unraveling the intricate relationships between the size, composition, and magnetic properties of nano-ferrites. Advanced characterization techniques, such as high-resolution imaging and spectroscopy, enable researchers to explore these materials at the nanoscale, providing unprecedented insights into their behavior. This deeper understanding lays the foundation for more precise tailoring of nano-ferrites for specific applications, from electronics to biomedicine [28]. In the realm of electronics, breakthroughs have been achieved in the development of spintronic devices using nano-ferrites. Research has demonstrated the potential for leveraging the unique spin properties of electrons in these materials for information processing and storage. The integration of nano-ferrites in spin valves and magnetic tunnel junctions opens new avenues for creating energy-efficient and high-performance electronic devices, contributing to the evolution of spintronics. Novel synthesis techniques have emerged as a key driver of progress in nano-ferrite research. One notable approach involves the use of green synthesis methods, utilizing plant extracts or other natural sources as reducing agents and stabilizers. This eco-friendly approach not only aligns with sustainable practices but also introduces unique surface functionalities to nano-ferrites, expanding their potential in biomedical and environmental applications. Additionally, advancements in sol-gel processes and co-precipitation techniques have enabled greater control over particle size, shape, and composition, addressing challenges related to uniformity and reproducibility in nano-ferrite synthesis [29].

19.5.1 Breakthrough Research

Breakthroughs in biomedical applications have been fueled by innovative research in surface engineering and functionalization of nano-ferrites. Researchers have developed sophisticated surface modifications to enhance the biocompatibility of these nanoparticles, enabling more effective targeted drug delivery and imaging applications. The development of multifunctional nano-ferrite platforms, capable of integrating therapeutic and diagnostic functionalities, represents a significant leap forward in the field of nanomedicine.

In environmental applications, recent advances include the use of nano-ferrites for efficient wastewater treatment and the removal of heavy metals. Breakthroughs in understanding the adsorption and catalytic properties of nano-ferrites have led to their enhanced performance in environmental remediation processes. These applications showcase the potential impact of nano-ferrites in addressing critical challenges related to water quality and pollution. Furthermore, the integration of nano-ferrites in energy-related technologies has witnessed significant strides. Breakthroughs in the development of nano-ferrite-based materials for energy storage and conversion devices, such as batteries and solar cells, hold promise for more efficient and sustainable energy solutions. Recent advances in nano-ferrite research encompass breakthroughs in understanding their properties and functionalities, as well as the development of novel synthesis techniques. These advancements pave the way for a new era of applications in electronics, biomedicine, environmental remediation, and energy technologies, positioning nano-ferrites as key players in driving innovation across diverse scientific and technological domains.

19.5.2 Novel Synthesis Techniques

Current trends in nano-ferrite research indicate a growing emphasis on sustainable synthesis methods and environmentally friendly applications. Researchers are increasingly exploring green synthesis techniques, utilizing natural sources and minimizing the use of hazardous chemicals [30]. This trend aligns with the broader movement toward sustainable and eco-friendly materials, reflecting a conscientious approach to the environmental impact of nanomaterial production. As the field advances, the integration of green synthesis methods not only contributes to the responsible development of nano-ferrites but also expands their applications in biomedicine and environmental science. Another notable trend involves the convergence of nano-ferrites with emerging technologies, such as the Internet of Things (IoT) and artificial intelligence (AI). The unique magnetic properties of nano-ferrites make them attractive for applications in sensors, actuators, and data storage devices within IoT systems [31]. Additionally, researchers are exploring how AI algorithms can optimize the synthesis and design of nano-ferrites for specific applications, accelerating the development of tailored materials with enhanced properties. This trend reflects the intersection of nanotechnology with cutting-edge technologies, paving the way for transformative advancements in various fields.

19.6 EMERGING TECHNOLOGIES

Emerging technologies are increasingly leveraging the unique properties of nano-ferrites, opening new frontiers in various scientific and technological domains. In the realm of electronics, the integration of nano-ferrites is contributing to the development of advanced sensors and actuators. Their size-dependent magnetic properties make them ideal for creating highly sensitive sensors capable of detecting minute changes in the surrounding environment [32]. Additionally, nano-ferrite-based actuators, influenced by external magnetic fields, offer precise control in MEMS, enabling the creation of responsive and miniaturized devices for diverse applications. In the field of energy storage, nano-ferrites are making significant strides. Research is focusing on utilizing these materials in batteries and supercapacitors to enhance energy storage capacity and charge-discharge rates. The tunable magnetic and electrical properties of nano-ferrites provide opportunities to design materials that address the challenges of energy storage systems, contributing to the development of more efficient and sustainable energy solutions. Biomedical applications are witnessing transformative advancements with the incorporation of nano-ferrites. Researchers are exploring their potential in theranostics, a field that combines therapy and diagnostics. Nano-ferrites with tailored magnetic and optical properties can serve as multifunctional platforms for simultaneous imaging and targeted drug delivery [33]. This convergence of imaging and therapeutic functionalities enhances precision in medical treatments, marking a significant leap forward in personalized medicine. The synergy between nano-ferrites and photonics is another emerging trend. Researchers are harnessing the optical properties of nano-ferrites for applications in photonic devices and telecommunications. The ability to modulate the optical characteristics of these materials opens avenues for developing novel photonic components, such as modulators and switches, with implications for advancing optical communication networks and information processing. Moreover, the integration of nano-ferrites into the burgeoning field of quantum computing is on the horizon. Quantum properties of nano-ferrites, such as spin states, make them intriguing candidates for qubits, the fundamental units of quantum information. Researchers are exploring ways to exploit the quantum behavior of nano-ferrites to enhance the performance of quantum computers, offering potential breakthroughs in computational capabilities and problem-solving.

Nano-ferrites are at the forefront of emerging technologies, influencing advancements in electronics, energy storage, biomedicine, photonics, and quantum computing. The versatility of these materials, coupled with ongoing research into their unique properties, positions nano-ferrites as key contributors to the transformative landscape of cutting-edge technologies.

19.6.1 Cross-Disciplinary Applications

Nano-ferrites exhibit versatile properties that make them well-suited for cross-disciplinary applications, fostering collaborations across various scientific domains. In the intersection of materials science and electronics, nano-ferrites play a pivotal role. Their magnetic properties make them essential in the development of sensors and data storage devices, impacting fields ranging from telecommunications to MRI in medical diagnostics [34]. This cross-disciplinary synergy is evident in the collaborative efforts to enhance the performance of electronic devices through the incorporation of nano-ferrites, pushing the boundaries of what is achievable in both material science and electronics. In the realm of nanomedicine, the convergence of materials science and biomedical research is prominent. Nano-ferrites find applications in targeted drug delivery systems [35], combining expertise in nanotechnology with pharmaceutical sciences. The surface modifications of nano-ferrites enable the attachment of therapeutic agents, guiding them precisely to diseased cells or tissues through external magnetic fields. This interdisciplinary approach holds immense potential for personalized medicine, where researchers draw on knowledge from both materials science and medicine to create innovative therapeutic solutions. The environmental sciences benefit from cross-disciplinary applications of nano-ferrites, particularly in areas such as water treatment and pollution control. Collaborations between materials scientists and environmental engineers explore the use of nano-ferrites for the efficient removal of heavy metals and contaminants from water sources. This collaborative effort not only addresses environmental challenges but also emphasizes the importance of interdisciplinary approaches in creating sustainable solutions for pressing global issues. Additionally, the convergence of nano-ferrites with fields like energy storage and photonics showcases the interdisciplinary nature of their applications. Researchers from materials science, physics, and engineering collaborate to develop nano-ferrite-based materials for advanced batteries, supercapacitors, and photonic devices. This collaboration fosters innovations at the intersection of these disciplines, contributing to the development of more efficient energy storage systems and novel optical components.

In the emerging field of quantum technologies, nano-ferrites offer interdisciplinary opportunities at the intersection of materials science and quantum physics. Quantum researchers explore the potential use of nano-ferrites in quantum information processing, capitalizing on their unique magnetic and quantum properties. This interdisciplinary synergy holds promise for the advancement of quantum technologies, contributing to the development of quantum computers and quantum communication systems. The cross-disciplinary applications of nano-ferrites exemplify the collaborative nature of scientific advancements. As these materials bridge the gap between materials science, electronics, medicine, environmental sciences, and quantum technologies, interdisciplinary collaborations become increasingly essential for unlocking the full potential of nano-ferrites in addressing complex challenges and driving innovation across diverse scientific domains.

19.7 ENVIRONMENTAL IMPACT

The environmental impact of nano-ferrites is a critical consideration in their widespread application. While these materials hold promise in various fields, from electronics to biomedicine, the synthesis methods and potential release of nano-ferrites into the environment raise environmental concerns (Figure 19.5). The production processes often involve the use of chemicals and energy-intensive methods, prompting the need for sustainable and eco-friendly synthesis techniques. Additionally, the environmental impact of nano-ferrites during their lifecycle, including disposal at the end of product life, requires careful assessment. Researchers and industry stakeholders are actively exploring greener synthesis approaches and conducting life cycle analyses to mitigate the environmental footprint of nano-ferrites, ensuring responsible development and use of these materials in technology and medicine [36, 37].

19.7.1 Sustainability Considerations

Sustainability considerations are paramount in the utilization of nano-ferrites, given their potential environmental impact throughout their lifecycle [37]. In the synthesis phase, there is a growing emphasis on adopting sustainable methods to reduce the environmental footprint. Green synthesis techniques, utilizing natural sources and minimizing hazardous chemicals, align with sustainability goals and promote eco-friendly production. Additionally, the recyclability and reusability of nano-ferrites play a crucial role in minimizing waste and resource depletion. Sustainable design principles are being incorporated into the development of nano-ferrite-based products, ensuring materials can be efficiently recovered or repurposed at the end of their lifecycle. As the field progresses, an overarching commitment to sustainable practices in both synthesis and application will be integral in realizing the full potential of nano-ferrites while minimizing their environmental impact.

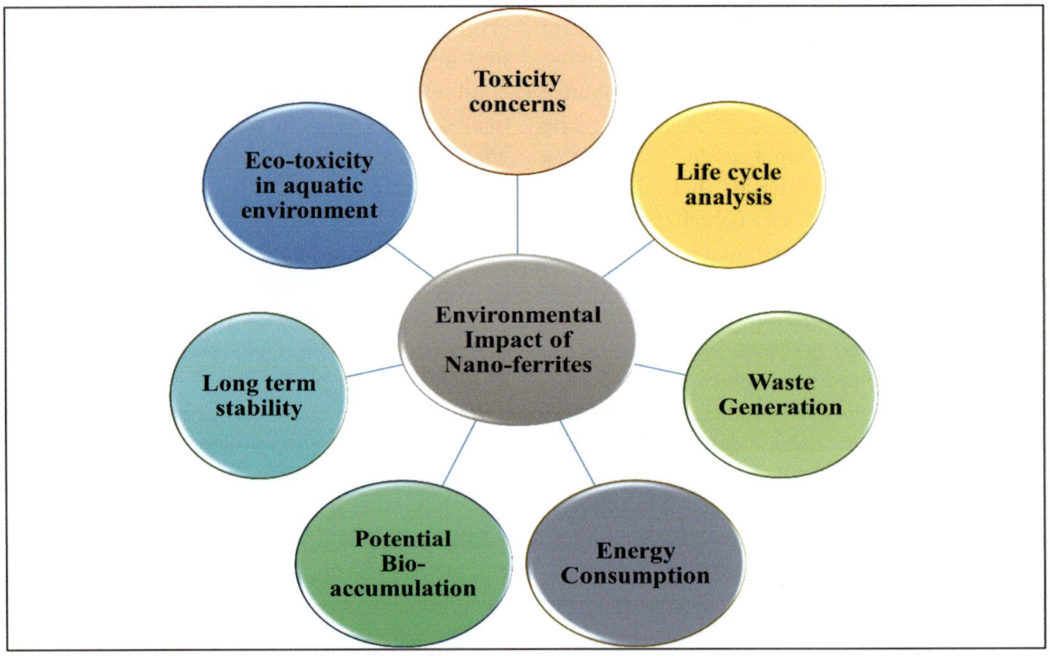

FIGURE 19.5 Environmental impact of nano-ferrites.

19.8 MARKET TRENDS: INDUSTRY ADOPTION AND ECONOMIC IMPLICATIONS

Market trends in nano-ferrites reflect an increasing industry adoption driven by the unique properties and versatile applications of these materials. The electronics industry, in particular, has witnessed a surge in the incorporation of nano-ferrites in various devices. The demand for high-performance electronics, coupled with the need for miniaturization, has fueled the adoption of nano-ferrites in components like sensors, data storage devices, and microwave devices [38]. As industry players recognize the potential of nano-ferrites to enhance device functionalities, their integration is becoming more commonplace, shaping market dynamics. The healthcare sector is another significant adopter of nano-ferrites, leveraging their applications in biomedicine. The use of nano-ferrites in targeted drug delivery systems and as contrast agents in medical imaging has gained traction. This adoption is driven by the desire for more effective and precise medical treatments, positioning nano-ferrites as valuable tools in advancing healthcare technologies. The growth in industry adoption is fostering collaborations between materials scientists, biomedical researchers, and healthcare professionals, leading to innovative solutions in diagnostics and therapeutics. The economic implications of the increasing adoption of nano-ferrites are noteworthy. The market for nano-ferrites is experiencing expansion, creating opportunities for businesses involved in their synthesis, manufacturing, and application. The demand for specialized expertise in materials science, nanotechnology, and related fields is on the rise, leading to job creation and economic growth. Additionally, the emergence of new applications for nano-ferrites, such as in energy storage and environmental remediation, contributes to diversifying revenue streams and fostering economic resilience. Investments in research and development play a pivotal role in driving market trends for nano-ferrites. Governments, private companies, and research institutions are allocating resources to explore novel synthesis techniques, improve material properties, and uncover new applications. This strategic investment fosters innovation accelerates technological advancements, and positions nano-ferrites as key contributors to economic progress.

However, challenges such as regulatory considerations and safety concerns also influence market trends. The responsible and ethical use of nano-ferrites requires adherence to regulatory frameworks to ensure the safety of both consumers and the environment. Industry stakeholders are increasingly mindful of these considerations, impacting market strategies and influencing the development of standards and guidelines for nano-ferrite applications. Furthermore, the global nature of the nano-ferrite market contributes to economic interdependencies. Supply chain dynamics, trade policies, and geopolitical factors can influence the availability and pricing of nano-ferrite materials. Industry players need to navigate these complexities, contributing to a dynamic market landscape that responds to economic shifts and global events. Market trends in nano-ferrites showcase increasing industry adoption across diverse sectors, driven by their unique properties and versatile applications. The economic implications include job creation, revenue diversification, and contributions to research and development, positioning nano-ferrites as key players in the evolving landscape of advanced materials and technologies.

19.9 SUMMARY

In this comprehensive chapter, we delved into the intricate world of nano-ferrites, exploring their fundamentals, properties, synthesis methods, and applications across diverse scientific and technological domains. The discussion initiated with an insightful overview of nano-ferrites, emphasizing their structural composition and magnetic behaviors. Understanding these fundamentals provided a solid foundation for subsequent explorations into the properties and characteristics that make nano-ferrites unique and valuable in various applications. The synthesis methods of nano-ferrites emerged as a pivotal aspect, showcasing the diversity of techniques employed to tailor their size, shape, and composition. From traditional approaches like sol-gel and co-precipitation to cutting-edge green synthesis methods, researchers navigate a spectrum of techniques to achieve precise control over nano-ferrite properties. This discussion illuminated the nuanced choices and considerations involved in the synthesis process, influencing the performance of nano-ferrites in specific applications. The chapter then transitioned into a detailed examination of the applications of nano-ferrites, shedding light on their pivotal role in magnetic storage, biomedicine, electronics, and environmental sciences. The size-dependent magnetic behaviors of nano-ferrites were harnessed in magnetic storage devices, contributing to increased data storage density. In biomedicine, their biocompatibility and targeted drug delivery capabilities positioned nano-ferrites as promising candidates for innovative therapies. The electronics industry embraced nano-ferrites in the development of high-performance devices, while their magnetic and electrical properties found applications in diverse fields, from sensors to catalysis.

A crucial theme throughout the chapter was the interdisciplinary nature of nano-ferrite research. Whether exploring their potential in quantum computing, collaborating on cross-disciplinary applications, or addressing sustainability considerations, researchers from various fields converged to unlock the full potential of nano-ferrites. This interdisciplinary approach was particularly evident in the discussion on eco-friendly synthesis methods and the careful considerations for sustainability in the production and application of nano-ferrites.

Market trends reflected the increasing industry adoption of nano-ferrites, driven by their unique properties and versatile applications. The economic implications of this adoption included job creation, revenue diversification, and contributions to research and development. However, challenges such as regulatory considerations and safety concerns were acknowledged as influencing factors in shaping market strategies and ensuring responsible industry practices.

This chapter provided a comprehensive exploration of nano-ferrites, from their fundamental aspects to advanced applications and market dynamics. The multidimensional nature of nano-ferrites was emphasized, highlighting their potential to revolutionize diverse fields and contribute to the evolving landscape of advanced materials and technologies. As researchers continue to push the boundaries of knowledge and innovation, nano-ferrites stand poised at the forefront of scientific and technological advancements, promising further breakthroughs and transformative applications.

REFERENCES

1. Thakur, P., & Thakur, A. (2022). Nanomaterials, their types and properties. In: Thakur, A., Thakur, P., Khurana, S. P. (eds) *Synthesis and Applications of Nanoparticles*. Springer, Singapore. https://doi.org/10.1007/978-981-16-6819-7_2
2. Sugimoto, M. (1999). The past, present, and future of ferrites. *Journal of the American Ceramic Society*, 82(2), 269–280. https://doi.org/10.1111/j.1551-2916.1999.tb20058.x
3. Tolani, S. C., Golhar, A. R., & Rewatkar, K. G. (2019). A review of morphological, structural behaviour and technological applications of ferrites. *Nucleation and Atmospheric Aerosols*. https://doi.org/10.1063/1.5100459
4. Ivanets, A., Prozorovich, V., & Kouznetsova, T. (2022). Advancement in water purification by nanomaterials. In: Thakur, A., Thakur, P., Khurana, S. P. (eds) *Synthesis and Applications of Nanoparticles*. Springer, Singapore. https://doi.org/10.1007/978-981-16-6819-7_18
5. Dhiman, P., Rana, G., Goyal, D., & Goyal, A. (2021). Basics of ferrites: Types and structures. In: Bhargava, G. K., Bhardwaj, S., Singh, M., Batoo, K. M. (eds) *Ferrites and Multiferroics. Engineering Materials*. Springer, Singapore. https://doi.org/10.1007/978-981-16-7454-9_1
6. Vidya, Y., Manjunatha, H., Sridhar, K., Seenappa, L., Munirathnam, R., & Chinnappareddy, B. (2023). Brief review on magnetic properties of nanoferrites. *Inorganic Chemistry Communications*, 158, 111408. https://doi.org/10.1016/j.inoche.2023.111408
7. Shirsath, S. E., Kolekar, Y. D., & Ramana, C. V. (2022). *Nanostructured Ferrites: Structure, Properties and Performance* (pp. 177–195). Elsevier eBooks. https://doi.org/10.1016/b978-0-12-815732-9.00082-6
8. Taneja, S., Punia, P., Thakur, P., & Thakur, A. (2022). Synthesis of nanomaterials by chemical route. In: Thakur, A., Thakur, P., Khurana, S. P. (eds) *Synthesis and Applications of Nanoparticles*. Springer, Singapore. https://doi.org/10.1007/978-981-16-6819-7_4
9. Shirsath, S. E., Wang, D., Jadhav, S. S., Mane, M. L., & Li, S. (2017). *Ferrites Obtained by Sol–Gel Method* (pp. 1–41). Springer eBooks. https://doi.org/10.1007/978-3-319-19454-7_125-3
10. Sajjia, M., Oubaha, M., Hasanuzzaman, M., & Olabi, A. G. (2014). Developments of cobalt ferrite nanoparticles prepared by the sol–gel process. *Ceramics International*, 40(1), 1147–1154.
11. Dippong, T., Levei, E. A., & Cadar, O. (2021). Recent advances in synthesis and applications of MFe_2O_4 (M = Co, Cu, Mn, Ni, Zn) nanoparticles. *Nanomaterials (Basel, Switzerland)*, 11(6), 1560. https://doi.org/10.3390/nano11061560
12. Thakur, P., Chahar, D., Taneja, S., Bhalla, N., & Thakur, A. (2020). A review on MnZn ferrites: Synthesis, characterization and applications. *Ceramics International*, 46(10), 15740–15763. https://doi.org/10.1016/j.ceramint.2020.03.287
13. Zhao, D., Wu, X., Guan, H., & Han, E. (2007). Study on supercritical hydrothermal synthesis of $CoFe_2O_4$ nanoparticles. *The Journal of Supercritical Fluids*, 42(2), 226–233.
14. Baykal, A., Kasapoğlu, N., Köseoğlu, Y., Başaran, A. C., Kavas, H., & Toprak, M. S. (2008). Microwave-induced combustion synthesis and characterization of $Ni_xCo_{1-x}Fe_2O_4$ nanocrystals (x = 0.0, 0.4, 0.6, 0.8, 1.0). *Central European Journal of Chemistry*, 6, 125–130.
15. Pandit, V. A., Repe, G. R., Bhamre, J. D., & Chaudhari, N. (2020). A review on green synthesis and characterization technique for ferrite nanoparticles and their applications. *Journal of Physics*, 1644(1), 012009. https://doi.org/10.1088/1742-6596/1644/1/012009
16. Valenzuela, R. (2012). Novel applications of ferrites. *Physics Research International*, 2012, 1–9. https://doi.org/10.1155/2012/591839
17. Kumari, N., Kour, S., Singh, G., & Sharma, R. K. (2020). A brief review on synthesis, properties and applications of ferrites. *Nucleation and Atmospheric Aerosols*. https://doi.org/10.1063/5.0001323
18. Hazra, S., & Ghosh, N. N. (2014). Preparation of nanoferrites and their applications. *Journal of Nanoscience and Nanotechnology*, 14(2), 1983–2000. https://doi.org/10.1166/jnn.2014.8745
19. Manori, S., Manori, A., & Shukla, R. K. (2023). Ferrite nanoparticles for energy storage applications. In: Sharma, P., Bhargava, G. K., Bhardwaj, S., Sharma, I. (eds) *Engineered Ferrites and Their Applications. Materials Horizons: From Nature to Nanomaterials*. Springer, Singapore. https://doi.org/10.1007/978-981-99-2583-4_10
20. Sharma, A., Thakur, R. C., & Sharma, R. (2023). Biomedical applications of ferrites. In: Sharma, P., Bhargava, G. K., Bhardwaj, S., Sharma, I. (eds) *Engineered Ferrites and Their Applications. Materials Horizons: From Nature to Nanomaterials*. Springer, Singapore. https://doi.org/10.1007/978-981-99-2583-4_13
21. Nandhini, G., & Shobana, M. (2022). Role of ferrite nanoparticles in hyperthermia applications. *Journal of Magnetism and Magnetic Materials*, 552, 169236. https://doi.org/10.1016/j.jmmm.2022.169236
22. Srivastava, R., & Yadav, B. C. (2012). Ferrite materials: Introduction, synthesis techniques, and applications as sensors. *International Journal of Green Nanotechnology*, 4(2), 141–154.
23. Bhatia, N., Kumari, A., Sharma, K., & Sharma, R. (2023). Ferrite nanoparticles for antimicrobial applications. In: Sharma, P., Bhargava, G. K., Bhardwaj, S., Sharma, I. (eds) *Engineered Ferrites and Their Applications. Materials Horizons: From Nature to Nanomaterials*. Springer, Singapore. https://doi.org/10.1007/978-981-99-2583-4_11
24. Kefeni, K. K., Msagati, T. A., & Mamba, B. B. (2017). Ferrite nanoparticles: Synthesis, characterisation and applications in electronic device. *Materials Science and Engineering: B*, 215, 37–55. https://doi.org/10.1016/j.mseb.2016.11.002

25. Hublikar, L. V., Ganachari, S. V., & Patil, V. B. (2023). Zn and Co ferrite nanoparticles: Towards the applications of sensing and adsorption studies. *Environmental Science and Pollution Research*, 30(25), 66994–67007. https://doi.org/10.1007/s11356-023-27201-z
26. Tamboli, Q. Y., Patange, S., Mohanta, Y. K., Sharma, R., & Zakde, K. R. (2023). Green synthesis of cobalt ferrite nanoparticles: An emerging material for environmental and biomedical applications. *Journal of Nanomaterials*, 2023, 1–15. https://doi.org/10.1155/2023/9770212
27. Cadar, O., Dippong, T., Senila, M., & Levei, E.-A. (2020). *Progress, Challenges and Opportunities in Divalent Transition Metal-Doped Cobalt Ferrites Nanoparticles Applications*. IntechOpen. https://doi.org/10.5772/intechopen.93298
28. Cruz, I., Freire, C., Araújo, J. P., Pereira, C., & Pereira, A. M. (2018). *Multifunctional Ferrite Nanoparticles: From Current Trends toward the Future* (pp. 59–116). Elsevier eBooks. https://doi.org/10.1016/b978-0-12-813904-2.00003-6
29. Rana, G. C., Dhiman, P., Kumar, A., Vo, D. N., Sharma, G., Sharma, S., & Naushad, M. (2021). Recent advances on nickel nano-ferrite: A review on processing techniques, properties and diverse applications. *Chemical Engineering Research and Design*, 175, 182–208. https://doi.org/10.1016/j.cherd.2021.08.040
30. Khargotra, R., András, K., Kalia, R., Himanshi, Prakash, J., & Verma, A. (2023). *Synthesis of Novel Hard/Soft Nanoferrite Composites* (pp. 15–35). Elsevier eBooks. https://doi.org/10.1016/b978-0-323-96115-8.00011-8
31. Shobana, M. (2021). Nanoferrites in biosensors – A review. *Materials Science and Engineering: B*, 272, 115344. https://doi.org/10.1016/j.mseb.2021.115344
32. Garg, J., Chiu, M. N., Krishnan, S., Kumar, R., Rifah, M., Ahlawat, P., … Gupta, P. K. (2023). Emerging trends in zinc ferrite nanoparticles for biomedical and environmental applications. *Applied Biochemistry and Biotechnology*. https://doi.org/10.1007/s12010-023-04570-2
33. Hussain, M., Xia, M., Xiao-NaRen, Akhtar, K., Ahmed, N., Sharma, S. K., & Javed, Y. (2020). Ferrite nanoparticles for biomedical applications. In *Nanomedicine and Nanotoxicology* (pp. 243–265). Springer. https://doi.org/10.1007/978-3-030-39923-8_7
34. Verma, R., Chauhan, A., Kalia, R., Batoo, K. M., & Kumar, R. (2023). *Magnetic Nanoferrites as an Alternative for Magnetic Resonance Imaging Application* (pp. 237–256). Elsevier eBooks. https://doi.org/10.1016/b978-0-323-96115-8.00013-1
35. Gautam, S., Thakur, V., & Goyal, N. (2023). *Nanoferrites as Drug Carriers in Targeted Drug Delivery Applications* (pp. 161–178). Elsevier eBooks. https://doi.org/10.1016/b978-0-443-18874-9.00016-3
36. Mandyal, P., Sharma, R., Sambyal, S., Fang, B., Sillanpää, M., Kumar, V., … & Shandilya, P. (2023). *A New Generation of Magnetic Nanoferrite-Based Nanocomposites for Environmental Applications* (pp. 257–293). Elsevier eBooks. https://doi.org/10.1016/b978-0-323-96115-8.00001-5
37. Gómez, P., Elduque, D., Pina, C., & Javierre, C. (2018). Influence of the composition on the environmental impact of soft ferrites. *Materials (Basel, Switzerland)*, 11(10), 1789. https://doi.org/10.3390/ma11101789
38. Goldman, A. (2006). *Modern Ferrite Technology*. Springer Science & Business Media Springer, New York, NY.

Index

3D imaging, 34
3D printing, 57
5G technology, 9

A

AC conductivity, 43
Actuators, 49
Adsorbents, 131–134
Adsorption, 111, 140–141
Aeration, 130
Agar disk, 71, 72
Agglomeration, 33
Airborne dispersal, 179
Antibacterial activity, 69–77, 177, 205–207
Antifungal activity, 78–89
Aptamers, 167
Atomic force microscopy (AFM), 35
Automotive industry, 63
Avoiding ingestion risks, 186

B

Barium ferrite, 62
Bioaccumulation, 173, 179
Biocompatibility, 28, 51, 79, 101, 112
Biodegradation, 178
Biofilm assays, 72
Biological methods, 98
Biomagnification, 173
Biomedical, 163–164, 216
Biosensing, 57
Biosensor, 204–205
Bioseperation, 168
Biotic disruption, 179
Bismuth ferrite, 203
Boiling, 130
Broth dilution assay (MIC/MBC), 72–74

C

Calcium ferrite, 203
Cancer diagnostics, 92–96, 168
Catalysis, 37, 65
Catalytic activity, 27, 37, 117, 119–122, 142
Cell membranes, 68
Cellular uptake, 51
Ceramic process, 21
Chlorination, 130
Citrate precursor, 22
Coagulation, 178
Coatings, 37
Cobalt ferrite, 74–75, 202
Coercivity, 6–8, 44, 62
Combination therapy, 86–87
Combustion method, 99
Computed tomography (CT), 52–53
Copper ferrite, 75–76, 120–121
Co-precipitation, 14, 82, 109, 118, 139, 153, 175, 215
Coral reefs, 179
Crystal structure, 27–32, 48–49, 63–64
Crystallography, 2, 27, 28, 37

D

DC conductivity, 42
DC resistivity, 43–44, 198
De Magnete, 1–2
Demagnetization, 8–9
Density Functional Theory (DFT), 172, 177, 180
Dentistry, 87–88
Diagnostic imaging, 87
Dielectric properties, 42–44
Dislocation, 32
DNA, 55, 167
DNA damage, 69–70
Drinking water, 126–129, 131, 138, 144, 149
Drug delivery system, 49, 86, 166
Drug loading, 111
Dyes adsorption, 132–133
Dynamic light scattering (DLS), 36

E

Electric motors, 3
Electrical conductivity, 27, 41, 42, 44
Electrochemical method, 110
Electromagnetic interference (EMI) suppression, 9
Electronics, 216–217
Emergency preparedness, 186
Encapsulation, 111
Energy conversion, 37
Energy efficiency, 10
Enhanced permeability and retention (EPR), 54
Environmental impact, 173
Environmental remediation, 37
Environmental safety, 186
Enzyme activity, 68–69
Extracellular, 18–19
Eye protection, 185–186

F

Face-centred cubic (FCC), 28–29, 48, 119
Feeders, 179
Fenton-like reactions, 141–142
Ferrimagnetic susceptibility, 5
Ferritic stainless steels, 2
Ferromagnetism, 2
Filtration, 130
Fourier transform infrared (FTIR), 31, 32
Fungal membranes, 83–84

G

Gas sensing, 37
Gas sensor, 203
Genotoxicity, 51
Glucose detection, 167
Grain size, 41, 42
Green synthesis, 20, 37, 172, 215, 218–222

H

Hard disk drives (HDDs), 5, 216
Heavy metals, 149–157
Hematite, 27–28, 200
High-strength low alloy (HSLA), 3
Hydrophobic, 112
Hydrothermal, 109, 118, 140
Hysteresis loop, 8, 46, 197

I

Immunoassay, 167
Immunological, 50
Implant coatings, 88
In vitro antifungal assays, 85
In vitro cytotoxicity, 49–50
In vivo biocompatibility, 50
Inorganic compounds, 101–103
Intracellular, 18–19, 51, 164
Inverse spinel ferrites, 64–65
Ion homeostasis, 84
Isotropic sensitivity, 30

L

Leaching, 70
Lithium ferrite, 194–199
Lodestone, 1–2, 61

M

Maghemite, 27, 93, 156, 202
Magnetic data storage, 4, 62
Magnetic domain wall, 4
Magnetic domains, 4, 5, 8
Magnetic field, 4–10
Magnetic fluids, 65
Magnetic hyperthermia, 37, 94, 206
Magnetic moment, 2–5, 45, 48, 139, 197–198
Magnetic permeability response, 196–197
Magnetic random-access memory (MRAM), 215–216
Magnetic recording, 5, 8, 65
Magnetic resonance imaging (MRI), 3, 9, 10, 28
Magnetic sensors, 10
Magnetic separation, 205
Magnetic storage, 8, 174, 216, 218
Magnetic susceptibility, 5–6, 205
Magnetic targeting, 53
Magnetite, 27, 50–52, 55, 61–65, 84, 93, 153, 178, 202, 213
Magnetization, 4–10, 45, 52, 175, 194, 197
Manganese ferrite, 64, 122
Material composition, 5
Mechanical characteristics, 28
Mechano-chemical, 99
Medical bioelectronics, 58
Medical imaging, 3, 5, 37, 52, 216
Metabolic pathway, 79
Microbial-assisted, 19
Microemulsion, 16, 109, 119
Micro-ferrites, 171, 174, 177, 179
Microfluidic chip, 168
Microplastics, 173–176
Microwave devices, 9–10, 64–65

Microwave-assisted, 19, 98, 140
Morphological, 32–35
MRI, 9, 203–204
Multifunctional materials, 146

N

NADPH oxidase (NOX), 70
Nickel Ferrite, 64, 202
Noise Suppression, 6

O

Organic polymers, 101
Osteogenesis, 56
Oxidation method, 119

P

Paramagnetic susceptibility, 5
Particle size, 32, 99
Permanent magnets, 63
pH-responsive loading, 112
Pharmaceutical, 144–145
Phase analysis, 30
Photocatalysis, 141
Photothermal therapy, 86
Pigments, 65
Plastic particles, 173–174
Point-of-care testing (POCT), 55
Polymerase chain reaction (PCR), 55
Polyol Process, 23
Power loss, 198
Prevention, 33

Q

Quantum Technologies, 4

R

Radar technology, 2, 8, 9
Radomes, 10
Reactive oxygen species (ROS), 65–67
Refrigerator door seals, 7
Remanence, 8
Renewable energy, 3
RFID (Radio-frequency identification), 9

S

Safe handling, 186
Scanning electron microscopy, 34, 35, 195
Sedimentation, 130
Silver ferrite, 76
Solar disinfection, 130
Sol-gel, 14, 82, 99, 108, 118, 140
Solid-state, 18, 99, 108
Solvothermal, 15, 98, 110
Sonochemical, 17, 110
Spintronics, 4
Stearic acid gel, 119
Strain, 30
Strontium ferrite, 62
Superparamagnetic, 54
Surface characterization, 27
Switch-mode power supplies (SMPS), 3
Synergistic effects, 82

T

T1 and T2 relaxation, 52
Targeted distribution, 206
Telecommunications, 10, 36, 63
Template-assisted synthesis, 83
Textile industry, 142–143
Theragnostic, 57
Three-pot method, 130
Time-kill assay, 70, 73–74
Topographical, 34
Toxicity, 179, 207
Transformer, 9
Transmission electron microscopy (TEM), 35

U

Ultraviolet (UV) radiation, 67

W

Waste disposal, 189–190
Wastewater treatment, 173–175
Water contaminants, 149, 152
Water purification, 128–132
Water treatment, 4, 37, 125–132
Wireless monitoring devices, 57
Wound dressing, 87, 168

X

X-ray diffraction, 29, 30, 195

Z

Zinc ferrite, 64, 121, 202

For Product Safety Concerns and Information please contact our EU representative GPSR@taylorandfrancis.com Taylor & Francis Verlag GmbH, Kaufingerstraße 24, 80331 München, Germany